Materials Processing and Design:

Grain-Boundary-Controlled Properties of Fine Ceramics II

Related titles published by the American Ceramic Society:

Ceramic Transactions, Volume 41: Grain Boundaries and Interfacial Phenomena in Electronic Ceramics
Edited by Lionel M. Levinson and Shin-ichi Hirano
© 1994, ISBN 0-944904-73-4

Handbook on Characterization Techniques for the Solid-Solution Interface
Edited by James H. Adair, Jon A. Casey, and Sridhar Venigalla
© 1993, ISBN 0-944904-67-X

Ceramic Transactions, Volume 33: Materials in Microelectronic and Optoelectronic Packaging
Edited by Hung C. Ling, Koichi Niwa, and Vishwa N. Shukla
© 1993, ISBN 0-944904-63-7

Ceramic Transactions, Volume 32: Dielectric Ceramics: Processing, Properties, and Applications
Edited by K.M. Nair, J.P. Guha, and A. Okamoto
© 1993, ISBN 0-944904-60-2

Phase Equilibria Diagrams, Phase Diagrams for Ceramists, Volume IX: Semiconductors and Chalcogenides
Edited by Gerald B. Stringfellow
© 1992, ISBN 0-944904-50-5

Phase Diagrams for High T_c Superconductors
Edited by John D. Whitler and Robert S. Roth
© 1991, ISBN 0-944904-41-6

Materials Processing and Design:

Grain-Boundary-Controlled Properties of Fine Ceramics II

▼▲▼

Edited by

Koichi Niihara
Osaka University

Kozo Ishizaki
Nagaoka University of Technology

Mitsuo Isotani
Japan Fine Ceramics Center

▲▼▲

Ceramic Transactions
Volume 44

Published by
The American Ceramic Society
735 Ceramic Place
Westerville, Ohio 43081

Proceedings of the Japan Fine Ceramics Center workshop entitled "Materials Processing and Design Through Better Control of Grain Boundaries: Emphasizing Fine Ceramics II," held in Koda-cho, Aichi, Japan, March 17–19, 1994.

Library of Congress Cataloging-in-Publication Data

JFCC International Workshop on Fine Ceramics '94 (1994 : Kōda-chō, Japan)
Materials processing and design : grain-boundary-controlled properties of fine ceramics II / edited by Koichi Niihara, Kozo Ishizaki, Mitsuo Isotani.
p. cm. -- (Ceramic transactions, ISSN 1042-1122 ; v. 44)
Includes index.
ISBN 0-944904-78-5
1. Ceramic materials. 2. Grain boundaries. 3. Electronic ceramics. I. Niihara, Koichi. II. Ishizaki, Kōzō1946- .
III. Isotani, Mitsuo. IV. Title. V. Series.
TA455.C43J43 1994
620.1'4--dc20 94-38725
CIP

For information on ordering titles published by the American Ceramic Society, or to request a publications catalog, please call 614-890-4700.

Printed in the United States of America.

1 2 3 4–98 97 96 95 94

ISSN 1042-1122
ISBN 0-944904-78-5

Contents

Functional Ceramics

Composite Materials

Synthesis and Sintering

Mechanical Properties

Poster Session

Preface

This volume contains selected papers presented at a workshop sponsored by the Japan Fine Ceramics Center, "Materials Processing and Design Through Better Control of Grain Boundaries: Emphasizing Fine Ceramics II," which was held March 17–19, 1994, in Koda-cho, Aichi, Japan.

This workshop brought together scientists and engineers, from all over the world and from a variety of disciplines, to focus on the application of grain boundary phenomena to materials processing and design. The topics covered included electronic materials, evaluation methods, structural materials, and interfaces. Also included is an illuminating overview of the current status of work on grain-boundary-assisted materials processing and design, particularly for fine ceramics. The attendance of nearly 100 participants from more than ten countries indicates the active interest in this topic.

The Japan Fine Ceramics Center had a two-day workshop two years ago on the same theme. However, since this topic is very important and covers a wide area, this was not enough time to discuss everything. Therefore, it was decided to have another workshop this year, which was extended by another day in order to discuss the topic sufficiently. In addition, the workshop was held at a remote location in traditional Japanese-style lodging, where attendees shared rooms. This encouraged informal discussions, some of which lasted well into the evening, after the oral presentations were over.

This collection of high-quality research papers presents current developments in the area of grain-boundary-related phenomena. These significant but little-understood phenomena generated much lively discussion during the workshop, as well as fruitful interaction among those attending.

Through the publication of this volume, the Japan Fine Ceramics Center hopes that many who were unable to attend this interesting and timely workshop will benefit from its contents.

Finally, we express our sincere appreciation to Laurel M. Sheppard of Lash Publications International and Ben Huybrechts of Nagaoka University of Technology for their devoted cooperation.

The workshop was subsidized by the Japan Keirin Association through its promotion funds from KEIRIN RACE.

KEIRIN

Koichi Niihara
Kozo Ishizaki
Mitsuo Isotani

JFCC International Workshop on Fine Ceramics '94

Organizing Committee

Committee Chairman

Masaaki Ohashi

Vice-Chairmen

Koichi Niihara
Kozo Ishizaki
Mitsuo Isotani

Members

Jun-ichi Kon
Jun-ichiro Tsubaki
Yuichi Ikuhara
Tsutomu Tsunooka
Mineo Mizuno
Hideaki Matsubara

Secretariat

Noriyuki Kosuge

Speakers

K. Niihara

K. Ishizaki

M. Isotani

Y.H. Choa

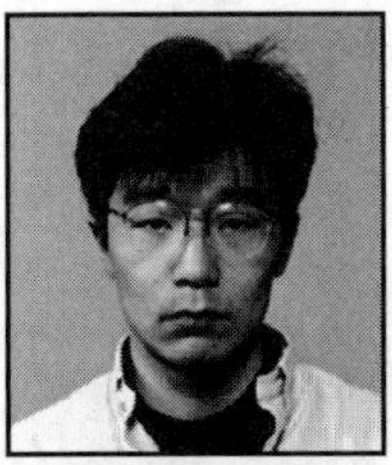
T. Hamasaki

K.S. Han

S. Hayashi

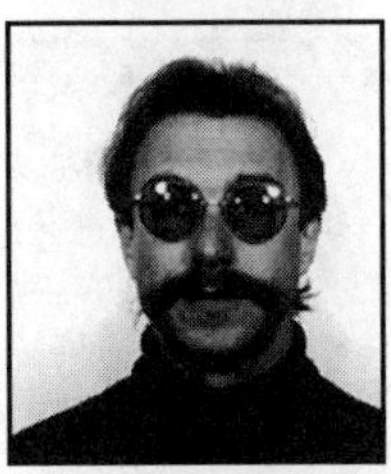
J.M. Heintz

T. Hirano

M. Hussain

B. Huybrechts

H.J. Hwang

O. Inoue

A. Ishida

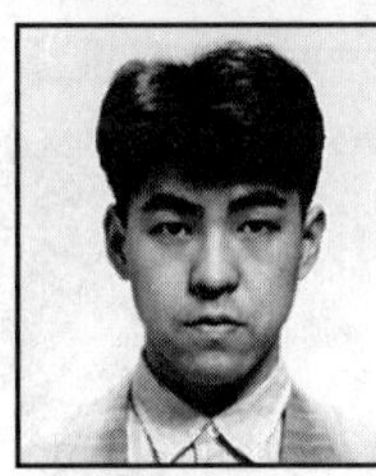
H. Kitayama

H.J. Kleebe

S.B.S. Kumar

M. Kusunoki

M.G.Q. Lu

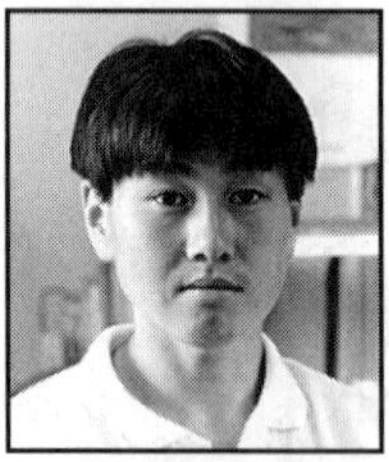
K. Morita

I. Nakamichi

J. Nowotny

P. Pirouz

T. Sekino

Y. Suzuki

I. Tanaka

S. Tanaka

H. Tsubakino

T. Watanabe

K. Watari

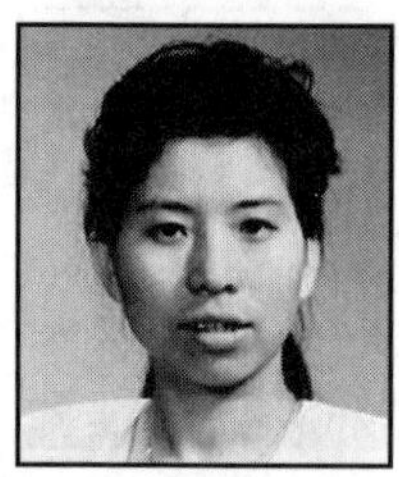
Y. Xu

T. Yanai

Z. Zhang

Y. Ikuhara

Electron Microscopy

ELECTRON MICROSCOPIC ANALYSIS OF Pd/MgO MISMATCHED INTERFACE

*Hideki Ichinose, Takashi Ichimori, Hideo Ishii, Yoichi Ishida and Werner Mader**

Department of materials Science, University of Tokyo
7-3-1 Hongo, Bunkyo-ku, Tokyo, JAPAN
**Anorganische Materialforschung,*
Institute fur Anorganische Chemie, Universitat Bonn,
Romerstrasse 164 D-53117 Bonn, GERMANY

ABSTRACT

A well defined Pd/MgO interface was produced by internal oxidation of dilute Mg-Pd alloy. High resolution electron microscopy was performed on the Pd/MgO interface in order to investigate structural features of the mismatched metal/ceramic interface. A dislocation network predicted by O-lattice theory was not recognized in the Pd/MgO interface by both dark field imaging and lattice imaging in spite of 7.7% mismatch at the interface. A periodic diffused lattice image that appeared along the interface was believed to correspond to the periodic Moire pattern.

INTRODUCTION

Many metallic[(1)(2)] and semiconductor [(3)(4)(5)(6)] grain boundaries were found to agree with the geometrical structure model[(7)], even though the observed atomic structure sometimes deviated significantly from the model at exact CSL orientation such as $\Sigma 3$ [(8)] in gold and aluminum. A similar type of deviation from the CSL model is also observed in silicon and germanium grain boundaries. In a covalent bond material such as α-SiC, the geometrical theory did not hold; the grain boundary energy was governed not by the number of coincident site but by the number of dangling bonds[(9)].

In the metal/ceramic interface it was pointed out that the dislocation network which accommodates the stress due to lattice mismatch at the interface shows various aspects. In the Pd/ZnO interfaces[(10)] no misfit dislocation network was observed by both conventional and high resolution TEM. In the V/Al_2O_3 interface a misfit dislocation network was scarcely detectable[(11)]. In the Nb/Al_2O_3 interface a dislocation network was actually present but was located at the stand off position from the chemical interface[(12)]. Previously reported results suggest that the interface structure can be different depending on both the properties of combined materials and the manufacturing process. In order to reach on accurate understanding further accumulation of experimental results due to HRTEM is required.

In the present paper, therefore, the interface structure of palladium/magnesium oxide is investigated by both conventional and high resolution electron microscopy in order to clarify the presence (or absence) of the misfit dislocation at the interface.

II. EXPERIMENTAL

(a) MgO/Pd interface production

A dilute Pd-Mg binary alloy is produced and is heated under oxygen atmosphere in an infrared furnace. Temperature, oxygen partial pressure of mixed gas and flow rate of the mixed gas are independently controlled by a computer aided system. A MgO precipitate is grown in the Pd matrix during heating. The growth rate of the precipitations is slow enough to form a well defined interface between the Pd matrix and oxides.

(b)Electron microscopy

A precipitated Pd plate is sliced and thinned by ion milling. A JEM-200CX high resolution electron microscope which is equipped with newly designed objective lens was used. The spherical and chromatic aberration constants of the lens are Cs=0.7mm and Cc=2.0mm respectively. A specimen is illuminated in the direction of the <110> zone axis of Pd. Diffraction waves up to (200) are allowed to pass through the objective aperture. A lattice image picture is taken at second Schelzer focus condition. At this condition the atomic position appears as a white spot.

III. RESULTS AND DISCUSSIONS

III-1 Orientation relationship

The orientation relationship between Pd and MgO was simple. The [110] axis of both crystals was parallel each other (Figure 1), so that two of the four (111) planes were parallel with the electron beam in both crystals. The lattice mismatch between the matrix and precipitates was estimated to be 7.7% from the diffraction pattern. The interface plane was parallel with the (111) plane of both Pd and MgO.

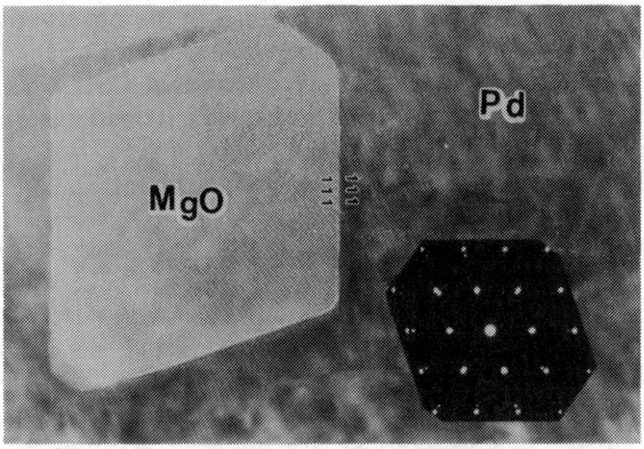

Fig.1. A MgO precipitate and diffraction pattern from the interface region.

III-2 Periodic contrast in dark & bright field observation

The dislocation network due to lattice mismatch, according to O-lattice theory, is expected to be present at the interface(Figure 2). In most cases, however, no image contrast which corresponds to the interface dislocation was observed by dark field imaging (Figure 3) as far as it is observed in the edge-on condition. The result is similar to that of the Pd/ZnO interface[10].

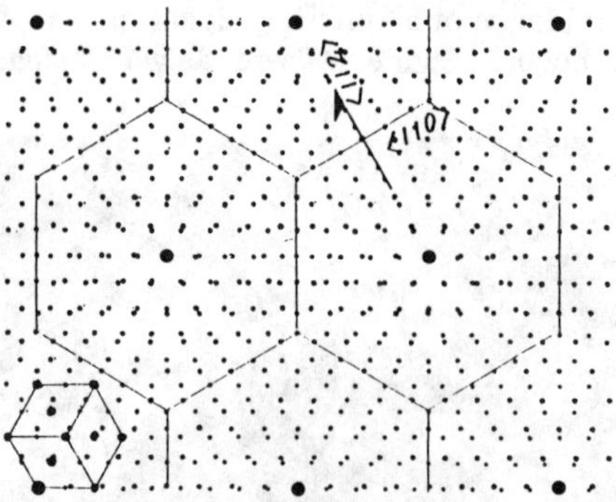

Fig.2. Dislocation network predicted by 0-lattice model.

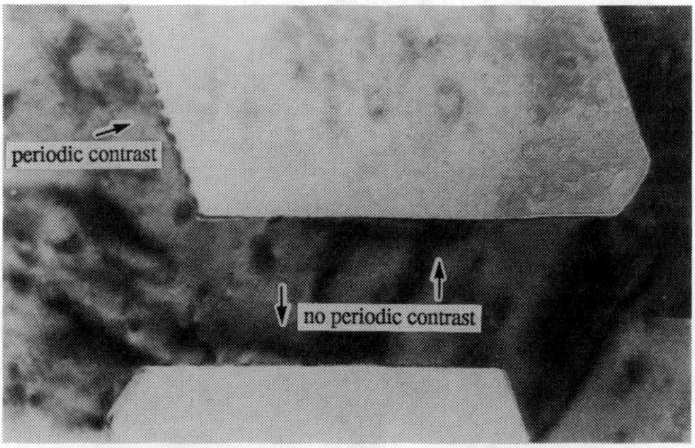

Fig.3. Bright field image of precipitations.

When the interface was inclined to the specimen surface the periodic contrast appeared(Figure 3). However no periodic contrast was observed when the interface was viewed at the edge-on condition. Absence of periodic contrast at the condition suggests that there is no three dimensional strain field due to misfit dislocation. The periodic contrast in the inclined interface must be attributed to another cause which is different from strain field due to dislocation.

Another example is shown in Figure 4. In this case, a periodic contrast is also visible. This periodic contrast, however, does not appear in the shape of a network. It is simply parallel lines. If the periodic contrast represents misfit dislocations it must be the shape of a network at the present illumination condition.

III-3 Atomic structure of Pd/MgO interface

In most interfaces no evidence of a misfit dislocation was obtained even by lattice imaging. An example is shown in Figure 5. There is a well matched

Fig.4. One dimensional periodic contrast in the MgO/Pd interface.

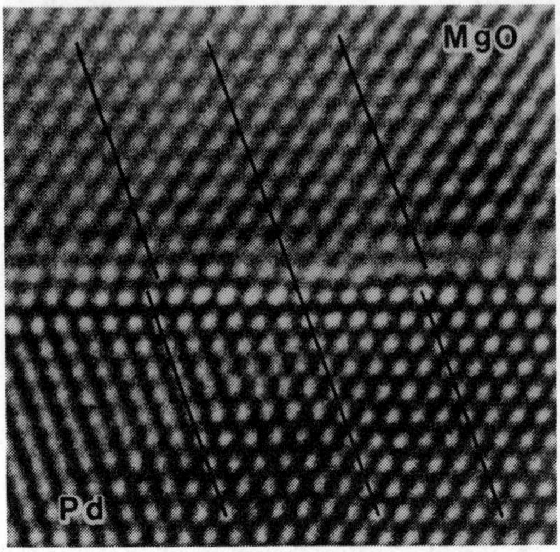

Fig.5. Lattice image of Pd/MgO interface.

region in the central part of the image whereas the mismatched regions are seen in the right and left parts of the image. The mismatching (matching) region is located at every 11 or 12 planes of (111) plane along the interface. This distance coincides well with the corresponding interval of the dislocation network segment which was predicted from 7.7% of lattice mismatch.

In the well matched region one can trace an atomic row on a straight line across the interface. In the mismatched region an atomic row can be traced also on a straight line toward the interface in both crystals. White dots in one plane that is parallel and is next to the interface deviated from its original position. In other words, the atomic displacement corresponding to long ranged strain field of dislocation is not detected. The result is supported by the results of dark field observation.

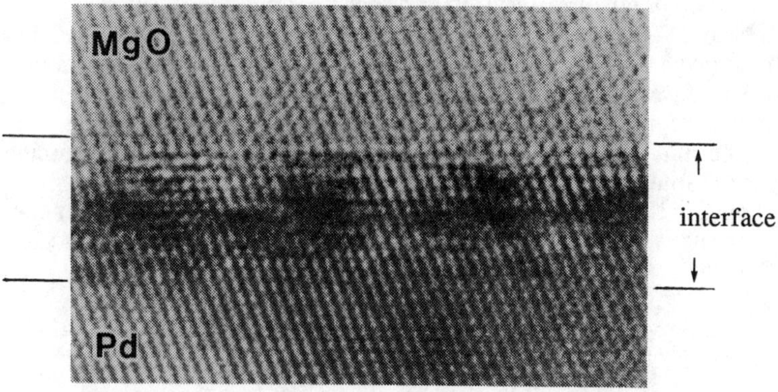

Fig.6. Moire pattern is produced along the interface.

Fig.7. Misoriented Pd/MgO interface.

The lattice image of the interface which showed periodic contrast in Figure 2 is shown in Figure 6. Lattice fringes look very straight and dark round contrasts are ranged along the interface. A misfit dislocation is usually expected to produce a similar contrast. But this is not the case. The dark contrast is too weak to be caused by the strain field of the dislocation so that the lattice image is still visible even in the dark contrast region. A moire pattern due to two parallel lattice fringes of MgO and Pd is more reasonable to explain the contrast.

The lattice image of another case that appeared in Figure 4 is shown in Figure 7. The interval of the linear contrast in the Figure 4 coincided with the termination of the $(111)_{Pd}$

plane at the interface in Figure 7. Since the two crystals were misoriented by a small angle to each other a weak strain field was produced at the termination of the $(111)_{Pd}$ plane.

III-4. Absence of misfit dislocation and thermal cracking
It can be said that the misfit dislocation network was absent in the Pd/MgO interface even though an image contrast which is similar to the dislocation is observed in minor occasion. The absence of the misfit dislocation must be a major cause of thermal cracking at the interface of metal/ceramic bonded materials. If the misfit dislocation network is absent, thermal stress at the interface must be shared by a limited number of atoms facing at the interface. The share of the stress for each atoms can easily exceed the atomic bonding force across the interface to break it. If the dislocation network is present, the thermal stress is dispersed along a large number of atoms included in the strain field of the dislocation and thus the interface remains intact.

Acknowledgment-The authors thank the Foundation for Promotion of Material Science and Technology of Japan for the financial support.

References

1. Y.Ishida and H.Ichinose, Proc. Int. Conf. HVEM(Kyoto),(1977)603, "Structure of [110] tilt boundaries of gold analyzed by high resolution HVEM"
2. Ichinose, H. and Ishida, Y., Phil.Mag.**A43**(1981)1253, "Observation of [110] tilt boundary structure in gold by high resolution HVEM"
3. Krivanek, O.L., Isoda, S. and Kobayasi, K.,Phil.Mag., **36**(1977)931, "Lattice imaging of a grain boundary in crystalline germanium"
4. Bourret, A., J.de Physique, **C4**(1985)27, "Atomic structure of <001> pure tilt grain boundaries in germanium and silicon"
5. Cunningham, B., Strunk, H.P. and Ast,D.G., Scripta Met.,**16**(1982)349, "High resolution electron microscopy of a Σ=27 boundary in silicon"
6. Ichinose,H. and Ishida,Y., Trans. J.I.M., **27**(1986)253, "Structure analysis of Σ3 CSL boundary in polysilicon by HRTEM lattice imaging"
7. Brandon, D.G., Ralph, B., Ranganathan, S. and Wald, M.S.,Acta.Met.,**12**(1964)813, "A field ion microscope study of atomic configuration at grain boundary"
8. Ichinose,H. and Ishida, Y., Phil.Mag.,**A52**(1985)51, "Lattice imaging analysis of Σ3 coincidence-cite-lattice boundaries in gold."
9. Ichinose,H.,Inomata,Y. and Ishida,Y.,-Ceramic Micro structure '86-, (Plenum Pub.Co.,1988)p.255, "High resolution electron microscopy of silicon nitride-metal bonded interfaces"
10. Ichinose H.,Mader W.,Kirchheim R. and Ishida Y., Abstracts of Int.Symp.on "New Directions and Future Aspects of HVEM"-1990, "High voltage high resolution electron microscopy of metal-ceramic interfaces segregated by hydrogen"
11. Ikuhara Y. and Pirouz P., Ultra microscopy **52**(1993)421, "A high-resolution electron microscopy study of vanadium deposited on the basal plane of sapphire"
12. Necker G. and Mader W., Phil.Mag.Letters **58**(1988)205, "Characterization of Ag/MgO interfaces"

INTERNAL INTERFACES IN SILICON NITRIDE STUDIED BY HIGH-RESOLUTION ELECTRON MICROSCOPY

H.-J. Kleebe
Universität Bayreuth, Lehrstuhl Keramik und Verbundwerkstoffe, Institut für Materialforschung, Ludwig-Thoma-Str. 36b, D-95440 Bayreuth, Germany

ABSTRACT

High-resolution electron microscopy (HREM) was utilized to investigate internal interfaces in Si_3N_4 ceramics. Typically, liquid-phase sintered Si_3N_4 materials reveal continuous, thin films of residual glass along interfaces. Depending on the sintering aid used, HREM observations, performed parallel to analytical electron microscopy (AEM), revealed differences in film thickness and chemical composition, respectively. Each material, with its given grain-boundary chemistry, showed a constant film width within ±0.1 nm. Crystallization of secondary phases can alter interface chemistry, changing intergranular film thickness. With increasing impurity-cation content at the interface, a widening of the films was observed. Moreover, experimental evidence on microstructural changes of a MgO-doped Si_3N_4 material, rapidly cooled from high temperatures, is presented. Quenching the material from 1350°C produced no observable variation in grain-boundary film thickness. However, quenching from temperatures of about 1420°C, a substantial increase in the amorphous film width was found. Specimens quenched from higher temperatures and/or longer holding times above the eutectic temperature revealed again rather thin intergranular films, which are comparable to those observed in the slowly cooled material. The observed changes in grain-boundary film thickness are related to (i) the eutectic temperature of the system, (ii) the solubility of Si_3N_4 and (iii) mass transport in the residual glass.

INTRODUCTION

Due to the high covalent bonding character and low self-diffusivity Si_3N_4 cannot be completely densified by solid-state sintering. Thus, the addition of sintering aids is required to promote liquid-phase sintering to achieve complete densification. Upon cooling the bulk of the secondary phase formed during sintering exists at multiple-grain pockets as devitrified or partially devitrified products of the residual liquid [1]. The remainder of the liquid solidifies as a thin, amorphous film with a constant film thickness, separating the crystalline phases. Large amounts of sintering additives result in detrimental effects on high-temperature mechanical properties, e.g. flexural strength and creep behavior, thus requiring a significant reduction in their volume fraction. Another approach to improve the high-temperature properties takes advantage of the refractoriness of rare-earth and certain transition element based oxides, which influence the transition temperature of the liquid phase formed [2,3]. In materials where crystallization of a stable, refractory secondary phase is achieved, the high-temperature behavior depends primarily on the properties of the residual amorphous films at the grain boundaries [4]. Therefore, a control of these films as well as a detailed characterization of the factors which influence the interganular grain-boundary films is required. Recent investigations have shown that the intergranular film width primarily depends on interface chemistry [5-7]. However, it should be noted that interface properties are mainly governed by interface chemistry rather than influenced by intergranular film thickness.

Moreover, the question of whether the Si_3N_4 microstructure is the same at room temperature as it is at high temperatures is of great interest. This question was first addessed by Clarke [8]. The microstructure of samples quenched from about 1450°C were different compared to the same material cooled under normal conditions. The Si_3N_4 grains were reported to have a more rounded morphology and the amorphous intergranular film was observed to be much thicker with 2-8 nm compared to about 1 nm for the normally cooled sample. In addition, the volume fraction of amorphous material was reported to have increased in the quenched material. With the development of more refractory Si_3N_4 materials, which are being developed for potential high-temperature applications, the question remains as to whether the microstructures of these materials are significantly different at room temperature than those of the materials at service temperature. Therefore, apart from factors influencing the resulting grain-boundary film thickness in Si_3N_4 ceramics at room temperature, which are discussed in the following sections, recent investigations regarding microstructural changes in Si_3N_4 ceramics quenched from elevated temperatures are presented.

EXPERIMENTAL PROCEDURES

Densification of the Si_3N_4 materials was performed by (i) hot pressing (NC-132, MgO-fluxed with a completely amorphous secondary phase, Norton Company, Northboro, MA), (ii) pressureless sintering of Si_3N_4 doped with 5 vol% Yb_2O_3+0.5 vol% Al_2O_3 (designated Yb/A) at 1800°C for 2 hrs under 0.1 MPa N_2, and (iii) by gas-pressure sintering of reaction-bonded Si_3N_4 (SRBSN) with various additives at 1925°C for 60 min under 10 MPa N_2.

Quenching thin specimens from elevated temperatures was performed by using a modifyed apparatus, originally designed for thermal shock testing of ceramics [9]. Quench rates sufficient to maintain the high-temperature microstructures were achieved, as described elsewhere [10]. Heating of the samples was perfomed with two 150-W tungsten halogen lamps positioned at the focal point of ellipsoidal gold-coated mirrors. Disc-shaped specimens of about 1 cm in diameter were thinned to about 200 μm thickness with the center dimpled to a final thickness of either 10 μm or 100 μm. Specimens, held between the tungsten lamps by a spring-loaded ZrO_2-insulated alligator clip, were heated to temperatures of 1350°C to 1550°C within 2 min and held at temperature for about 5 s. In addition, one experiment was performed at 1470°C for about 300 s. The temperature at the center of the specimens was monitored by an IR pyrometer. A water bath was used as the quenching medium. The compressed spring of the aligator clip was released, allowing the specimen to enter the water quenching bath at room temperature within 0.005 s, as determined by a light barrier [10].

TEM foils of the non-quenched and quenched samples were prepared using standard methods of grinding, dimpling, and ion milling to electron tranparency. Conventional transmission electron microscopy (TEM) and analytical electron microscopy (AEM) were performed using a JEM-2000FX, JEOL, Inc., Tokyo, Japan, operating at 200 kV, equipped with an energy-dispersive x-ray spectroscopy (EDS) ultra-thin window germanium detector. Internal interfaces in Si_3N_4 were studied by HREM using a JEM-4000EX, JEOL, Inc., Tokyo, Japan, microscope operating at 400 kV.

RESULTS AND DISCUSSION

Grain Boundaries at Room Temperature

In general, the overall Si_3N_4 microstructure consists of elongated β-Si_3N_4 grains surrounded by equiaxed fine-grained Si_3N_4 particles. Secondary phases are located at triple-grain pockets of the Si_3N_4-matrix grains. Apart from the secondary-phase pockets, which can be crystallized by additional heat treatments, all internal interfaces in these liquid-phase sintered materials consist of a thin layer of residual glass. The presence of such continuous intergranular films is independent of additive composition. The amorphous films were observed at interpenetrating grains as well as along curved and stepped boundaries. Figure 1 shows a HREM micrograph of a Si_3N_4-material doped with pure SiO_2 (note the presence of a continuous, thin intergranular film).

Although a number of different sintering aids were used for densification, the overall microstructure described above showed only minor variations. However, at close inspection, differences in grain-boundary film thickness were observed, depending on sintering aid composition. All liquid-phase sintered Si_3N_4 materials investigated, with different interface chemistry, revealed a characteristic intergranular-film thickness.

In order to verify the preliminary observations mentioned above, different TEM techniques for determining the intergranular-film thickness as well as a statistical approach by studying a number of different grain boundaries in one material were empolyed. Amorphous intergranular films can be detected by three different TEM techniques: (i) diffuse dark-field imaging, (ii) the Fresnel fringe technique, and (iii) by high-resolution lattice-fringe imaging [11-13]. It was shown that the most accurate results are obtained by the HREM-imaging technique [14].

Statistical analyses of a number of grain-boundary films in the Yb/A-doped material provided experimental verification of a theoretically proposed equilibrium film thickness [15-17]. Fifteen grain boundaries were investigated and the HREM imaging technique applied to each boundary by averaging three to four measurements from each HREM micrograph. The intergranular film thicknesses were evaluated at four locations at each interface. A Gaussian-like distribution of the film-thicknesses, with a standard deviation of about 0.05 nm, was observed. The mean value to the thickness measurements from a single grain boundary was similar to the one evaluated by statistical means. The results seemingly suggest a constrant film thickness for all high-angle grain boundaries. Moreover, the results give confidence to quantitatively determine grain-boundary film thickness from single HREM micrographs, with an accuracy better than 0.1 nm, and can be seen as experimental evidence for the existence of an equilbrium film thickness in Si_3N_4 ceramics.

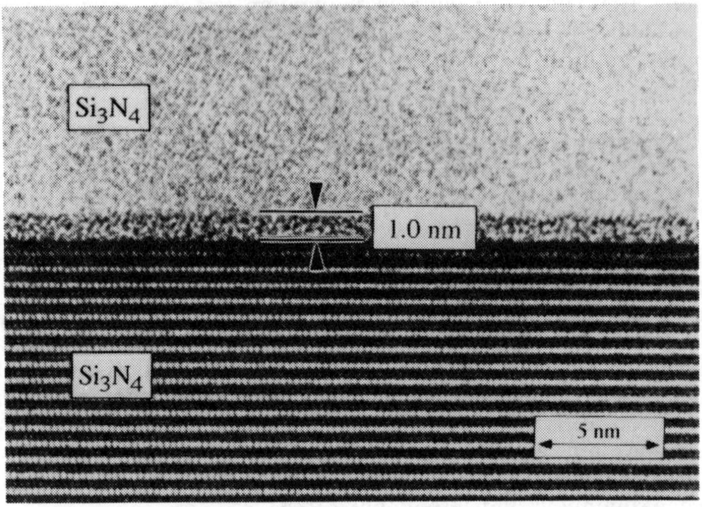

Figure 1. HREM micrograph of a Si_3N_4-material doped with 4.0 vol% SiO_2 as the only sintering aid. Note the presence of a continuous, thin intergranular film of 1.0 nm thickness.

In the following sections, experimental results of β-Si_3N_4-β-Si_3N_4 grain-boundary film thickness measurements, depending on additive-volume fraction, interface chemistry, impurities, and secondary-phase crystallization will be briefly summarized. It should be borne in mind that all measurements were performed by the aforementioned HREM-imaging technique with a given accuracy of about 0.1 nm. Experimental results from HREM imaging of interfaces at room temperature are given in the schematic drawing shown in Figure 2.

Volume Fraction: The intergranular secondary phase in liquid-phase sintered Si_3N_4 is silicate based, since SiO_2, which is present on the Si_3N_4 particle surfaces, reacts with the sintering aids during densification. Three reference materials were investigated, which contained SiO_2 as the only additive, however, the secondary-phase volume fractions differed (3.6, 4.0, and 10.5 vol% SiO_2). An increasing amount of the sintering aid results in larger triple-point regions containing the surplus material. In addition, HREM studies confirmed a continuous, amorphous film along the interfaces [5]. The determined film thickness was 1.0 nm for all three materials (see also Figure 1). It should be emphasized that for the pure SiO_2-doped materials, grain-boundary film widths were found to be independent of the volume fraction of the secondary phase.

Interface Chemistry: Opposite to the results for the SiO_2-doped materials, Y_2O_3+Al_2O_3-doped SRBSN materials showed an increase in grain-boundary film width with increasing secondary-phase content. The thickness of the intergranular films varied from 1.0 nm to 1.5 nm for the addition of 1.2 and 12.0 wt% Y_2O_3+Al_2O_3, respectively. Hence, a distinction between metal-oxide ratio and chemical composition of the intergranular phase is required. Although the Y_2O_3/Al_2O_3 ratio is constant for both compositions, the $(Y_2O_3+Al_2O_3)/SiO_2$ ratio differs for about a factor of 10, since the SiO_2 volume fraction is constant. The material with lower additive content forms a silica-rich secondary phase, giving a grain-boundary film thickness of 1.0 nm, as observed for pure SiO_2-doped materials. It is concluded that the secondary phase volume fraction has no major effect on grain-boundary film thickness. In this context, it should be noted that each material, with its given interface chemistry, reveals a characteristic film width. The addition of 5 wt% of unstabilized ZrO_2 as a sintering aid stabilized the thinnest intergranular films observed in liquid-phase sintered Si_3N_4 materials, with a film thickness of 0.5 nm. However, Sc_2O_3-doped SRBSN was shown to result in continuous grain-boundary films of 1.5 nm width [6]. Hence, resulting film thickness predominantly depends on interface chemistry, i.e., the sintering aids employed as well as impurities, which may have been incorporated into the material during processing.

Impurities and Secondary Phase Crystallization: Moreover, Yb_2O_3/Al_2O_3-fluxed Si_3N_4 materials were prepared with the addition of small amounts of CaO, i.e., 0.07 and 0.25 percent by volume [7]. CaO was selected since no solid-solution between CaO and Si_3N_4 is known. The CaO was expected to segregate to the internal interfaces, thus changing both interface chemistry and grain-boundary film width. HREM measurements confirmed an increase in film thickness from 1.0 nm to 1.3 nm with increasing CaO-addition.

Clarke et. al. reported on a theoretical approach rationalizing the presence of an equilibrium film thickness [15,16]. Segregation of Ca^{2+}-cations to the interface is thought to change the surface and space charge at the boundary as well as the dielectric properties of the residual glass. Therefore, the attractive van der Waals dispersion force and the repulsive forces (steric force and electrical double layer) across the interface are changed. This force balance stabilizes a characteristic film thickness, which is predicted to be wider for a high impurity level [16].

The experimental results showed that impurity cations can segregate to the interfaces and thereby change both interface chemistry and grain-boundary film thickness. Moreover, crystallization of secondary phases can also modify the chemical composition of the residual glass and hence influence film width. Different crystalline secondary phases were identified in the materials doped with

0.07 and 0.25 vol% CaO. For the low CaO addition, only $Yb_2Si_2O_7$ was found as the crystalline secondary phase after heat treatment. TEM investigations of the annealed 0.25 vol% CaO-containing specimen revealed, apart from the $Yb_2Si_2O_7$ phase, a second crystalline phase $CaYb_9(SiO_4)_6ON$ (Ca- and N-stabilized apatite). The incorporation of CaO into the crystal lattice strongly reduces the Ca^{2+}-cation concentration at the interface. However, if no impurities can be incorporated, the impurity level is greatly increased at the interface, owing to the reduced volume of residual glass after crystallization of triple-grain pockets. Therefore, the two materials doped with 0.07 and 0.25 vol% CaO showed a film thickness of 1.5 nm and 1.4 nm, respectively. It should be emphasized that even small changes in interface chemistry can strongly affect the intergranular film thickness, in particular if the impurities are not incorporated into the crystalline secondary phases formed upon heat treatment and thus can lead to a marked degradation in high-temperature properties [2-4].

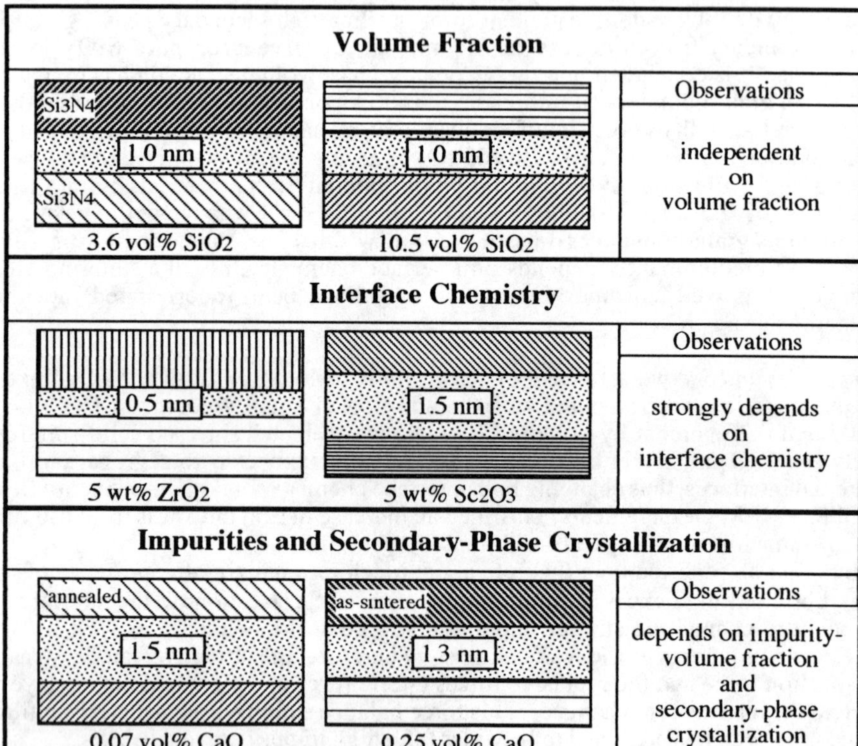

Figure 2. Schematic of grain-boundary film thickness development in various Si_3N_4 materials obtained by HREM imaging [5-7]. Note that the main influencing factors such as volume fraction, interface chemistry, impurities, and secondary-phase crystallization are indicated.

Grain Boundaries at Elevated Temperatures

In order to study high-temperature microstructures, disc-shaped samples of MgO-doped Si_3N_4 (NC-132) were quenched from 1350°C, 1420°C, 1500°C and 1550°C with a holding time at temperature of about 5 s [10]. In addition, one sample was quenched from 1470°C, held at temperature for about 300 s. The specimens entered the cooling medium within 0.005 s, which is rapid enough to freeze-in the high-temperature microstructure. It is assumed that cooling occured by linear heat transfer through the lateral surface by convection. It is worth mentioning that the overall microstructures of the materials quenched from high temperatures are similar to the microstructure of the sample cooled under normal processing conditions. No observable increase in the volume fraction of the secondary phase was found after quenching.

Below Eutectic Temperature: Previous studies of internal interfaces in the NC-132 material, using the diffuse dark-field and lattice-fringe imaging techniques, have reported a film thickness between <1nm to 5 nm [18,19]. In the present study, the intergranular film thickness of the as-received, normally cooled material was evaluated to be 0.8 nm (see Figure 3), which is consistent with the measurements reported by Clarke and Thomas (<1 nm) [19].

Figure 3. HREM micrographs of amorphous grain-boundary films in MgO-doped Si_3N_4 (NC-132), (a) at room temperature (RT) showing a film width of 0.8 nm, (b) quenched from 1420°C revealing a film thickness of 2.8 nm. Note that the film thickness in the as-received material and the sample quenched from 1350°C are identical, however, above the eutectic temperature a pronounced widening of the film thickness occurs.

HREM imaging of β-Si_3N_4-β-Si_3N_4 grain boundaries in the material quenched from 1350°C revealed no noticable changes in the intergranular films compared to those observed in the normally cooled material (compare Figure 3a). The intergranular film thickness was measured to also be 0.8 nm. Since no change in grain-boundary film thickness was observed up to a service temperature of 1350°C, it can be concluded that for short exposure times no microstructural changes occur below the eutectic temperature.

Above Eutectic Temperature: Although no changes in the overall microstructures were observed after quenching, HREM imaging of grain boundaries quenched from temperatures >1350°C revealed marked changes in intergranular film widths. Grain-boundary film thicknesses were observed ranging from 1.8 nm to 7 nm and from 1.5 nm to 3.5 nm in the samples quenched from 1420°C and 1500°C, respectively, implying a nonequilibrium state (Figure 3b). These film thicknesses are of the same magnitude as those reported by Clarke [8] in NC-132 quenched from 1450°C, using conventional bright-field imaging. The widening of the intergranular films, while exposed to temperatures above the eutectic temperature, can be rationalized by the solution of a small amount of Si_3N_4 into the newly formed liquid, assuming there is only little diffusion out of the grain-boundary region into the triple-grain pockets at such short exposure times. Assuming that the amount of Si_3N_4 equivalent to five atomic planes ({10-10} with an interplanar spacings of 0.66 nm) dissolves into the liquid, the film thickness would increase to about 4 nm. Therefore, it is concluded that the widening of the intergranular films is caused by rapid solution of Si_3N_4 into the liquid, which forms at the eutectic temperature.

In contrast to the aforementioned results, grain boundaries observed after quenching from 1550°C/5s as well as from 1470°C/300s revealed intergranular film thicknesses of 1.0 nm, similar to those in the normally cooled sample (compare schematic shown in Figure 4). Quenching from a higher temperature as well as a longer time at temperatures above the eutectic temperature allows for diffusional processes to occur such that an equilibrium film thickness can again be attained. While enhanced solution of Si_3N_4 into the newly formed liquid is expected slightly above the eutectic temperature, which causes the initial widening of the grain-boundary films, at higher temperatures (1550°C/5s) as well as longer holding times above the eutectic temperature (1470°C/300s) allows for diffusional processes to occur. Therefore, an equilibrium film thickness can be achieved by redistribution of liquid from two-grain junctions into triple-grain pockets. The observed small difference in film thickness of 0.2 nm between the as-received material and the sample quenched from 1550°C can be attributed to small changes in interface chemistry owing to the solution of Si_3N_4 into the glass.

Although an equilibrium state is again attained while exposing the material to rather high temperatures, it should be borne in mind that at these temperatures a liquid phase is present at the interface. Hence, potential applications of liquid-phase sintered Si_3N_4 materials at high service temperatures is strongly limited by the interface properties. Further improvement of the formation of more refractory grain-boundary phases is required.

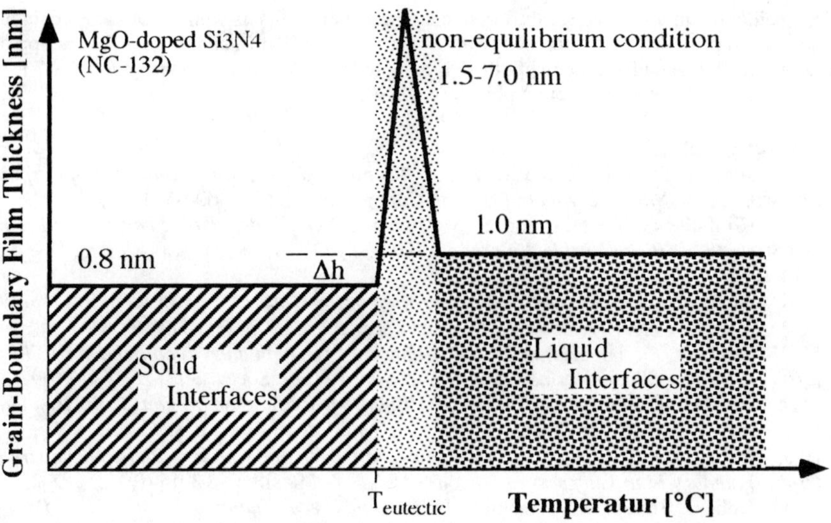

Figure 4. Schematic showing grain-boundary film thickness values obtained from HREM imaging before and after quenching of MgO-doped Si_3N_4 (NC-132).

SUMMARY

With a given accuracy of better than 0.1 nm, small changes in β-Si_3N_4-β-Si_3N_4 grain-boundary film thicknesses were evaluated employing the HREM-imaging technique. Intergranular film thickness was found to be independent of additive-volume fraction for SiO_2-doped materials. Si_3N_4 materials doped with various sintering aids revealed a characteristic film width for each material. Segragation of impurities to the interface typically results in a widening of the film. Moreover, crystallization of secondary phases can alter interface chemistry and hence change grain-boundary film width. Intergranular film thickness can be seen as a fingerprint of the corresponding interface chemistry.

High-temperature microstructures of an MgO-hot pressed Si_3N_4 (NC-132) were obtained by quenching thin samples from high temperatures. Specimens quenched at a temperature of 1350°C revealed no change in microstructure from those cooled under normal conditions. Specimens quenched from temperatures of 1420°C and 1500°C exhibited a pronounced widening of the amorphous film, indicating a non-equilibrium state. Upon quenching from 1550°C/5s as well as 1470°C/300s the intergranular films again adopted a near equilibrium thickness, which is similar to those observed in the as-received material.

Acknowledgements

I am greatly indebted to Prof. M. Rühle for supporting this work, which was performed during my stay at the Max-Planck-Institute in Stuttgart. I also would like to thank R. Mager and Dr.

G.A. Schneider for their assistance during quenching experiments as well as Dr. M.K. Cinibulk for his strong support with the microscopy involved. Dr. M.J. Hoffmann and Dr. E. Hampp are gratefully acknowledged for processing the Yb/A materials. Moreover, I thank the BMFT for financial support under the contract number NTS 0230/0.

References

1. D. A. Bonnell, T.-Y. Tien, and M. Rühle, "*Controlled Crystallization of the Amorphous Phase in Silicon Nitride Ceramics*," J. Am. Ceram. Soc., **70** [7] 460-65 (1987).
2. M. K. Cinibulk, G. Thomas, and S. M. Johnson, "*Grain-Boundary-Phase Crystallization and Strength of Silicon Nitride Sintered with a YSiAlON Glass*," J. Am. Ceram. Soc., **73** [6] 1606-12 (1990).
3. L. K. L. Falk and G. L. Dunlop, "*Crystallization of the Glassy Phase in an Si_3N_4 Material by Post-Sintering Heat Treatments*," J. Mater. Sci., **22** 4369-76 (1987).
4. M. K. Cinibulk, G. Thomas, and S. M. Johnson, "*Strength and Creep Behavior of Rare-Earth Disilicate - Silicon Nitride Ceramics*," J. Am. Ceram. Soc., **75** [8] 2044-49 (1992).
5. I. Tanaka, H.-J. Kleebe, M.K. Cinibulk, J. Bruley, M. Rühle, "*Amorphous Grain Boundary Films in SiO_2 Containing Si_3N_4 Ceramics*," J. Am. Ceram. Soc., in review.
6. H.-J. Kleebe, M. J. Hoffmann, and M. Rühle, "*Influence of Secondary Phase Chemistry on Grain Boundary Film Thickness in Silicon Nitride*," Z. Metallkd., **83** [8] 610-17 (1992).
7. H.-J. Kleebe, J. Bruley, and M. Rühle, "*HREM and AEM Studies of Yb_2O_3-Fluxed Si_3N_4 Ceramics with and without CaO Additions*," J. Europ. Ceram. Soc., (1994) in press.
8. D. R. Clarke, "*High-Temperature Microstructure of a Hot-Pressed Silicon Nitride*," J. Am. Ceram. Soc., **72** [9] 1604-09 (1989).
9. G. A. Schneider and G. Petzow, "*Thermal Shock Testing of Ceramics - A New Testing Method*," J. Am. Ceram. Soc., **74** [1] 98-102 (1991).
10. M.K. Cinibulk, H.-J. Kleebe, M. Rühle, "*Amorphous Intergranular Films in Silicon Nitride Ceramics Quenched From High Temperatures*," J. Am. Ceram. Soc., **76** [11] 2801-808 (1993).
11. D.R. Clarke, "*On the Detection of Thin Intergranular Films By Electron Microscopy,*" *Ultramicroscopy*, **4** 33-44 (1979).
12. J.N. Ness, W.M. Stobbs, and T.F. Page, "*A TEM Fresnel Diffraction-Based Method for Characterizing Thin Grain-Boundary and Interfacial Films,*" Phil. Mag. A, **54** [5] 679-702 (1986).
13. O.L. Krivanek, T.M. Shaw, and G. Thomas, "*Imaging of Thin Intergranular Phases By High-Resolution Electron Microscopy*," J. Appl. Phys., **50** 4223-27 (1979).
14. M. K. Cinibulk, H.-J. Kleebe, and M. Rühle, "*Quantitative Comparison of TEM Techniques for Determining Amorphous Intergranular Film Thickness*," J. Am. Ceram. Soc., **76** [2] 426-32 (1993).
15. D. R. Clarke, "*On the Equilibrium Thickness of Intergranular Glass Phases in Ceramic Materials*," J. Am. Ceram. Soc., **70** [1] 15-22 (1987).
16. D. R. Clarke, T. M. Shaw, A. P. Philipse, and R. G. Horn, "*Possible Electrical Double Layer Contribution to the Equilibrium Thickness of Intergranular Glass Phases in Polycrystalline Ceramics*," J. Am. Ceram. Soc., **76** [5] 1201-204 (1993).
17. H.-J. Kleebe, M. K. Cinibulk, R. M. Cannon, and M. Rühle, "*Statistical Analysis of the Intergranular Film Thickness in Silicon Nitride Ceramics*," J. Am. Ceram. Soc., **76** [8] 1969-77 (1993).
18. D.R. Clarke and G. Thomas, "*Grain Boundary Phases in a Hot-Pressed MgO-Fluxed Silicon Nitride*," J. Am. Ceram. Soc., **60** [11-12] 491-95 (1977).
19. O. L. Krivanek, T. M. Shaw, and G. Thomas, "*The Microstructure and Distribution of Impurities in Hot-Pressed and Sintered Silicon Nitrides*," J. Am. Ceram. Soc., **62** [11-12] 585-590 (1979).

GRAIN-BOUNDARY STRUCTURE AND HIGH-TEMPERATURE MECHANICAL RESPONSE IN LOW Ca-DOPED Si_3N_4

Isao TANAKA
Department of Materials Science and Engineering, Kyoto University,
Sakyo, Kyoto 606-01 JAPAN

ABSTRACT

High-purity Si_3N_4 (with 2.5 wt% glassy SiO_2) materials doped with 0 to 450 at.ppm of Ca, were prepared as a model system in order to understand the relationship between the impurity-modified structure of the grain boundary (gb) film and its behavior at high temperatures. Doped Ca was found to be localized in the amorphous gb film. The thickness of the gb film at two grain junctions depended sensitively on the Ca content, but was independent of grain misorientation in all materials within a scattering of ± 1 Å. Internal friction of the doped materials exhibited the superposition of a gb relaxation peak and high-temperature background. Apparent viscosity of the gb film at 1000 ~ 1200 °C determined from the peak was about 4 orders of magnitude greater than that of the bulk glass with the same composition, indicating the presence of structural ordering in the gb film. This is consistent with the "equilibrium thickness" model by Clarke: In this model, repulsive forces that originate from structural ordering in the gb film acts to balance the van der Waals attractive force, resulting in constant gb film thickness.

INTRODUCTION

Continuous siliceous glassy phase is present in most of polycrystalline Si_3N_4-ceramics. Since the intergranular glassy film acts as a short diffusion path at high temperatures, the overall diffusion-related response of the material, such as high-temperature strength and creep is determined by the properties of the intergranular phase. In order to improve these properties, efforts to manipulate the volume fraction as well as composition of the glassy phase (grain boundary chemistry) have been attempted for more than two decades. In the course of these investigations, the intergranular glassy film at two grain junctions was recognized to be very difficult to crystallize. The stability of the intergranular glassy film was first theoretically discussed by Raj and Lange[1]. Clarke[2,3] first pointed out the existence of a stable intergranular film having an equilibrium thickness based on a continuum approach. The presence of such an equilibrium gb film was experimentally confirmed by Kleebe, Rühle and coworkers[4-7] through systematic high-resolution electron microscopy (HREM) investigations of various Si_3N_4 -based materials. They also found that the equilibrium thickness depends only on the gb chemistry, and is independent of the volume fraction of the glassy phase. In other words, the amount of the glassy phase at two grain junctions of two different materials are the same when the gb chemistry and grain size distribution are identical.

High temperature strength shows little dependence on the volume fraction of glass, when the gb chemistry is the same. This is a natural consequence of the fact that the amount of glass at two grain junctions is constant when the volume fraction of the glass is changed. These relationships were most clearly demonstrated in a model system, high-purity Si_3N_4+glass-SiO_2 materials with different SiO_2 contents. Both high-temperature strength[8], and intergranular glassy phase thickness[9] in materials with different SiO_2 glass content were found to be the same.

Although both the equilibrium gb film thickness and high temperature strength are determined by the gb chemistry, there is no direct correlation between the thickness itself and high tempertaure strength according to the data by Kleebe et al[4-7]. It is still an open question how the grain boundary chemistry affects the grain-boundary structure as well as affects the high-temperature diffusion-related response.

The effects of low concentration of dopants on the high-temperature strength of Si_3N_4 have been systematically investigated by the present author and coworkers using undoped, high-purity Si_3N_4 material as a reference[10-12]. Doping of the order of 100 ppm Ca was found to clearly degrade the high-temperature properties[13]. Ca is a typical grain boundary segregant in β-Si_3N_4 and it is expected to modify the structure of the intergranular film significantly. HREM and nano-beam analytical electron microscopy was utilized to systematically study the effects of Ca in a series of low Ca-doped materials[14]. Results of these studies are joined in this paper and the observed high-temperature response is discussed based on the knowledge obtained through the HREM study.

EQUILIBRIUM INTERGRANULAR FILM THICKNESS

HREM observations of the intergranular glassy phase were systematically made for undoped and three Ca-doped materials. All materials were hot-isostatic-press sintered up to the relative density of > 99.5 %, and contained 2.5 wt% of SiO_2 delived from the surface oxide of the starting powder. Ca concentration of the four

samples were determined after sintering by ICP (inductively-coupled plasma spectrometry) to be 450 ppm, 220 ppm, 80 ppm and <1 ppm (undoped). Other cation impurity contents were less than 50 ppm. Figure 1 shows the TEM image of the undoped Si_3N_4 material (containing 2.5 wt% of SiO_2). The TEM specimen was tilted so as to orient the grain boundary film "edge-on" and both adjacent grains in good diffracting conditions. An intergranular glassy film, as exemplified in Fig. 1, always existed at high-angle grain boundaries in all materials. The film thickness was constant at all boundaries and independent of grain misorientation. Appearance of constant thickness even in a curved boundary maintaining "edge-on" condition confirms these observations. These results support the "equilibrium gb thickness" model by Clarke[2]. In this model, the intergranular film is determined by the repulsive steric force, which acts to balance the attractive van der Waals dispersion force when other impurities are absent.
Effects of grain boundary chemistry on the thickness value were systematically investigated using a series of Ca doped materials. Their HREM images are shown in Fig. 2. The thickness was measured for more than six different grain boundaries in each material, and the measured thickness was plotted as a function of Ca concentration in Fig. 3. The thickness of the intergranular film increased with the increase of Ca concentration above 80 ppm. It almost doubled by an increase from 80 ppm to 450 ppm. It is interesting that the thickness in the undoped material is larger than that in the 80ppm Ca material, suggesting that the thickness decreased with the increase of Ca concentration in the very dilute region.

Elemental analyses of the intergranular phase were made at two grain junctions by an STEM with a probe size of 8 Å. Ca was detected even in an 80 ppm Ca specimen in which the film thickness was only 7 Å as shown in Fig. 2. No impurities other than oxygen were detected in the undoped specimen in the intergranular phase. Ca was also not detected in the Si_3N_4 phase in any of the materials. The composition of the intergranular film can be estimated from the intensity ratio of Ca and O peaks assuming that all Ca and O are localized in the intergranular phase. The content at two grain junctions was found to be 1.5±0.3 mol% CaO / ($CaO+SiO_2$) for 80 ppm doped material. This value shows good agreement with a simple estimate that the glass contains 1.0 mol% CaO, by assuming that all Ca atoms in the 80 ppm material are localized in the intergranular glass, which is 3.6 vol% of the material.

The observed variation in thickness of the intergranular film with concentration of added Ca can be understood using Clarke's model[2,3]. Ca^{2+} is expected to disrupt the network structure of the silica intergranular phase, decreasing the structural correlation length and hence the magnitude of the steric repulsive force. Since

Fig. 1 HREM image of undoped Si_3N_4-SiO_2 material. Curved intergranular glassy phase tilted to the edge-on condition exhibits the constant film thickness of 10 Å.

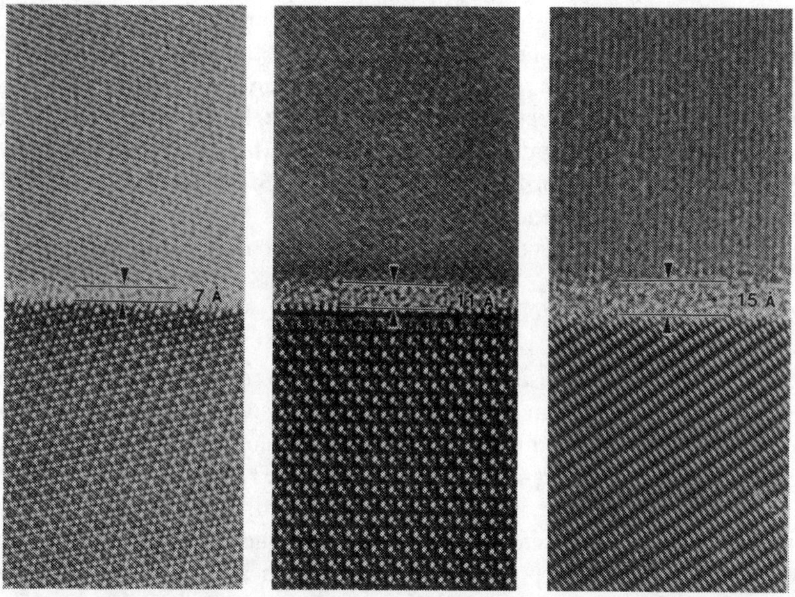

Fig. 2 HREM images of grain boundaries in a high-purity Si_3N_4 containing an SiO_2 intergranular phase with 80 ppm Ca, 220 ppm Ca, and 450 ppm Ca, from left to right, respectively.

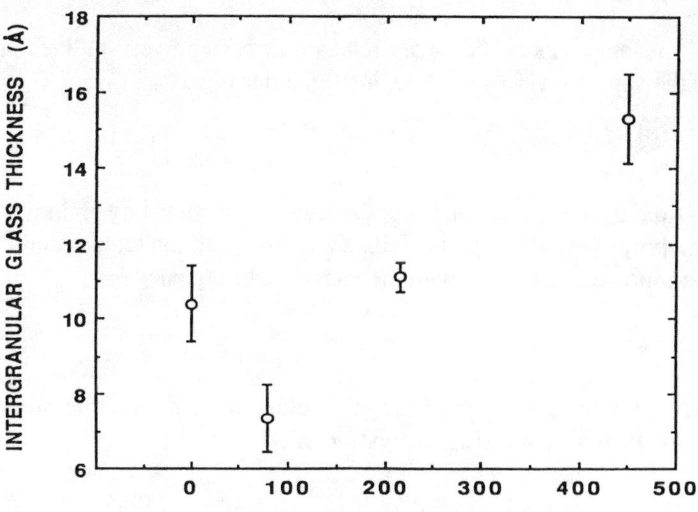

Fig. 3 Plot of equilibrium film thickness in high-purity Si_3N_4 as a function of Ca content.

Ca^{2+} is known to be a silica network modifier, minor Ca^{2+} additions may decrease the steric force and cause a reduction in the film thickness. The addition of Ca^{2+} also provides a charged species to the material which may result in the increase of the repulsive electrical double layer force, leading to further increases in film thickness. A minimum in thickness of the intergranular film at a low concentration of a network modifier, as observed at ~80 ppm Ca, is thus consistent with a competition between the effects of disrupting the network structure and the development of a force due to an electrical double layer.

APPARENT VISCOSITY OF THE INTERGRANULAR FILM

Subcritical crack growth resistance as well as creep resistance of these undoped and Ca-doped materials were evaluated at 1400 °C, and both of them were found to be significantly degraded with an increase in Ca content[13]. In order to extract the contribution of the intergranular glassy film to the high temperature properties, the internal friction of these materials were measured. Figure 4 shows the results measured by the torsional pendulum method at a frequency of 10 Hz. Three doped materials show the superposition of a peak with an exponential-like high-temperature background increase. Peak-top temperature is plotted in Fig. 5 as a function of Ca concentration. Although a peak was not found in the undoped material, the presence of a peak was assumed above the highest measurable temperature in our equipment, i.e., >1350 °C.

The time constant for grain-boundary relaxation, τ, in polycrystalline materials is related to the apparent viscosity of the intergranular film, η_{ap},

$$\tau = \alpha\, \eta_{ap}\, d/\, \delta, \qquad (1)$$

where δ is the film thickness and α is a constant determined by grain morphology and elastic properties of the grains with size d. At peak top temperature, τ should satisfy the following condition with the vibrational frequency f_0,

$$2\, \pi f_0\, \tau = 1 \qquad (2)$$

For polycrystals with idealized hexagonal close pack microstructure, Raj and Ashby[15] obtained the following expression of α,

$$\alpha = 1.14\, (1-\nu^2)\, /\, E \qquad (3)$$

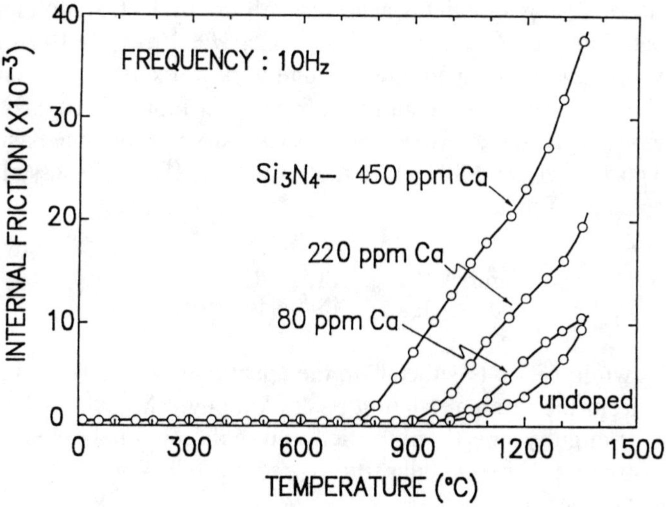

Fig. 4 Internal friction of undoped and low Ca-doped Si_3N_4 materials.

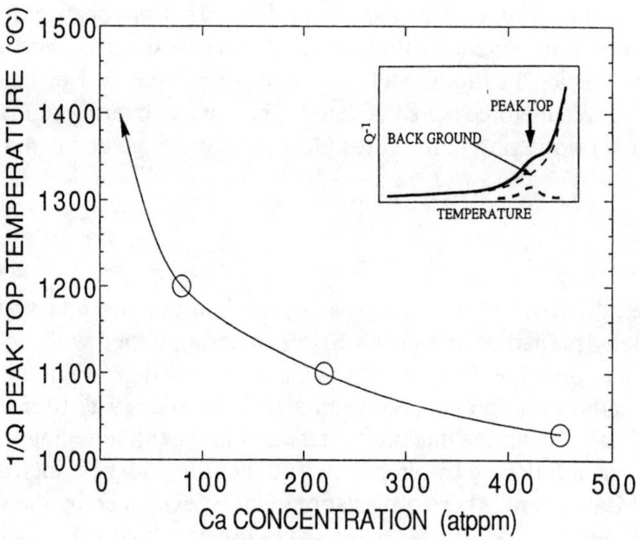

Fig. 5 Peak top temperature of internal friction as a function of Ca concentration.

where E and ν are Young's modulus and Poisson's coefficient. Adopting data of the present materials, i.e., $E = 300$ GPa, $\nu = 0.24$, $d = 1$ μm, $\delta = 10$ Å and $f_0 = 10$ Hz, η_{ap} at the peak top temperature is found to be 4.4×10^6 Pa s. Within the authors knowledge, no viscosity data for <u>bulk</u> SiO_2 - x mol% CaO glass system with $x < 38$ mol% is available. Therefore, the viscosity was obtained using data for pure SiO_2 glass[16], η_0, and SiO_2-38.8 mol%CaO glass[17], $\eta_{38.8}$, assuming the following simple law of mixing:

$$\frac{1}{\eta(x)} = \frac{38.8-x}{38.8}\frac{1}{\eta_0} + \frac{x}{38.8}\frac{1}{\eta_{38.8}} \quad . \qquad (4)$$

Results are shown in Fig. 6 together with the apparenmt viscosity of the glassy film obtained from the internal friction peak. Although the data of bulk glass contain some ambiguity due to simplified extrapolations, it is clear that the apperent viscosity of the intergranular film is aproximately 4 orders of magnitude greater than that of the bulk glass.

Greater viscosity of the intergranular glass can be interpreted to be due to the presence of structural ordering in the glassy film. This is consistent with the model of equilibrium intergranular film thickness used for the interpretation of the present HREM results. In this model, two repulsive forces act to balance the attractive van der Waals force are considered. The repulsive term originates from the higher degree of ordering in the silica glass structure at gb compared with the bulk glass.

CONCLUSION

High resolution microscopy investigations of the intergranular glassy film in a series of specially prepared high-purity Si_3N_4 material, doped with 0~450 ppm Ca, have shown that the film thickness is strongly dependent on the Ca concentration in the film. In undoped material, the thickness was 10 ± 1 Å. With the increase of Ca content, the thickness decreased in the dilute region (≤ 80 ppm Ca), which was then followed by a region where the film thickness increased with the increase of Ca content. These observations can be explained by the structural disjoining and electrical double layer repulsive forces which act to balance the attractive van der Waals force. Apparent viscosity of the gb film at 1000 ~ 1200 °C determined from the internal friction peak was about 4 orders of magnitude greater than that of the bulk glass with the same composition, indicating the presence of structural ordering in the gb film. This is consistent with the model used to explaine the thickness results.

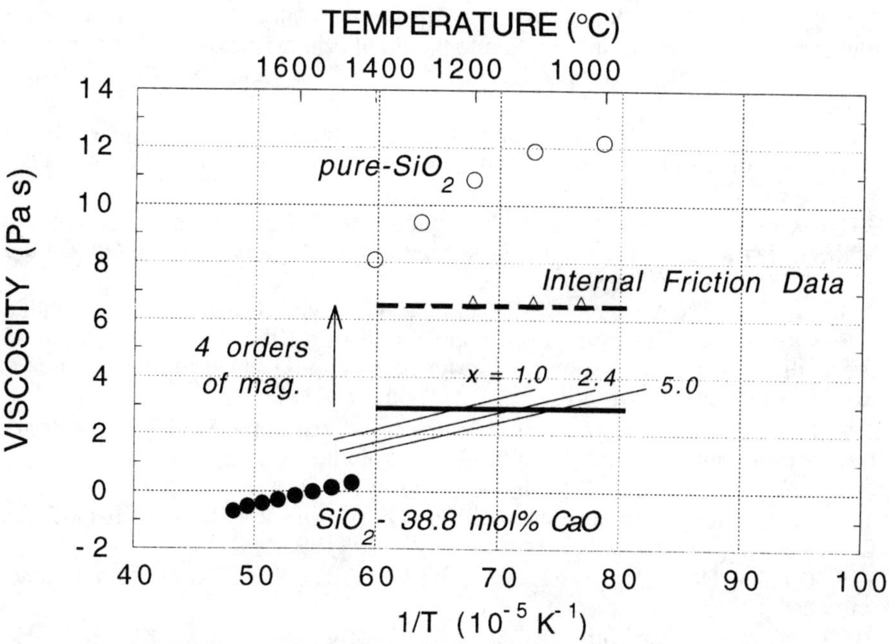

Fig. 6 Arrhenius plots of viscosity for bulk SiO_2 glass, SiO_2-38.8mol%CaO glass and SiO_2- x mol%CaO glass with x=1.0, 2.4, 5.0 obtained using eq(4). Apparent viscosity of the intergranular glass obtained from the present internal friction data were shown for comparison.

ACKNOWLEDGMENTS: The author is grateful to D.R. Clarke, R.M. Cannon, M.K. Cinibulk, H-J. Kleebe, M. Rühle and J. Bruley for helpful discussion. Internal friction measurements were done by K. Matsushita.

REFERENCES

1. R. Raj and F.F. Lange , "Crystallization of Small Quantities of Glass Segregated in Grain-Boundaries in Ceramics," *Acta Metall.*, 20 1993-2000 (1981).

2 D.R. Clarke, "On the Equilibrium Thickness of Intergranular Glass Phases in Ceramic Materials," *J. Am. Ceram. Soc.*, 70[1]15-22 (1987) .

3 D.R. Clarke, T.M. Shaw, A.P. Philipse, A.P., and R.G. Horn, "Possible Electrical Double-Layer Contribution to the Equilibrium Thickness of Intergranular Glass Films in Polycrystalline Ceramics," *J. Am. Ceram. Soc.* 76[5]1201-204 (1993) .
4 H-J. Kleebe, M.J. Hoffmann, and M. Rühle, "Influence of Secondary Phase Chemistry on Grain Boundary Film Thickness in Silicon Nitride," *Z. Metallkd.* 83 610-617 (1992).
5 H-J. Kleebe, M.K. Cinibulk, R.M. Cannon and M. Rühle, "Statistical Analysis of the Intergranular Film Thickness in Silicon Nitride Ceramics," *J. Am. Ceram. Soc.*, 76[8] 1969-77(1993) .
6 H-J. Kleebe, J.S. Vetrano, M.K. Cinibulk, I. Tanaka, J. Bruley, and M. Rühle, "High Resolution Electron Microscopy Studies on Silicon Nitride Ceramics", Proceedings of the Workshop on "Tailoring of High-Temperature Properties of Si_3N_4-Ceramics (Schloss Ringberg 1993), in press.
7 H-J. Kleebe, "High Resolution Electron Microscopy Studies on Internal Interfaces in Silicon Nitride Ceramics", This volume.
8 J. Zeng, I. Tanaka, Y. Miyamoto, O. Yamada, and K. Niihara, "High-Temperature Strength and Cavitation Threshold of Silicon-Nitride-Silica Ceramics," *J. Am. Ceram. Soc.*, 75[1] 195-200 (1992).
9 I. Tanaka, H-J. Kleebe, M.K. Cinibulk, J. Bruley, M. Rühle, unpublished work 1993.
10 I. Tanaka, G. Pezzotti, T. Okamoto, Y. Miyamoto and M. Koizumi, "Hot Isostatic Press Sintering and Properties of Silicon Nitride without Additives," *J. Am. Ceram. Soc.*, 72 1656-60 (1989).
11 I. Tanaka, G. Pezzotti, K. Matsushita, Y. Miyamoto and T. Okamoto, "Impurity Enhanced Cavity Formation in Si_3N_4 at Elevated Temperatures," *J. Am. Ceram. Soc.*, 74 [4] 752-59(1991).
12 I. Tanaka, G. Pezzotti, "Delayed Failure Resistance of High-Purity Si_3N_4", *J. Am. Ceram Soc.*, 75 1023-25(1992) .
13 I. Tanaka, K. Igashira, T. Okamoto, K. Niihara, and R.M. Cannon,"High-Temperature Fracture Mechanism of Low Ca Doped Silicon Nitride, *J. Am. Ceram. Soc.*, in press.
14 I. Tanaka, H-J. Kleebe, M.K. Cinibulk, J. Bruley, D.R. Clarke and M. Rühle, " Calcium Concentration Dependence of the Intergranular Film Thickness in Silicon Nitride", *J. Am. Ceram. Soc.*, in press.
15 R. Raj and M.F. Ashby, "On Grain-Boundary Sliding and Diffusional Creep" *Met. Trans.* 2 1113-1127(1971).
16 G. Hetherington, K.H. Jack and J.C. Kennedy," The Viscosity of Vitereous Silica," *Phys. Chem. Glasses,* 5[5] 130-136 (1964).
17 J. O'M. Bockris and D.C. Lowe, "Viscosity and the Structure of Molten Silicates," *Proc., Roy. Soc.*, A226 423-35 (1954).

THE $15R \rightarrow 3C$ STRESS-INDUCED TRANSFORMATION IN SiC

Jinwei Yang, Nils Nordell*, and Pirouz Pirouz
Department of Materials Science and Engineering, Case Western Reserve University, Cleveland, OH 44106-7204, U.S.A.
*Now at Industrial Microelectronics Center, P. O. Box 1084, S-164 21 Kista, Sweden.

ABSTRACT

A recently developed dislocation model for polytypic transformation in SiC is briefly described. Some experiments carried out on the basis of this model are then presented. The experiments consist of indenting single crystals of the $15R$ α-SiC polytype in the temperature range 1100-1300°C in vacuum with a load of 300 g. The plastic zone around the indents have then been investigated by transmission electron microscopy (TEM). Within the plastic zone of the indents, bands of cubic β-SiC ($3C$) are found indicating that a $\alpha \rightarrow \beta$ polytypic transformation has taken place. This stress-induced transition is consistent with the proposed dislocation model for polytypic transformation.

1. INTRODUCTION

In addition to being an important material as a structural ceramic, SiC is also a potential electronic material as a wide band-gap semiconductor. The applications of SiC for the former purpose go back to the beginning of this century and, presently, SiC is perhaps the most widely used reinforcement in structural composites. The electronic applications of SiC are more recent, basically because the crystal quality requirements in electronic applications are much more stringent. It is only in the last 10-15 years that new crystal growth techniques have been developed to a stage where single crystals of SiC with a quality sufficiently good to be employed in electronic devices have become available. A serious limitation in the employment of SiC as an electronic material has always been that this material frequently grows as a mixture of many different phases. These phases, called polytypes, occur quite readily in SiC while they are absent in most other semiconductors. Each polytype has different electronic properties (e.g., band gap and charge carrier mobility). The occurrence of various different phases with different structures and different electronic properties is a major problem because it means that the material is electronically heterogeneous. The simultaneous presence of different polytypes also entails the existence of interphase interfaces. Such boundaries can scatter charge carriers and act as regions with a high density of recombination centers.

In general, one may look at polytypism in SiC from two different points of view: (i) the thermodynamic stability of SiC at a given pressure and temperature, and (ii) the mechanism by which one polytype transforms to another.

Despite extensive work, the thermodynamic stability of different polytypes is still not clear (see, however, ref. 1). Clearly, at a particular temperature, only one polytype can be thermodynamically stable; this implies that the other phases in a mixed polytype material are in a metastable state. However, presumably because the free energy of the different polytypes is very close to each other, many of them can occur simultaneously at any temperature. Not only is the stability of different polytypes not known, but the mechanism by which one polytype transforms to another is not understood either. Clearly, the capabilities of SiC can only be realized when single polytype materials with a very low density of defects can be routinely grown. This is only possible when conditions under which a particular polytype occurs and transforms to another is understood.
One of the first discussions of the stability of a polytype was by Jagodzinski who proposed that the stability of a polytype was predominantly determined by the vibrational states of the crystal[2]. A different approach was taken by Knippenberg[3], and later Inomata *et al.*[4], who built up stability diagrams for SiC based on empirical observations. These diagrams, however, have not been very successful in predicting the occurrence of a particular polytype under a given set of thermodynamic conditions. This has been partly because the occurrence of a polytype depends on many different factors, such as the effect of non-stoichiometry and different impurities which tend to stabilize different phases. The most recent approach to the stability regimes of different polytypes is to perform total energy calculations of different polytypes using the *ab initio* methods of quantum mechanics which are now possible given powerful

computers. Using such methods, Heine and collaborators[1] have had some success with theoretical comparison of the energy, and thus the stability, of various polytypes at different pressures and temperatures. The experimental build-up of a stability diagram, or the theoretical calculation of the minimum energy phase, provides little information about the mechanism of transformation, although they do tell something about the driving force for the phase transformation.

In discussing the mechanism of polytypic transformation, two routes are available: (a) the transformation occurs by a solid state diffusional process whereby atoms change their configuration under some thermodynamic driving force, or (b) it occurs by a dislocation mechanism whereby the glide of partial dislocations changes the stacking sequence of the crystal. A proper diffusional model has never been proposed in detail, although many people have mentioned such a possibility. A dislocation mechanism for polytypic transformation was first proposed by Frank[5], which was itself based on his dislocation theory of crystal growth[6]. These models have been widely reviewed (see, e.g. ref. 7) and will not be discussed here. A new dislocation model for polytypic transformations based on a very different set of hypotheses than Frank's model was proposed a few years ago[8]. This has been recently reviewed[9] and its details will not be described here. Instead, a very brief outline of the model is given in section 2, and then some recent experiments that have been performed to test the validity of this model are presented (section 3). It should be mentioned that the problem of polytypic transformation is complex and no existing model can claim to provide an explanation for every observable aspect of it.

2. MECHANISM OF POLYTYPIC TRANSFORMATION

The model proposed in Ref. 8 is based on a number of points and assumptions each of which is described in more detail in Ref. 9 to which reference should be made for clarification.

a. A screw dislocation in SiC dissociates into two partials which have cores consisting of atoms of one type only. The core of one partial consists of silicon atoms only, while the core of the other partial consists of carbon atoms only.

b. Partial dislocations in SiC belong to the "glide set" rather than the "shuffle set". This is important in two respects: first the motion of a glide partial dislocation changes a SiC tetrahedron to its twinned variant as it should. Because of this, its motion is probably much easier than a shuffle dislocation which severely distorts the bonding configuration in a SiC tetrahedron. Second, the bonds in the core of glide partial dislocations can reconstruct with relative ease in order to eliminate the broken bonds. This is not the case in the core of shuffle partial dislocations. Using the convention proposed in Ref. 10, we call the two partials of a screw dislocation Si(g) and C(g).

c. Because of reconstruction in the core of the Si(g) and C(g) partial dislocations, their movement necessitates the breakage of Si-Si and C-C bonds, respectively. This is because the motion of dislocations in SiC, which has a high Peierls energy, involves the formation of kink pairs and kink migration. From the comparative strength of Si-Si and C-C bonds, it is concluded that the activation energy for the motion of the two partials are different and thus they can have very different mobilities in a certain range of temperature. It also follows that the Si(g) partial can probably move at lower temperatures and under lower stresses than the C(g) partial.

d. It is now assumed that a screw dislocation segment, that is dissociated into a Si(g) and a C(g) partial, is pinned and experiences a resolved shear stress on the primary glide plane, τ_{pgp}. Further it is assumed that the Si(g) partial is the leading and the C(g) is the trailing dislocation. Under the action of τ_{pgp} each partial bends in a manner similar to the behavior of a dislocation in a Frank-Read source. However, since, in the range of temperature under consideration, the mobility of the two partial dislocations is widely different, the fast Si(g) partial forms a loop before the slow C(g) has a chance to bend to the critical radius.

e. The loop that forms by the Si(g) partial is faulted and within it the stacking sequence has changed. The important point here is that unlike a perfect dislocation, a partial dislocation cannot repeat its action of forming a loop a second time. In fact, by the Friedel-Escaig mechanism[11-13], the screw dislocation automatically cross-slips after one loop has formed[14].

f. Because of the geometry of dissociation on the primary glide plane *vs.* that on the cross-slip plane, the core of the partials switches in the cross-slip process, e.g. a Si(g)/C(g) dissociated screw dislocation on the primary glide plane becomes a C(g)/Si(g) screw dislocation on the cross-slip plane. As a result, and again by the Friedel-Escaig mechanism, there is an immediate tendency for the screw dislocation to double-cross-slip back onto the primary glide plane.

g. In the cubic phase, i.e. β-SiC, the sequence of faulted loop formation followed by double-cross-slip* can take place on each neighboring plane and the stacking sequence changes to that of a twinned material. In this case, a stress on the cross-slip plane, τ_{csp}, is not necessary for the flipping of the screw dislocation from one plane to a neighboring one.

h. In a non-cubic crystal, i.e. α-SiC, double cross-slip can take place only on those primary glide planes on which the stacking rule is not violated. This automatically transforms the crystal to one with the stacking sequence of cubic SiC and is the basis of the $\alpha \rightarrow \beta$ polytypic transformation.

i. Consideration of the back stress, τ_{bs}, of loops surrounding the source dislocation in a cubic crystal leads to the continuation of cross-slip in order to 'escape' the back stress such that it can get back onto the primary glide plane. This results in missing of a number of planes before double cross-slip occurs. Here the resolved shear stress, τ_{csp}, on the cross-slip plane does play an important role. The final result is the sequential formation of a nd_{111}-thick twin band followed by cross-slip through a md_{111} thick untwinned band, where d_{111} is the $\{111\}$ interplanar spacing in β-SiC and n and m are integers. The values of n and m, which depend on the ratio τ_{pgp}/τ_{csp} determine the particular α polytype that forms. This is the basis of the $\beta \rightarrow \alpha$ polytypic transformation in the present model.

From the above summary, a polytypic transformation is stress-driven although the required stresses are quite moderate because the transformation is primarily carried out by the motion of the Si(g) partial dislocation. This is in contrast to slip in SiC where the motion of both Si(g) and C(g) partials are necessary and the process is controlled by the

---

* In this case, it is more a re-configuration of the screw dislocation core from one plane to a neighboring one rather than an actual double-cross-slip of the dislocation.

slow C(g) partial dislocation. Thus, thermal stresses in an annealing process may be quite sufficient to drive the Si(g) partial to form faulted loops and change the stacking sequence leading to a polytypic transformation. In fact, according to this model, a polytypic transformation depends on the relative contribution of resolved shear stresses on the primary glide plane (τ_{pgp}) and on the cross-slip plane (τ_{csp}). Temperature enters the problem because it is this parameter which primarily determines the asymmetry in the mobility of the partial dislocations. Impurities, in particular donor or acceptor impurities, can be important in a polytypic transformation because they are known to affect dislocation velocity in covalent materials.

This model is presently being verified by performing various experiments involving deformation of SiC under different conditions, e.g. temperature, crystal orientation, and strain rate[15]. Some of the deformation tests in SiC require low temperatures where plastic deformation of SiC is very difficult. In fact, the only way that such deformations can be carried out are those in which a hydrostatic component is present so that brittle fracture of the material is inhibited. The simplest types of such deformations involving high temperature indentation tests have already produced some very interesting results on the $6H \rightarrow 3C$ polytypic transformation of SiC[16]. One of the problems with indentation tests is that the stress field is very complicated and not liable to analysis. The conditions are much clearer if uniaxial compression tests under confining pressures could be carried out.

In the remainder of this paper, some new experiments are described which were designed to check the validity of a stress-induced polytypic transformation. In these experiments, specimens of 15*R*-SiC crystals were indented in the range 1000-1300°C to induce a transformation to the 3*C* polytype. These experiments are similar to those which were carried out on 6*H*-SiC polytype in Ref. 16 where a partial $6H \rightarrow 3C$ phase transformation was observed in the plastic zone of the indents.

3. EXPERIMENTAL

The SiC crystal examined here was prepared in Kyoto University by the modified sublimation technique[17]. Using the transmission mode in the optical microscope, the crystal consisted of regions with three different colorations corresponding to the three polytypes 15*R*, 4*H*, and 6*H*. After optical examinations, thin TEM foils from the crystal were prepared in the <11$\bar{2}$0> orientation. Six of the thin foils were examined by electron diffraction: five of the foils were exclusively 15*R*, while the sixth foil was exclusively 4*H*. In all the foils, the spots in the diffraction patterns were sharp with little streaking. This indicates that the crystal was of a relatively good quality; in fact, only a few faults were observed all lying on the basal planes.

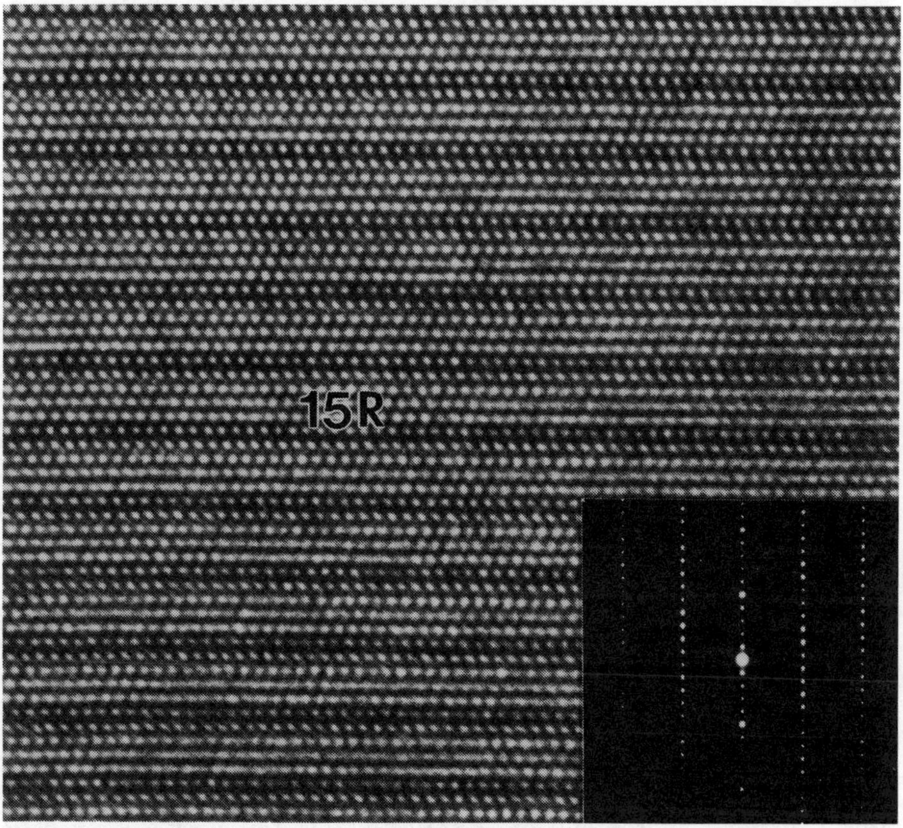

Fig. 1. <11$\bar{2}$0> HREM micrograph of the as-received 15*R*-SiC crystal. Inset is the diffraction pattern. Areas a few microns away from the indents in an indented crystal (e.g. Fig. 2) also look the same.

A few 3 mm discs parallel to the $\{11\bar{2}0\}$ plane were next polished to a thickness of ~500 μm using different grits of sand paper followed by 1 μm diamond paste. These discs were indented at 1100°C, 1200°C, and 1300°C with a load of 300 g and a dwell time of 30 s. Subsequent to this the discs were thinned to electron transparency from the back side (the side opposite to the indented face) and examined in a JEOL 200CX for conventional strain contrast TEM, and a JEOL 4000EX for HREM.

4. RESULTS AND DISCUSSION

Fig. 1 shows a typical high resolution image of an as-received 15*R* region before indentation; regions a few microns away from an indentation look very similar. This region is defect-free and its 15*R* structure is clear from the inset $<11\bar{2}0>$ diffraction pattern: four spots are clearly visible between the transmitted 0-beam spot and the 0 0 0 15 (strong first) spot.

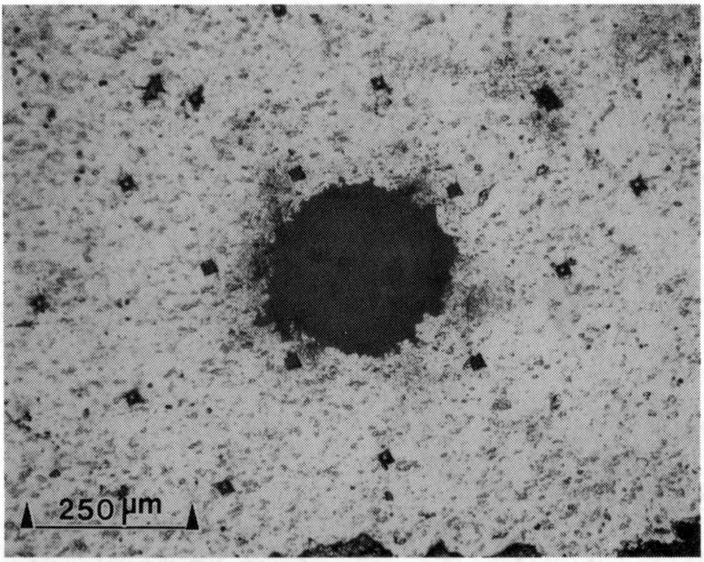

Fig. 2. Optical micrograph of a $<11\bar{2}0>$ TEM foil with an array of indents. The dark patch at the center of the figure is the hole produced by ion beam thinning.

The high resolution image shows a $(32)_3$ sequence, or alternate 3 and 2 layers repeated three times, corresponding to a stacking sequence of ...*ABCB'A'BCAC'B'CABA'C'*... (the letters indicate Si-C tetrahedra at different stacking positions, and the primed letters denote the twinned variants of the corresponding tetrahedra at the same sites; for more detail, see Ref. 9).

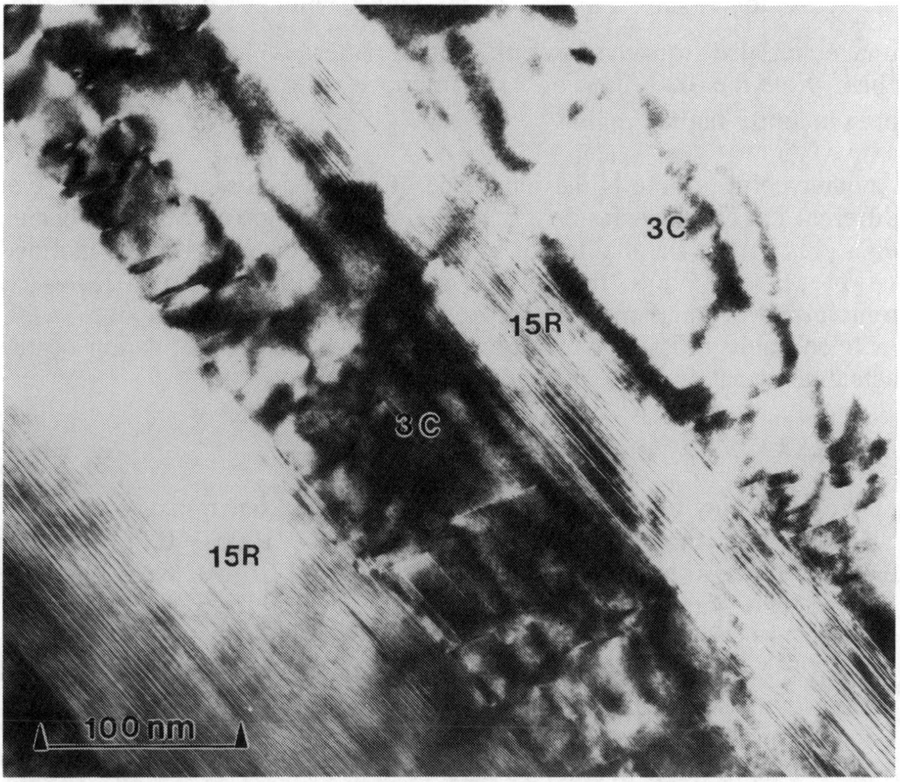

Fig. 3. A bright-field multi-beam TEM micrograph from an edge of an indent in Fig. 2. Note the interleaving bands of 3*C*- and 15*R* -SiC. The 3*C*-SiC bands have been produced by a stress-induced polytypic transformation of the 15*R*-SiC crystal.

Fig. 2 is an optical micrograph of the array of indents on a TEM foil; the dark patch in the center of the figure has been thinned away during specimen preparation. It is the regions around this hole that have been investigated by TEM.

A higher magnification of part of an indentation plastic zone at the edge of the hole is shown in Fig. 3. This region consists of a mixture of long (>1 μm) bands of 15*R*- and 3*C*-SiC with widths of about 100 nm. The different polytypes in these regions were characterized by electron diffraction and also by high resolution TEM. A HREM micrograph of the 3*C* region is shown in Fig. 4; the cubic symmetry is clear from the image and also from the inset diffraction pattern which is along a $<1\bar{1}0>$ zone. Note the $(0001)_{15R}//(111)_{3C}$ interface which extends over a ~5-10 nm randomly faulted region.

Another point to note is the incidence of single faults occurring on two different {111} planes inside the 3*C* region; while one of the faults occurs on a {111} plane parallel to the interface, the other occurs on an inclined {111} plane. While the former faults are related to the 15*R*→3*C* transformation, and may be the residues of this transition, the latter, inclined, faults must have clearly occurred after the transformation by the actual deformation of the 3*C* band.

5. CONCLUSION

We have now observed a non-cubic→cubic partial transformation of SiC after deformation in the temperature range 1000-1500°C in three different types of experiments. Single crystals of 6*H*-SiC were observed to undergo partial transformation to the cubic 3*C* phase after compression[15] and also after high-temperature indentation[16]. In the present experiments, 15*R* single crystals have also been observed to undergo stress-induced partial transformation to the cubic phase in the same temperature range. On the other hand, some recent compression experiments in the temperature range 1650-1920°C by Tsurekawa *et al.*[18] on polycrystalline β-SiC show that, in agreement with many other experiments, a partial $\beta\to\alpha$ transformation takes place. A significant point in the work of Tsurekawa *et al.*[18] is that these authors could not find a $\beta\to\alpha$ transformation just by annealing the same β-SiC polycrystal at 1920°C suggesting that deformation plays an important role in polytypic transformation of SiC. In brief, the results appear to suggest that the $\beta\leftrightarrow\alpha$ transformation is stress-induced and that the direction of transformation depends on the temperature: below ~1500°C for $\beta \rightarrow \alpha$ and above ~1500°C for the $\alpha \rightarrow \beta$ transformation. These temperature ranges are also consistent with the stability diagrams of Knippenberg[3] and Inomata *et al.*[4].

ACKNOWLEDGMENT

We are grateful to Professor H. Matsunami for providing us with the 15*R*-SiC crystal. This work has been supported by grant numbers DE-FG02-93ER45496 from DoE and NAG 3-758 from NASA.

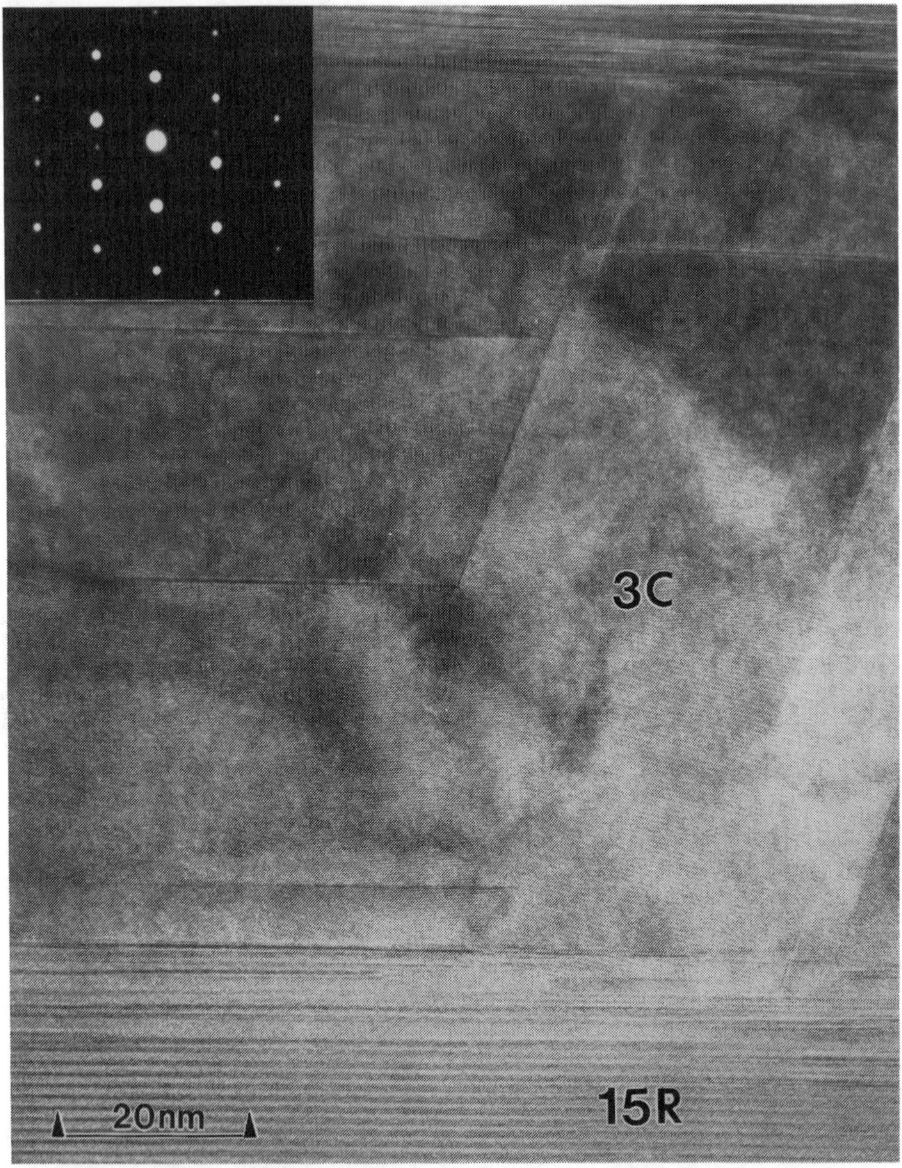

Fig. 4. A <11$\bar{2}$0> HREM micrograph of one of the 3C-SiC bands in Fig. 3. Note the inclined and the horizontal faults in the center of the band. The horizontal faults are parallel to the $(0001)_{15R}//(111)_{3C}$ interface.

REFERENCES

[1]V. Heine, C. Cheng, G. E. Engel and R. J. Needs, "The Origin of Polytypes in SiC and ZnS," *Mat. Res. Soc. Symp. Proc.*, **242** 507-518 (1992).

[2]H. Jagodzinski, "Wachstums-bzw. Umwandlungspolytypie beim SiC und ZnS," *N. J. Min. Monatsch*, **3** 209-225 (1954).

[3]W. F. Knippenberg, "Growth Phenomena in Silicon Carbide," *Philips Research Reports*, **18** 161-274 (1963).

[4]Y. Inomata, A. Inoue, M. Mitomo and H. Suzuki, "Relation Between Growth Temperature and the Structure of SiC Crystals Grown by the Sublimation Method," *Yogyo-Kyokai-Shi*, **76** 313-319 (1968).

[5]F. C. Frank, "The Growth of Carborundum: Dislocations and Polytypism," *Phil. Mag.*, **42** 1014-1021 (1951).

[6]F. C. Frank, "The Influence of Dislocations on Crystal Growth," *Disc. Faraday Society*, **5** 48-54 (1949).

[7]N. W. Jepps and T. F. Page, "Polytypic Transformations in Silicon Carbide," *Crystal Growth and Characterization*, **7** 259-306 (1983).

[8]P. Pirouz, "Dislocation Mechanisms for Twinning and Polytypic Transformations in Semiconductors," *Inst. Phys. Conf. Ser.*, No. **104** 49 (1989).

[9]P. Pirouz and J. W. Yang, "Polytypic Transformations in SiC: The Role of TEM," *Ultramicroscopy*, **51**(6) 189-214 (1993).

[10]H. Alexander, P. Haasen, R. Labusch and W. Schröter, Foreword to *J. Physique*, **40** Colloque C6 (1979).

[11]J. Friedel: "Regarding Seeger's Paper on Work Hardening," in *Dislocations and Mechanical Properties of Crystals.*, edited by J. C. Fisher. (John Wiley: New York, 1957).

[12]B. Escaig, "Sur le Glissement Dévié des Dislocations dans la Structure Cubique a Face Centrées," *J. Physique*, **29** 225-239 (1968).

[13]J. Rabier and P. Boivin, "Dislocation Mobilities and Low-Temperature Macroscopic Plasticity of III-V Compound Semiconductors," *Philos. Mag. A*, **61**(4) 673-683 (1990).

[14]P. Pirouz and P. M. Hazzledine, "Cross-Slip and Twinning in Semiconductors," *Scripta Met.*, **25**(5) 1167-1172 (1991).

[15]J. W. Yang, T. Suzuki, P. Pirouz, J. A. Powell and T. Iseki, "Stress-Induced Polytypic Transformation in 6*H*-SiC," *Mat. Res. Soc. Symp. Proc.*, **242** 531-536 (1992).

[16]J. W. Yang and P. Pirouz, "The $\alpha \rightarrow \beta$ Polytypic Transformation in High-Temperature Indented SiC," *J. Mater. Res.*, **8**(11) 2902-2907 (1993).
[17]Y. M. Tairov and V. F. Tsvetkov, "Investigation of Growth Processes of Ingots of Silicon Carbide Single Crystals," *J. Crystal Growth*, **43** 209-212 (1978).
[18]S. Tsurekawa, Y. Hasegawa, K. Sato, Y. Sakaguchi and H. Yoshinaga, "Effect of Crystal Structure on High-Temperature Deformation Behavior of Silicon Carbides," *Materials Transactions, JIM*, **34**(8) 675-681 (1993).

Evaluation

EFFECT OF GRAIN BOUNDARY STRUCTURE AND SEGREGATION ON ELECTRICAL RESISTIVITY IN ALUMINUM

Isao Nakamichi

Cryogenic Center and Faculty of Science, Hiroshima University, Higashi-Hiroshima 724, Japan.

Recent results of the electrical resistivity measurement of a single grain boundary using Superconducting Quantum Interference Devices (SQUID) on zone-refined Al and dilute Al-Ag are reviewed. On pure aluminum, the resistivity has been found to depend on the boundary type and structure systematically. This is well explained by the boundary dislocation model. The measurement on Al-Ag has shown that migrating boundaries cause a large amount of segregation. It suggests also that the resistivity of solute segregating around boundaries becomes an order of magnitude smaller.

1. INTRODUCTION

Electrical resistivity has two possibilities of giving information on grain boundaries. One is that the resistivity now depends on the boundary structure and gives the information on boundary structure. The other is that the resistivity gives information on the impurity segregation, because the electrical resistivity is very sensitive to the impurities.

Metals can give the intrinsic electrical resistivity due to boundary structure which is not affected by impurities, because super-pure material of 99.9999% purity is relatively easily obtainable on many metals. This enables us to separate the resistivity contributions due to boundary and impurities. However, the resistivity due to boundaries is orders of magnitude smaller in metals than in the other materials such as ceramics and is difficult to be measured. Thus, such kind of study has not been done for a long time.

Recently, the measurements with Superconducting Quantum Interference Devices (SQUID) have given the resistivity due to a single boundary in aluminum[1-5]. One set of experiments has revealed that the electrical resistivity depends systematically on the boundary structure[1,2,4]. The other set of experiments has given the information on the solute segregation mechanism and the effect of segregation on resistivity[2,3,5]. Here, these experiments are reviewed together.

2. EXPERIMENTAL

2.1 Specimen preparation

Two types of specimens were prepared. One is to obtain the intrinsic boundary resistivity. Zone-refined Al bicrystals with the cross section 1x3 mm^2 were used for this. Their bulk residual resistance ratio was 30,000 (99.9999% pure). The other is to study both the boundary-segregation effect on resistivity and the segregation mechanism. Al-60 and -106 at.ppm Ag bicrystals with the cross section 1x2 mm^2 were prepared from silver of purity 99.999% and zone-refined Al for this.

Both types of specimens were prepared by the strain-annealing method using a moving furnace in which the highest temperature was 873 K; grains were grown to a few centimeters in length. The speed of moving furnace, which would be nearly equal to the grain growth speed, was 12 mm/h for Al-Ag specimens.

Before electrical measurements, zone-refined Al specimens were annealed at 823 K for two hours and Al-60 at.ppm Ag specimens were annealed at 853 K for 10 hours. They were followed by the cooling in the furnace.

2.2 Resistivity measurement of a single boundary

The electrical resistivity due to a single boundary was obtained by the following method. As shown in Fig.1, resistances were measured for a small part with a boundary (B part) and two grain parts (G1 and G2) contiguous to the boundary on a bicrystal, at 4.2 K with a SQUID and at 300 K with a potentiometer. From these resistances, the residual resistivities ρ_B, ρ_{G1} and ρ_{G2} were calculated for the three parts.

The specific boundary resistivity ρ_{gb} at 4.2 K was obtained by the following equation

$$\rho_B - (\rho_{G1} + \rho_{G2})/2 = \rho_{gb}S/V, \qquad (1)$$

where S is the cross-sectional area of the specimen and V is the volume producing ρ_B.

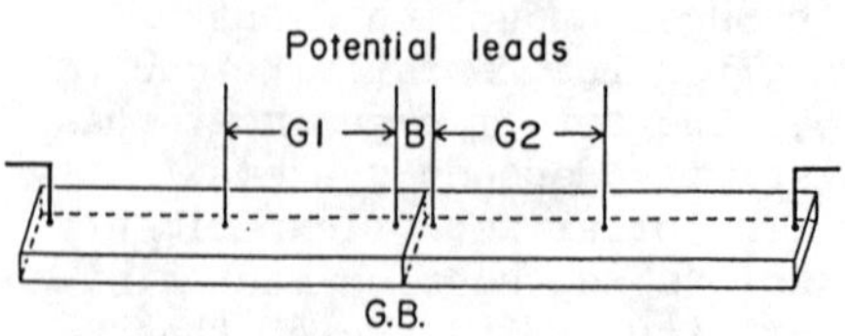

Fig. 1. Specimen used for resistivity measurement.

The boundary parameters, namely the rotation axis, the rotation angle, etc. were determined using x-ray diffraction analysis with a precision better than 1 degree. Further experimental details have been described elsewhere[1-3].

3. RESULTS AND DISCUSSION

3.1 Pure Aluminum

Figure 2 is the specific boundary resistivity ρ_{gb} plotted against the rotation angle ω, where the rotation axes are random. Closed symbols are general boundaries (small-angle and large-angle boundaries) and full symbols are coincidence-related boundaries. We see that the values for general boudaries are larger than those for coincidence-related boundaries when the rotation angles are fixed. The value for coherent twin boundary is $1.5\times10^{-17}\,\Omega\,m^2$, which is one order of magnitude smaller than that for general boundaries. However, this figure does not clearly show the angle-dependence of boundary resistivity.

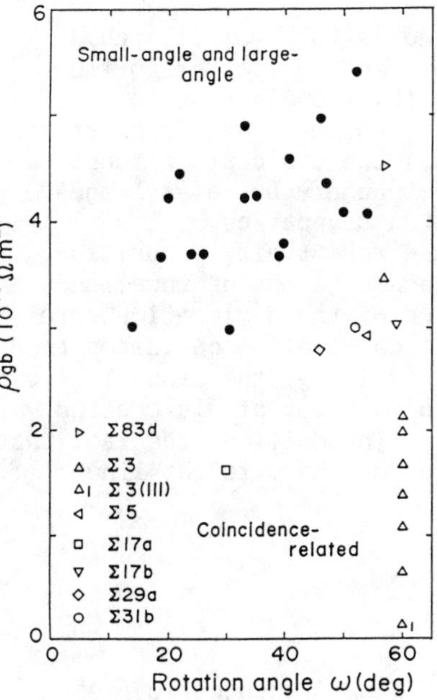

Fig. 2. The ρ_{gb} plotted against the rotation angle ω.

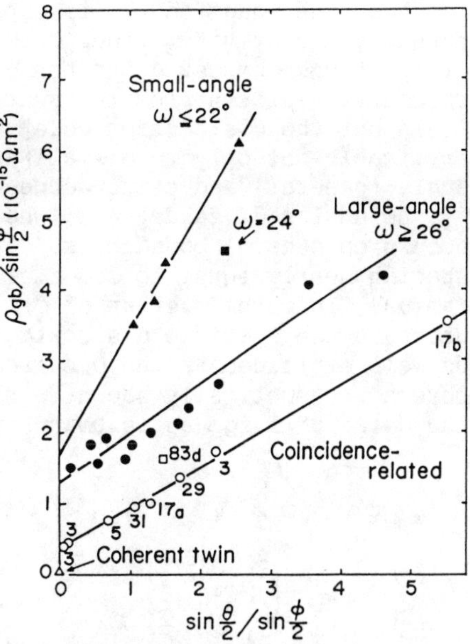

Fig. 3. The plot of $\rho_{gb}/\sin(\phi/2)$ versus $\sin(\theta/2)/\sin(\phi/2)$. θ, tilt angle; ϕ, twist angle.

The figure is then plotted in a different manner, using the tilt angle θ and the twist angle ϕ. Figure 3 is the plot of $\rho_{gb}/\sin(\phi/2)$ against $\sin(\theta/2)/\sin(\phi/2)$. Numbers beside the data points are Σ values. We see that the sets of data for small-angle boundary, large-angle boundary and coincidence-related boundary lie on respective straight lines. In addition, the data for coherent-twin boundary deviates downwards from these lines, and general boundaries have a transition region of resistivity at the angles between 22 and 26 degrees.

This shows that the specific boundary resistivities for small-angle, large-angle and coincidence-related boundaries are expressed by the equation

$$\rho_{gb} = A\sin(\theta/2) + B\sin(\phi/2), \quad (2)$$

where constants A and B depend on the boundary types.

According to the dislocation model, which describes that a boundary consists of primary/lattice dislocations, the densities of these dislocations are $(2/b)\sin(\theta/2)$ and $(4/b)\sin(\phi/2)$ for edge dislocations and screw dislocations, respectively[6]. Here, b is the magnitude of Burgers vector.

Thus, the equation (2) is explained as follows: (1) the boundary resistivity arises from the electron scattering from dislocations composing the boundary; (2) the dislocation resistivity arises from the scattering not the long-range strain field but the dislocation core[7]; (3) the dislocation model is applicable not only to a small-angle boundary but also large-angle (general) and coincidence-related boundaries.

The angles 22-26 degrees, where the resistivity transition occurs on general boundaries, corresponds to the dislocation spacing nearly equal to 2b, i.e. twice of the dislocation core radius. Since the overlap of dislocation core is considered to decrease the specific dislocation resistivity, the transition can be well explained as the occurrence of overlap of dislocation core as schematically shown in Fig. 4. In addition, the fact that the data for large-angle boundary lies on the straight line,

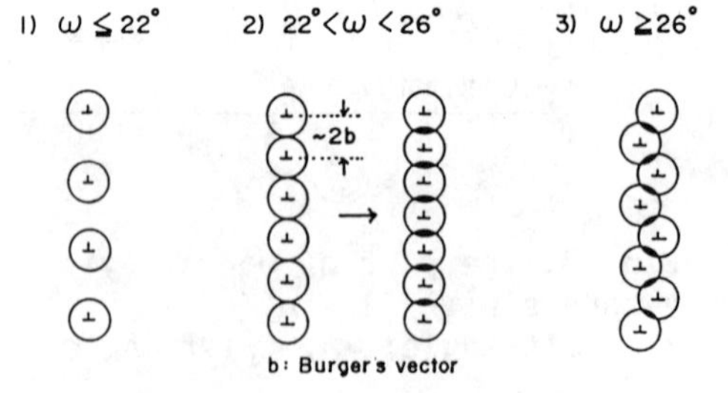

Fig. 4. Schematic presentation for the change in dislocation array as the rotation angle ω increases. Circle, dislocation core.

suggests that the overlap of dislocation core does not proceed further. The dislocation arrays may take a zigzag arrangement to avoid the overlap.

Therefore, the equation (2) is equivalent to

$$\rho_{gb} = \rho_d(\text{edge})(2/b)\sin(\theta/2) + \rho_d(\text{screw})(4/b)\sin(\phi/2), \quad (3)$$

where ρ_d(edge) and ρ_d(screw) are the specific resistivity for edge and screw dislocations, respectively. Comparing eq.(3) with Fig. 3, we obtain the specific resistivity of dislocations in boundaries and the results are shown in Table I.

Table I. The values of A and B in eq.(2) determined from experiment and the specific dislocation resistivities ρ_d(edge) and ρ_d(screw) in grain boundaries estimated from these

type of boundary	A	B	ρ_d(edge)	ρ_d(screw)
	($f\Omega m^2$)	($f\Omega m^2$)	($10^{-25}\Omega m^3$)	($10^{-25}\Omega m^3$)
small-angle	1.8	1.6	2.6	1.2
large-angle(general)	0.69	1.3	1.0	0.91
coincidence-related	0.57	0.37	0.80	o.26

The value for small-angle boundaries nearly agrees with the experimental value for dislocations in lattice, $(1.5\pm0.3)\times10^{-25}$ Ωm^3[8,9], because this value is for the dislocations randomly oriented and should be multiplied by 3/2 to compare with that for dislocations in boundaries. In addition, the small value for the screw dislocations in coincidence-related boundary suggests that the dislocation core is much overlapped on the screw dislocations in coincidence-related boundary. This seems to be consistent with the fact that the rigid-body translation of two grains along a boundary is small on twist boundaries compared with tilt boundaries[10].

As noted in the above, the specific boundary resistivity for a coherent twin boundary ($\Sigma3$, {111}) is one order of magnitude smaller than others. This is consistent with the fact that there is no deviation of lattice site and no dislocations on a coherent twin boundary. The resistivity for near-coherent twin boundaries increases when the boundary plane deviates slightly from {111}. This increase is proportional to the angle γ between the boundary normal and the rotation axis as shown in Fig. 5. This is explained by the model that near-coherent twin boundaries are formed by the adding of dislocations on {111} plane as shown in Fig. 6. The slope of line in Fig. 5 gives the Burgers vector near (1/6)<112> and agrees with the electron-microscopy observation in gold[11].

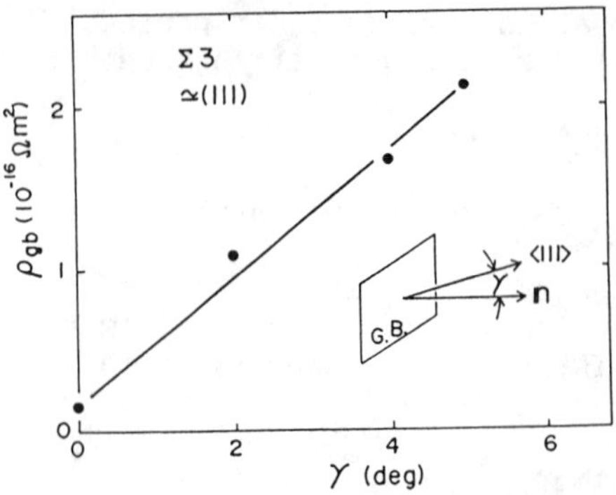

Fig. 5. The ρ_{gb} versus the angle γ between the boundary normal **n** and <111> for near-coherent twin ($\Sigma 3$, $\simeq(111)$) boundaries.

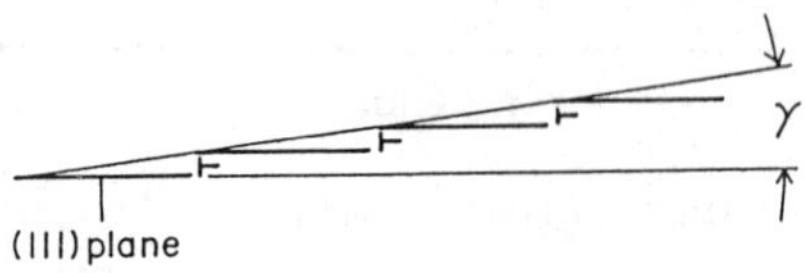

Fig. 6. Model for near-coherent twin boundaries.

3.2 Al-Ag alloys

Figure 7 is the plot of specific boundary resistivity in Al-60 at.ppm Ag against the rotation angle. The data for zone-refined Al (open symbols) are included for a comparison. The ρ_{gb} in Al-Ag has two features: (1) ρ_{gb} is 2 to 30 times larger than that in zone-refined Al; and (2) in contrast to the zone-refine Al data, ρ_{gb} increases with the decrease of the rotation angle and is large on twin (incoherent) boundaries. Thus, eq.(2) is not directly applicable to the data on Al-Ag. This shows that the boundary resistivity in this alloy is mainly due to the Ag segregating around the boundary.

The segregation can be estimated by taking the value of ρ_{gb}(Al-Ag)-ρ_{gb}(Al) as equal to the amount of segregated Ag multiplied by the specific solute resistivity. Using the specific resistivity equal to 10.0 nΩm(at.%)$^{-1}$ for Ag in Al[12], and the ρ_{gb}(Al) estimated by eq.(2) and Table I, the amount of Ag segregated is obtained from the ρ_{gb}(Al-Ag) in Fig.7. This corresponds to a length of 0.7 nm (3 layers) of Ag for large-angle boundaries and to that of 2-3 nm (10 layers) of Ag for small-angle and twin boundaries. Such a large amount of segregation is beyond the equilibrium segregation.

Figure 8 is the ρ_{gb}(Al-Ag) - ρ_{gb}(Al) plotted against $(\sin\theta/2)^{-1}$. This suggests that the segregation is proportional to the extent of strain field of dislocation array, because the strain field of the edge dislocation wall extends as $(\sin\theta/2)^{-1}$

[6,13]. This is consistent with the large segregation on incoherent twin boundaries, because they also have a relatively wide strain-field.

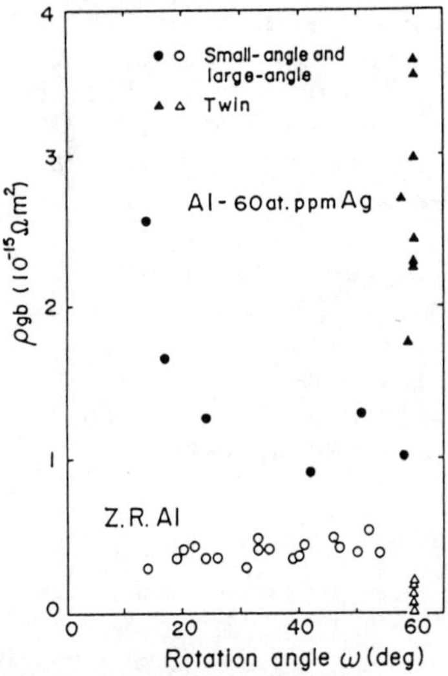

Fig. 7. The ρ_{gb} versus for Al-60 at.ppm Ag.

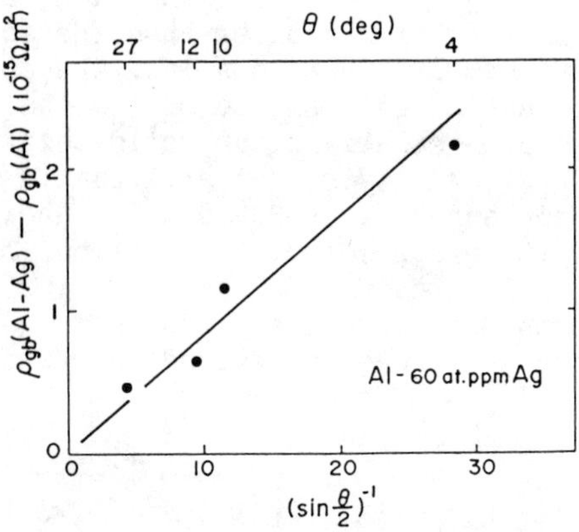

Fig. 8. The plot of ρ_{gb}(Al-Ag)-ρ_{gb}(Al) against $(\sin\theta/2)^{-1}$ for Al-60at.ppm Ag. θ, tilt angle.

The above segregation features are much different from those of equilibrium segregation. Two kinds of nonequilibrium segregation mechanisms are considered for the observed segregation. One is the vacancy-drag segregation, which is due to the dragging of solute atoms by vacancies annihilated at boundaries during cooling[14]. The other is the segregation by a migrating boundary during grain growth[15].

The former segregation is expected to increase with the decrease of cooling rate[14]. The latter segregation should be decreased by the long-time anneal at high temperatures. Therefore, an Al-106 at.ppm Ag specimen shown in Fig. 9 was annealed in vacuum at 873 K for 0.5, 10, 106 and 227 hours(total) and followed by the two types of cooling - rapid(18 K min^{-1}) and slow (=furnace cooling, 2.4 K min^{-1}) ones. The rapid cooling is aimed at reducing vacancy drag.

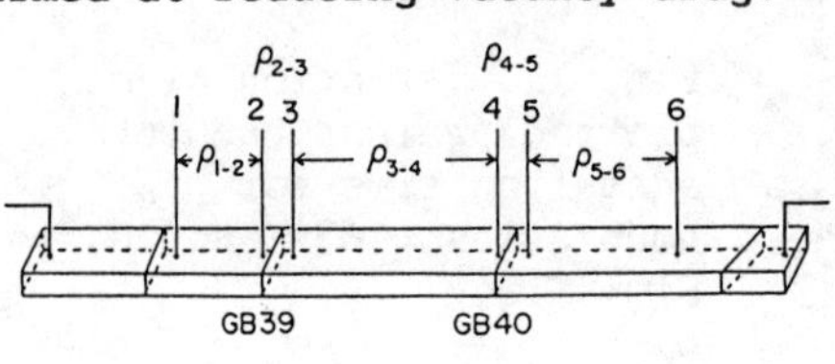

Fig. 9. The Al-106 at.ppm Ag specimen with boundaries GB39 and GB40. 1,2,...6 are potential leads.

First, the segregated-solute distribution varied by the solute diffusion during anneals is calculated with Shewmon's equation, and shown schematically in Fig. 10[3,5]. The figure shows that the segregated solute diffuse out from B parts to grains by the anneal for 106 hours, where the width of B parts is about 1.6 mm for the specimen shown in Fig. 9. In addition, since the solute staying within B part on the anneals for 0.5 and 10 hours has the concentration smaller than the solubility equal to about 200 at. ppm at 300 K in Al-Ag, the amount of initial segregation can be estimated from the ρ_{gb} on the two anneals, using the specific solute resistivity of solid solution.

Figure 11 is the variation in ρ_{gb} due to heat treatments for two boundaries (GB39 and 40) shown in Fig. 9. The width calculated for segregated solute distribution, $4\sqrt{Dt}$, is shown at the top of the figure. Within the annealing time of 10 hours, ρ_{gb} is almost constant and corresponds to the 39 and 50 layers of silver for the general boundary and the twin boundary, respectively.

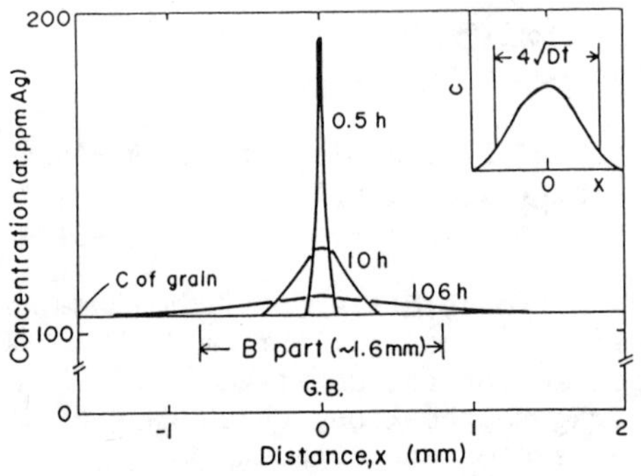

Fig. 10. Estimation for the segregated-solute distribution varied by the Ag diffusion during anneals.

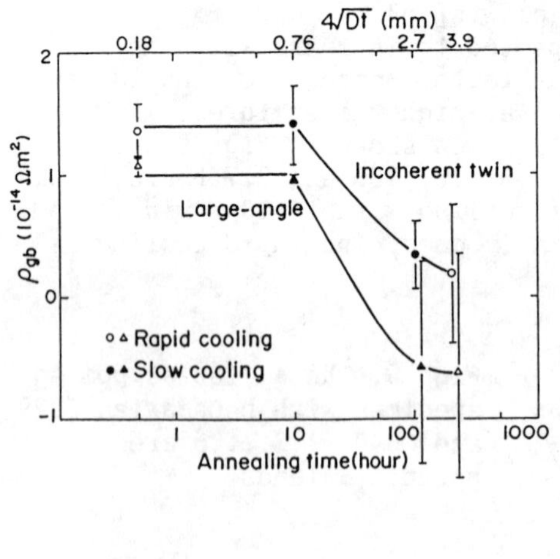

Fig. 11. Variation of ρ_{gb} plotted against the total annealing time for boundaries (GB39 and 40) in Fig. 9.

The ρ_{gb} decreases significantly when the annealing time becomes hundreds of hours, irrespective of cooling types. This demonstrates that the large amount of segregation observed is not due to the vacancy drag during cooling, but mainly to the segregation by the migrating boundary.

The observed segregation is beyond Kasen's proposal that moving boundaries interact thermodynamically with the impinging solute field in a manner similar to that of a stationary boundary in a steadily increasing solute field[15]; the maximum segregation by this mechanism is a monolayer of solute. On the other hand, it has been shown that migrating boundaries absorb and generate vacancies[16,17]. The boundary-vacancy interaction may be stimulated by the long-range strain field, and we have observed the increase in segregation with the extent of strain field on small-angle boundaries. Moreover, the Ag-vacancy binding energy is known to be positive in Al[18]. It is thus suggested that migrating boundaries interact with solute atoms through the vacancies associated with boundary migration and cause a large segregation when the solute-vacancy binding energy is positive.

In addition, another type of resistivity decrease was observed on the initial heat treatment. As shown in Fig. 12, the resistivities of B parts (ρ_{2-3} and ρ_{4-5}) decrease significantly when they are furnace-cooled after the one-hour anneal (2nd heat treatment), and recover again by the 10-hours anneal. Since the distance diffusing to a boundary during furnace cooling is estimated to be nearly same as that diffusing out from the boundary during the one-hour anneal for Ag in Al, vacancy-drag is expected to return much of the silver to the boundaries[3]. These

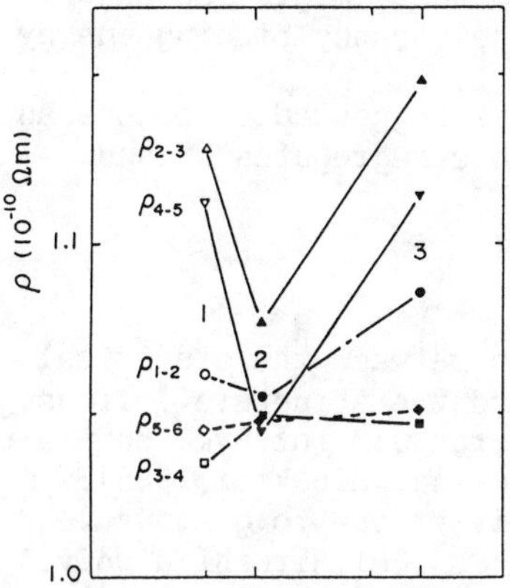

Fig. 12. Variation in the residual resistivity ρ due to heat treatments (1, 2 and 3) for the five parts of specimen shown in Fig. 9. Open symbols are rapid cooling and full symbols are furnace (slow) cooling.

suggest that the specific solute resistivity becomes an order of magnitude smaller when the solute segregates around the boundaries. This can be attributed to the interference of electron scattering waves between the segregated solute atoms[19].

CONCLUSIONS

The electrical resistivity measurement of a single boundary has revealed the followings.

(1) The boundary electrical resistivity depends on boundary types; they are classified into four: small-angle, large-angle general, coincidence-related and coherent twin boundaries.

(2) The specific boundary resistivity is expressed with the equation (2), which is based on the primary/lattice dislocation model, except coherent twin boundaries. This suggests that the dislocation model is applicable to not only a small-angle boundary but also large-angle general and coincidence-related boundaries.

(3) The boundary resistivity has a transition region at the rotation angles between 22 and 26 degrees, where the dislocation core begins to overlap each other.

(4) The analysis with equation (2) shows that the dislocation core is overlapped in large-angle boundaries and most strongly on the screw dislocations in coincidence-related boundaries.

(5) The resistivity of twin boundary is an order of magnitude smaller when the boundary plane is {111} (a coherent twin boundary), and increases with the deviation from {111}.

(6) Migrating boundaries cause a large amount of segregation which is beyond the equilibrium segregation in Al-Ag. It is proposed that migrating boundaries interact with solute atoms through the vacancies associated with boundary migration and cause a large segregation when the solute-vacancy binding energy is positive.

(7) The specific solute resistivity is suggested to become an order of magnitude smaller when the solute segregates around boundaries.

REFERENCES

1. I.Nakamichi and T.Kino, "Correlation between the electrical resistivity due to grain boundary and the structure", Trans. Japan.Inst.Met.Suppl.27(1986)1013 [Proc.JIM Int.Sym. on Grain Boundary Structure and Related Phenomena, Minakami,1985].
2. I.Nakamichi, "The electrical resistivity due to grain boundary and the boundary structure", J.Sci.Hiroshima Univ.A 54(1990)49.
3. I.Nakamichi, "Strong segregation by a migrating boundary and

the decrease in electrical resistivity", J.Phys.: Condens. Matter 5(1993)4967.
4. I.Nakamichi, "Electrical resistivity and the structure of coincidence-related and twin boundaries in pure aluminum", Conf.on Advanced Materials, Tokyo(1993)(in press).
5. I.Nakamichi and T.Kino, "Grain boundary segregation and electrical resistivity on dilute Al-Ag alloys", Conf.on Advanced Materials, Tokyo(1993)(in press).
6. S.Amelinckx and W.Dekeyser, "The structure and properties of grain boundaries", in Solid State Phys. vol.8 ed.by F.Seitz and D.Turnbull(Academic,New York 1959)p325.
7. B.R.Watts, "Conduction electron scattering in dislocated metals", in Dislocations in Solids vol.8 ed F.R.N.Nabarro (North Holland,Amusterdam 1989)p175.
8. J.G.Rider and C.T.B.Faxon, "An experimental determination of electrical resistivity of dislocations in aluminum", Phil. Mag. 13(1967)289.
9. T.Kino, T.Endo and S.Kawata, "Deviations from Matthiessen's rule of the electrical resistivity of dislocations in aluminum", J.Phy.Soc.Jpn. 36(1974)698.
10. V.Vitek, A.P.Sutton, D.A.Smith and P.C.Pond, "Atomistic studies of grain boundaries and grain boundary dislocations", in Grain Boundary Structure and Kinetics, ed.by American Society for Metals,(Metals Park,Ohio 1979)p115.
11. W.Krakow and D.A.Smith, "Core structure of grain boundaries in gold" Trans.Japan.Inst.Met.Suppl.27(1986)277.
12. I.Nakamichi, "Enhancement of surface resistivity due to anisotropic Guinier-Preston zone scattering", J.Phys.: Condens.Matter 1(1989)8887.
13. W.W.Webb, "The interaction of solutes with dislocation walls", Acta Metall.5(1957)89.
14. P.Doig and P.E.J.Flewitt, "Segregation of chronium to prior austenite boundaries during quenching of a 2.25%Cr1%Mo steel", Acta Metall.29(1981)1831.
15. M.B.Kasen, "Solute segregation and boundary structural change during grain growth", Acta Metall.31(1983)489.
16. C.Gottshalk, K.Smidoda and H.Gleiter, "The generation of migrating boundaries by vacancies", Acta Metall.28(1980) 1653.
17. K.Lücke and G.Gottstein, "Grain boundary motion-1. Theory of vacancy drag during grain boundary motion", Acta Metal.29 (1981)779.
18. J.Takamura, M.Koike and K.Furukawa, "The vacancy-solute interaction in aluminum", J.Nucl.Mater.69-70(1978)738[Proc. Conf.on Properties of Atomic Defect in Metals, Argonne,1976].
19. P.L.Rossiter, The Electrical Resistivity of Metals and Alloys (Cambridge University Press, Cambridge 1987).

PREDICTION OF FRACTURE IN CERAMIC COMPOSITES BY MEASUREMENT OF ELECTRICAL RESISTIVITY

Atsumu Ishida, Masaru Miyayama, Hiroaki Yanagida*, and Hideaki Matsubara+

Research Center for Advanced Science and Technology,
University of Tokyo, Tokyo 153, Japan

*Department of Industrial Chemistry,
University of Tokyo, Tokyo 113, Japan

+Integrated Materials Research Division,
Japan Fine Ceramics Center, Nagoya 456, Japan

ABSTRACT

Prediction of fracture by measuring electrical resistivity in ceramic composites was investigated. Ceramic composites forming an electric conduction path were fabricated with calcium fluoride as the insulating matrix and silicon carbide whiskers as the conducting additive. Applying the bending load to the composites, the connectivity of the conduction path changed, and the electrical resistance increased before fatal fracture occurred. Thus, the measurement of electrical resistivity can be used to predict the fracture in ceramic composites. The movement of whiskers disconnecting the conduction path and leading to an increase in electrical resistance was confirmed by scanning electron microscopy.

INTRODUCTION

Ceramic composites have been used in many applications. The interface between the matrix and additive in ceramic composites plays an important role in achieving high performance. In structural ceramics, composites with whisker or fiber reinforcements have been widely investigated in order to overcome the brittleness problem.[1,2] In the case of ceramic composites with whiskers, bonding conditions at the interface between the matrix and the whisker are known to affect the debonding- and pullout-behaviors of whiskers under applied loads. This, in turn influences a toughening behavior.[3,4]

Prediction of fracture in structural ceramics is desired to maintain a reliability of the system and to increase the practical applications. Composites consisting of an insulator matrix with electric conducting additives are known to show a large variation in electrical resistivity with changes of composition and microstructure. A large decrease in electrical resistivity occurs through the formation of conduction paths.[5,6] In the present study, measurements of electrical resistivity before and after applying a load were conducted on semiconducting composites consisting of insulating CaF_2 and conducting SiC whiskers. In such ceramic composites, prediction of fracture would be possible by measuring the change of electrical resistivity. This change results from microdeformation or a change of connectivity of the conduction paths when subjected to an external load. Based on the results, the possibility of predicting fracture in ceramic composites is discussed.

EXPERIMENTAL

CaF_2 and SiC were chosen as insulator and conductor ceramics with a low reactivity during sintering. Whisker morphology of SiC was chosen for this study as the whisker has possibilities of producing a conduction path at a low concentration, and of enhancing toughness in the composites. CaF_2 powder with a purity over 99.9% and a particle size of under 2 μm (Kojundo Chemical Laboratory Co., Ltd., Japan), and SiC whiskers from 0.1 to 1 μm in diameter and from 50 to 200 μm in length (Tokai Carbon Co., Ltd., Tokyo, Japan) were used as the starting materials.

The starting materials were mixed with ethanol in an ultrasonic bath and then dried. The powder was pressed into pellets of 15 mm in diameter and about 1.8 mm in thickness at 220 MPa. The pellets were sintered by hot pressing at a pressure of 40 MPa at 900° to 1200°C for 30 min. Hot-pressed samples were then cut and machined to a proper shape and evaluated. Density, microstructure observation by scanning electron microscopy, composition by X-ray diffraction, electrical resistivity, and fracture strength were evaluated in this study.

The electrical resistivity of the composites was measured in disk pellet samples with circular platinum electrodes sputtered at both sides of the disk pellets. The measurements were done by the two-probe direct-current method using an ultrahigh-resistance meter at room temperature. The sample was then cut into test pieces of 15mm x 2mm x 1mm for the measurement of fracture strength with a three-point bending apparatus having a span of 10 mm and crosshead speed of 0.2 mm/min.

The measurement of electrical resistivity before and after applied loading was conducted on the test piece used in the measurement of fracture strength, with platinum sputtered at both ends of the test piece as electrodes and connected to an electrical resistance meter. Electrical resistance, applied load, and deflection were measured. SEM observation was conducted to measure the grain size of the composites and to examine the microstructure after applying the bending load.

RESULTS AND DISCUSSION

Density and Microstructure

Figure 1 shows the SEM micrographs of the sintered CaF_2 and the composite with 10-vol%-SiC-whisker hot-pressed at 1000°C. Both samples have high relative density of 97%. In the composites, aggregation of the SiC whiskers was not observed, and the grain growth of CaF_2 seemed to be inhibited by the addition of SiC whiskers. From X-ray diffraction measurements, no peaks except for CaF_2 and SiC were observed.

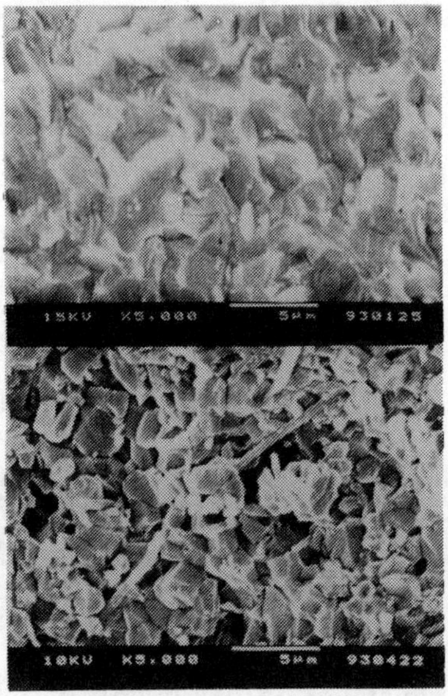

Figure 1. SEM micrographs of CaF_2 and CaF_2/10-vol%-SiC-whisker composite.

Electrical Resistivity

Electrical resistivity plotted as a function of SiC-whisker content in the composites hot pressed at 1000°C is shown in Figure 2. A large decrease of electrical resistivity from 10^{12} to 10^5 Ω·cm for SiC-whisker contents above 2 to 3 vol% shows that the conduction path has been formed in this region of whisker concentration. As the composites were fabricated with whisker-type conductive material, a clear, sharp decrease of electrical resistivity was achieved for a relatively small amount of conductive material.

Applying the percolation theory to the results, the critical volume fraction for the percolation threshold is 2 vol% of SiC whiskers. Holm and Cima[7] have developed a computer model which treats ceramic powder matrix ceramic whisker composites as a percolative network, and reported that a critical volume fraction depends on the whisker aspect ratio. The critical volume fraction of the present composites was calculated to be from 0.9 to 2 vol%, using an aspect ratio of around 200 to 500 for the SiC whiskers. The calculated value is in good agreement with the experimental results.

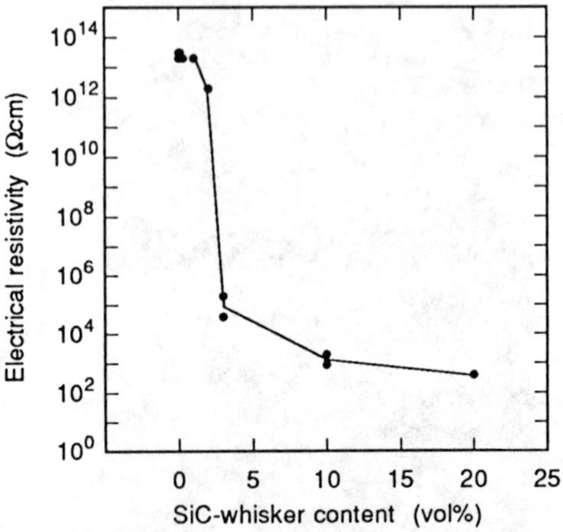

Figure 2. Electrical resistivity as a function of SiC-whisker content in CaF_2/SiC-whisker composites hot pressed at 1000°C.

Fracture Strength

Figure 3 shows the fracture strength of the composites as a function of SiC-whisker content. The fracture strength of the composites with 3 vol% of SiC whiskers had the minimum value, about half of that of single phase CaF_2 ceramics. The fracture strength of the composites decreased with SiC-whisker content of 1 to 3 vol%, similar to the change of electrical resistivity shown in Figure 2. Although the fracture strength increased with increasing SiC-whisker content of more than 3 vol%, it was still less than the fracture strength of single phase CaF_2 ceramics. As a result, addition of SiC whiskers up to 20 vol% seems to have negligible effect on improving the fracture strength of the composites.

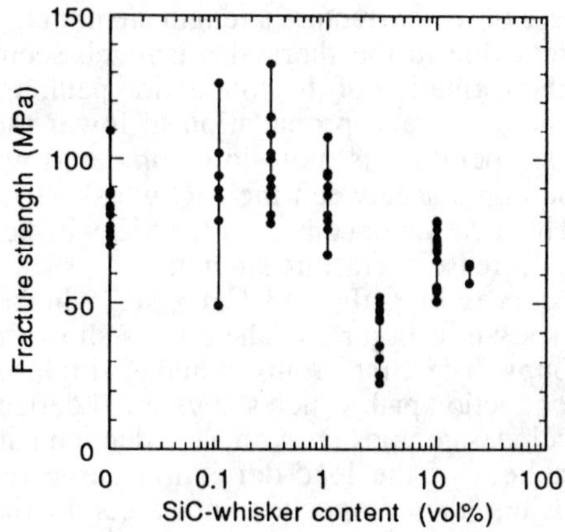

Figure 3. Fracture strength as a function of SiC-whisker content in CaF_2/SiC-whisker composites hot pressed at 1000°C.

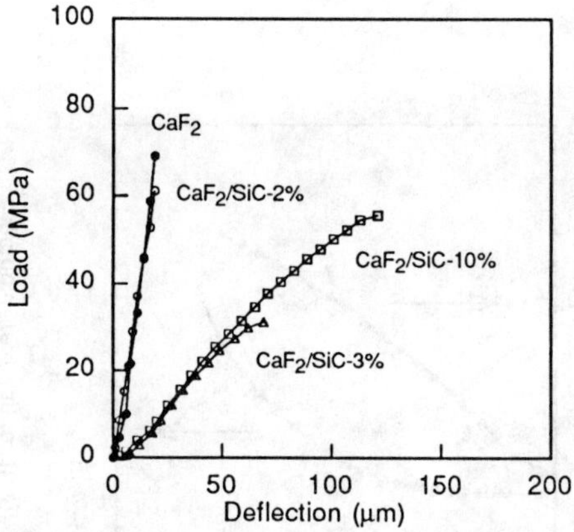

Figure 4. Load-deflection curves of CaF_2 and CaF_2/SiC-whisker composites hot pressed at 1000°C.

Although the causes of lower fracture strength are not clear in this system, the high tensile matrix stress due to the thermal mismatch seems to be one of the reasons. Furthermore, the formation of the conduction path, as shown in Figure 2, may have some influence on crack propagation to lower the fracture strength. Because the sintering temperature is quite low compared with the dissociation temperature of SiC, the contact between the SiC whiskers in the composites is presumed to be weak. The increase in contact of the SiC whiskers in the composites appears to influence the decrease of fracture strength.

Load-deflection curves of CaF_2 and CaF_2/SiC-whisker composites hot-pressed at 1000°C are shown in Figure 4. The curves shown in the figure can be classified into two groups. The first group includes single-phase CaF_2 and the composites without a conduction path which shows small deflections up to fracture, and the second group includes composites with a conduction path which show large deflections. As the gradient of the load-deflection curve represents an elastic modulus of the composites, the composites with a conduction path have lower elastic moduli than single-phase CaF_2 and the composites without a conduction path. A low elastic modulus seems to be related to the increase in contact of the SiC whiskers in the composites, as in the case of decreasing the fracture strength.

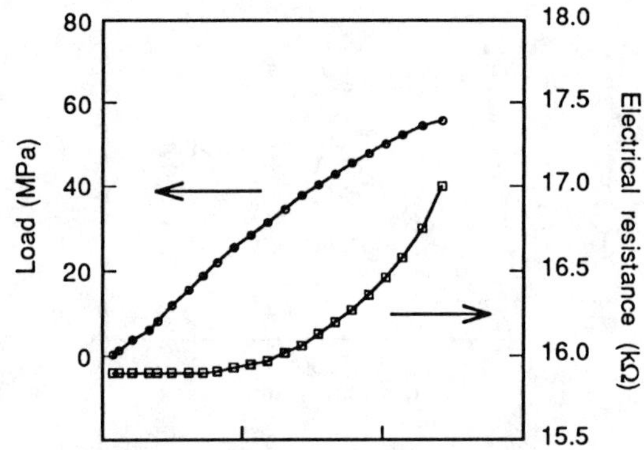

Figure 5. Load-deflection curve and change of electrical resistance under bending load to CaF_2/10-vol%-SiC-whisker composite hot pressed at 1000°C with initial electrical resistivity of 1×10^3 Ω·cm.

Change of Electrical Resistance by Applying Load

The relation between applied load, deflection, and electrical resistance of the CaF_2/10-vol%-SiC-whisker composite hot pressed at 1000°C is shown in Figure 5. The increase of electrical resistance was observed to average about 7% above the initial electrical resistance, after applying less than half of the fractured load and increased until fatal fracture occurred. This result indicates that rupture of the conduction path has occurred by applying the load before fracture, and that the change of the electrical resistance can be used to predict the fracture of the composites.

The increase in electrical resistance under loads was observed in other conducting composites as well. Figure 6 shows the increased ratio of electrical resistance at the time of fracture to initial electrical resistance plotted as a function of SiC-whisker content. The ratios vary widely from 1% to 10% and show no clear dependence on the SiC-whisker content or on the hot-pressing temperature. Compared with CaF_2/10-vol%-SiC-whisker composite, a large increase of the electrical resistance was not observed in CaF_2/3-vol%-SiC-whisker composite, which is close to the critical volume fraction of forming the conduction path. An ambiguity results in adopting the bend test, which gives rise to both compression and tension in the different part of the test piece by applying a load. More precise measurements are needed to clarify the relation between the microstructure and the change in electrical resistance under applied load.

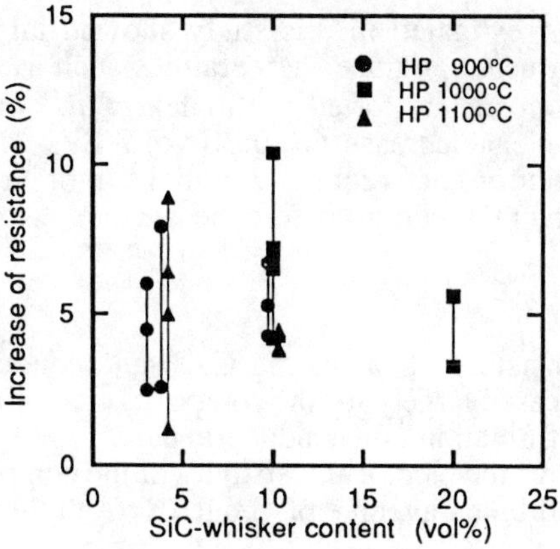

Figure 6. Ratio of the increase in electrical resistance to the initial electrical resistance as a function of SiC-whisker content in CaF_2/SiC-whisker composites.

Figure 7 shows a SEM micrograph of the fracture surface of CaF_2/10-vol%-SiC-whisker composite. Many SiC whiskers are observed extending from the fracture surface, indicating the sliding of whiskers during deformation of the composites. It is suggested that the increase in electrical resistance under load occurred not only by the breaking of the SiC whiskers, but also by the sliding of the SiC whiskers to disrupt the conduction path.

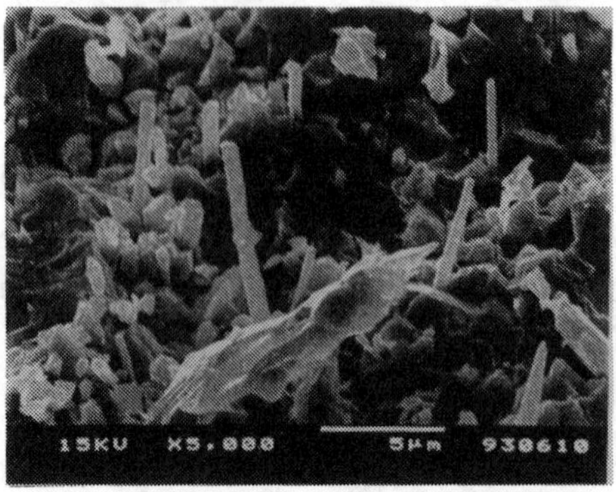

Figure 7. SEM micrograph of CaF_2/10-vol%-SiC-whisker composite hot pressed at 1000°C.

The composites investigated in this study showed large deflections until fracture, compared with usual engineering ceramics such as Si_3N_4 or Al_2O_3. In Si_3N_4/SiC-whisker composites reported by Shalek *et al.*,[8] elastic moduli of the composites showed no clear decrease up to 30 vol% SiC-whisker content. It is assumed that the prediction of fracture would be more reliable in ceramic composites that show larger deflections before the fracture.

CONCLUSIONS

In composites consisting of insulating CaF_2 and conducting SiC whisker, electrical resistivity decreased largely for the compositions of more than 3 vol% SiC whiskers, indicating the formation of a conduction path.

Under bending loads, the electrical resistivity of the composites with 10 vol% of SiC-whisker increased by an average of about 7% from the initial value before fracturing.

The debonding of whiskers leading to an increase in electrical resistivity was confirmed by SEM observation. Possibilities for the prediction of fracture and for fatigue detection are suggested for ceramic composites possessing electric conduction.

References

[1] P. F. Becher, "Advances in the Design of Toughened Ceramics," *J. Ceram. Soc. Jpn.*, **99**[10]993-1001(1991).

[2] L. M. Sheppard, "Enhancing Performance of Ceramic Composites," *Am. Ceram. Soc. Bull.*, **71**[4]617-631(1992).

[3] N. Claussen, K. L. Weisskopf, and M. Ruhle, "Tetragonal Zirconia Polycrystals Reinforced with SiC Whiskers," *J. Am. Ceram. Soc.*, **69**[3]288-292(1986).

[4] G. C. Wei and P. F. Becher, "Development of SiC-Whisker-Reinforced Ceramics," *Am. Ceram. Soc. Bull.*, **64**[2]298-304(1985).

[5] S. Kirkpatrick, "Percolation and Conduction," *Rev. Mod. Phys.*, **45**[4]574-588(1973).

[6] D. S. McLachlan, M. Blaszkiewicz, and R. E. Newnham, "Electrical Resistivity of Composites," *J. Am. Ceram. Soc.*, **73**[8]2187-2203(1990).

[7] E. A. Holm and M. J. Cima, "Two-Dimensional Whisker Percolation in Ceramic Matrix-Ceramic Whisker Composites," *J. Am. Ceram. Soc.*, **72**[2]303-305(1989).

[8] P. D. Shalek, J. J. Petrovic, G. F. Hurley, and F. D. Gac, "Hot-Pressed SiC Whisker/Si_3N_4 Matrix Composites," *Am. Ceram. Soc. Bull.*, **65**[2]351-356(1986).

SURFACE INTERACTION AND INTERFACIAL CHEMISTRY IN MULTICOMPONENT NANOCRYSTALLINE MATERIALS

Jackie Y. Ying
Department of Chemical Engineering
Massachusetts Institute of Technology
Cambridge, MA 02139, USA

ABSTRACT

Nanocrystalline processing offers a new route for synthesizing multicomponent systems with unique surface interaction and interfacial chemistry. This paper discusses three approaches to create unique multicomponent materials with ultrahigh dispersion of active phases. The objective is to create catalytic materials with unusual surface reactivity, and to modify ceramic properties through a better control of interfacial chemistry. Inert gas condensation was employed to synthesize nanocrystals of a Cu-Ce alloy. Through controlled post oxidation, Cu_2O was segregated to the surface of the CeO_{2-x} nanocrystals to form a model supported oxide catalyst. A uniform ultrafine mixture of Pt nanocrystals and TiO_{2-x} nanocrystals can also be obtained by inert gas condensation with simultaneous or sequential processing. Finally, molecular-level packaging of mixed oxide phases was achieved through pillaring of layered structures. The novel processing approaches lead to unique surface and interfacial interaction between different phases. The structure and thermal stability of these multicomponent systems were examined by X-ray diffraction and X-ray photoelectron spectroscopy.

INTRODUCTION

Multicomponent systems may exist as a solid solution or as a mixture of different components. In a number of materials applications, it is desirable to have an intimate mixture of different components or phases to engineer the surface interaction and interfacial chemistry in the multicomponent systems. Conventionally, a mixed component system is obtained through physical mixing of different constituents, or through chemical co-precipitation of the corresponding metallic salt precursors. As expected, homogeneous mixtures of multicomponent systems are not easily achieved through these conventional routes, and undesirable bulk phase separation may occur upon heat treatment.

The objective of this study is to explore novel means of processing multicomponent mixtures to tailor the interfacial characteristics critical to catalytic and ceramic applications. In heterogeneous catalysis, reacting species would typically adsorb on the surface sites of an active component such that surface reaction could proceed. The active component may be a metal or a metal oxide, which is usually present in a small quantity. It relies on a relatively inert support component to disperse it over the large support area, and to stabilize it as a reactive phase. The interaction between the active component and the oxide support component directly influences the catalytic activity, and the long-term resistance of the supported catalyst against sintering and poisoning effects. In ceramic applications, a second component is commonly added to the major ceramic component for several functions: (i) as a sintering aid to promote low-temperature sintering; (ii) as a 'pinning' secondary phase that suppresses grain growth during heat treatment to facilitate the final stage of densification; and (iii) as a reinforcing agent that interacts synergistically with the matrix component for greater structural strength and toughness. Here again the dispersion of the secondary component in the matrix component is extremely important. The interfacial chemistry governs the structure of the grain boundaries, and impacts on the overall mechanical, thermal, electrical, and optical behavior of the ceramic materials.

In both catalytic and ceramic applications, it would be ideal to achieve an ultrahigh dispersion of the different components. This would ensure that the resulting multicomponent material has a uniform property, with maximum surface contact between the different phases such that the desirable interfacial characteristics can be attained with minimum use of the secondary component. The secondary component may be a noble metal in a supported metal catalyst, such that spreading out the use of this expensive constituent by fine dispersion in the major component would be highly desirable. In a ceramic system, the secondary component may be a sintering aid such as Y_2O_3 needed in the liquid phase sintering of Si_3N_4 or AlN. The oxide dopants enable the covalent nitride ceramics to densify at a more reasonable temperature, but their presence leads to glassy phases in the final ceramics. Such glassy phases are detrimental to the high-temperature strength of Si_3N_4 in structural applications, and significantly deteriorate the thermal conductivity of AlN in electronic packaging applications. These examples illustrate a common need for finding new approaches to add the secondary component in a controlled small quantity to maximize its impact in sintering and structural reinforcement, while minimizing its undesirable side effects on the properties of the main ceramic component.

Nanocrystalline processing offers the possibility to design model multicomponent systems with an ultrahigh level of mixing that provides nanometer-scaled uniformity. Fig. 1 depicts three types of mixing that would give rise to unique surface interaction between different components. In Fig. 1(a), each grain has a surface coating which consists of a different component from that of the core crystallite. The amount of interaction between the surface component and the core component is maximized when the crystallite size is minimized. The complexity involved in direct deposition of a coating layer on fine crystallites is rather formidable, so unique means of materials engineering has to be developed to generate such surface dispersion features.

Fig. 1(b) illustrates two possible types of uniform mixing of grains of different composition. The first type involves a random mixing of different crystallites, which is difficult to achieve mechanically without alloying effects, or chemically without being compromised by agglomeration. The second type of mixing shown in Fig. 1(b) involves an alternate packing of grains of different compositions. It cannot be achieved by physical/mechanical mixing or by chemical routes in non-self-assembled systems. The two types of mixing in Fig. 1(b) can however, be achieved by vapor synthesis routes through simultaneous or sequential materials generation from different sources. Ideally, the level of mixing would be highest if the grain sizes are of an ultrafine nature. In Figs. 1(a) and (b), the grains may be consolidated to form a bulk solid compact. If the grains are synthesized as non-agglomerated nanocrystals, we achieve not only an ultrahigh level of mixing between different components, but also a nanostructured compact with a high volume fraction of grain boundaries and a microporous network. The significant grain boundary component provides a further opportunity to tailor the interfacial chemistry of the multicomponent system. The pores provide high surface area for enhanced catalytic activity, and their small diameter may offer catalytic selectivity like the zeolitic structures. A ceramic green body with such fine porosity has greater mechanical reliability associated with small flaw sizes, and may be sintered at much lower temperatures due to the shorter diffusion distances involved in the nanostructure.

Another way of packaging different components is shown in Fig. 1(c). Here one composition is presented as a two-dimensional lamellar structure, the second component is introduced as a 'pillaring agent' or robust intercalant between the layered structures. The layered structures may be held together by weak van der Waals forces, their pillaring may be driven by charge balance between the layers and the pillaring agents. In the process of introducing the pillaring agents between the layered structures, we not only create an intimate mixture of different components, but also a porous structure whose pore dimensions are governed by the intercalant's size and spatial density within the interlayer region. When the layered structures and the pillaring agents are of molecular dimensions in the *c* axis, the interlayer spacing would typically be ≤ 2 nm. This would correspond to a molecular-level packaging of the layered and pillar components, yielding a microporous structure and a high level of surface interaction between different constituents.

SYNTHESIS OF Cu_2O/CeO_{2-x} SUPPORTED CATALYSTS

The model mixed oxide systems described in Figs. 1(a) and (b) may be generated experimentally through nanocrystalline processing by inert gas condensation. Inert gas condensation with Joule-heating [1-4] has been used to form nanoclusters that were subsequently consolidated to make nanostructured solids. In this synthesis method, the precursors of the nanocrystalline materials are placed in a refractory metal boat which is resistively heated (see Fig. 2(a)). Collision between the precursor vapor and the cooling inert gas results in the nucleation of crystallites, which then undergo particle or cluster growth by a coalescence mechanism [5]. The aerosol of particles generated by thermal evaporation and inert gas condensation may be collected via natural gas convection and thermophoresis on a liquid-nitrogen cooled rotating cold-finger. After evaporation, the particles can be scraped from the cold-finger in vacuum and compacted *in situ* in a piston-and-anvil device.

The nanostructured materials synthesis has been successfully employed to generate a variety of metallic, ceramic, intermetallic, alloy, and composite materials. It yields high-purity nanocrystals with a high degree of control of grain sizes through parameters such as type of gas, gas pressure, evaporation rate, and source to substrate distance. Nanoclusters of $\leq$ 10 nm diameter can be synthesized and *in situ* assembled to form nanostructured materials with clean interfaces. The Joule-heating synthesis of materials however, is governed by the melting point and vapor pressure of the precursors. It has inherent limitations in generating materials, such as many ceramics which have high melting temperatures and low vapor pressures. It also faces difficulties in producing alloys and composites with a consistent composition when the different components involved have dissimilar vapor pressures and thermodynamic activities. Contaminants may also be introduced from the crucibles in Joule-heating when reactive precursors are used. In deriving the ideal mixing in a multicomponent system as described in the introduction section, Joule-heating may not always provide sufficient flexibility and uniformity. To extend the inert gas condensation to synthesizing a greater variety of materials with controlled composition and purity, dc- or rf-magnetron sputtering, a technique conventionally used for thin film deposition [6,7], was developed to yield fine particles [4,8-12].

During sputtering, ions of a suitable substance (such as Ar or Kr) accelerated to high energies are directed toward a surface, from which atoms and clusters, both neutral and ionic, are ejected. This technique has a number of distinct advantages compared to Joule-heating: (i) low contamination, since crucibles are not used; (ii) greater flexibility in the choice of source materials since metals, ceramics, and intermetallics with high melting point can be readily sputtered; (iii) preservation of the source composition for alloys and composites; and (iv) easily controllable process parameters. For particle formation, a higher gas pressure of > 1 Pa is needed compared to the pressure range of 10^{-1}-10^{-2} Pa typically employed in thin film deposition processes. We have found that particle yield by magnetron sputtering in the usual line-of-sight deposition was rather low, whereas the deposition in the plane of the sputter gun above the ground shield was significant at the relatively high pressures for particle formation [13]. Consequently, the ground shield of the sputter gun was modified as illustrated in Fig. 2(b), such that it may be used as an effective cooled substrate that can be rotated for *in situ* scraping of particles collected.

The synthesis of the nanocrystalline multicomponent system in Fig. 1(a) was accomplished with the availability of the modified magnetron sputtering technique [13]. The system of interest is nanocrystalline Cu_2O/CeO_{2-x} catalyst for selective SO_2 reduction by CO [14,15]. The core material is made of cerium oxide since this fluorite-structured oxide has an intrinsic high oxygen vacancy mobility needed for enhancing the redox reaction mechanism,

$$SO_2 + 2\,CO \rightarrow S + 2\,CO_2.$$

The oxygen vacancy concentration in the catalytic materials is also extremely important to enable the complete reduction of SO_2 to elemental S. This high defect concentration can be tailored by the nanocrystalline processing through generation of metallic Ce clusters, followed by controlled oxidation. The Cu_2O component in such a catalyst should ideally be well-dispersed over CeO_{2-x} to maximize the surface contact between the two components to form a Schottky junction that may lower the enthalpy of formation of oxygen vacancies. The interaction between Cu_2O and CeO_{2-x} may also enhance the adsorptivity of reacting species with the catalyst through spill-over

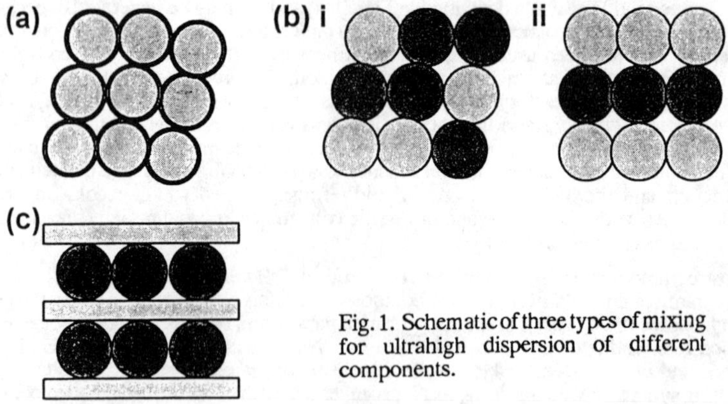

Fig. 1. Schematic of three types of mixing for ultrahigh dispersion of different components.

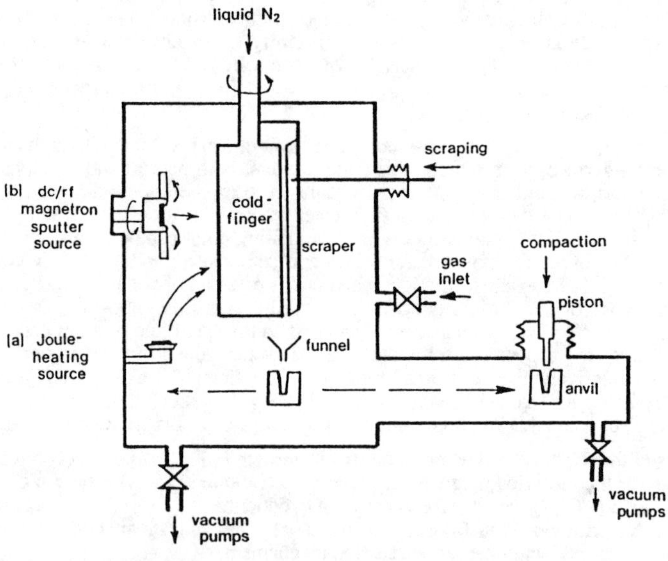

Fig. 2. Schematic drawing of a gas-condensation chamber for the synthesis of nanometer-sized clusters by: (a) Joule-heating and/or (b) modified dc-/rf-magnetron sputtering.

effects from Cu_2O to CeO_{2-x}. We set out to create the optimum dispersion of Cu_2O on CeO_{2-x} through deriving a mixed multicomponent system depicted in Fig. 1(a). A coating of Cu_2O on CeO_{2-x} nanocluster cannot be easily achieved in a direct vapor synthesis process. Our idea is to make use of surface segregation effects to achieve the Cu_2O coating on CeO_{2-x} [13]. By starting with a mixed 15 at% Cu/85 at% Ce target, Cu-Ce nanocrystals with a uniform alloy composition were generated by dc-magnetron sputtering at a power of 160 W in an argon pressure of 20-30 Pa. The alloy nanoclusters were then carefully post-oxidized by slowly back-filling the ultrahigh vacuum chamber with 500 Pa of oxygen. As Cu and Ce were gradually oxidized, they became no longer miscible in their respective oxide forms, and we anticipated that the minor component, copper oxide, would be segregated to the surface of the CeO_{2-x} crystallites.

Characterization by X-ray diffraction (XRD) and X-ray photoelectron spectroscopy (XPS) (Perkin Elmer PHI-5500) confirms the structure of model Cu_2O/CeO_{2-x}. Fig. 3(a) illustrates the XRD pattern of nanocrystalline Cu-Ce sample after annealing in an oxidizing atmosphere at 500 °C. The pattern only displays the diffraction peaks associated with the face-centered cubic structure of the cerium sublattice for the fluorite-structured cerium oxide. Cu peaks were not obtained presumably because: (i) Cu was in solid solution with Ce, and/or (ii) Cu was highly dispersed in the grain boundaries or on the surfaces of the nanocrystallites. Separate copper phase in the form of CuO was found only after heating at 550 °C. This phase became more distinct upon further heat treatments to 600 °C and 650 °C. The mean crystal size of the cerium oxide phase was determined from the (111) and (222) reflection peak broadening by Kochendorfer analysis. The as-prepared sample has a grain size of 7 nm. Annealing at 600 °C led to a minor grain growth to ~10 nm. Significant grain coarsening did not initiate till the sample was heat treated to ≥ 700 °C.

XPS illustrated the surface stoichiometry and chemical composition of the samples (see Fig. 4). Both Ce-3d lines (870-920 eV) and Cu-2p lines (920-960 eV) were found in the XPS spectrum. The as-prepared sample has two special peaks u' and v', which correspond to the $4f^1$ initial state of Ce^{3+} [16,17] (Fig. 4(a)). These components became less distinct for samples oxidized at increasing temperatures. They illustrated the significant oxygen deficiency particularly in the as-prepared CeO_{2-x} phase. The position of the Cu-2p line and LMM Auger line, after correcting for peak position using C-1s, shows that Cu was in a +1 oxidation state [18]. There was a distinct increase in the Cu component in the XPS spectrum when the sample was heated to 450 °C. Since XRD did not reflect separate copper phase formation until treatment >500 °C, the large increase in Cu-2p line in XPS at 450 °C was interpreted to be associated with surface segregation of Cu_2O in the CeO_{2-x} crystallites.

The different analytical sensitivity to crystalline and surface structure of XRD and XPS could be complemented to demonstrate the following structural evolution in nanocrystalline Cu/CeO_{2-x}: (i) copper-doped CeO_{2-x} exists as a solid solution up to at least 300 °C (Fig. 5(a)); (ii) copper was segregated to the surface of the CeO_{2-x} nanocrystals when heated between 300 °C and 450 °C, producing a surface copper enrichment as illustrated in Fig. 5(b); (iii) copper existed as Cu^{+1} to at least 450 °C, and substantial Ce^{3+} was still detected in the sample at this temperature; (iv) copper was finely dispersed below 550°C, so that only cerium oxide crystalline phase but no bulk copper oxide phase was detected by XRD; (v) distinct copper oxide phase (CuO) was developed when the sample was treated between 500 °C and 550 °C (Fig. 5(c)). The change in copper oxidation states and Ce^{3+} peak intensities could be attributed to the extent of oxidation. This experiment illustrates that through a controlled post-oxidation process, one may tailor different level of dispersion of one component in another for model supported oxide catalyst. The Cu_2O/CeO_{2-x} nanocrystals derived by surface segregation of Cu between 300 °C and 450 °C in O_2 enabled unusually high catalytic activity for SO_2 reduction and CO oxidation. The novel catalyst achieved full conversion of SO_2 to S and of CO to CO_2 at a remarkably low temperature of 420 °C and 80 °C, respectively [15].

SYNTHESIS OF Pt-TiO_{2-x} NANOCOMPOSITES

Using ultrafine particles of different materials to make novel composites represents a potentially fertile field of materials research. The possibility for *in situ* reacting, coating, and mixing various types, sizes, and morphologies of clusters presents a significant capacity for synthesizing new multicomponent composites with nanometer-sized microstructures and engineered properties.

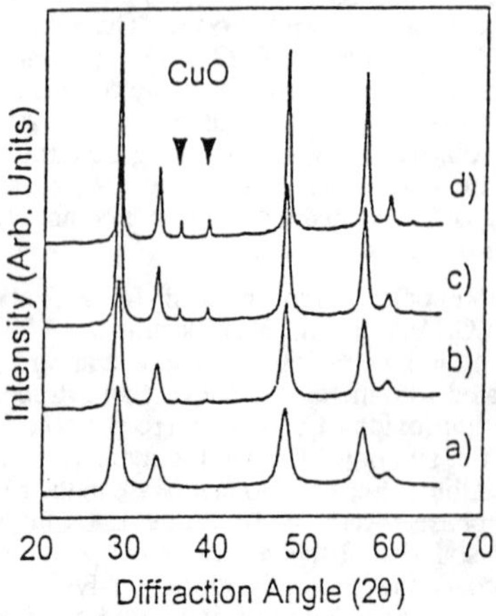

Fig. 3. X-ray diffraction pattern for Cu-doped CeO_{2-x} after annealing at (a) 500°C, (b) 550°C, (c) 600°C, and (d) 650°C. The main peaks are due to the cerium fcc sublattice whereas the smaller peaks at 35.5° and 39° correspond to CuO. Notice that the Ce peak after annealing at 500 °C is still broadened as compared to higher temperatures.

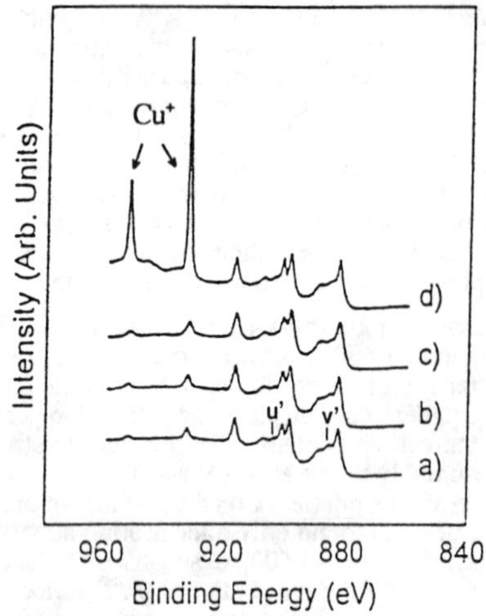

Fig. 4. XPS spectrum of Cu-doped CeO_{2-x}: as prepared (a), and after annealing at (b) 150°C, (c) 300 °C, and (d) 450°C. The two distinct lines in curve (d) are Cu-2p and their increase in relative intensity with respect to Ce-3d indicates Cu segregation.

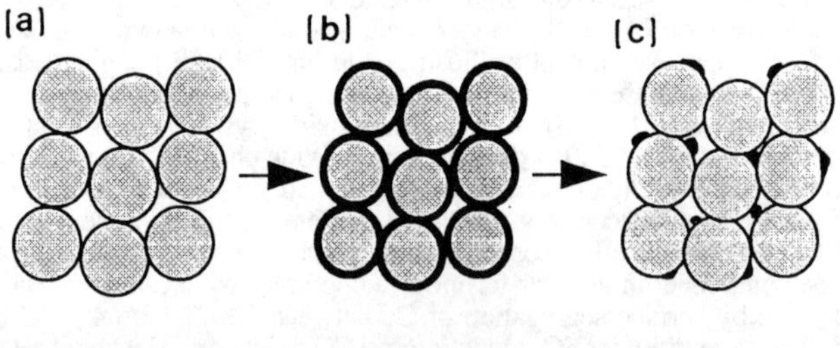

Fig. 5. Synthesis of Cu_2O/CeO_{2-x} nanocrystalline supported catalyst: (a) Cu-Ce nanocrystalline alloy, (b) surface segregated Cu_2O in CeO_{2-x} nanocrystals, and (c) phase separated CuO in CeO_{2-x} nanocrystalline matrix.

Nanostructured composite processing offers several advantages: (i) nanostructured cermets and ceramic-matrix composites with components well dispersed on the nanometer-scale is possible for enhanced structural and catalytic behavior, and can be formed at relatively low temperatures; (ii) the nanostructured composite may exhibit low-temperature ductility to facilitate net-shape forming of materials; and (iii) the nanostructured composite with ultrafine pore sizes may be densified at significantly lower temperatures.

Nanostructured composites, such as Pt-TiO_{2-x} and Pd-Al_2O_3, were studied as a model system for supported metal catalysts and cermet systems [4,19]. Composite processing by the nanocrystalline route might be achieved by simultaneous evaporation of the corresponding metallic precursors, followed by post-oxidation to obtain the desired oxide ceramic (note: noble metal does not oxidize as easily and may be reduced fairly readily compared to the ceramic support component). For the effective processing of ultrafine-grained and stoichiometric oxides, reactive sputtering and direct rf-sputtering were developed. These techniques offer the possibility of simultaneous sputtering of the ceramic and noble metal components. This type of simultaneous generation of clusters from sources of different components would produce the random dispersion of different types of nanocrystals depicted in Fig. 2(b)i.

Quite frequently, the conditions (e.g. gas pressures) for good sputtering or evaporation yields of two different materials might be dissimilar. We have devised a sequential processing method that would provide both reasonable yields and a unique dispersion of the components in nanostructured cermets [4,19]. This method also allows a combination of techniques such as Joule-heating and magnetron sputtering to be utilized. In the processing of Pt-TiO_{2-x} for example, the sequence involved: (i) evaporation of Ti by Joule-heating in He, (ii) oxidation of Ti clusters, and (iii) sputtering of Pt in Ar. This sequence was repeated several times, and the alternating deposition of TiO_{2-x} and Pt in this sequential processing creates the kind of dispersion of nanocrystals illustrated in Fig. 2(b)ii. The composition, degree of dispersion, and particle size of each component could be flexibly controlled by factors such as the evaporation/sputtering periods, evaporation/sputtering rates, and He/Ar gas pressures. Fig. 6 illustrates the XRD pattern of a nanostructured Pt-TiO_{2-x} sample with ~10 wt% Pt. TiO_{2-x} was present mainly in the rutile phase, and trace quantity of the anatase phase was also found. By the Scherrer formula, the average crystal size was determined to be 11 nm for TiO_{2-x}, and 8 nm for Pt. This particular cermet system has potential applications as oxide supported metal catalysts. It is anticipated that by this processing the ultrafine clusters of Pt and TiO_{2-x} would be dispersed effectively on the nanometer scale, demonstrating superior metal-support interactions. The grain boundary component of nanostructured composites could additionally be varied in volume and structure by the size and composition of the grains. The porosity and surface reactivity related to the unique microstructure and grain boundary component of nanostructured composites might be engineered in an analogous fashion as the cage-like zeolites for tailoring of catalytic selectivity.

The future for synthesizing nanostructured composites appears very exciting. Whether by the solid-state reaction of neighboring particles in a nanostructured compact, the infusion of dopants and/or reactants through the grain-boundary network, or the mixing of a variety of powders with differing and complementary properties [20], the synthesis of nanostructured composite materials might have the greatest technological impact in this new field of research.

PILLARING OF LAYERED STRUCTURES

The term pillaring is used to describe the process whereby a layered solid undergoes permanent expansion in the *c* axis due to the intercalation of a robust, bulky molecule or ion. This process provides a molecular-level packaging of two different phases, and offers an unique dispersion of different components in a scheme illustrated in Fig. 1(c). The interlayer region of the host material is typically increased to heights comparable to the van der Waals dimension of the guest through pillaring. This phenomenon introduces porosity into the layered material. If the spatial density of the pillaring species can be carefully manipulated to control the lateral distance between the pillars, the potential exists for generating a class of materials with a two-dimensional, intracrystalline microporous network for size-selective adsorption and catalysis.

In general, ionic lamellar solids are preferred hosts for the preparation of pillared intercalates through simple ion-exchange reactions. Two classes of layered materials, layered double hydroxides (LDHs) and transition metal oxide gels, have been successfully synthesized and pillared by us. The family of LDHs has a permanent positive layer charge, which may be compensated by anionic pillars in the interlayer region. Transition metal oxide gels, on the other hand, have a negative layer charge, which may be balanced by introducing cationic guests between the layered structures.

LDHs [21] or the so-called anionic clays, represent a potentially useful class of ionic solids for forming microporous pillared derivatives. The structure of the LDH is very similar to that of brucite, $Mg(OH)_2$, where octahedra of Mg^{2+} (6-fold coordinated to OH^-) share edges to form infinite sheets. These sheets are stacked on top of each other, and are held together by hydrogen bonding [22]. Isomorphous substitution of a divalent cation in the lattice by a trivalent cation having a similar radius results in a positive charge generated in the hydroxyl sheet. This net positive charge is balanced by having hydrated anions in the interlayer region. The compositions of the LDHs are thus typically represented by the general formula $[M^{II}_{(1-x)}M^{III}_{x}(OH)_2]^{x+}[A^{n-}_{x/n} \cdot yH_2O]$, where A^{n-} is the gallery anion, and M^{II} and M^{III} are the divalent and trivalent cations, respectively. A large number of compositions are possible depending on the choices of M^{II}, M^{III}, A^{n-}, and the layer cation stoichiometry (typically in the range of $0.17 \leq x \leq 0.33$).

We have successfully synthesized and pillared novel Mg/Ni/Al layered double hydroxides with heptamolybdate anions [23]. The parent LDHs were prepared with terephthalate dianions as the intermediate interlayer anions. Protonation of the organic anions under mildly acidic conditions facilitated the subsequent exchange with the highly charged pillaring anions, such as the desirable heptamolybdate ions. Fig. 7(a) illustrates the XRD pattern of the Mg/Ni/Al LDH with the carbonate anion in the interlayer region. The *d* spacing for this LDH was increased from 8.0 Å to 12.2 Å upon pillaring with $Mo_7O_{24}^{6-}$ through the terephthalate dianion intermediate (Fig. 7(b)).

Transition metal oxides may be synthesized in unique lamellar structures through a novel single-step ion-exchange sol-gel approach [24-26]. These gels consist of a net negative charge in their layered structure due to the presence of mixed valence states. Vanadium pentoxide xerogel, $V_2O_5 \cdot nH_2O$ ($n < 2$), synthesized by this technique has been successfully pillared with polyhydroxyaluminum polymers. Analysis of the dry xerogel by wide-angle X-ray scattering indicated a stacking periodicity of 11.4 Å (Fig. 8(a)). After pillaring, the xerogels showed up to 9 orders of *00l* reflections (Fig. 8(b)), with a mean basal spacing of 17.7 Å. The effect of temperature on the original and pillared xerogels was also examined by XRD. The unpillared xerogel transformed to the orthorhombic crystal structure by 350 °C, while the pillared xerogel maintained its layered structure at this temperature. The Al_{13}-pillaring agent sustained the layered structure of the xerogel to 450 °C. The pillared materials are being examined as catalysts for partial oxidation and hydrodesulfurization reactions.

SUMMARY

Nanocrystalline processing and pillaring of layered structures were employed in the derivation of novel mixing in multicomponent systems. Intimate surface interaction between different components could be achieved through: (i) surface coating of one component over the nanocrystallite of another composition, (ii) random mixing of different types of nanocrystallites, (iii) alternating arrangement of nanocrystallites of different compositions, and (iv) packaging of different constituents as layered structures and pillaring agents stabilized by charge balance. These unique ways of dispersing different components may allow us to achieve enhanced catalyst-support interaction in catalytic applications, and better control of interfacial chemistry in engineering of advanced ceramics.

ACKNOWLEDGMENT

This project is sponsored by National Science Foundation (CTS-9257223 and DMR-9022933) and Sloan Funds. The author acknowledges the contribution of her research group, particularly Andreas S. Tschöpe and Doron Levin, in the synthesis of nanocrystalline oxides and pillared structures discussed in this paper.

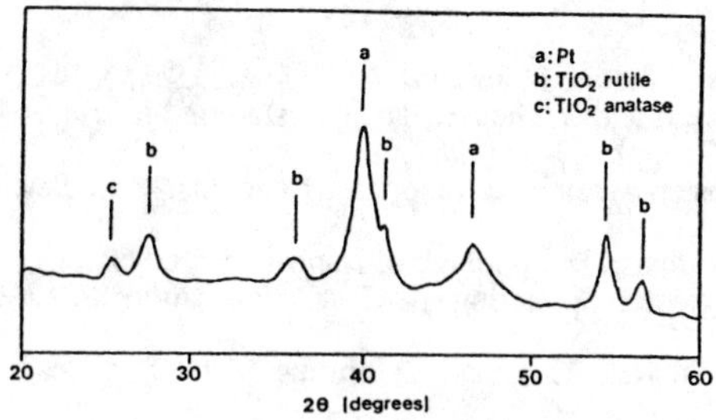

Fig. 6. XRD pattern of a cermet nanocomposite of Pt-TiO_{2-x} produced by sequential processing [4,19].

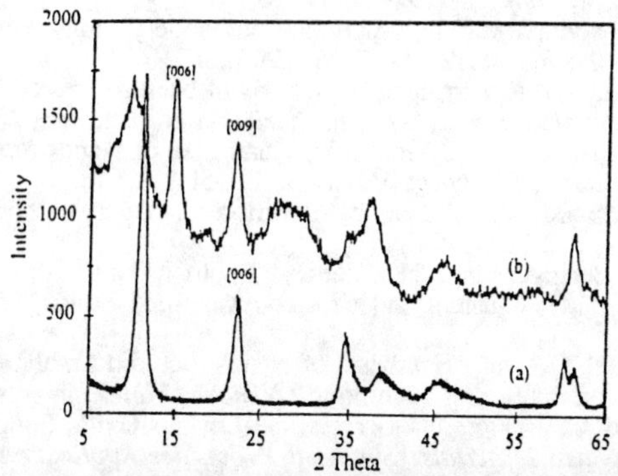

Fig. 7. XRD patterns of Mg/Ni/Al LDH: (a) as-prepared with carbonate anion in the interlayer region, (b) after pillaring with $Mo_7O_{24}^{6-}$ through the terephthalate dianion intermediate [23].

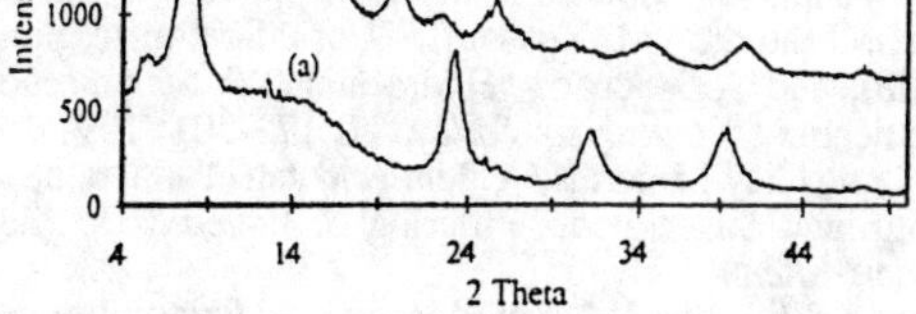

Figure 8. XRD patterns of vanadium pentoxide xerogel: (a) before pillaring, and (b) after pillaring [26].

REFERENCES

[1] H. Gleiter, "Nanocrystalline Materials," *Prog. Mater. Sci.,* **33**[4], 223-315 (1989).

[2] R. Birringer, U. Herr, and H. Gleiter, "Nanocrystalline Materials - A First Report," *Trans. Jpn. Inst. Met.,* **27** suppl., 43-52 (1987).

[3] R.W. Siegel, "Cluster-Assembled Nanophase Materials," *Ann. Rev. Mater. Sci.,* **21,** 559-78 (1991).

[4] J.Y. Ying, "Structure and Morphology of Nanostructured Oxides Synthesized by Thermal Vaporization/Magnetron Sputtering and Gas Condensation," *J. Aerosol Sci.,* **24**[3], 315-38 (1993).

[5] C.G. Granqvist and R.A. Buhrman, "Ultrafine Metal Particles," *J. Appl. Phys.,* **47**[5], 2200-19 (1976).

[6] L.I. Maissel and R. Glang, Eds., *Handbook of Thin Film Technology.* McGraw-Hill, New York, 1970.

[7] K. Reichelt and X. Jiang, "The Preparation of Thin Films by Physical Vapour Deposition Methods," *Thin Solid Films,* **191,** 91-126 (1990).

[8] H. Hahn and R.S. Averback, "The Production of Nanocrystalline Powders by Magnetron Sputtering," *J. Appl. Phys.,* **67**[2], 1113-15 (1990).

[9] V. Haas, H. Gleiter, and R. Birringer, "Synthesis of Nanostructured Materials by the Use of a Thermophoretic Forced Flux System," *Scripta Metall. Mater.,* **28,** 721-24 (1993).

[10] V. Haas and R. Birringer, "The Morphology and Size of Nanostructured Cu, Pd and W Generated by Sputtering," *Nanostr. Mater.,* **1,** 491-504 (1992).

[11] G.M. Chow and A.S. Edelstein, "Sputter Deposition of Nanocrystals and Nanocomposites," *Nanostr. Mater.,* **1,** 107-11 (1992).

[12] J.Y. Ying, J.B. Benziger, and H. Gleiter, "Photoacoustic Infrared Spectroscopy of Nanometer-sized Al_2O_3 Clusters and Cluster-Assembled Solids," *Phys. Rev. B,* **48**[3], 1830-36 (1993).

[13] A.S. Tschöpe and J.Y. Ying, "Synthesis of Nanostructured Catalytic Materials Using A Modified Magnetron Sputtering Technique," *Nanostr. Mater.,* in press.

[14] A.S. Tschöpe and J.Y. Ying, "Nanocrystalline Cerium Oxide Catalytic Materials"; pp. 781-84 in *Nanophase Materials: Synthesis-Properties-Applications.* Edited by G.C. Hadjipanayis and R.W. Siegel. Kluwer, Netherlands, 1994.

[15] A.S. Tschöpe, J.Y. Ying, W. Liu, and M. Flytzani-Stephanopoulos, "Nanocrystalline Doped Cerium Oxide as a Catalyst For SO_2 Reduction by CO"; in *Materials and Processes for Environmental Protection.* Materials Research Society, Pittsburgh, PA, in press.

[16] F. Le Normand, L. Hilaire, K. Kili, G. Krill, and G. Maire, "Oxidation State of Cerium in Cerium-Based Catalysts Investigated by Spectroscopic Probes," *J. Phys. Chem.,* **92,** 2561-68 (1988).

[17] F. Le Normand, J. El Fallah, L. Hilaire, P. Légaré, A. Kotani, and J.C. Parlebas, "Photoemission on 3d Core Levels of Cerium: An Experimental and Theoretical Investigation of the Reduction of Cerium Dioxide," *Solid State Comm.,* **71**[11], 885-89 (1989).

[18] *Handbook of X-Ray Photoelectron Spectroscopy,* Perkin Elmer Corp.

[19] J.Y. Ying and A.S. Tschöpe, to be submitted.

[20] R.W. Siegel and H. Hahn, "Nanophase Materials"; pp. 403-419 in *Current Trends in the Physics of Materials.* Edited by M. Yussouff. World Scientific, Singapore, 1987.

[21] F. Cavani, F. Trifirò, and A. Vaccari, "Hydrotalcite-Type Anionic Clays: Preparation, Properties and Applications," *Catalysis Today,* **11,** 173-301 (1991).

[22] W.T. Reichle, "Anionic Clay Minerals," *Chemtech,* Jan. 1986, p. 58-63.

[23] D. Levin, J.Y. Ying, and S.L. Soled, "Pillaring of Layered Double Hydroxides," to be submitted to *J. Mater. Chem.*

[24] J.Y. Ying and D. Levin, "Pillaring of Layered Vanadium Pentoxide Gels," Gordon Research Conference on 'Zeolitic and Layered Structures,' New Hampshire, June 1993.

[25] J.Y. Ying and D. Levin, "Pillaring of Sol-Gel Derived Vanadium Pentoxide," American Institute of Chemical Engineers Annual Meeting, St. Louis, Nov. 1993.

[26] D. Levin and J.Y. Ying, "Pillaring of Vanadium Pentoxide Gels with Polyoxoaluminum Cations," to be submitted to *J. Am. Chem. Soc.*

CHARACTERIZATION OF THE SURFACE CHEMISTRY OF SILICON NITRIDE POWDERS.

Lennart Bergström
The Institute for Surface Chemistry
P.O. Box 5607
S-114 86 Stockholm, Sweden

INTRODUCTION

Silicon nitride is a ceramic material with a potential for high strength at room as well as elevated temperatures, good thermal shock resistance and relatively good oxidation resistance. However, optimal properties can only be realized when the microstructure and the grain boundary composition of the sintered body is controlled. Several studies have shown that the surface properties of the starting powder, together with the choice and addition of sintering aids, play a decisive role in determining the final properties [1-3].

However, the final properties of the sintered material depend on many different parameters. These parameters include the green-body characteristics, the physical and bulk chemical powder properties and the sintering conditions together with the powder surface chemical properties. Hence, it has been difficult to define the relationship between only the surface chemical properties and materials properties more precisely. Also, it is not clear how the relevant surface properties of silicon nitride should be characterized and quantified.

This paper is a part of a review on the surface chemistry of silicon nitride [4]. The focus will be on an overview of the methods that are available to characterize relevant surface properties of ceramic powders, together with a summary on the present understanding of the surface chemistry of silicon nitride.

SURFACE CHEMICAL CHARACTERIZATION OF POWDERS

There is some crucial information, vital for optimal processing as well as for sintering, that a thorough surface analysis of ceramic powders should provide. The surface chemistry of the starting powder plays an important role in many of the powder processing steps. These steps include the mixing and dispersing of the starting powders, controlling the interparticle forces and the rheology and behavior during forming of casting suspensions. The surface chemistry also influences the composition and phases of the grain-boundary phase. Hence, there is a great need to characterize the different features of the surface properties of the starting powders. It has been suggested that the information wanted can be phrased into an attempt to answer the following questions [5]:

What is the chemical composition and crystallographic phase of the surface film (stoichiometry, extent of oxidation, thickness,surface impurities etc.)?

What is the chemistry of the surface functional groups? How is their stability and reactivity to thermal and chemical treatments?

What is the behavior at the solid-liquid or solid-gas interface?

These questions relate to an analysis with different information depths. The characterization of the chemical composition of the surface film typically involves information from the top 1-5 nm, while characterization of the chemistry of the surface functional groups deals with the top molecular layer, 0.3-0.6 nm. In the case of characterizing the solid-gas or solid-liquid interface, the measurements necessarily involves characterizing the *interactions* at the interface. Available surface characterization methods are summarized in Table I and briefly described below. A more detailed description is given in ref. [5].

The surface chemical composition can be measured using different types of high vacuum techniques such as AES, ESCA and SIMS. Transmission electron microscopy (TEM) can also give useful information regarding the thickness and crystallinity of the surface film. The vibrational spectrum (IR spectroscopy) of inorganic powders is the traditional, and still mainly used, source of information about the chemistry of functional surface groups. Some other techniques, such as ESR and NMR, have also been used to study surface species.

More information regarding the surface species can be obtained by studying the interactions at the solid-gas interface. Adsorption onto oxides has been extensively studied in relation to heterogeneous catalysis where the main method has been IR spectroscopy. Inverse gas

Table I. Surface Chemical Characterization Methods for Ceramic Powders

Type of Characterization	Methods*
Elemental composition of powder surfaces	AES, ESCA, SIMS TEM
Chemistry of functional surface groups	IR spectroscopy (NMR, ESR, Raman)
Interaction at the solid-gas interface	IR spectroscopy (IGC, TPD)
Interaction at the solid-liquid interface	Electrokinetic methods Potentiometric titration Calorimetry, IR spectroscopy Adsorption isotherms

* Methods within parentheses are considered to be of less importance.

chromatography (IGC) is also a promising and simple method for characterizing the surface properties. Characterization of the interactions at the solid-liquid interface are of particular importance to optimal suspension processing. In aqueous media, the surface charge and double layer potential can be estimated by potentiometric titrations and electrokinetic measurements, respectively. In non-aqueous media, the interaction with the solvent and the adsorption of different additives can be characterizad using calorimetry (measuring the heat of immersion and heat of adsorption), adsorption isotherms to measure the amount adsorbed, and IR spectroscopy to obtain molecular information on the sorbate-solid interaction.

Hence, there are a wide variety of methods available to characterize the surface chemical properties of ceramic powders. Unfortunately, it is not possible to find one universal method which can provide all the information wanted. As will be demonstrated in the following section, it is necessary to combine a variety of experimental techniques, preferably from the different groups in Table I, to obtain a deeper understanding of the surface chemistry of ceramic powders.

SURFACE CHEMISTRY OF SILICON NITRIDE

Surface composition

Various techniques such as TEM, ESCA, AES, SIMS and IR have been used to characterize the surface structure and composition of Si_3N_4 powders and thin films. Electron microscopy studies have shown that a typical commercial Si_3N_4 powder is coated with an amorphous oxide film with a thickness of about 1-5 nm [6,7]. There has been some controversy regarding the composition of this oxide layer. ESCA studies on a plasma-derived Si_3N_4 powder with a high oxygen content (3.1 wt%) [8] have indicated that the oxide layer consists of SiO_2. The conclusion drawn from other ESCA studies on commercial Si_3N_4 powders [7,9,10] has been that the oxide layer composition corresponds to nonstoichiometric silicon oxynitride.

The different conclusions drawn are essentially related to the shape of the Si(2p) peak in the ESCA spectrum. This peak is always quite broad when oxidized Si_3N_4 is being investigated with a half-width above 2 eV. Studies [8] indicating a SiO_2 composition of the surface film have been able to deconvolute this peak into one component corresponding to Si in Si_3N_4 (at~101.5-102 eV) and one component corresponding to Si in SiO_2 (at ~103.7 eV). The studies [7,9,10] indicating a non-stoichiometric silicon oxynitride composition have not deconvoluted the Si(2p) peak and have thus obtained only one peak energy usually between 102 and 102.5 eV.

Recent ESCA and Bremsstrahlung-excited AES (XAES) studies have also included measurements of the Si KLL Auger peak [11,12,15]. Since this peak has a much larger energy separation between the Si-N and Si-O species (Δ(B.E)~3.4 eV) than the Si(2p) peak (Δ(B.E.)~1.5 eV) it was possible to resolve the two components. This information has been used to estimate the thickness of the oxidized surface layer on silicon nitride powders [11,12,25]. By assuming an uniform, dense SiO_2 layer on a Si_3N_4 powder containing 4 wt% Y_2O_3, Wang et al [11] estimated the oxide film thickness to be 0.84 nm. Using the same model, Hackley et al [12] calculated oxide thicknesses ranging from 0.1 to 0.7 nm for five commercial silicon nitride powders.

However, these calculated thicknesses of the oxide film should only be used as a rough estimate. Results from angle-dependent ESCA measurements of an oxidized silicon nitride film showed that the oxygen content varied with depth [13]. Hence, it is not possible to assume a homogeneous composition of the oxide layer. Futhermore, the spatial distribution of oxygen species, the variation in oxide film thickness over the powder surface, and the density of the surface film has not been characterized. The descripancy between the oxide thickness estimated from TEM (t= 1-5 nm) and the estimations using XAES (t= 0.1-0.8 nm) may be related to the existance of such a nonuniform or discontinous surface layer with a varying stoichiometry.

Manipulation of the surface film.

There are several different possibilities in manipulating the thickness and composition of the oxide film. Oxidation in air at elevated temperatures always results in an increased oxygen content on the surface [2]. It has been shown that the thickness of the oxide film on Si_3N_4 powders can be reduced by leaching in anhydrous hydrofluoric acid or treatment in a mixture of argon/hydrogen at 1300°C [7]. Another viable method to remove part of the oxide film is to leach oxidized Si_3N_4 powders in an aqueous medium [6,10,12,14]. Chu [6] used a Soxhlet extraction unit which continuously subjected the powder to fresh, hot distilled water. She found that the thickness of the oxide film decreased from 2-4 nm to less than 1 nm during Soxhlet extraction for 24 h and that the thickness of the oxide film could be controlled by changing the extraction time. Similarily, Bergström et al [10,14] showed that by controlling the pH and solids loading in a silicon nitride suspension, the kinetics and total amount of dissolved silica could be controlled.

Surface functional groups.

Infrared studies on silicon nitride materials can be divided into two groups: those investigating grown or deposited amorphous thin films, and those investigating synthesized, high surface area (>50 m^2/g) powders. Reliable IR studies on commercially available silicon nitride powders are practically non-existant, probably due to experimental difficulties using powders of low (~10 m^2/g) surface area.

Work on thin films [16,17] shows a consistent picture regarding the functional groups in amorphous silicon nitride or silicon oxynitride materials. Examining silicon nitride films, Claesson *et al.* [16] found strong absorption bands at 850 cm^{-1} (Si-N stretching), 3300 cm^{-1} (N-H stretching) and 2180 cm^{-1} (Si-H stretching). Increasing the oxygen content in the film resulted in a shift of the N-H band to higher wave numbers, up to 3400 cm^{-1}, when all the nitrogen atoms have been exchanged for oxygen attached to the silicon tetrahedron. The IR work by Tsu et al. [17] on thin films of silicon nitride and silicon diimide produced by remote plasma-enhanced chemical vapour deposition showed similar features. They also found a N-H stretching band centered at about 3335 cm^{-1} and a S-N stretching band between 835 and 885 cm^{-1}. In addition, they found a N-H bending vibration at 1175 cm^{-1} and a shoulder on the high wave number side of 3335 cm^{-1} for the films grown at the lowest temperature. This was interpreted as being a N-H_2 stretching vibration.

Work on synthesized silicon nitride powders showed a similar picture [18-20]. Nielsen et al. [18] investigated an ultrafine laser-synthesized silicon nitride powder. Powder unexposed to ambient atmosphere showed two overlapping peaks in the IR spectrum at 3490 cm^{-1} and 3395 cm^{-1}, interpreted as representing asymmetric and symmetric N-H stretching modes. When the powder was exposed to humidity, a sharp peak at 3760 cm^{-1} appeared, related to O-H stretching. This peak is caused by silanol (Si-OH) groups created on the surface. The N-H peaks changed to a broad, unresolved peak at 3350 cm^{-1}. Studies on silicon nitride produced by laser synthesis or ammonia nitridation of silica [19,20] provide more detailed information on the nature and interactions of the surface groups. The most important features of the IR spectrum were a very strong band at 3742 cm^{-1} (O-H stretching of surface hydroxyl groups), a band at 3350 cm^{-1} (N-H stretching of Si_2-NH groups), and a band at 1550 cm^{-1} (in-plane N-H deformation). The existence of Si-NH_2 was not well established with some studies reporting its presence [20] and others not [19].

In summary, IR studies have shown that secondary amino groups (Si_2-NH) and silanol groups (Si-OH) exist on silicon nitride materials exposed to an atmosphere containing hydrogen and oxygen (Fig. 1). Primary amino groups (Si-NH_2) are probably too reactive to exist for any extended period of time under ambient conditions.

Behavior at the solid-liquid interface.

A large number of studies have dealt with the electrokinetic properties and surface charge behaviour of silicon nitride powders in aqueous media [6,10,12,14,21-23]. These studies report a wide range of pH of the isoelectric points (pH_{iep}), between $pH_{iep} \approx 3$ to $pH_{iep} \approx 9$. Although it was early recognized that the wide variation in the isoelectric point of silicon nitride was probably related to the degree of surface oxidation, relatively few studies have combined measurements of the electrokinetic or charge behaviour with a thorough surface characterization.Bergström and Pugh [10] investigated three different silicon nitride powders using ESCA and electrophoretic mobility. The

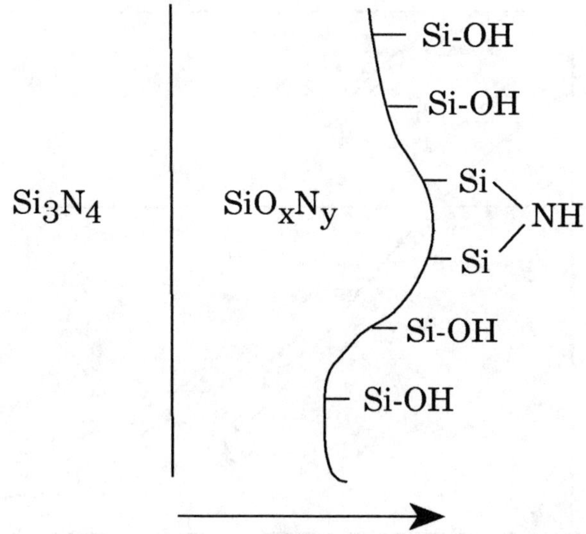

Figure 1. Model of the oxidized silicon nitride surface consisting of a graded, nonstoichiometric silicon oxynitride layer with silanol (Si-OH) and secondary amine (Si_2-NH) functional surface groups.

variation in isoelectric point was found to be correlated to the N/Si ratio in the surface film determined by ESCA. A high pH_{iep} was accompanied by a high N/Si ratio. In a later study [14], the surface composition of the powders was systematically altered by leaching in water or annealing in vacuum and the effect on the isoelectric point was investigated. Fig. 2 illustrates the change in iep as a function of conditioning time in an aqueous solution at pH≈9. The change in iep is caused by a continuous leaching of the oxidized surface layer. It was found that the shifts in pH_{iep} could be related to the atomic ratios N/Si and O/Si. There is a general trend for the N/Si ratio to increase and the O/Si ratio to decrease with increasing pH_{iep}. An increase in pH_{iep} is also associated with a decrease in the total oxygen content. In a study on the effect of different milling conditions, the same consistent correlation between pH_{iep}, bulk oxygen content and surface composition was found [23].

These results together with other studies [6,12,21] clearly show the strong link between the surface composition and the electrokinetic behaviour. In an attempt to explain this behaviour, a model of the charge behaviour of silicon nitride was presented [14]. The model is

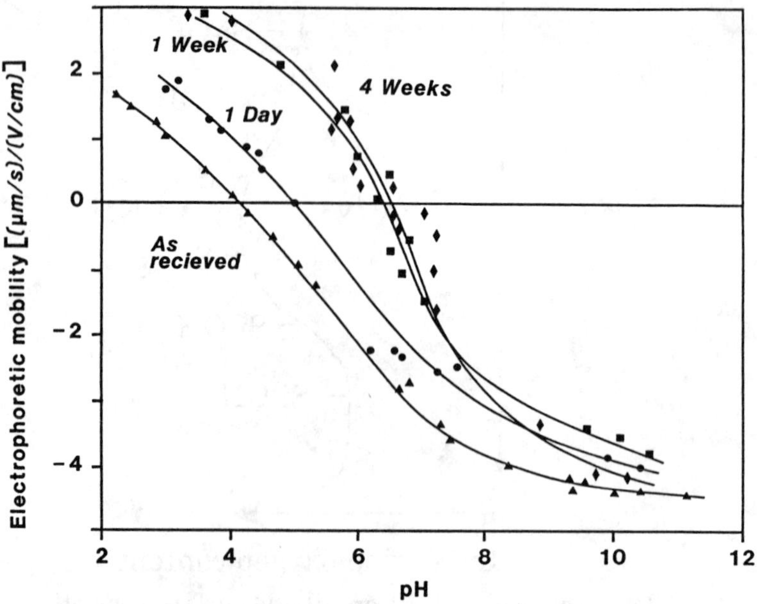

Fig. 2. Electrophoretic mobility of a silicon nitride powder in 10^{-2}M NaCl as a function of conditioning time and pH. From ref [14].

based on the assumption that the silicon nitride surface contains two types of ionizable surface groups; silanol (Si-OH) and secondary amine (Si_2-NH) groups (fig. 1). The potential-determining reactions of these amphoteric (silanol) and basic (amine) surface groups can be written as

$$\text{Si-OH}^+ \Longleftrightarrow \text{Si-OH} + \text{H}^+$$

$$\text{Si-OH} \Longleftrightarrow \text{Si-O}^- + \text{H}^+$$

$$\text{Si}_2\text{-NH}^+ \Longleftrightarrow \text{Si}_2\text{-NH} + \text{H}^+$$

Each of these reactions can be assigned an equilibrium constant and it was shown that it is possible to relate the ratio of basic amine groups to amphoteric silanol groups to the pH of zero charge (or pH_{iep}) [14]. Hence, electrokinetic measurements can be used as a sensitive tool to detect changes in the composition of the functional surface groups.
In a recent paper, a method was developed to investigate the effect of surface properties on the adsorption behavior in organic media [24].

Adsorption isotherms of organic probe molecules covering a wide spectrum of Lewis acidity/basicity was determined and analyzed using a Langmuir-Freundlich model. Measurements on three different silicon nitride powders with large differences in pH_{iep}, showed that the relative surface concentration of amine and silanol groups could be correlated to the maximum amount adsorbed (Γ_M) of some selected probes. Using the previously described model to calculate the relative site density of the surface groups, it was found that Γ_M of benzoic acid (an acidic molecule) increased linearly while Γ_M of pyridine (a basic molecule) decreased with an increase in the relative amount of amine groups on the surface. Hence, adsorption measurements on suitable probe molecules in organic media can also be used to study the surface composition of silicon nitride and other ceramic powders.

SUMMARY

The surface chemistry of silicon nitride has been investigated in a large number of studies, characterizing both the surface film and the functional surface groups together with the interactions at the solid-liquid interface using a variety of methods. By combining the data obtained from these thorough studies it is possible to obtain a rather consistent picture regarding the chemical composition of the surface film and the nature of the functional surface groups.

Silicon nitride powders, which have been subjected to air or water, are always coated with an oxidized film. Most of the ESCA work on powders and thin films suggests that this surface film has a composition which corresponds to an intermediate state between silica and silicon oxynitride. Most probably, the oxidized surface film consists of a graded, nonstoichiometric silicon oxynitride film which is oxygen-rich at the surface. However, the estimations of the thickness of the oxidized film has been found to be strongly method dependent. ESCA and XAES measurements indicate a film thickness between 0.1-0.8 nm, while TEM shows a thickness between 1 and 5 nm. Thorough studies on single crystals or very smooth thin films are needed to elucidate the detailed structure of the surface film on silicon nitride. In addition, information on the spatial variation of film thickness or oxygen rich areas is needed.

There are several ways of altering the composition of the surface film, e.g. by leaching in anhydrous HF, by treatment in Ar/H_2 gas at high temperature, milling in different media or by leaching in aqueous media. Leaching silicon nitride powders in water in a controlled way was found to be an effective method of controlling the surface composition.

Infrared studies on synthesized, high surface area powders and amorphous thin films show that two types of functional surface groups; secondary amino groups (Si_2-NH) and silanol groups (Si-OH), always exist on silicon nitride materials exposed to ambient atmosphere.

Experimental studies have established a strong link between the electrokinetic properties and the surface composition. A low oxygen content in the surface film is generally reflected in a high pH of the isoelectric point (pH_{iep}). A model has been developed which relates the relative site density of the two ionizable surface groups to the pH_{iep}. In a similiar fashion has the adsorption behaviour of small probe molecules in non-aqueous media been correlated to the surface composition using the Lewis acid/base concept.

REFERENCES.

[1] G. Ziegler, J. Heinrich, and G. Wötting, "Relationships between processing, microstructure and properties of dense and reaction-bonded silicon nitride", J. Mater. Sci., 22, 3041-86, 1987
[2] S. Natansohn, A.E. Pasto, and W.J. Rourke, "Effect of Surface Modifications on the Properties of Silicon Nitride Ceramics", J. Am. Ceram. Soc., 76 [9], 2273-84 (1993)
[3] A.J. Pyzik, and D.R. Beaman, "Microstrucure and Properties of Self-reinforced Silicon Nitride", J. Am. Ceram. Soc., 76 [11] 2737-44 (1993)
[4] L. Bergström, "Surface and Colloid Chemistry of Silicon Nitride", in preparation
[5] L. Bergström, "Surface Chemical Characterization of Ceramic Powders", pp 71-125 in *Surface and Colloid Chemistry in Advanced Ceramics Processing*, eds. R.J. Pugh and L. Bergström, Marcel Dekker, New York, 1994

[6] S-M. Chu, "Dispersion of Silicon Nitride in Aqueous Media", M.Sc. Thesis, Massachusets Institute of Technology, 1990
[7] M.N. Rahaman, Y. Boiteaux, and L.C. de Jonghe, "Surface Characterization of Silicon Nitride and Silicon Carbide Powders", Am. Ceram. Soc. Bull., 65 [8] 1171-76 (1986)
[8] I. Bertoti, G. Varsanyi, G. Mink, T. Szekely, J. Vaivads, T. Miller, and J. Grabis, "XPS Characterization of Ultrafine Si_3N_4 Powders", Surf. Interface Anal., 12, 527-30 (1988)
[9] M. Peuckert and P. Greil, "Oxygen distribution in silicon nitride powders", J. Mater. Sci. 22, 3717-20 (1987)
[10] L. Bergström and R.J. Pugh, "Interfacial Characterization of Silicon Nitride Powders", J. Am. Ceram. Soc., 72 [1], 103-109 (1989)

[11] P.S. Wang, S.M. Hsu, S.G. Malghan, and T.N. Wittberg, "Surface oxidation kinetics of Si_3N_4-4%Y_2O_3 powders studied by Bremsstrahlung-excited Auger spectroscopy", J. Mater. Sci., 26, 3249-52 (1991)
[12] V.A. Hackley, P.S. Wang, and S.G. Malghan, "Effects of soxhlet extraction on the surface oxide layer of silicon nitride powders", Mater. Chem. Phys, 36, 112-118 (1993).
[13] S.I. Raider, R. Flitsch, J.A. Aboaf, and W.A. Pliskin, "Surface Oxidation of Silicon Nitride Films", J. Electrochem. Soc., 123, 560-565 (1976)
[14] L. Bergström, and E. Bostedt, "Surface Chemistry of Silicon Nitride Powders: Electrokinetic Behaviour and ESCA Studies", Coll. Surf., 49, 183-97 (1990)
[15] K. Fukuyama, and K. Okada, "XPS Analysis of Surface State of various Silicon Nitride Powders", in *Third Euro-Ceramics*, eds. by P. Durán and J.F. Fernández, Faenza, Spain, 1993, pp 143-148.

[16] W.A.P. Claassen, H.A.J.Th. v.d.Pol, A.H. Goemans and A.E.T. Kuiper, "Characterization of Silicon-Oxynitride Films Deposited by Plasma-Enhanced CVD", J. Electrochem. Soc., 133, 1458-64 (1986).
[17] D.V. Tsu, G. Lucovsky and M.J. Mantini, "Local atomic structure in thin films of silicon nitride and silicon diimide produced by remote plasma-enhanced chemical-vapor deposition", Phys. Review B, 33, 7069-76 (1986).
[18] K.J. Nilsen, R.E. Riman and S.C. Danforth, "The Effect of Moisture on the Surface Chemistry and Dispersion Properties of Silicon Nitride in Imidazoline-hexane Solutions", in *Ceramic Powder Science II*, G.L. Messing, E.R. Fuller Jr., H. Hausner, eds., The American Ceram. Soc. Westerville, 1988, p. 469-76.
[19] G. Busca, V. Lorenzelli, M.I. Baraton, P. Quintard, and R. Marchand, "FT-IR Characterization of Silicon Nitride and Silicon Oxynitride Surfaces", J. Molec. Struc. 143, 525-28 (1986).
[20] G. Ramis, G. Busca, V. Lorenzelli, M. I. Baraton, T. Merle-Mejean, and P. Quintard, "FT-IR Characterization of High Surface Area Silicon Nitride and Carbide", in *Surfaces and Interfaces of Ceramic Materials*, eds. L.-C. Dufour et al, Kluwer Academic Publishers, Amsterdam 1989, pp. 173-184.

[21] P.Greil, R. Nietsche, H. Friedrich, and W. Hermel, "Evaluation of Oxygen Content on Silicon Nitride Powder Surface from the Measurements of the Isoelectric Point", J. Europ. Ceram. Soc., 7, 353-359 (1991)
[22] P.K. Whitman, and D.L. Feke, "Comparison of the Surface Charge Behavior of Commercial Silicon Nitride and Silicon Carbide Powders", J. Am. Ceram. Soc., 71 [12], 1086-93 (1988)
[23] L. Bergström, M. Ernstsson, B. Gruvin, R. Brage, B. Nyberg, and E. Carlström, "The Effect of Wet and Dry Milling on the Surface Properties of Silicon Nitride Powders", in *Ceramics today- Tomorrows Ceramics*, ed. P. Vincenzini, Elsevier, Amsterdam 1991, pp 1005-14.
[24] L. Bergström, "Surface Chemistry of Silicon Nitride Powders: Adsorption from Non-Aqueous Solutions", Coll. Surf., 69, 53-64 (1992)
[25] P.S. Wang, S.G. Malghan, S.M. Hsu, and T.N. Wittberg, "Effects of α-silicon nitride powder processing on surface oxidation kinetics", J. Mater. Res., 8, 3168-3175 (1993)

Grain Boundary Control and Design

On Wetting and Segregant Films at Ceramic Grain Boundaries

R. M. Cannon

LBL
University of California
Berkeley, CA 94720

Interfacial impurites in ceramics are widely seen directly or inferred based on large effects of additives on properties or upon processing. Understanding the structure and behavior of such impure interfaces is crucial to controlling microstructure evolution during sintering as well as subsequent properties. The amounts involved are often much more than can be accounted by simple segregation models that apply up to monolayer levels. Often such behavior entails amorphous films of 1 to many nanometers thick. Such films have been described in terms of wetting by a second phase, but such a view is also insufficient. Specifically, nanometer thick boundary films exist which often do not satisfy the condition for complete wetting, but are properly viewed as multilayer adsorbate in a thermodynamic sense. In other systems grain boundary films behave as though fully wetting. Examples will be discussed in terms of grain boundary characteristics observed in structural (Si_3N_4, Al_2O_3, ZrO_2) and electrical ceramics (ZnO, $SrTiO_3$, ruthanates). Parallels noted to films between colloidal particles have motivated further experiments to provide alternative insights. Recent experiments of this type will be described. Aspects of important factors controlling particle interactions which must be incorporated into theories for such behavior will be discussed briefly.

If the auther send the manuscript after the dead line, we will publish it on next edition.

STRATEGIES FOR CONTROLLING THE WETTING BEHAVIOR OF INTERGRANULAR FILMS IN LIQUID PHASE SINTERED CERAMICS

David R. Clarke
Materials Department, College of Engineering, University of California at Santa Barbara, Santa Barbara, CA 93106-5050

ABSTRACT

By considering the thermodynamic and geometric factors that determine grain boundary and surface wetting of a liquid in a polycrystalline solid, generic strategies for eliminating a remnant intergranular film in liquid phase sintered ceramics are described. Two that appear to have the greatest viability are a) changes in the surface wetting angle by changing the sintering atmosphere during the latter stages of sintering, and b) the preferential vaporization of one or more components of the liquid phase.

I. INTRODUCTION

With the exception of certain ceramics, the majority are densified by liquid phase sintering. Although the formation of a liquid phase promotes rapid densification, the liquid phase often remains in the microstructure after sintering and adversely affects the physical properties of the ceramic. Thus, a remnant liquid phase, particularly at the two-grain junctions, can decrease the thermal conductivity of AlN ceramics, cause enhanced high-temperature creep in structural ceramics such as silicon nitride ceramics, and increase the optical scattering in transparent ceramics such as yttria. Although the presence of a remnant liquid phase can be advantageous, in general it is not. The use of liquid phase sintering thus poses a processing dilemma; the liquid phase is necessary for densification but is undesirable in the final material. To overcome this processing dilemma, the liquid phase should wet the grains during sintering but becoming dewetting in some post-densification process. In this short contribution, we consider the thermodynamics of grain boundary wetting with emphasis on the variables that can, at least in principle, be altered to cause dewetting. This provides a basis for possible processing strategies for removing the liquid phase after densification.

For ease of exposition as well as simplicity, we consider in the following the representative microstructure illustrated in figure 1. Although two-dimensional rather than three-dimensional, it is nevertheless capable of representing the essential parameters involved. The repeating microstructural unit is indicated by the dashed square and consists of four identical units shown in figure 1. The liquid is principally located at the multiple, four-grain junctions in square pockets of side length, l, and depending on the wetting conditions may extend as a thin film of thickness, 2h, along the two-grain junctions (figure 2). This representation allows the geometric parameters to be expressed in terms of the volume fraction, V_L, of liquid phase in the microstructure.

Only the macroscopic aspects of wetting are described here. Qualitatively similar arguments can be developed to describe microscopic aspects of wetting but the mathematics is considerably more complex. Nevertheless, the conclusions concerning the strategies for grain boundary dewetting are essentially the same.

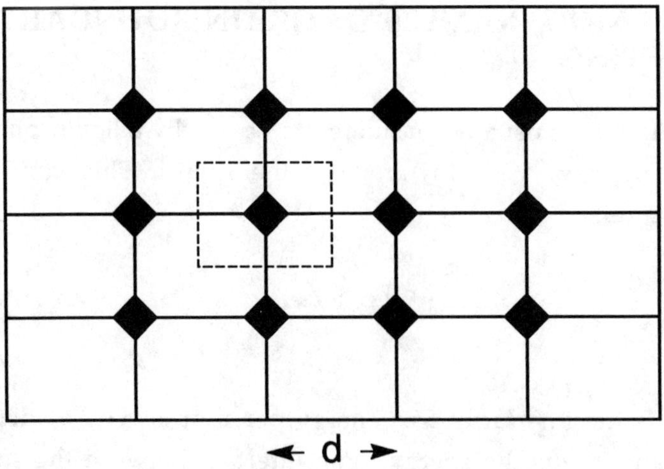

Fig. 1. Representative microstructure of a liquid phase sintered ceramic. The liquid phase is located in pockets at the grain corners and as a thin film along the grain boundaries. The microstructural motif is outlined by the dashed lines.

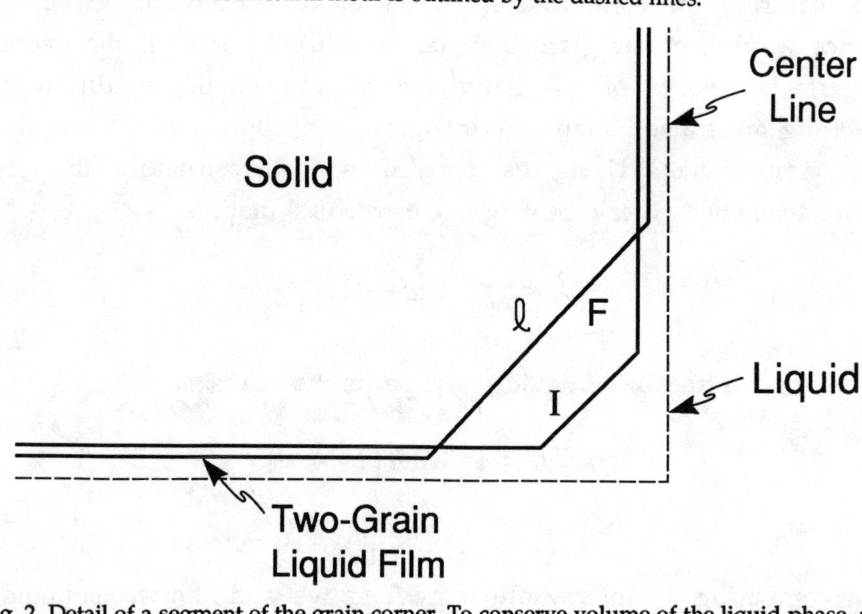

Fig. 2. Detail of a segment of the grain corner. To conserve volume of the liquid phase, a decrease in the thickness of the liquid film is accompanied by a growth of the facet from position I to F.

II. THERMODYNAMICS OF GRAIN BOUNDARY WETTING

A grain boundary is usually considered to be wet by a liquid phase if its interfacial energy, γ, is lowered by the liquid. This condition is normally expressed as:

$$2\,\gamma_{SL} < \gamma_{SS}. \qquad [1]$$

where γ_{SS} is the interfacial free energy per unit area of a "dry" grain boundary and γ_{SL} is the energy of an interface between the solid and liquid phases. Although this describes whether the interfacial energy per unit area of boundary is lowered, it does not take into account the net contribution to the total interfacial energy in the microstructure. Since wetting of the grain boundaries will only occur if the overall interfacial energy of the material is minimized, the condition for wetting must also include the interfacial contribution of the liquid at the grain pockets. Using the representative microstructure, the total interfacial energy per repeating microstructural motif is:

$$U_S = l\,\gamma_{SL} + \left(d - \sqrt{2}\,l\right)\gamma_{gb} \qquad [2]$$

The volume fraction of liquid phase per unit motif is:

$$V_L\,d^2 = l^2 + 2\,h\left(d - \sqrt{2}\,l\right) \qquad [3]$$

By comparing the interfacial energy for a wet and non-wet boundary subject to conserving the volume fraction of liquid, grain boundary wetting will only be favored when:

$$2\,\gamma_{SL}\left[d\left(1-\sqrt{V_L}\right)-l\left(\sqrt{2}-1\right)\right]<d\left(1-\sqrt{2\,V_L}\right)\gamma_{SS} \qquad [4]$$

and is seen to be dependent on both the volume fraction of liquid phase and the grain size. This result is a direct consequence of the constraint that the volume of liquid phase is conserved. When the grain size is much larger than the size of the pockets and the volume fraction of liquid is small, then the wetting condition approaches that of equation 1.

III. GRAIN BOUNDARY DEWETTING

The liquid phase will retract from along the grain boundaries if the interfacial energies are altered from the wetting condition. Two possible situations can occur. One is if the relative grain boundary and interfacial energies are altered by, for instance, a decrease in temperature such that the wetting condition of equation 4 is reversed. The liquid phase will then withdraw to the grain pockets leaving the grain boundaries free of any liquid film. The total interfacial energy per microstructural motif is then:

$$U_s=2\left(d-2\sqrt{2}\,l\right)\gamma_{SS}+4d\,\sqrt{V_L}\,\gamma_{SL} \qquad [5]$$

The other possibility is that the relative grain boundary and surface energies are altered, causing the liquid phase to be drawn to the free-surface. The role of the important parameters, the relative interfacial energies, are illustrated by the schematic diagram in figure 3 representing an intermediary situation in which the liquid phase is partially wetting a portion of the free-surface and a grain boundary. The liquid forms a contact angle θ_S with the surface and a contact angle, θ_g, along the grain boundary At equilibrium, the distribution of the liquid

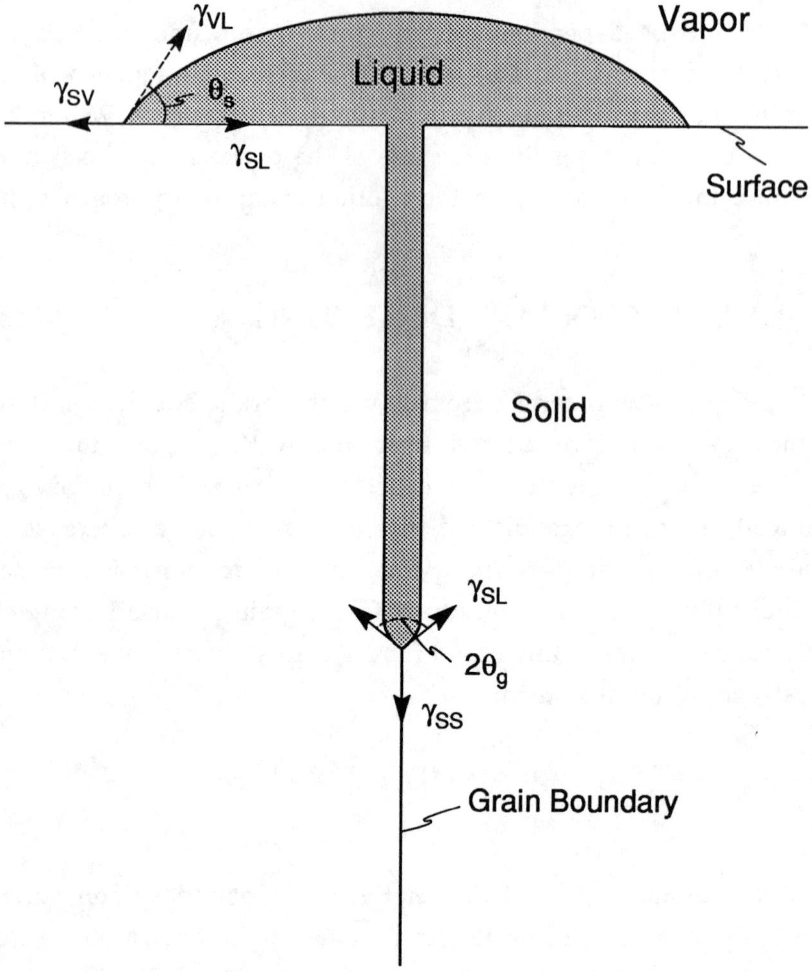

Fig. 3. Schematic diagram illustrating a volume of liquid phase partially wetting both the free-surface and a grain boundary.

phase along the free-surface and the grain boundary is the one that minimizes the total interfacial free energy. However, when equilibrium is perturbed, for instance when the temperature is changed, the liquid phase will redistribute until the interfacial energy is again minimized. Two limiting situations can be considered. One is when the grain boundary contact angle is finite and the surface contact angle goes to zero. Then, the liquid will preferentially wet the free-surface and will be drawn from the grain boundaries until none remains at the grain boundaries. The opposite behavior results in the other limiting situation in which the surface contact angle is finite but the grain boundary contact angle goes to zero. Thus, changing the two contact angles provides a means of re-distributing the liquid phase and can, in principle, lead to grain boundary dewetting. This is discussed in the following section.

IV. PROCESSING STRATEGIES

The simple thermodynamic arguments concerning macroscopic wetting behavior outlined above provide a basis for discussing processing strategies for the removal of the liquid phase from the microstructures of polycrystalline ceramics.

The most straightforward strategy is to use an additive that leads to the formation of a fugitive liquid phase, one that can be vaporized from the material by a heat treatment at high temperatures. Whilst this is believed to be the basis of the use of LiF as a fugitive liquid phase in the densification of, for instance, MgO [1] and $BaTiO_3$ [2], the rate of vaporization must be carefully controlled. If it is too fast, porous channels will remain in the body or liquid pockets will be trapped within the ceramic. If it is too slow, extensive grain growth may occur during the evaporation which may be undesirable.

An alternative processing strategy might be to lower the surface contact angle at the end of the liquid phase densification stage in order to provide a driving force for the liquid phase to migrate to the surface. The obvious means by which this might be achieved is through a change in the sintering atmosphere to increase the surface interfacial energy, γ_{SV} and/or lowering the liquid interfacial energy, γ_{LV} . There are two possible means by which the appropriate atmosphere might alter the liquid interfacial energy. One is directly. The other is indirectly by promoting the preferential vaporization of one or more of the component species in the liquid phase which in turn changes the liquid interfacial energy, γ_{LV} . This would correspond to an alteration of the interfacial energy by the Marangoni effect [3], much as the vaporization of alcohol from water-alcohol mixtures causes its interfacial energy to increase until all the alcohol has evaporated.

Although changing the atmosphere after densification is straightfoward, only two examples of its use to remove a remnant liquid phase have been reported as far the the author is aware. One is, apparently, in the densification of transparent AlN ceramics in which a highly reducing atmosphere is used in conjunction with a Y_2O_3 densification aid to form YN phase on the surface of the ceramic [4]. The surface YN phase is subsequently dissolved away after cooling leaving the material free of the grain boundary phase. The other example is the densification of Mg-PSZ with SrO additions in different partial pressures of oxygen [5]. When fired in pure oxygen, the liquid phase remains at the grain boundaries. However, under strongly reducing conditions the liquid phase is drawn to the free-surfaces where it can again be removed on cooling. In neither case is it clear which interfacial energy is altered by the change in atmosphere, although in the Mg-PSZ case it is apparently associated with the loss of MgO from the liquid phase.

Although dewetting of grain boundaries may be thermodynamically favored it may nevertheless be prevented from actually occurring, or prevented from reaching completion, if the necessary mass transport is kinetically limited. Trying to estimate the length of time required for dewetting is complex because of the variety of processes by which the dewetting might occur. The intergranular film may, for instance, simply thin uniformly or undergo a morphological instability. Detailed microstructural observations will be required to establish the particular processes of dewetting. Nevertheless, a lower bound, albeit crude, estimate of the time for the liquid to withdraw to the grain pockets can be made by considering just the removal of the final few monolayers of liquid and equating the effective diffusivity to the grain boundary diffusivity. The lower bound time, t_{min}, is then:

$$t_{\min} = \frac{d^2}{D_b\,\delta_b} \quad [6]$$

If the liquid phase is glass-forming, as most silica-based liquid phases probably are, then an additional kinetic limitation is the freezing-in of mass transport on cooling through the glass-transition temperature. Taken together, these considerations suggest that processes that promote dewetting at the sintering temperature, for instance, by changing the sintering atmosphere after densification are to be preferred.

CONCLUSIONS

Removal of the liquid phase from the grain boundaries in polycrystalline ceramics is dependent on altering the thermodynamics of wetting through changes in the interfacial energies of the grain boundaries, the grain pockets and the free surfaces. By using a representative microstructure, the conditions for grain boundary wetting are shown to be dependent on the volume fraction of the liquid phase and the grain size as well as on the interfacial energies per-se. The grain boundaries will dewet if the liquid phase is fugitive or the grain boundary wetting angle increases with decreasing temperature. Alternatively, the grain boundaries may dewet by changing the chemical potential during sintering. This may be accomplished by changing the sintering atmosphere provided it decreases the surface contact angle.

REFERENCES

1. P. E. Hart, R. B. Atkin and J. A. Pask, "Densification Mechanisms in Hot-Pressing of Magnesia with a Fugitive Liquid Phase", Journal of the American Ceramic Society, **53** (2) 83-86 (1970).

2. J. M. Haussonne, G. Desgardin, PH. Bajolet and B. Raveau, "Barium Titanate Perovskite Sintered with Lithium Fluoride", Journal of the American Ceramic Society, **66** (11) 801-807 (1983).

3. L. E. Scriven and C. V. Sternling, "The Marangoni Effects", Nature, **187,** 186 (1960).

4. F. Ueno and A. Horiguchi, "Grain Boundary Phase Elimination and Microstructure of Aluminium Nitride", in "Euro-Ceramics", edited G. de With, R.A. Terpstra and R. Metselaar, Elsevier Applied Science, New York, p 1.383-1.387.

5. J. Drennan, D. R. Clarke, R. H. J. Hannink and T. M. Shaw, "Effect of Oxygen Partial Pressure During Sintering of the Microstructure of Mg-PSZ with Strontia and Silica Additions", Journal of the American Ceramic Society, In Press.

GRAIN BOUNDARY STRUCTURES IN COVALENT-BONDED CERAMICS

K. Morita*, S. Tsurekawa, H. Nakashima** and H. Yoshinaga****

*Graduate Student, Kyushu University, Kasuga, Fukuoka, 816, Japan
**Department of Materials Science and Technology, Graduate School of Engineering Sciences, Kyushu University, Kasuga, Fukuoka, 816, Japan

ABSTRACT

An amorphous-like layer of *ca.* 0.5 nm in thickness is observed along the grain boundaries in SiC with B+C, and SiC without any sintering aids. Since these materials do not show a remarkable stress decrease at high temperatures, Ikuhara *et al.* presumed that the amorphous-like layer is not a second phase of sintering aids or impurities, but a relaxed structure formed to reduce its high energy. This structure was called an extended boundary. The concept of the extended boundary, however, has not yet been generally accepted for lack of convincing evidence. In the present work, the chemical composition of the boundary layer was determined in order to inspect the extended boundary concept.

From the viewpoint of the extended boundary assumption, TiC and Si are also expected to have grain boundary structures similar to that observed in SiC because of its high covalency and directionality in atomic bonding. In this paper, the grain boundary structures were revealed by means of high resolution electron microscopy, and then the difference in the structure in different materials will be discussed based on these observations in connection with the difference in covalency and polarity.

INTRODUCTION

Non-oxide ceramics such as silicon carbide (SiC) and silicon nitride (Si_3N_4) are expected to be useful as structural materials at high temperature because of their high thermal resistance, high chemical stability, and excellent high temperature strength. It is well known that mechanical properties are markedly affected by the presence of grain boundaries; *e.g.*, grain boundary sliding at high temperature should induce stress concentration at triple points and ledges of grain boundaries, resulting in a strength decrease [1]. Such behavior depends strongly on the grain boundary character and extrinsic factors such as additives, impurities, sintering procedure and so on [1]. The main points obtained in our previous studies concerning the relationship between high temperature strength and grain boundary structure in SiC are summarized as follows. (1) The fracture strength of SiC with AlN added as a sintering aid decreases remarkably at high temperature, while the strengths of sintered SiCs with simultaneous addition of B and C, and without any additives do not decrease at high temperature [1][2]. (2) Nevertheless, whether the strength decreases or not at high temperature, a similar amorphous-like layer in appearance is observed at random boundaries [1]. (3) The misorientation dependence of the grain boundary strength in SiC [3] is much weaker than that in Mo [4][5].

Based on these results, Ikuhara *et al.* made an extended boundary assumption[1]; that is, the boundary layer observed in SiC with B + C and without any additives is not a second phase mainly composed of sintering aids or impurities, but of a relaxed structure to decrease its high energy. This structure is amorphous-like and has a chemical composition similar to that of the grain interior. Therefore, the strength of these materials do not decrease at high temperature. However, some questions arose against this extended boundary assumption [2]: Is the high energy grain boundary preferentially amorphorized during Ar ion-thinning? Is the grain boundary layer a second phase (such as B_4C), which has a high strength at high temperature? Thus, we carried out some collateral verifications for the extended boundary assumption [2]. However, the concept of the extended boundary has not yet been generally accepted for lack of direct evidence. Hence, in the present work, the chemical composition of the boundary layer was determined using electron energy loss spectroscopy (EELS), and energy dispersive X-ray spectroscopy (EDS) using the micro-beam method. This allowed the analysis of an extremely small area. As a result, it is found that the chemical composition of the boundary layer is not much different from that of the grain interior, although a minimal segregation of oxygen is observed [2].

From the viewpoint of the extended boundary assumption, TiC and Si are also expected to have grain boundary structures similar to that observed in SiC because of their high covalency and directionality in atomic bonding. In the present work, therefore, grain boundary structures in TiC and Si are revealed by means of high-resolution electron microscopy. The difference in grain boundary structures in these different materials will be discussed.

EXPERIMENTAL PROCEDURES

The materials used for the chemical analysis of boundary layer were pressureless sintered α-SiC with additions of 0.6mass% B and 3.0mass% C and hot isostatic pressed β-SiC without any sintering aids. Thin foils suitable for the analytical electron microscopy (AEM) were prepared by the following processes: slices about 1 mm thick were cut out from the sintered materials with a low speed diamond cutter, and disks (3 mm in diameter) were trepanned from the slices with a ultrasonic machine, and then polished mechanically to less than 50 μm. Further, these disks were Ar ion-milled into foils. The chemical composition analyses of the grain boundary layer were conducted with a JEOL JEM-2010F, which allowed the convergence of the electron beam to 0.4 nm in diameter. The measurements were carried out under edge-on conditions at areas of about 10 nm in thickness for which high-resolution TEM observation was possible.

Silicon and titanium carbide bicrystals with <110> symmetric tilt boundaries of misorientation angle ϕ, which is defined as the angle made by the two (001) planes of adjacent crystals, were prepared by a hot-press technique. The details of the preparation method of the bicrystals are as follows. Oriented single crystals were cut out with a low speed diamond cutter, and were mechanically and chemically polished to obtain flat surfaces to be joined and to remove the probable oxide film at the surfaces. Then two crystals were put together and joined in a vacuum of *ca.* 3 mPa at 1700 K under 30~40 MPa for 1.8 ks for Si, and at 2273 K under 25~30 MPa for 1.8 ks for TiC using an electrically controlled hydraulic machine. After joining the bicrystal specimens were annealed *in-situ* at the same temperatures for 3.6 ks without load in order to make the grain boundary structure in equilibrium.

Thin specimens for HRTEM observation were prepared by the ion-milling technique for Si and the twin-jet electropolishing technique for TiC. The details of the preparation method of thin foil were almost the same as those described elsewhere [6][7]. The HRTEM observations of grain boundary structure were performed with a JEOL JEM-2000EX/T.

RESULTS AND DISCUSSION

Analysis of Grain Boundary Layer in SiC

The analytical transmission electron microscope (AEM) used here is JEM-2010F equipped with a field emission gun, which allows the EELS and the EDS analyses with a convergent beam of 0.4 nm in diameter. Since measurements were conducted at thin regions less than 10 nm in foil thickness, even considering the beam scattering through the foil, it is believed that information from a small area less than 1 nm in diameter should have been obtained.

Figure 1 shows an EELS spectrum obtained from a grain boundary layer in α-SiC with B+C [2]. At the peak position of B indicated by an arrow in this figure, no EELS peak of B is observed. Further, the peak height of C at the boundaries was almost the same as that in the grain interior [2]. Hence, the chemical composition of an amorphous-like layer observed in sintered α-SiC with B+C should not be much different from that of the grain interior. The amount of aids or impurities segregation at boundaries is beyond the detectable power of the apparatus.

Figure 2 shows results of the EDS analyses for the hot isostatic pressed β-SiC without any additives [2]. It is found that the EDS peak height of O in the grain boundary layer is about three times as high as that of the grain interior. Although oxygen segregated at the boundaries, the amount of O segregation was small and

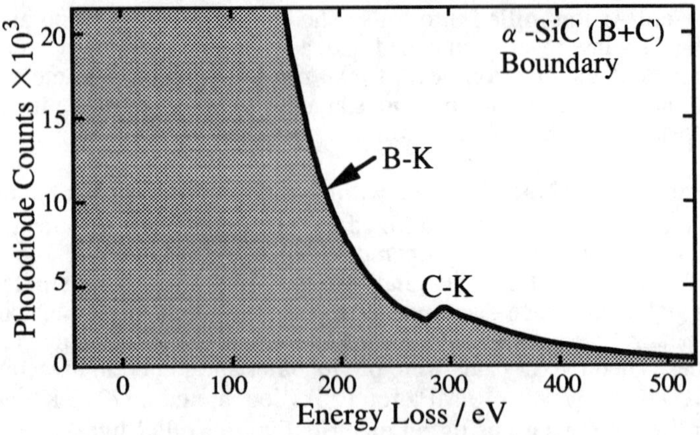

Fig. 1 Electron energy loss spectrum obtaind from a grain interior and a grain boundary in sintered α -SiC with B+C[2].

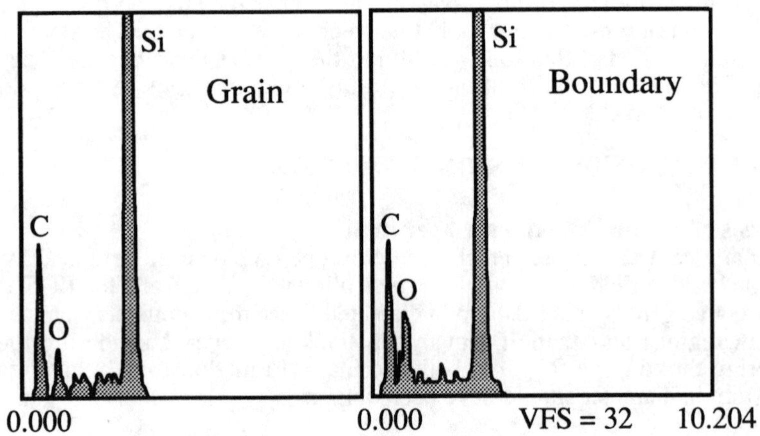

Fig. 2 The EDS spectra obtaind from a grain interior and a grain boundary in the hot isostatic pressed β -SiC without any additives[2].

no oxide phase was recognized. These results strongly support the extended boundary concept.

The results of the EELS and the EDS analyses verify that the grain boundary layer observed in SiC sintered materials with B+C additions and without any sintering aids is not a second phase mainly composed of sintering aids or impurities, but is of a relaxed structure to decrease the high energy, though a little segregation is observed. Consequently, we concluded that the extended boundary concept is valid.

Grain boundary structures in TiC and Si

Figures 3 (a) and (b) show HRTEM images of the (122) $\Sigma 9$ symmetric tilt boundary of TiC [6] and Si [7], respectively. Contrary to the expectation, any sign of amorphous-like layer similar to that observed in SiC is not observed either in Si or in TiC. For TiC, some periodic structures are observed along the boundary; *i.e.*, a longer periodic structure of 9.4 nm and a shorter one of 0.9 nm coexist. Since the misorientation angle of this boundary is $1.0 \pm 0.5^\circ$ off from the exact (122) $\Sigma 9$ coincidence misorientation, $\phi = 141.06^\circ$, the longer periodic structure, which is accompanied with grain boundary steps, may correspond to a DSC dislocation array accommodating the small deviation from the exact coincidence. A schematic explanation of the DSC dislocation model for (122) $\Sigma 9$ coincidence boundary is shown in Fig. 4 [6]. Assuming $\vec{b}_1$ dislocations ($\vec{b}_1 = (a/9)\langle 22\bar{1}\rangle$), which are perpendicular to the grain boundary plane and produce a deviation of tilt angle $\Delta\phi$ from the exact $\Sigma 9$, the predicted dislocation spacing is 8 nm, which is very close to the observed spacing of 9.4 nm. Inversely, the value of $\Delta\phi$ estimated from the observed period of 9.4 nm becomes 0.85°, which agrees with the misorientation within the experimental error of a misorientation angle measurement. Moreover, the predicted step height of 0.56 nm from the DSC dislocation model is in good agreement with the observed value of 0.46 nm. The shorter periodic structure of *ca.* 0.9 nm can be well described as a lattice dislocation ($\vec{b} = a[110]$) array based on the O-lattice theory [6]. The observed periodic grain boundary structures in Si also agreed well with the expectation from the geometric models mentioned above [7].

Figures 5 (a) and (b) show HRTEM images of near (115) $\Sigma 27$ ($\phi = 31.6^\circ$) symmetric tilt boundaries of TiC [6] and Si [7], respectively. The expected boundary layer is not recognized even at the less-coherent boundaries either in Si or in TiC. Two kinds of periodic structures similar to those observed at near $\Sigma 9$ coincidence boundaries are observed along the boundaries in Si. Again, the structures can be explained by the DSC dislocation model and the O-lattice model [7]. Further, as seen in Fig. 5 (b), the atomic positions in Si are slightly disturbed over a range of a few interatomic distances at the boundary and the structure looks to have some boundary extension. According to Papon and Petit [8], the atomic bonding should be reconstructed along $\langle 110 \rangle$ tilt boundaries so as to remove dangling bonds. It is believed, therefore, that the observed slight extension should be caused by the reconstruction.

On the other hand, at the boundary in TiC, the shorter periodic structure of 0.85 nm, which is described by the O-lattice model, is only detected, probably because the boundary has the exact (115) $\Sigma 27$ misorientation. Further, such disturbance as that observed in Si was not recognized in TiC over a wide range of $\phi = 26.53^\circ$ to 172.0° [6].

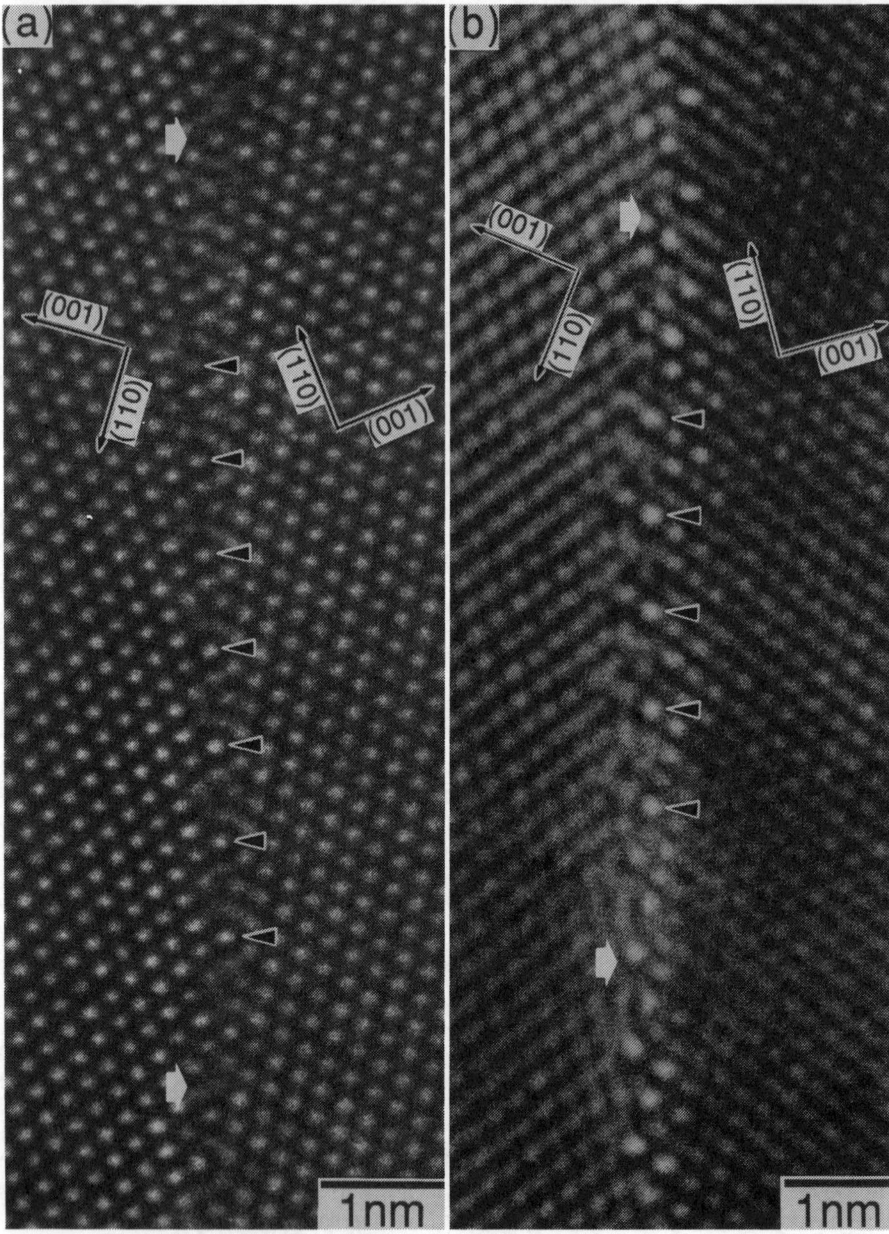

Fig. 3 HRTEM images of (221) $\Sigma 9$ symmetric tilt boundary; (a) TiC [6] and (b) Si [7].

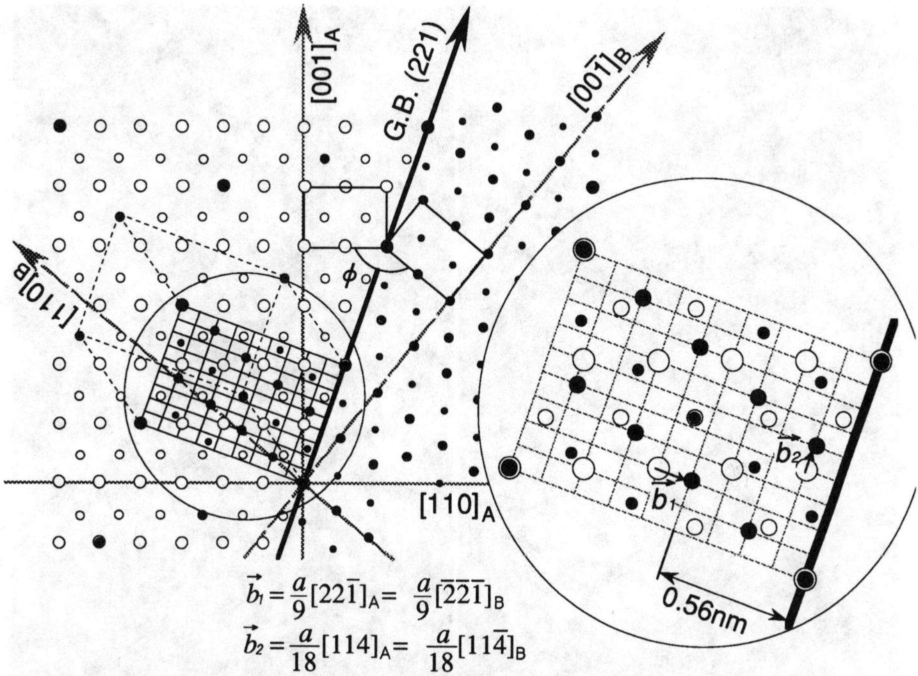

Fig. 4 Coincidence site and DSC lattice for (221) $\Sigma 9$ symmetric tilt boundary showing Burgers vectors of DSC dislocations [6].

Differences in the grain boundary structure of different materials

As mentioned above, the grain boundary structures in TiC and Si can be described by the dislocation models irrespective of the misorientation, and the grain boundary structures are not greatly different from those in metals such as Al [9] and Mo [10][11], though in Si a slight sign of boundary extension is observed. This slight difference probably results from the reconstruction of dangling bonds.

In the case of SiC, Kohyama *et al.* [12] reported that both bond-angle and bond-length distortions are fairly large as compared with those in Si as a result of Si-Si and C-C wrong bonds. The majority of the total boundary energy is due to these distortions. More distortion in SiC should give a wider boundary extension when observed by TEM. Consequently, grain boundary structures in SiC are considered to be closely related to the wrong bonds [12][13].

Contrary to the expectation, there is no definite sign of boundary extension in TiC. Moreover, the atomic displacement in the vicinity of observed boundaries is much smaller than that in Si. In this respect, the grain boundary structure in TiC is more metal-like than that in Si. This is probably attributed to the atomic bonding

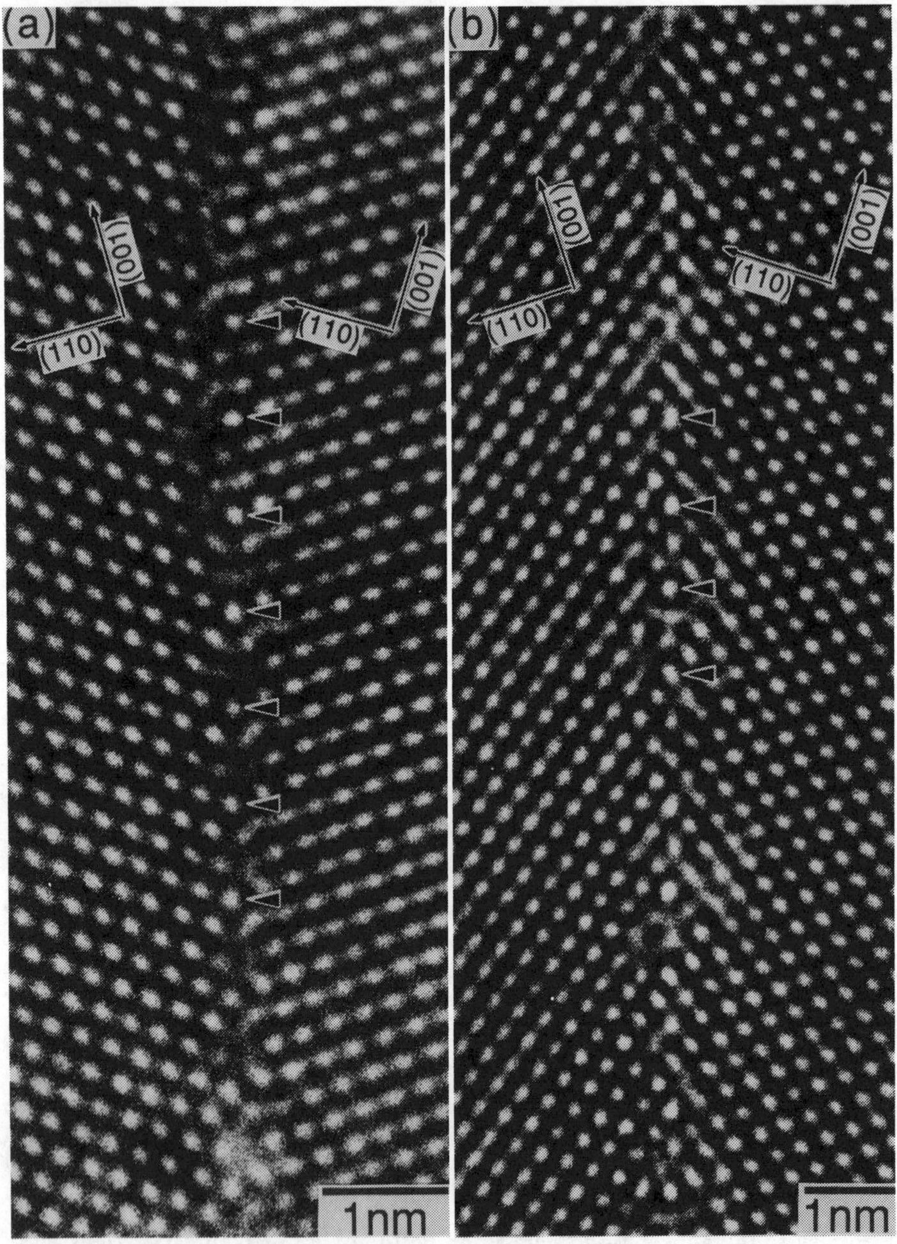

Fig. 5 HRTEM images of near (115) $\Sigma 27$ symmetric tilt boundary; (a) TiC [6] and (b) Si [7].

in TiC, which has a crystal structure like that of rock salt. Different from ionic crystals, the structure is described as the carbon atoms occupy the octahedral interstitial sites in the fcc titanium sublattice. The metal-like boundary structure suggests that the directionality in atomic bonding and the bond strength in TiC having that structure should be relatively weak as compared with those in Si and SiC, and the effect of the wrong bonds on the boundary structure should be less than that in SiC. Further, the effect of nonstoichiometry at grain boundaries must be insignificant because TiC has a wide range of nonstoichiometry without changing the crystal structure [14].

SUMMARY

In order to verify the extended boundary assumption proposed by Ikuhara *et al.*, the chemical composition analyses of the boundary layer in SiC were conducted. Further, HRTEM observations of grain boundaries in Si and TiC were conducted in order to understand the characteristics of grain boundary structure in covalent bonded materials. The main results obtained are as follows:

1) No EELS peak of B was observed in the grain boundary layer of α -SiC with B+C. The peak height of C in boundaries was almost the same as that in the grain interior. The EDS peak height of O in the grain boundary layer was about three times as high as that in grain interior for β -SiC without any additives. The amount of O segregation was not enough to form an oxide phase. These results show that the grain boundary layer observed in SiC sintered materials with B+C and without any additives is not a second phase of sintering aids or impurities, but is a relaxed structure to decrease its high energy. This means that the extended boundary concept is valid.
2) High-resolution observation revealed that no amorphous-like layer existed either in Si or in TiC, irrespective of the misorientation. Instead of the layer, two kinds of periodic structures were observed both in Si and in TiC. The periodic structures were well described by the dislocation models based on the CSL (Coincidence site lattice) and O-lattice theories.
3) In Si, although the amorphous-like layer was not found, a slight disturbance of atom positions was observed within a range of a few interatomic distances along a less-coherent boundary, whereas such a disturbance was not observed in TiC.

Acknowledgment

The authors would like to express their thanks to Dr. Y. Ikuhara, the Japan Fine Ceramics Center, for the supply of pressureless sintered SiC and the chemical analysis of impurities in SiC, to Dr. I. Tanaka, the Institute of Scientific and Industrial Research, Osaka University (present address; Factuly of Engineering, Kyoto University), for the supply of hot isostatic pressed SiC, to Drs. T. Oikawa and M. Kawasaki, JEOL Ltd., for their help for the EELS and the EDS measurements and to Mr. T. Seguchi, Graduate Student, Kyushu University, for his help for the experimental work. The present work was financially supported by the Grant-in-Aid for Scientific Research, the Ministry of Education, Science and Culture, Japan. The support is also very much appreciated.

References

[1] Y. Ikuhara, H, Kurishita and H. Yoshinaga, "Grain Boundary and High Temperature Strength of Sintered SiC," *Yogyo-Kyokai-Shi*, **95,** 638-645 (1987).

[2] S. Tsurekawa, S. Nitta, H. Nakashima and H. Yoshinaga, "Grain Boundary Structures in Silicon Carbides," submitted to *Interface Science*.

[3] Y. Ikuhara, H. Miyazaki, H. Kurishita and H. Yoshinaga, "Grain Boundary Structure and Strength of SiC Bicrystals," *Nippon Seramikkus Kyokai Gakujyutsu Ronbunshi*, **97**, 1511-1516 (1989).

[4] H, Kurishita, A. Oishi, H. Kubo and H. Yoshinaga, "Grain Boundary Fracture in Molybdenum Bicrystals with $\langle 110 \rangle$ Symmetric Tilt Boundaries," *J. Jpn. Inst. Met.* **47**, 546-554 (1983): *Trans. Jpn. Inst. Met.* **26,** 341-352 (1985).

[5] S. Tsurekawa, T. Tanaka and H. Yoshinaga, "Grain Boundary Structure, Energy and Strrength in Molybdenum," *Mater. Sci. Eng*, in press.

[6] K. Morita, S. Tsurekawa, H. Nakashima and H. Yoshinaga, "Grain Boundary Structures in Titanim Carbide," will be submitted to *J. Jpn. Inst. Met.*

[7] S. Tsurekawa, T. Seguchi, S. Nitta and H. Yoshinaga, "Microstructure of <110> Symmetric Tilt Boundaries in Silicon," will be submitted to *J. Jpn. Inst. Met.*

[8] A. M. Papon and M. Petit, "A Survey of Geometrical Reconstruction of [011] Defects in Semiconductors: Grain Boundary and Dislocations," *Scripta Met.*, **19**, 391-396 (1985).

[9] M.J. Mills, M. S. Daw, G. J. Thomas and F. Cosandey, "High-resolution Transmission Electron Microscopy of Grain Boundies in Aluminum and Correlation with Atomistic Calculations," *Ultramicroscopy* , **40**, 247-257 (1992).

[10] S. Tsurekawa, T. Tanaka and H. Yoshinaga, "Microstructure of $\langle 110 \rangle$ Symmetric Tilt Boundaries in Molybdenum," *J. Jpn. Inst. Met.* **56**, 609-618 (1992).

[11] S. Tsurekawa, T. Tanaka, H. Nakashima and H. Yoshinaga, "Microstructure of a Σ17b Grain Boundary in Molybdenum," *J. Jpn. Inst. Met.* **58**, 377-381 (1994).

[12] M. Kohyama, S. Kose and R. Yamamoto, "The Atomic and Electronic Structures of Grain Boundaries in Semiconductors and Ceramics," *Proc. of Inter. Conf. on Computer- assisted Materials Design and Process Simulation*, 1993, Tokyo, ISIJ.

[13] K. Sato, H. Miyazaki, Y. Ikuhara, H. Kurishita and H. Yoshinaga, "Structure and Strength of Grain Boundaries in Si Bicrystals with $\langle 111 \rangle$ Twist Misorientations," *Mater. Trans., JIM*, **31**, 865-871 (1990).

[14] E. K. Storms; *The Refractory Carbides*, Academic Press , New. York. (1967), p. 1.

STRESS REDISTRIBUTION AROUND A CERAMIC/ METAL INTERFACE

Shun–ichiro Tanaka
TANAKA Project, ERATO, JRDC & Toshiba R&D Center
Yokohama High–Tec Center, 1–1–1, Fukuura,
Kanazawa–ku, Yokohama 236, Japan

ABSTRACT

Stress redistribution after debonding of a ceramic/metal joined interface was studied as an actual model of a heterogeneous interface. Localized interfacial debonding was successfully accomplished in Si_3N_4/Cu/steel by propagated cracks from an indentor, which could only be detected sensitively by a scanning acoustic microscope. Stress redistribution by debonding was measured quantitatively using the collimated X–ray method. Debonding induced a change of the tensile residual stress into a large compressive stress. Stress along the interface was released to zero at the center of the debonded portion, whereas it reached maximum compression at a singular point of the debonded edge.

INTRODUCTION

Among the basic knowledges on various interfaces in materials, there seems to be a lack of knowledge on stress related phenomena such as the relation between the strengthening mechanism and the stress/strain distribution around fillers in composites. Microscopic interfacial stress information should be studied with a simple plane interface. As a model of heterogeneous interfaces, the residual stress distribution around a ceramic/metal joined interface has been studied using a ϕ0.1 mm collimated X–ray beam by the author [1][2] for the past 6 years to clarify its relation with the bonding strength. Evaluation of interfaces has also used scanning acoustic microscopy [3].

The features of the stress distribution were a large normal stress around both sides of the anomaly point, a steep stress drop just inside there, and a stress maximum at the center region. These profiles were first obtained experimentally and were compared with FEM analysis [2]. It has already been reported that the stress was redistributed to maintain such symmetrical stress profiles even after cutting the specimen across the interface [4].

The actually joined parts are usually exposed to various dynamic conditions such as applied stress or heat loading. Under static or dynamic stress/heat conditions, the properties change at and around the interface and the bonding strength seems to be affected by the above–mentioned residual stress redistribution. The detailed behavior of stress redistribution by an external factor has not yet been understood either experimentally or theoretically.

The purposes of this paper are as follows: first to create a localized artificial debonded interface as a result of external loading; and secondly to know the characteristics of the redistributed residual stress around such an interface.

EXPERIMENTAL

(1) Specimens

The ceramic/metal joined specimen used in this experiment was a Si_3N_4/Cu/steel plate having a dimension specified by JIS (4 X 3 X 40 mm), which was prepared through Ag–Cu–Ti brazing and a cut/ground process [1]. The cross–sectional ultrasonic SAM (scanning acoustic microscope) views, **Fig. 1**, clearly showed its characteristics of a brazed rugged interface, and showed that no debonding existed before external loading.

(2) Detection of interface debonding

A Vickers indentor was used to introduce localized debonding at the joined interface. The load level was 20 kg and the center of the pyramid was located 0.3 mm from the interface. One crack propagated from one corner of the pyramid and reached the interface. An attempt was made to detect interface debonding by optical microscopy (OM), scanning electron microscopy (SEM), and scanning acoustic microscopy (SAM). For SAM, the frequency of the inlet ultrasonic wave was 400 MHz and a point focus lens was used. The location of debonding was confirmed by a 30 MHz ultrasonic C–scan mode after cutting.

(3) Measurement of stress redistribution

Stress redistribution was measured by the $\sin^2\psi$ method using a 10 kW collimated X–ray beam [2], which had already been applied to many interfaces [4]. Cu–Kα was collimated to 0.1 mm and the irradiated area was confined to less than 0.03 mm^2. The measured stress was two normal stresses, σy and σx, and the averaged stress, σ_R, obtained by rotation.

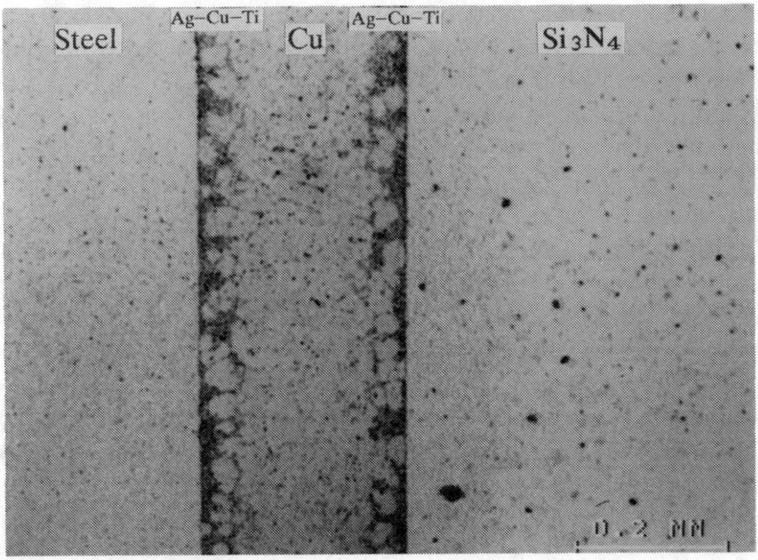

Fig. 1 Scanning acoustic micrograph of Si_3N_4/Cu/steel brazed interface using Ag–Cu–Ti filler

LOCALIZED DEBONDING BY THE INDENTOR

An attempt was made to detect the location and the area of the interface debonding by the two conventional methods of OM and SEM. SEM micrographs around the interface are shown in **Fig.** 2. Four cracks were initiated from the corner of the pyramid and it showed the following typical features in the case of the existence of residual stress ; (1) the lengths of the cracks parallel to the interface were longer than these perpendicular to it, which is direct evidence of a large normal stress perpendicular to the interface, (2) the bowing of the longer cracks was induced by the mutual interaction between both stress fields of the interface and the crack. These facts lead to the demerits of the stress evaluation method.

One of the cracks propagated and reached the interface, and so localized debonding should have been induced, but neither OM nor SEM could reveal its location even when a magnified photograph was used, as shown in **Fig.** 2. On the contrary, SAM revealed well the debonding at the joined interface of Si_3N_4/Cu. As shown in **Fig.** 3, the debonded portion was clearly detected as a black stripe, and the open length was 0.4 mm. The principle of SAM imaging is scattering or absorption of the surface acoustic wave, and so discontinuities in the structures and materials can be sensitively detected. The shape of the debonding was detected as a white region in **Fig.** 4 and was confirmed as one half of a disk, which was checked by the 30 MHz ultrasonic C–scan technique after cutting.

The whole view around the indented joined interface was very colorful as shown in **Fig.** 5, which was composed artificially by three sliced images. **Figure** 5 includes important information on related phenomena caused by the indentation, which contained a protrusion around the pyramid and a more important plastic deformation in the Cu layer induced by interface debonding. These indirect phenomena affected the stress redistribution around the debonded interface.

Fig. 2 Scanning electron micrographs around Vickers indented joined interface; interface debonding could not be detected even in the magnified image

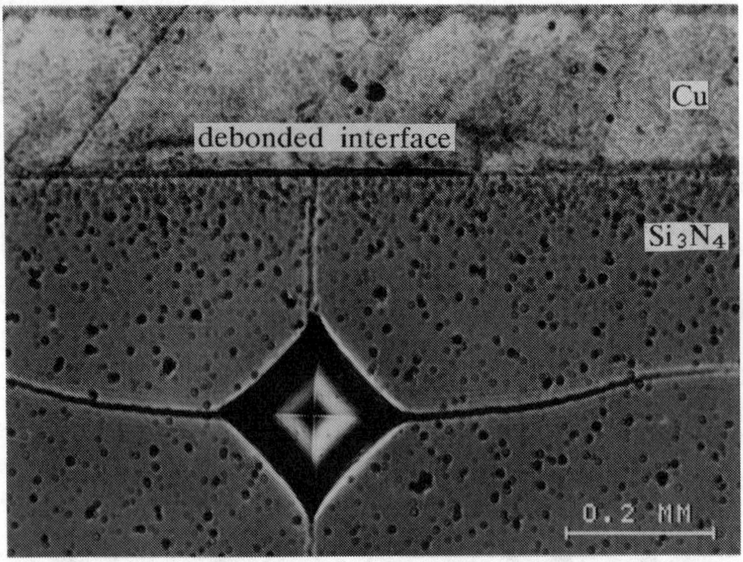

Fig. 3 Scanning acoustic micrograph around Vickers indented joined interface using a 400 MHz point focus lens (depth Z=7 μm)

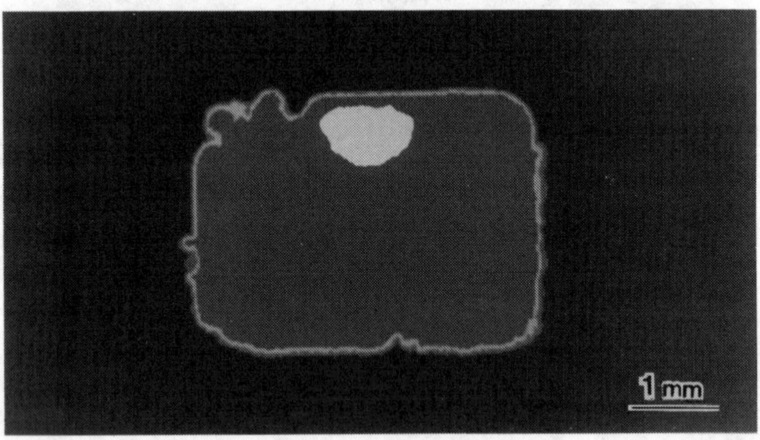

Fig. 4 Ultrasonic C–scan image of debonded interface obtained after steel cutting taken at 30 MHz

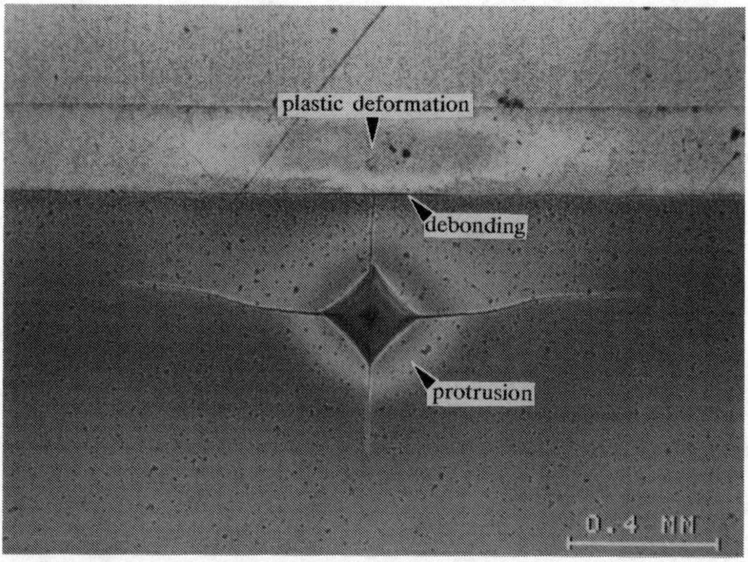

Fig. 5 3Z mode of scanning acoustic micrograph around Vickers indented joined interface in which the protrusion in Si_3N_4 and plastic deformation in Cu were observed in addition to interface debonding (original:color)

STRESS REDISTRIBUTION BY LOCALIZED DEBONDING

The direct effect of the Vickers indentor is its application of a compressive stress onto the ceramics. During indentation of a Si_3N_4 plate, the compressive stress value reached 850–900 MPa at the side of the pyramid and 600 MPa at the crack tip, which was measured by the same collimated X–ray method. Since the indentor added a compressive stress to the joined interface, the stress was redistributed from tensile to compressive.

The stress redistributions were measured for the Si_3N_4 side of the Si_3N_4/Cu debonded interface and the induced cracks, which positions are shown in Fig. 6. The row of the measured position was laid on the lines of y = 0.1 and 0.45 mm from the interface. The normal and the averaged stress distribution around the

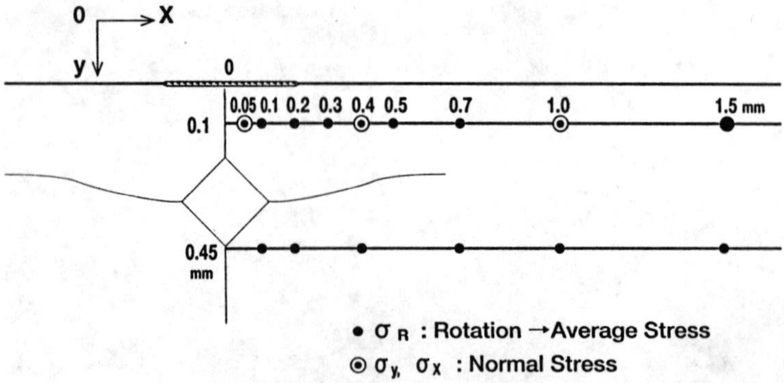

Fig. 6 Measured positions of residual stress redistribution around debonded interface, on the lines y = 0.1 and 0.45 mm

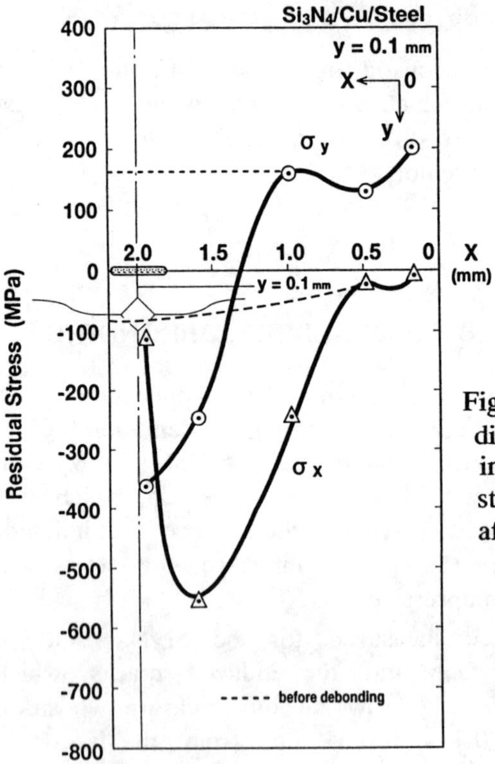

Fig. 7 Residual stress re-distributions around debonded interface; change in normal stress distribution before and after indentation

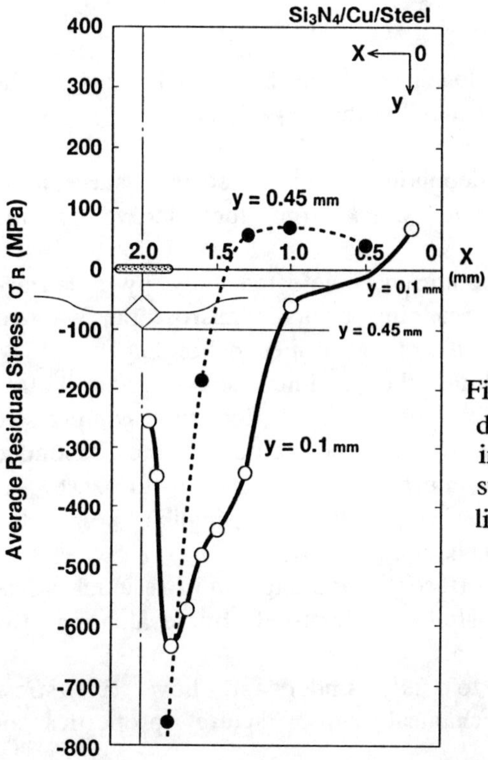

Fig. 8 Residual stress re-distributions around debonded interface; change in average stress distribution on the lines of y=0.1 and 0.45 mm

debonded interface are shown in **Fig. 7** and **Fig. 8**, respectively. In **Fig. 7**, indentation caused the stress distribution change from the dotted line to the solid line. By the Vickers indentation, the normal stress value, σy, changed from a tensile strength of 200 MPa to a compressive strength of 300 MPa, whereas σx, along the interface, changed from a compressive strength of 100 MPa to a maximum compressive strength of 550 MPa. The stress value, σx, reached a maximum at a singular point of the interface and the stress was relaxed to zero at the center of debonding. The average stress distribution fell with the same tendency on the lines of y=0.1 and 0.45 mm. The effects of interfacial debonding by indentation was mainly stress redistribution to compressive largely concentrating at the debonded edge, the singular point.

CONCLUSIONS

Regarding the effects of localized interface debonding, the following features have been revealed in this experiment ;

(1) Localized interface debonding has been successfully accomplished by using a propagated crack from the concave pyramid corner of Vickers indentation.
(2) A localized interface debonding was detected only by a surface acoustic wave generated by a scanning acoustic misroscope, whereas an optical microscope or SEM did not reveal such debonding.
(3) Around localized debonding, the residual stress redistribution was largely compressive which was induced by the compressive stress field around the indentation. At the center of the debonded portion, the stress along the interface was released to zero. At the debonded edge of Si_3N_4/Cu interface, a singular point, the residual stress reached maximum compression.
(4) The relation between interface debonding and residual stress redistribution has been successfully understood by evaluating the stress values of the confined region.
(5) Further work is needed to fully understand how this stress redistribution affects the mechanical and fracture properties of the joint.

ACKNOWLEDGMENT

The author is indebted to Dr. Chiaki Miyasaka of the Olympus Optical Co. Ltd for his helpful assistance in scanning acoustic microscope operation.

REFERENCES

[1] Shun–ichiro Tanaka," Measurement of Residual Stress Distribution around a Si_3N_4/steel Joined Interface Using a Collimated X–ray Beam," Residual stresses–III, Science and Technology, (1992) p.887, ed., H.Fujiwara et al, Elsevier Applied Science, London and New York

[2] Shun–ichiro Tanaka and Yumiko Takahashi," Effects of X–ray Beam Collimation on the Measurement of Residual Stress Distribution in a Si_3N_4/steel Joint," ISIJ International **30** (1990) 1086

[3] Shun–ichiro Tanaka and Chiaki Miyasaka," Residual Stress Distribution in a Ceramic/Metal Joint Evaluated from V(Z) Curves of a Scanning Acoustic Microscope," Residual stresses–III, Science and Technology, (1992) p.278, ed., H.Fujiwara et al, Elsevier Applied Science, London and New York

[4] Shun–ichiro Tanaka," Measurement of Residual Stress Distribution around Si_3N_4/steel Joined Interface Using ϕ 0.1 mm Collimated X–ray Beam," Bulletin of Japan Institute of Metals **29** (1990) 924 (in Japanese)

THE IMPORTANCE OF GRAIN BOUNDARY CHARACTER DISTRIBUTION (GBCD) TO GRAIN BOUNDARY DESIGN AND CONTROL IN CERAMICS

Tadao Watanabe

Laboratory of Materials Design and Interface Engineering
Department of Machine Intelligence and Systems Engineering
Faculty of Engineering, Tohoku University, Sendai, 980, Japan.

ABSTRACT

In order to understand, design and control boundary-related bulk properties in polycrystalline materials, the grain boundary character distribution (GBCD) has been introduced which can corelate properties of individual boundaries to the bulk properties. Possible applications of GBCD to future development of advanced ceramics are discussed.

KEYWORDS: Grain boundary design and control, grain boundary character distribution (GBCD), grain growth, fracture, superconductors.

INTRODUCTION

The important roles of grain boundaries have been widely recognized in designing and processing advanced materials with high performance [1]. Grain boundaries and interphase boundaries are preferential sites for metallurgical phenomena occurring in polycystalline materials. It has only recently been established that the type and structure of the boundaries can strongly affect the properties of the boundaries [2]. The nature and the magnitude of their influence depend on the boundary type and structure. The bulk boundary-related properties and performance of polycrystalline materials have also been found to be controlled by the type and the frequency of grain boundaries, termed the grain boundary character distribution (GBCD) [3]. Therefore, it is important to design and control grain boundaries and interphase boundaries in order to obtain high performance or desired properties in a polycrystalline material. In particular, there seems to be the high potential for grain boundary design and control of both structural and functional ceramics, as has been proved with electroceramics [4] and superconducting ceramics [5]. The present paper will first discuss structure-dependent boundary properties which affect the boundary-related bulk properties. The importance of a new microstructural parameter, i.e. the grain boundary character distribution (GBCD), to design and control the boundary-related bulk properties and performance in polycrystalline advanced materials will then be described. Future prospects of interface engineering through the design and control of GBCD and other boundary parameters are also presented.

STRUCTURE-DEPENDENT BOUNDARY PROPERTIES IN CERAMICS

Most ceramic materials are produced by sintering compacted particles because of their high melting temperatures. Grain boundary diffusion plays an important role in sintering. It has been shown that grain boundary diffusion strongly depends on the type and structure of the grain boundary [6]; as seen from the observations on the effect of boundary misorientation, low-angle boundaries and some special boundaries are the less efficient sites and paths of atomic diffusion than high-energy general boundary, so called random boundary. It is also well known that grain boundary migration velocity strongly depends on the boundary type and misorientation [7]. Grain boundary diffusion and migration are involved in the densification of compacted particles which is a key processing of ceramic materials. Moreover, grain boundary deformation which must involve the interaction between the grain boundary and lattice dislocations from the grain interior, also depends on grain boundary type and structure [8]. Therefore, it is important to understand how and how much the grain boundary type and structure can affect grain boundary phenomena occurring in a polycrystal. This is an important basis for understanding structure-dependent boundary related properties in polycrystalline ceramic materials.

GRAIN BOUNDARY CHARACTER DISTRIBUTION (GBCD) IN POLYCRYSTALS

Recently it has become possible to characterize individual grain boundaries (more than four or five hundred) existing in a polycrystal, by using the scanning electron microscopy-electron channelling pattern (SEM-ECP) technique or more recently the SEM-electron back scattering pattern (SEM-EBSP) technique [9]. Grain boundaries are classified into three categories; low-angle boundaries with the boundary misorientation angle smaller than 15 ° , coincidence boundaries (specified by Σ plus numeral) and high-angle random boundaries. In fact, the type and the density of grain boundaries, termed the grain boundary character distribution (GBCD), is an important microstructural parameter which was recently introduced by the present author [10], and is increasingly being used in the understanding of boundary-related properties and phenomena in polycrystalline materials, as shown by Aust and Palumbo [11]. GBCD is a new microstructural parameter which can bridge a gap between the structure-dependent property of individual boundaries and the bulk property of a polycrystal, in a quantitative manner. The grain size, which is widely used for understanding and controlling boundary-related properties, is a parameter which describes the density of grain boundaries. However it does not include any information on the boundary type and structure. Until recently it was not certain whether the type and structure of grain boundary can be correlated to the grain size. The present author has revealed that GBCD can change with the grain size, depending on processing of polycrystalline materials [9]. It was found that polycrystalline materials produced by sintering (sintered iron polycrystal or yttria-stabilized zirconia polycrystals) contain a high frequency of random boundaries [12]. However, the effect of processing on

GBCD has only recently been investigated and there still exists many open questions.

RELATIONSHIP BETWEEN GBCD, TEXTURE AND PROCESSING

It may be useful to look at recent works on the GBCD in various materials. Thermomechanical processing is the most popular processing method for metallic materials and it has been widely utilized. Recent investigation into the GBCD in polycrystalline metals and alloys produced by this method has revealed that the frequency of occurrence of low-Σ coincidence boundaries (including Σ1, called low-angle boundary) increases with decreasing the grain size [9]. However, the observed grain size dependence of GBCD can not be general for every polycrystalline materials. Quite opposite grain size dependence of GBCD has been observed on iron-silicon polycrystalline ribbons which were produced by rapid solidification and subsequent annealing [13]. Again, the fabrication process is important. It has been found that the frequency of low- Σ coincidence boundaries increases with the sharpness of texture. Moreover, only specific Σ coincidence boundaries can occur in relation to the type of texture; for example, a polycrystalline iron-6.5mass%silicon ribbon sample with a sharp {110} -texture contains high frequencies of Σ1, 3, 9, 11, 17, 19 in descending order [14]. The inverse cubic root Σ dependence of coincidence boundaries has been found and seems to be general at least for BCC metals and alloys, but not for FCC metals and alloys. Thus, GBCD in polycrystalline materials can be affected not only by the processing method but also by the crystal structure, as expected from the knowledge of atomistic and electronic structure of the grain boundary. Quite recently it has been confirmed that microalloying can drastically change GBCD in polycrystalline materials produced by thermomechanical processing [11]. It is reasonable to expect that grain boundary migration can be affected by the grain boundary segregation of elements introduced by microalloying, resulting in the production of grain boundary microstructure different from that in pure material.

POSSIBLE EFFECTS OF GBCD ON BOUNDARY-RELATED PHENOMENA

(1) Sintering

Liquid phase sintering is often used for metallic or ceramic materials with high melting temperature. The penetration of liquid phase between particles, i.e. along grain boundaries is essential to this type of sintering. In fact it is known that the penetration of liquid phase along grain boundaries strongly depends on the type of the grain boundary. High-angle random boundaries are preferential paths for liquid phase penetration while low-angle boundary and low-Σ coincidence boundaries are very difficult paths, as observed on liquid metal-induced intergranular fracture [3]. It is expected that more efficient densification can be achieved by controlling GBCD at different stages of sintering. This has never been taken into account and attempted, to the author's knowledge.

(2) Grain Boundary Segregation

It is common to use some additive to enhance densification during sintering in ceramic materials. For example, MgO additions are required to obtain successful densification of Si_3N_4. Such additive exist as either by forming a grain boundary liquid phase or by being segregated to grain boundaries. The presence of impurities at the grain boundaries can strongly affect mechanical or electromagnetic properties of ceramics. Therefore it is important to reveal the segregating species and the amount of segregants in order to control the effect of segregants on bulk properties of polycrystalline ceramics. It is known that the amount of segregant depends on the solid solubility and the grain boundary misorientation [15]. It is known that high-energy random boundaries can accomodate segregants more effectively than low-energy boundaries. The effect of additives may be more effective when sintered compact contains more high energy random boundaries. However it is necessary to know what is the final GBCD for sintered material and what is the best GBCD to confer a desired bulk property. We should remember that it is not always true that the best GBCD for processing is the best GBCD for material performance. Quite opposite case may be possible depending on fabrication method of material.

(3) Grain Growth

Grain growth is a key factor in controlling microstructure during sintering in ceramics. Often abnormal grain growth occurs during sintering, generating the heterogeneity of microstructure. Such heterogeneity of microstructure is known to cause serious problems in development of advanced ceramics. In the case of yttria-stabilized zirconia, abnormal grain growth degradates mechanical properties (e.g. fracture toughness). The performance of electroceramics can also be affected by abnormal grain growth [16]. Superplasticity disappears when abnormal grain growth occurs. Quite recently it has been found that abnormally grown grains are often surrounded by more random boundaries [17]. The importance of GBCD to grain growth has been discussed by the present author in detail [18]. However the origin of abnormal grain growth has not been clarified yet.

(4) Superplasticity

In recent years, it is known that fine-grained ceramic materials can be superplastically deformed at high temperatures [19]. Quite recently more than 1000% elongation was obtained in SiO_2 containing ZrO_2 polycrystal [20]. Superplasticity is an important deformation mode for hot working, particularly for normally so called "brittle materials" like ceramics or intermetallics. As well known, possible mechanism of superplasticity is grain boundary sliding [21]. It is well known that grain boundary sliding strongly depends on the type and structure of grain boundary, taking place easily at random boundary, but hardly at low-angle boundary or low- Σ coincidence boundaries [22]. It has been reported that the formation of high angle boundaries is related to superplastic deformation in Al-Li alloy [23]. Therefore it is reasonable to expect that the introduction of a high frequency of high angle random boundaries may be one of requirements for superplasticity.

(5) Intergranular Fracture and Brittleness

It is general recognition that the brittleness of ceramics is associated with intergranular fracture. Recent experimental work on the effect of grain boundary type and structure has revealed that high energy random boundaries are preferential sites for crack nucleation and propagation while low-angle boundaries and low-Σ coincidence boundaries are very resistant to fracture in polycrystalline metals and alloys [3], and also intermetallics [24]. Similar effect has been reported on alumina polycrystals [25]. It has been proposed and partly confirmed by experiments that the introduction of low-energy boundaries which are resistant to fracture can control the brittleness associated with intergranular fracture [3, 10]. From modelling fracture behaviour in polycrystals with different GBCD, it was found that the fracture toughness can increase with increasing the fraction of low-energy and fracture-resistant boundaries [26].

GRAIN BOUNDARY DESIGN AND CONTROL FOR STRUCTURAL CERAMICS

The role of grain boundaries is particularly important in high temperature plasticity and fracture. More intergranular phenomena are involved and the activity of intergranular phenomena which are associated with processing and also during service increases with increasing temperature. Grain boundary diffusion, segregation, migration and sliding are involved together with crystal deformation and concomitant interaction between crystal slip and the grain boundary in a polycrystal. Quite recently, the present author has discussed the grain boundary design and control for high temperature materials in detail [22]. Recent achievement of the control of intergranular brittleness in Ni_3Al (the most popular intermetallic material) has been reported. The introduction of a high frequency of low-Σ coincidence boundaries (particularly $\Sigma 1$ and $\Sigma 3$) can dramatically control the intergranular brittleness and improve the ductility of "intrinsically brittle" polycrystalline Ni_3Al. Unfortunately, up to now, no attempt has been made for the control of intergranular brittleness in ceramics by designing and controlling GBCD according to the concept of "grain boundary design and control for strong and ductile polycrystalline materials" [10].

High thermal stability of microstructure must be one of requirements for high temperature materials. Microstructural modification can brought about during service through grain growth. In particular, abnormal grain growth is the major source of the heterogeneity of the microstructure. It is expected that GBCD may be related to the thermal stability of the microstructure in a polycrystal. There has been little work on the relationship between the thermal stability of the microstructure and GBCD. Future work is strongly required.

GRAIN BOUNDARY DESIGN AND CONTROL FOR SUPERCONDUCTING CERAMICS

Recently development of high T_c superconducting ceramics has drawn much interest of an increasing number of material scientists and engineers. In particular, $YBa_2Cu_3O_{7-x}$ has been extensively investigated in relation to the effect of grain boundary. It was found that the grain boundary critical density, J_c is strongly correlated to the tilt angle in <001> tilt bicrystals and it varies by almost 2 orders of magnitude over the range of the tilt angle $\theta = 0° - 35°$, having a saturation value of about 1/50 at a tilt angle of about 20° [27]. The presence of low-angle boundaries and low-Σ coincidence boundaries appears to be favorable to produce high Jc polycrystalline $YBa_2Cu_3O_{7-x}$. Since the evolution of a sharp texture can introduce a high frequency of low-angle boundaries and low-Σ coincidence boundaries, several investigators have already attempted to control grain boundaries through texture control in this material and obtained some success [28]. Unfortunately, up to now, the new microstructural parameter, GBCD, which can bridge between individual boundaries and the bulk properties of a polycrystal, has not been fully discussed in connection with the grain boundary connectivity, the importance of which was pointed out by the present author for fracture control [10] and recently by Nichols and Clarke for superconductivity [29]. This kind of approach may be fruitful for development of high T_c and J_c superconducting ceramics in the future.

CONCLUSION

There is a high potential for grain boundary design and control of structural and functional ceramics, by using a new microstructural parameter, termed the grain boundary character distribution (GBCD) which can bridge the properties of individual grain boundaries and the bulk properties of a polycrystal. Until recently some successes have been obtained for advanced metallic or intermetallic materials. The application of grain boundary design and control to advanced ceramics is widely open.

ACKNOWLEDGEMENTS

The author wishes to record his gratitude to coworkers who have collaborated with him inside and outside of Japan. This work was partly supported by a Grant-in-Aid for Research from the Ministry of Education, Science and Culture of Japan.

REFERENCES

1 M.H.Yoo, W.A.T.Clark and C.L.Briant (ed.), Interfacial Structure, Properties and Design, MRS Proc., **122**, Materials Research Society, Pittsburg, (1988).

2 G.A.Chadwick and D.A.Smith (ed.), Grain Boundary Structure and Properties, Academic Press, London, (1976).

3 T.Watanabe, "Grain boundary design for the control of intergranular fracture", Materials Science Forum, **46**, 25-48(1989).

4 L.M.Levinson (ed), Grain Boundary Phenomena in Electronic Ceramics, Advances in Ceramics, **1**, American Ceramic Society, Columbus, (1981).

5 T.E.Mitchell, D.R.Clarke, J.D.Embury and A.R.Cooper, "Processing Ceramic Superconductors", Journal of Metals, **41**[1], 6-10(1989).

6 R.W.Balluffi, "Grain boundary diffusion mechanisms in metals", Metallurgical Transactions, **13A** [12], 2069-2095, (1982).

7 K.T.Aust and J.W.Rutter, "Grain boundary migration in high-purity lead and dilute lead-tin alloys", Trans. AIME., **215** [2], 119-127(1959).

8 D.A.Smith, "Interaction of dislocations with grain boundaries" J.de Physique, Colloque C6, **43**, 225-237(1982).

9 T.Watanabe, "Grain boundary design for advanced materials on the basis of the relationship between texture and grain boundary character distribution (GBCD)", Textures and Microstructures, **20**[1-4], 195-216(1993).

10 T.Watanabe, "An approach to grain boundary design for strong and ductile polycrystals", Res Mechanica, **11**[1], 47-84(1984).

11 K.T.Aust and G.Palumbo, "Interface control in materials", Proc. Intern.Conf.on Advanced Structural Materials, D.S.Wilkinson (ed) Pergamon Press, New York, pp.215-226(1989).

12 T.Watanabe, unpublished work.

13 T.Watanabe, H.Fuji, H.Oikawa and K.I.Arai, "Grain boundaries in rapidly solidified and annealed Fe-6.5mass%Si polycrystalline ribbons with high ductility", Acta Met., **37**[3], 941-952(1989).

14 T.Watanabe, K.Arai, K.Yoshimi and H.Oikawa, "Texture and grain boundary character distribution (GBCD) in rapidly solidified and annealed Fe-6.5mass%Si ribbons", Phil.Mag.Let., **59**[2], 47-52, (1989).

15 T.Watanabe, S.Kitamura and S.Karashima, "Grain boundary hardening and segregation in alpha iron-tin alloy", Acta Met., **28**[4], 455-463(1980).

16 M. F. Yan, "Microstructural control in the processing of electronic ceramics", Materials Science and Engineering, **48**, 53-72(1981)

17 T. Watanabe, "Prediction of change in grain boundary energy during grain growth", Scripta Met., **21**[4], 427-432(1987).

18 T. Watanabe, "Grain boundary character distribution (GBCD) analysis of grain growth-related phenomena in polycrystalline materials", Materials Science Forum, **94-96**, 209-220(1992).

19 T. G. Nieh and J. Wodsworth, "Superplastic ceramics", Annu. Rev. Mater. Sci. **20**, 117-140(1990).

20 K. Kajihara, Y. Yoshizawa and T. Sakuma, "Superplasticity in SiO_2-containing tetragonal zirconia polycrystal", Scripta Met. Mater., **28**[5], 559-562(1993).

21 T. G. Langdon: "The physics of superplastic deformation", Materials Science and Engineering, AL A137, 1-11(1991).

22 T. Watanabe, "Grain boundary design and control for high temperature materials", Materials Science and Engineering, **A166**[1-2], 11-28(1993).

23 L. Qing, H. Xiaoxu, Y. Mei and Y. Jinfeng, "On deformation-induced continuous recrystallization in a superplastic Al-Li-Cu-Mg-Zr alloy", Acta metall. mater., **40**[7], 1753-1762(1992).

24 H. Lin and D. P. Pope, "The influence of grain boundary geometry on intergranular crack propagation in Ni_3Al", Acta metall. mater., **41**[2], 553-562(1993).

25 S. Lartigue and L. Priester, "Analysis of grain boundaries in fine-grained Mg-doped aluminas with different microstructures", Trans. JIM., suppl., **27**, 205-212(1986).

26 L. C. Lim and T. Watanabe, "Fracture toughness and brittle-ductile transition controlled by grain boundary character distribution (GBCD) in polycrystals", Acta metall. mater., **38**[12], 2507-2516, (1990).

27 D. Dimos, P. Chaudhari, J. Mannhart and F. K. LeGoues, "Orientation dependence of grain-boundary critical currents in $YBa_2Cu_3O_{7-d}$ bicrystals", Phys. Rev. Let., **61** [2], 219-222, (1988).

28 Y. Zhu, H. Zhang, H. Wang and M. Suenaga, "Grain boundary in textured $YBa_2Cu_3O_{7-d}$ superconductor", J. Mater. Res., **6** [12], 2507-2518, (1991).

29 C. S. Nichols and D. R. Clarke, "Critical currents in inhomogeneous triangular Josephson arrays: A model for polycrystalline super conductors", Acta metall. mater., **39** [5], 995-1002(1991).

COMPUTER SIMULATIONS OF THERMAL EXPANSION ANISOTROPY INDUCED MICROCRACKING IN BRITTLE CERAMICS.

Narayanaswamy Sridhar and David J. Srolovitz
University of Michigan, Ann Arbor, MI 48109

Edwin R. Fuller, Jr.
National Institute of Standards and Technology, Gaithersburg, MD 20899

ABSTRACT

Fracture processes in brittle, granular materials, such as fine ceramics, are profoundly influenced by grain-boundary properties and microstructure. The microcracking in singlephase ceramics which results from the intrinsic thermal expansion anisotropy (TEA) of the grains is such a process. Typically, these phenomena are modelled by extensions of linearelastic fracture machanics either as "regular anays" of microstructural features or from a "continuum perspective." Indeed, such models have contributed to many advances in understanding in recent years. However, an important aspect of the fracture process is ignored by these models: the stochastic nature of the microstructure. Computer simulations offer an excellent way to examine these phenomena. In this paper TEA-induced microcracking in single-phase ceramics has been simulated by using a simple mechanics model hased upon a regular lattice of brittle, elastic springs. Microcracks preferntially from at grain boundaries and propagate either along grain boundaries or into the grain, depending on the relative toughness of the grain interiors. The present results show that TEA-induced microcracking has a qualitatively different dependence on grain size depending on the damage measure employed. Although microcracking is first initiated in the large grain microsturactures, the small grain microstructures for most of the TEA misfit strain regime have more total microcrack length per area of cracking. The large grain-size microstructures, however, undergo a more detrimental strength degradation as a result of the larger microcrack sizes compared to those in the small grain samples. Unlike in unconstrained sample, when a sample is constrained during a temperature excursion, the stress created by the overall thermal expansion can directly lead to fracture of the entire sample.

[ERF was funded by the U.S.DOE, Advanced Industrial Concepts (AIC) Division, AIC Materials Program under Interagency Agreement No. DE-A105-900R21941]

If the auther send the manuscript after the dead line, we will publish it on next edition.

Functional Ceramics

INTERFACE PROPERTIES OF OXIDE CERAMIC MATERIALS

J. Nowotny
Australian Nuclear Science & Technology Organisation, Advanced Materials Program, Lucas Heights Research Laboratories, Menai, NSW 2234, Australia

ABSTRACT

Segregation results in an enrichment of interfaces in certain lattice elements and, consequently, leads to the formation of chemical and electrical potential gradients within the interface region. These gradients have substantial effect on the gas/solid kinetics in general and on processing of ceramics in particular.

This paper illustrates the effect of segregation on properties of ceramic oxides. Several applied aspects of interfaces in the preparation of functional ceramic materials, such as ceramic gas sensors, dielectrics and superconductors, are considered.

INTRODUCTION

The properties of ceramic materials are to a large extent controlled by interfaces, such as external surfaces and grain boundaries. Awareness is growing that the preparation of ceramics with enhanced properties is dependent on better understanding the effect of interfaces on materials properties.

Properties of the bulk and grain boundaries of solids are entirely different in respect to chemical composition, structural ordering and resulting electrical properties. Unfortunately, in contrast to much information which has been accumulated on bulk properties of the materials, little is known about interfaces and their impact on processing and properties of ceramic materials.

This paper will consider several aspects of ceramic interfaces and their impact on the preparation of sensors, dielectrics and superconductors.

SEGREGATION

Segregation can be considered as an adsorption from the solid phase. The process of segregation, which is rate controlled by lattice diffusion, requires elevated temperatures to occur. Figure 1 illustrates schematically the segregation of lattice defects to the gas/solid interface resulting in enrichment of the interface.

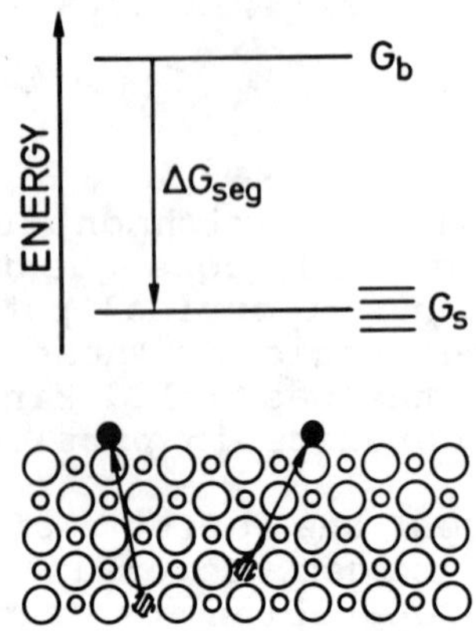

Fig. 1. Schematic illustration of surface segregation of defects and related change in surface energy

A theory of segregation has been developed for metallic solids [1]. Since there is no simple analogy between metals and compounds, current studies aim to develop an adequate theory of segregation for ionic compounds. There is thus an increasing interest in the determination of surface properties of these compounds.

SURFACE VS. BULK PROPERTIES

Metal oxides are basic components of many industrial ceramic materials such as sensors, high T_C superconductors, and dielectrics. The properties of these materials are determined by chemical composition of interfaces. However, because little is known about this dependence, the ceramic materials have been prepared empirically rather than using knowledge of interface properties.

Surface properties of solids can be assessed by many

surface-sensitive methods. However, the application of methods such as SIMS, XPS, LEED and ISS, is limited to high vacuum and lower temperature environment.

Electrical methods, such as thermopower and electrical conductivity, have found wide application in studies of ionic solids. These methods are very sensitive to defect-related properties and may be applied at elevated temperatures under controlled gas atmosphere. These two methods, however, are mainly sensitive to the bulk phase rather than to interfaces. In contrast the work function method (WF) [2, 3] is extremely sensitive to both chemical composition and structure of the outer surface layer.

Figure 2 illustrates the change of the Fermi energy (E_F) determined independently by thermopower (curve 1) and WF (curve 2) for Li-doped NiO [4].

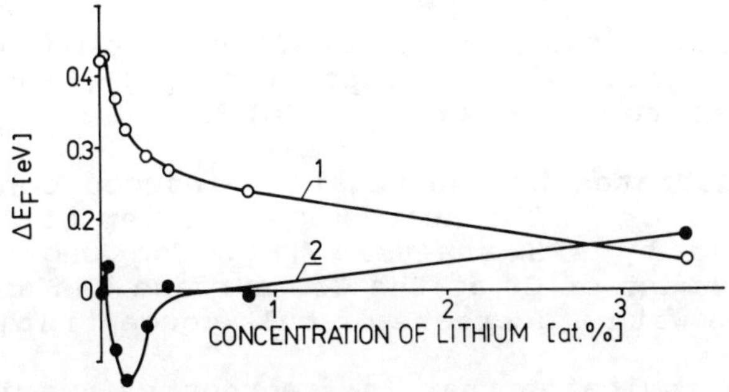

Fig. 2. Changes of the Fermi energy (E_F) for Li-doped NiO at 400°C determined by thermopower (1) and work function (2) [4]

Lithium addition results in a continuous decrease of the Fermi energy as determined by thermopower (curve 1). Since thermopower is a bulk sensitive method, this dependence indicates that incorporation of Li results in the formation of acceptors in the bulk phase of NiO. On the other hand, one may see that at higher concentrations Li forms donor centers at the surface. These experimental data indicate that the same ion may be incorporated by the lattice according to two different mechanisms, into the surface layer on one hand and the bulk on the other hand. It is interesting to note that the activation energy of CO to CO_2 oxidation over Li-doped NiO as a catalyst exhibits the same dependence as that of ΔE_F vs. Li-content based on work function measurements (Figure 3).

This correlation between the work function data and catalytic properties illustrates that catalytic properties of metal oxides are entirely dependent on surface rather than bulk properties.

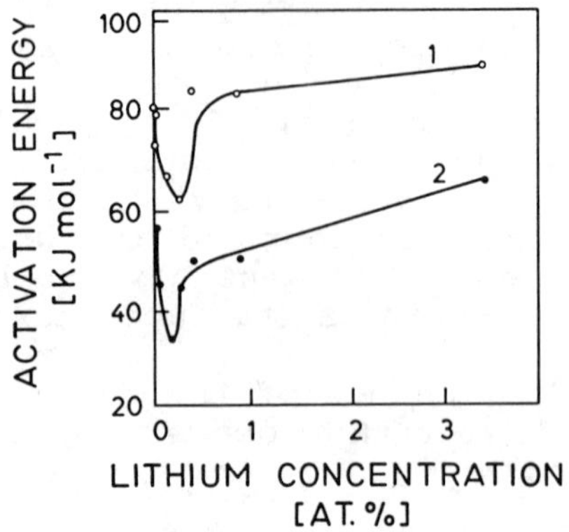

Fig. 3. The activation energy of CO oxidation over Li-doped NiO as a catalyst vs. composition [5] (1-stoichiometric $CO+O_2$ mixture, 2-excess of O_2)

Figure 4 illustrates the segregation-induced concentration profile of Cr in the near-surface layer of Cr-doped CoO determined by XPS analysis [6]. The segregation-induced enrichment of Cr at the CoO surface has a tendency to increase with a decrease of bulk concentration [6].

This tendency indicates that (a) segregated species have a tendency to reach the same surface concentration independently of the bulk composition, and (b) even impurities present at a very low concentration in the bulk may lead to a high surface concentration if their segregation driving force is high enough.

Zirconia is an important ceramic material with a wide application in electrochemical devices, such as sensors and fuel cells. The effect of segregation on its grain boundary properties has been reported in several publications [7-11]. AES studies have shown that the surface composition of yttria in yttria-stabilized zirconia is about 30-35 at% and is practically independent of the bulk composition [10].

Recent work function and SIMS studies have shown that the surface of yttria-doped zirconia has a sandwich-type character.

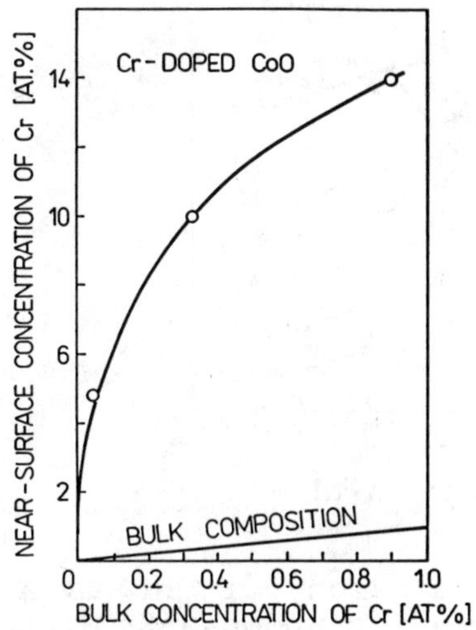

Fig. 4. Segregation-induced enrichment of Cr in $CoO-Cr_2O_3$ solid solution as a function of composition [6]

The overlayer is predominantly enriched in CaO, which is present in the bulk as an impurity at a very low level, and the sublayer enriched with yttria [10]. From a comparison of the work function and AES studies it appears that the thickness of the surface layer enriched with CaO is too small to be detectable by AES.

It is interesting to note that work function changes of crystals measured at elevated temperatures in the gas/ /solid equilibrium may be considered in terms of defect chemistry of the surface layer. Figure 5 illustrates work function changes for undoped zirconia, determined by thermionic emission, as a function of oxygen activity [12]. The measured slopes of these dependencies are consistent with the defect model of zirconia and related electrical properties [13]. Incorporation of yttria into zirconia results in the work function vs. $p(O_2)$ dependence which again is consistent with the bulk defect model of yttria-stabilized zirconia (Figure 6) [13, 14].

According to the above illustrated examples the $p(O_2)$ exponent resulting from work function measurements is the same as that expected for the bulk defect disorder. However, work function changes of zirconia after prolonged annealing is inconsistent with the bulk defect structure [10, 15].

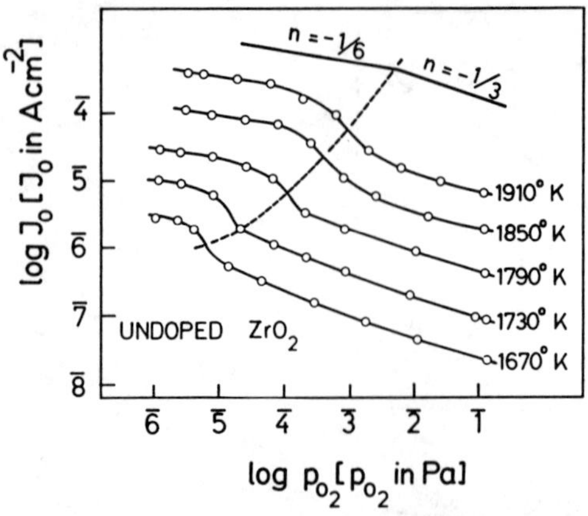

Fig. 5. The thermal emission current as a function of oxygen partial pressure for undoped ZrO_2 [12]

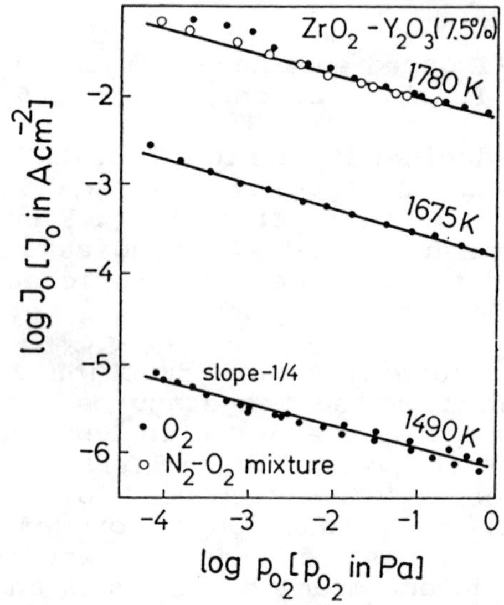

Fig. 6. The thermal emission current as a function of $p(O_2)$ for yttria-stabilized zirconia [14]

Figure 7 illustrates changes in the contact potential difference (CPD) between zirconia and Pt at 780°C during reduction and oxidation runs. These changes are determined by work function changes of the zirconia surface. As seen there is a continuous change of measured values with time reaching relatively stable values after 120 hours. Figure 8 illustrates changes of related values of reciprocal of oxygen pressure exponent which varies between 4, as predicted by theory, to almost 2.

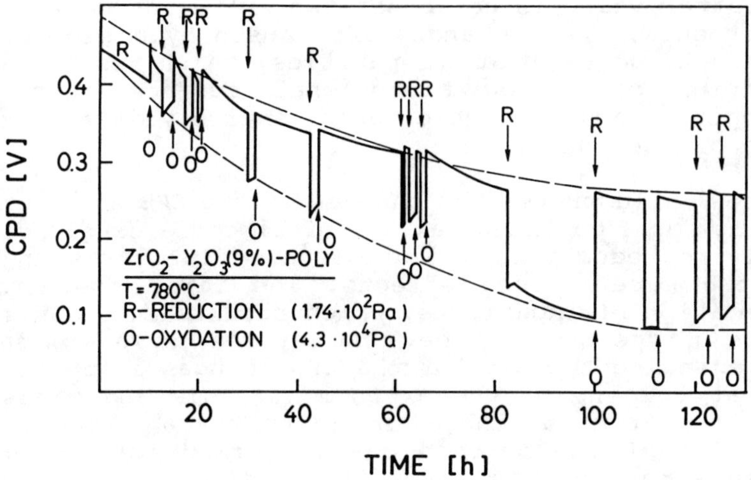

Fig. 7. Changes in the contact potential difference (CPD) of yttria-stabilized zirconia during subsequent reduction and oxidation runs vs. time of annealing at 780°C [10]

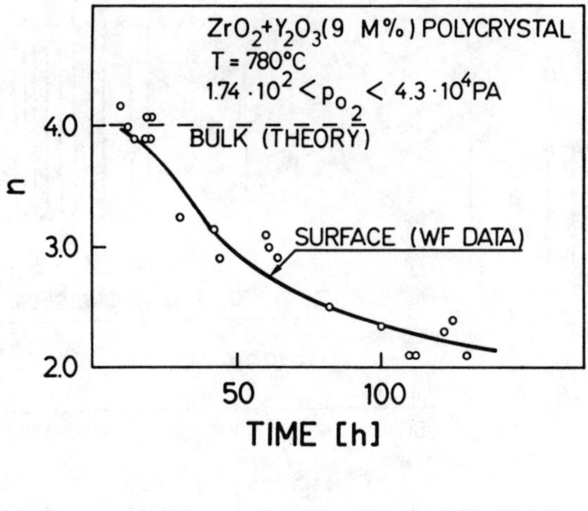

Fig. 8. Changes of the reciprocal of the $p(O_2)$ exponent of work function vs. time of annealing for yttria-stabilized zirconia [10]

The latter value is not consistent with any defect model of zirconia. These changes are caused by a slow process of defects segregation (impurities) which results in the formation of a low dimensional surface layer with composition and properties being entirely different from those of the bulk phase.

Figure 9 illustrates the changes of the CPD as a function of time for a polished surface of zirconia single crystal involving reductions and oxidations [15]. As seen there are two superimposed effects. The first one involves segregation of impurities, which results in a continuous change of the CPD in time. One may see, however, that at the beginning the sign of the CPD changes is opposite to that at the end of the experiment. This indicates that after some time of annealing (70h at 780°C) there is an inversion of the surface charge formed as a result of oxygen sorption. The formation of the positive surface charge as a result of oxygen sorption in the initial period results from the formation of dipoles localized at mechanical defects formed during polishing [15].

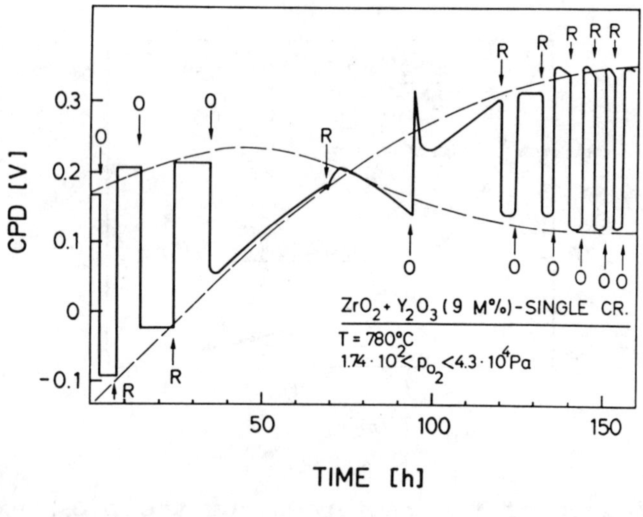

Fig. 9. Changes in the contact potential difference (CPD) of a zirconia single crystal (9 %mol Y_2O_3) during subsequent reduction and oxidation runs vs. time of annealing at 780°C [15]

Thus, we can see that the surface and the near-to-surface layer of yttria-doped zirconia are the subject of complications involving the bulk phase of well defined properties, as well as the sublayer enriched in yttria and the overlayer enriched in impurities such as Ca [13].

Barium titanate is another compound with wide technological application. Several properties of this compound, such as dielectric properties, are controlled by grain boundaries. However, the grain boundary properties of $BaTiO_3$ are not well understood. Likewise, both the chemical composition and the resulting electrical properties are different for the bulk phase and the grain boundary region.

It has been reported that an overlayer of $BaTiO_3$ exhibits the cubic structure while the bulk phase remains tetragonal (Figure 10) [16]. Thus, $BaTiO_3$ cannot assume the tetragonal structure below a certain grain size. Consequently, fine-grained $BaTiO_3$-based ceramic materials exhibit different properties, such as temperature characteristics of resistance, than bulk materials.

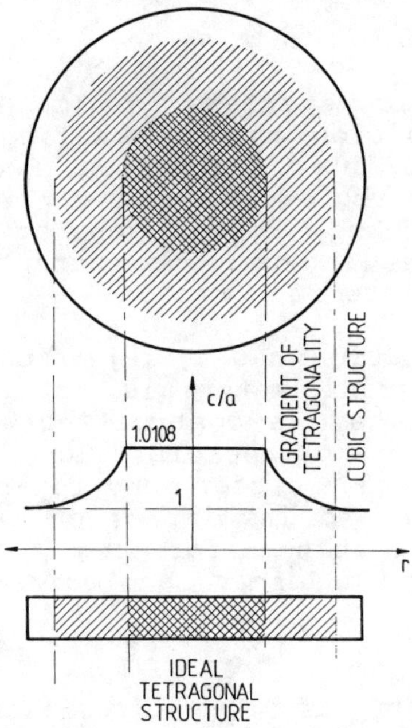

Fig. 10. Tetragonality gradient within a $BaTiO_3$ grain

There has been an accumulation of data indicating that the critical parameters of superconducting oxide cuprates are strongly influenced by their grain boundaries [17].

There is a disagreement as to the mechanisms involved and as to whether their effects on properties are positive or negative. It seems that the controversy is related to both the impurity level and the preparation conditions of individual specimens. This issue is discussed in detail in ref. [18]. Duan et al. [19] have reported a spectacular experiment showing that the ac susceptibility of a freshly prepared $YBa_2Cu_3O_x$ sample differs markedly from one with the overlayer removed. It has also been shown that grain size has a substantial effect on the superconducting properties of oxide cuprates [20, 21]. Dimos et al. [22] have shown a substantial effect of the grain boundary misorientation angle on the critical current.

Current literature reports confirm that properties of grain boundaries are of a strategic importance for the development of oxide ceramic materials with optimum or enhanced properties.

CONCLUSIONS

Segregation in ceramic materials results in changes in both chemical composition and crystalline ordering of interfaces and of the interface region. Because properties of the interface region are entirely different from those of the bulk phase, correct characterization of ionic solids requires independent characterization of both bulk phase and interface region.

The effect of segregation has a significant impact on processing of oxide ceramic materials. Interfaces such as grain boundaries may have a controlling effect on the mechanism of sintering and, therefore, on final material properties. Therefore, better understanding of the effect of segregation and its impact on grain boundary chemistry is of importance for the development of advanced materials for modern technologies.

ACKNOWLEDGEMENTS

The financial support of the Australian Government through the Department of Industry, Technology and Regional Development (DITARD) as well as the Australian

Nuclear Science and Technology Organisation is gratefully acknowledged.

REFERENCES

1. P. Kofstad, "Nonstoichiometry, Diffusion and Electrical Conductivity of Binary Metal Oxides", Wiley & Sons, New York, 1972
2. J. Nowotny and M. Destriau, "Application des Mesures D'extraction Electronique a l'Etude de la Chimisorption et de la Catalyse sur les Oxydes Semiconducteurs", Bull.Soc.Chim.France 91-106 (1976)
3. J. Nowotny, J.Mater.Sci., "Surface Equilibration Kinetics of Nonstoichiometric Oxides" 12, 1143-60 (1977)
4. J. Deren, J. Nowotny and J. Ziolkowski, "Surface and Bulk Electronic Properties of Lithium-Doped Nickel Oxide, Bull.Polon.Sci., Ser.Sci.Chim., 16, 45-50 (1968)
5. A. Bielanski and J. Deren, "Relation Between the Electronic and Catalytic Properties of Semiconducting Oxide Catalysts", in: "Electronic Phenomena in Chemisorption and Catalysis on Semiconductors", F.F. Volkhenstein and K. Haufe, Eds., W. de Gruyter, Berlin, 1969, pp. 149-65
6. J. Haber, J. Nowotny, I. Sikora and J. Stoch, "Segregation of Cr in $CoO-Cr_2O_3$ Solid Solutions", Appl.Surf.Sci., 17, 324-30 (1984)
7. K.O. Axelsson, K.E. Keck and B. Kasemo, "Surface Compositional Changes of ZrO_2 in H_2O, H_2 and Atomic Hydrogen, Investigated by AES and EELS", Appl.Surf.-Sci., 25 217-30 (1986)
8. A.J. Winnubst, P.J. Kroot and A.J. Burggraaf, "AES/STEM Grain Boundary Analysis of Stabilized Zirconia Ceramics", J.Phys.Chem.Solids 44 955-60 (1983)
9. A.J. Burggraaf, M.Van Hemert, D. Scholten and A.J.A. Winnubst, "Chemical Composition of Oxide Interfaces in Relation with Electric and Electrochemical Properties", in: "Materials Science Monographs" # 28, P. Barret and L.C. Dufour, Eds., Elsevier, Amsterdam, 1985, p.797-802
10. J. Nowotny, M. Sloma and W. Weppner, "Surface Relaxation of Y_2O_3-Stabilized ZrO_2", Solid State Ionics 28-30, 1445-50 (1988)

11. J. Nowotny, M. Sloma and W. Weppner, "Surface Reactivity of Yttria-Doped Zirconia with Oxygen", Solid State Ionics 32/33, 709-713 (1989)
12. P. Odier, J.C. Rifflet and J.P. Loup, "Defect and Surface Phenomena at High Temperatures for Oxides by Thermal Emission of Electrons", in: "Materials Science Monographs" #10, Elsevier, Amsterdam, 1982, pp. 458-66
13. J. Nowotny and M. Sloma, "Work Function of Oxide Ceramic Materials", in: "Surface and Near-Surface Chemistry of Oxide Materials", Elsevier, Amsterdam, 1988, pp. 281-343
14. P. Odier and J.P. Loup, "Thermal Emission of Electrons: A Sensitive Probe for Investigations in Non-Stoichiometry of Oxides at High Temperature", in: "Materials Science Monographs" # 15, Elsevier, Amsterdam, 1982, pp. 393-408
15. J.C. Niepce, "Some Aspects of the Influence of Particle Size on Properties and Behaviour of a Dielectric Material: Example of Barium Titanate", in: 'Surfaces and Interfaces of Ceramic Materials' Kluwer, Dordrecht, 1989, pp. 512-33
16. J. Nowotny, M. Rekas, D.D. Sarma and W. Weppner, "High T_c Oxide Superconductors. Possible Effect of Interfaces", in Ref. 13, p. 13, 669-699
17. Duan Hongmin, Lu Liand and Zhang Dian Lin, "The Role of Surface Shellof the $YBa_2Cu_3O_{7-x}$ Single Crystals in High T_c Superconductivity", Solid State Comm., 67 [8]809-813 (1988)
18. Y.M. Chiang, J.A.S. Ikeda and A. Roshko, "Grain Boundary Superconductors", in: "Ceramic Superconductors II. Research Update", M.F. Yan, Ed., Amer.-Ceram.Soc., Westerville, 1988, pp. 607-18
19. Duan Hongmin, Lu Li and Zhang Dianlin, "$YBa_2Cu_3O_{7-x}$: A Bulk Superconductor?", Chinese Phys.Lett., 5 [6] 253-256 (1988)
20. D. Dimos, P. Chaudhari, J. Mannhardt and F.K. Le Goues, "Orientation Dependence of Grain Boundary Critical Current in $YBa_2Cu_3O_{7-x}$ Bicrystals", Phys.-Rev.Lett., 61 [2] 219-222 (1988)

GRAIN BOUNDARY OXIDATION RATE OF SEMICONDUCTIVE Mn-DOPED $BaTiO_3$.

Ben Huybrechts, Masasuke Takata, and Kozo Ishizaki
Nagaoka University of Technology-(Nagaoka Gijutsu Kagaku Daigaku)
Kamitomioka 1603-1, 940-21 Nagaoka
Niigata, Japan

ABSTRACT

The relationship between the Mn concentration, the grain boundary oxidation after sintering, and the positive temperature coefficient of resistivity-properties were examined. Semiconductive $BaTiO_3$ was doped with 0.02 and 0.1 mol% Mn. The 0.1 mol% doped samples showed a bigger difference between the minimum and maximum resistivity. Surprisingly this was not caused by an increase in electron trap density (N_s) at the grain boundaries, but by an increase in the energy difference between the conduction band and the electron traps, combined with a decrease in charge carrier density, N_d. The oxidation rate did not increase with increasing Mn concentration, contrary to what was expected. This is explained by the fact that the electron traps in the 0.02 % doped samples are not caused by Mn but by intrinsic electron traps, such as barium vacancies and adsorbed oxygen.

INTRODUCTION

Since it was discovered in 1955 that properly doped $BaTiO_3$ shows a positive temperature coefficient resistance (PTCR) above its Curie-Point, there have been many studies. PTCR-barium titanate is now a common electronic device with many applications, such as fuses, heaters, demagnitizers of color TV-screens and computer monitors, and other devices. The PTCR-effect was modeled satisfactorily by Heywang in 1961 [1]. Heywang assumed electron traps at the grain boundaries, which attract electrons from the inside of the grains thereby forming a double Schottky barrier. This barrier increases abruptly with increasing temperature above the Curie temperature because of a decrease in permittivity, which follows the Curie-Weiss law. The increase in the barrier height in turn causes the increase in resistance. Although the Heywang model is well accepted,

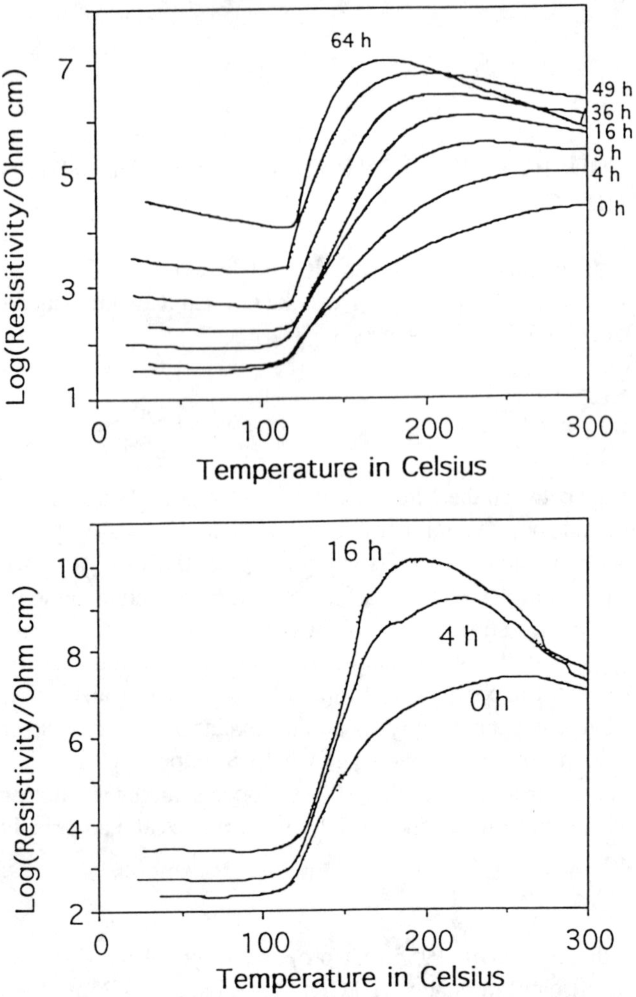

Figure 1. The resistivity temperature graphs for 0.02 % Mn-doped (a) and 0.1 % Mn-doped (b) samples. Notice that the PTCR-jump for the 0.01 % Mn doped samples is much larger than for 0.02 % Mn-doped samples. The resistivity-temperature curves for the 0.02 % Mn doped samples after 1 h, 3 h, 6 h, and 25 h annealing were omitted for clarity reasons. The curves of the 0.1 % Mn-doped samples after 1 h, and 9 h of annealing were omitted for the same reasons.

the exact nature of the electron traps, the relationships between the density of trapped electrons (N_s), the energy difference between the conduction band and the electron traps (E_s), and the charge carrier density (N_d) are still is not well understood. The authors proposed earlier a model to explain the relation between the grain boundary oxidation after sintering and the resistivity properties. This model is based upon a model proposed by Ueoka [2]. He assumes that there is an acceptor enriched region at the grain boundaries, which seems to be reasonable considering the segregation results of Chiang et al. [3,4]. The acceptors, usually Mn, are in the 2+ oxidation state during sintering. The Mn substitutes Ti, and the charge difference between the Mn^{2+} and the Ti^{4+}, is compensated by double charged oxygen vacancies. Because Mn^{2+} does not cause an electron trap, no PTCR is observed after quenching from the sintering temperature. The oxidation of the Mn to higher oxidation states, which act as electron traps, takes place during cooling or annealing at temperatures below the sintering temperature. The oxidation is caused by the diffusion of double charged oxygen vacancies towards the grain boundary, which acts as a sink for the vacancies. Ueoka did not specify whether the Mn was in the 3+ or the 4+ oxidation state after normal furnace cooling. However, recently the present authors have found that E_s increased after oxidation at high P_{O2} or after long annealing time [5,6]. The rise in E_s was attributed to the oxidation of Mn to its 4+ oxidation state. Based on the above, one would expect that the oxidation rate increases with higher Mn concentration at the grain boundaries, as proposed by the authors in ref. 5, This is because the density of the oxygen vacancies compensating the Mn at the sintering temperature increases.

Although the diffusion constant of the oxygen vacancies does not change, the number of oxygen vacancies annihilated during the same oxidation time increases, if the Mn concentration at the grain boundary increases. Therefore the number of electron traps created in a fixed oxidation time should increase with higher Mn concentration at the grain boundary. This is equivalent with a higher oxidation rate. The current research has been carried out to check if the oxidation rate changes with the Mn concentration at the grain boundaries.

EXPERIMENTAL

$BaTiO_3$ powder (Central Glass Comp. Ltd., purity 99.9 %, BET 10.1 m^2/g) was ball milled for 16 h with 0.15 mol% Sb_2O_3, 1 mol% Al_2O_3 (Soekawa Chemicals, 99.9 %, particle size 2-3 μm), 1 mol% TiO_2 (Soekawa Chemicals, 99.9 %, paritcle size 1-2 μm), and 1.5 mol% SiO_2 (High Purity Chemistry Laboratory, >99.9%, particle size 0.8 μm) in a monomaron pot using Al_2O_3 balls and deionized water. After drying, the powder was mixed with the appropriate amount of $Mn(NO_3)6H_2O$ solution in order to obtain 0.02 mol% Mn and 0.1 mol % Mn doped mixtures. The mixing was done using a magnetic stirrer. After drying, the powder was calcined at 750°C for 4 h. The calcined powder was sieved and subsequently uniaxially pressed (47 MPa) in 2 gram pellets with a diameter of 20 mm. The samples were placed on a bed of powder presintered at 1360°C for 1 h. Sintering and annealing took place in a tubelar furnace. The samples were heated at 20°C/min up to 1330°C. Sintering time was 1 h. After sintering the samples were cooled to 1183°C, and annealed for 0 to 64 h. After annealing the samples were furnace cooled to room temperature. The resistance temperature characteristics were measured after rubbing In-Ga paste on both sides of the samples, using a computer controlled 2 probe technique. The complex impedance was measured using a Hewlett-Packard 4192A-LF Impedance Analyzer in the range 5 Hz–13 MHz.

RESULTS

Figure 1 shows the resistivity as function of temperature for different annealing time, and demonstrates the well known fact that the maximum resistivity (ρ_{max}) increases with increasing annealing time. The same phenonema is found by many authors [7-11]. It is also clear that ρ_{max} for the Mn doped samples is much higher than that for the 0.02 % doped samples. Although at first sight these results seem to indicate an increase in electron trap density, it will be shown in the discussion that this is not the case.

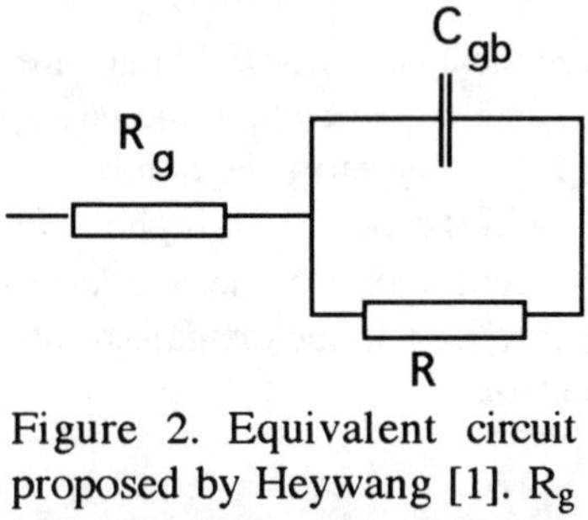

Figure 2. Equivalent circuit proposed by Heywang [1]. R_g indicates the grain resisitance, C_{gb} the grain boundary capacitance caused by the double Schottky barrier, and R_{gb} the grain boundary resistance.

The grain size of the 0.02 % doped samples was 9 μm and 12 μm for the 0.1 % Mn doped samples. The grain size did not increase with annealing time.

DISCUSSION

One of the important factors which influence the resistivity is the charge carrier density, which can be easily obtained from the grain resistivity. The grain resistivity in its turn can be obtained by impedance spectroscopy, using the equivalent circuit proposed by Heywang, see figure 2 [1]. The proposed equivalent circuit should give a semicircle in the impedance plane [12], however for PTCR-ceramics one usually obtains a depressed semicircle; i.e. a circle with its center below the x-axis, as is shown in figure 3. The depression of the semicircle is caused by distributed elements, which actually means that the elements in the equivalent circuit are not one fixed value but are spread around a mean value, as explained in reference [12].

To check if the obtained experimental data can be approximated by a depressed semicircle the following calculations are carried out. From the point of maximum imaginary resistivity, the value y_0 is obtained, as defined in figure 3a. x_0 is obtained using a grain resistivity of 9 ohm cm for the 0.02 % samples and 35 ohm cm for the 0.1 % doped samples. The radius of the depressed circle is obtained from:

$$r=\frac{x_0^2+y_0^2}{2y_0} \qquad (1)$$

The center of the circle is located a distance $r-y_0$ below the x-axis.The calculated lines in figure 3, are plotted using the radius and the position of the center point. Although only two semicircles are shown here, all depressed semicircles could be well approximated by the calculated circles. This leads to the conclusion that the

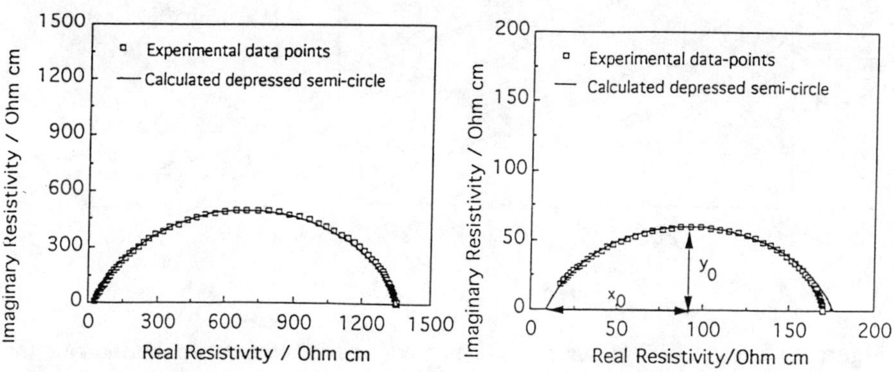

Figure 3. The impedance plane for a sample doped with 0.02 % Mn (a) and 0.1 % Mn doped (b). Both samples are annealed for 16 h. Remark that the experimental data can be very well approximated by a "*depressed*" semicircle. This shows that the elements in the equivalent circuit are distributed around a mean value. The values x_0 and y_0 are values used to calculate the depressed semicircle their meaning is explained in detail in the text.

circuit proposed by Heywang is valid, if we allow the elements to be distributed.

The grain resistivity did not change with increasing annealing time, which means that also N_d is constant as a function of annealing time. The grain resistivity is, however, quite different for the samples with a higher Mn concentration. N_d is 9 ohm cm for the 0.02 % Mn doped samples and 35 ohm cm for the 0.1 % Mn doped samples. This difference can be explained by a partial compensation of the donors by the Mn addition as is explained by Peng et al. [13].

A second parameter which greatly influences the resistivity behavior is the depth of the electron traps below the conduction band, E_s. This depth can be easily obtained from the plot of the maximum resistivity (ρ_{max}) versus the temperature of maximum resitivity (T_{max}), as is shown by Jonker [14]. The curves for the samples under discussion are shown in figure 4. The calculated lines are iteratively obtained using the equations proposed by Heywang. The lines of 0.224

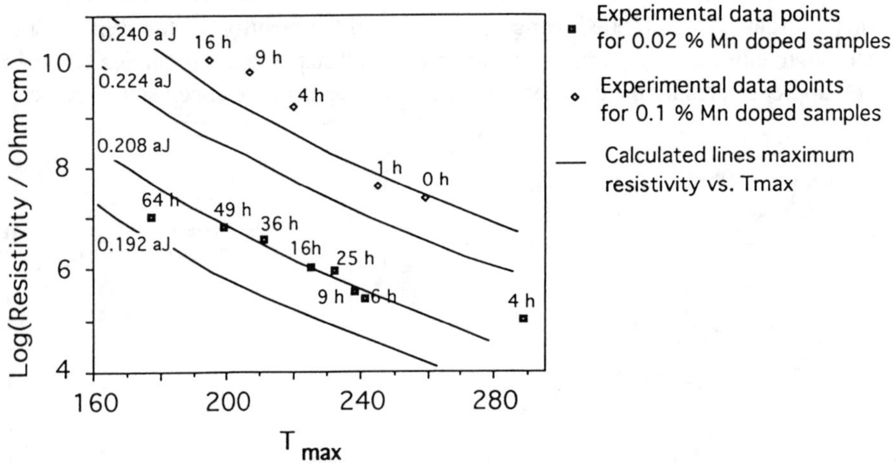

Figure 4. The ρ_{max}-T_{max} curves. The numerals besides the data-point indicate the annealing time at 1183°C. The numerals on the left side of the lines indicate the energy gap (E_s) between the conduction band and the electron trap. The calculated lines are obtained from the equations proposed by Heywang. Although all the samples are doped with Mn, E_s increases with increasing Mn concentration. This shows that the type of electron trap depends not only on the fact if there is doping or not, but also on the dopant concentration.

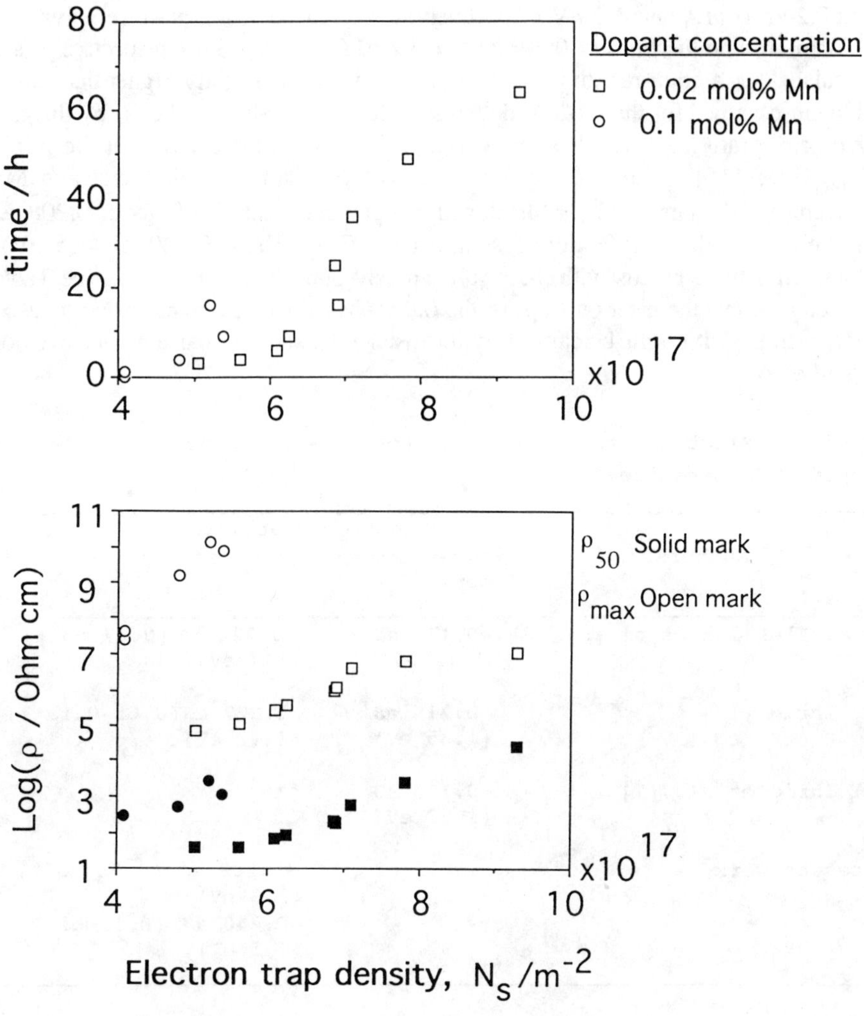

Figure 5. Experimental PTCR-chart. Figure (a) shows the electron trap density as a function of the annealing time. No significant difference in oxidation rate could be found between the 0.02 % and the 0.1% Mn-doped samples. Figure (b) shows the resistivity at 50°C (ρ_{50}), and the maximum resistivity (ρ_{max}) versus the electron trap density. Notice that (a) and (b) are associated. By plotting the data in this way one can see better the relation between the electron trap density, annealing time, ρ_{50}, and ρ_{max}.

and 0.240 aJ (1.4 and 1.5 eV respectively) are calculated for a grain resistivity of 35 ohm cm and the lines of 0.208 and 0.192 aJ (1.2 and 1.3 eV respectively) are calculated for a grain resistivity of 10 ohm cm, which is slightly higher than the 9 ohm cm obtained for the 0.02 % doped samples. As is already shown by Ihrig, a change in grain resistivity has almost no influence on the position of the ρ_{max}-T_{max} lines [15]. Surprisingly we find that E_s changes with increasing Mn concentration. The 0.02 % Mn-doped samples have an E_s of about 0.208 aJ (1.3 eV) and the 0.1 % doped samples an E_s of about 0.240 aJ (1.5 eV). Comparing these results with the results already published in literature, see Table 1, indicates that the electron traps in the 0.02 %Mn-doped samples are not caused by Mn-traps but might be caused by intrinsic traps such as barium vacancies or adsorbed oxygen.

Table I. Electron trap levels for non-acceptor doped and acceptor doped PTCR-ceramics

Authors	Electron trap level non-acceptor doped	Mn-doped
H.M. Al-Allak et al.[8]	~0.179 aJ (1.12 eV)	~0.224 aJ (0.07 mol%) (1.4 eV)
H. Ihrig [15]	~0.216 aJ (1.35 eV)	~0.227 aJ(0.05-0.1at%) (1.42 eV)
H. Ihrig et al. [7]	~0.195 aJ (1.22 eV)	***
Present work		~0.208 aJ (0.02 mol%) (1.3 eV) ~0.240 aJ (0.1 mol%) (1.5 eV)

In order to make any conclusion about the oxidation rate, N_s as a function of time has to be obtained. Using the Curie Weiss constant C=1.5x10^5 K, and the Curie-Weiss temperature θ=383 K, N_s is obtained from S_{arrh}, minus the slope in the reciprocal temperature plot log ρ-1/T in the region from low ρ to high ρ. Until trapped electrons start to jump to the conduction band this slope is constant and can be expressed by the following equation [16]:

$$S_{arrh}=\left(\frac{e^2\theta}{8\ \ln 10\ k\varepsilon_0 C}\right)\frac{N_s^2}{N_d} \tag{2}$$

where e is the charge of an electron, and k the Boltzmann constant. Calculation of N_s for 0.02 % Mn-doped samples annealed for 0, 1 and 3 h was not possible due to the broad peak and consequently small linear region. The exact reason why there is no distinct linear region is still unclear, but it might be caused by shallow electron traps (i.e., traps with a small E_s). Figure 5 shows the PTCR-chart obtained using these results. The upper part shows that the oxidation rate does not change considerably with the increase in Mn-dopant concentration. This is opposite to what was expected, but it can be explained by the fact that the electron traps in the 0.02 % Mn-doped samples are not caused by Mn at the grain boundaries. As suggested by the results in Figure 4 and Table I, the character of the electron traps is different in the 0.02 % and the 0.1 % doped samples.The difference in PTCR-jump, shown in the lower part of the figure, is caused by two factors, an increase in E_s and a decrease in N_d, and not by an increase in N_s.

CONCLUSIONS

1. Surprisingly semiconductive $BaTiO_3$ doped with 0.02 % Mn only showed a trap level of 0.208 aJ (1.3 eV), which was lower than the 0.240 aJ (1.5 eV) trap level for 0.01 % Mn-doped $BaTiO_3$. An increase in E_s with increasing Mn concentration is believed not to be reported before. Probably the 0.02 % Mn is too low to form Mn-electron traps at the grain boundaries, and only intrinsic electron trap levels, i.e. barium vacancies and or adsorbed oxygen, are present.

2. Increasing the Mn concentration increased the ρ_{max} and the PTCR-jump. This is caused by an increase in E_s, combined with a decrease in N_d. It is **not** caused by an increase in N_s.

ACKNOWLEDGEMENTS

The authors would like to thank Central Glass Company Ltd. for providing the raw $BaTiO_3$ powder, and Mr. Yoshiaki Kinemuchi for the assistance in calculating the ρ_{max} -T_{max} curves.

REFERENCES

1. W. Heywang, "Barium Titanate as a Semiconductor with Blocking Layers," *Solid State Electron.*, **3** 51-58 (1961).

2. H. Ueoka, "The Doping Effects of Transition Elements on the PTC Anomaly of Semiconductive Ferroelectric Ceramics," *Ferroelectrics*, **7** 351-353 (1974).
3. Y.M. Chiang and T. Takagi, "Grain Boundary Chemistry of Titanate and Strontium Titanate: I,High-Temperature Equilibrium Charge," *J.Am.Ceram.Soc.*, **73** [11] 3278-3285 (1990).
4. Y.M. Chiang and T. Takagi, "Grain Boundary Chemistry of Barium Titanate and Strontium Titanate: II, Origin of Electrical Barriers in Positive Temperature Coefficient Thermistors," *J.Am.Ceram.Soc.*, **73** [11] 3286-3291 (1990).
5. B. Huybrechts, K. Ishizaki, and M. Takata, "A Proposed Phenomenological PTCR-model and an Accompanying Phenomenological PTCR-chart," *J.Am.Ceram.Soc.*, **76** [11] 286-288 (1993).
6. B. Huybrechts, K. M Ishizaki, and Takata, "Improved Positive Temperature Coefficent of Resistivity Properties of $BaTiO_3$ by O_2-HIPing," *Hot Isostatic Pressing '93*, 451-458, Elsevier Publ., Editors L. Delaey, H. Tas, and W. Kaysser, (1993).
7. H. Ihrig and W. Puschert, "A Systematical and Theoretical Investigation of the Grain Boundary Resistivities of N-doped $BaTiO_3$ Ceramics," *J.Appl.Phys.*, **48** [7] 3081-3088 (1977).
8. H.M. Al-Allak, A.W. Brinkman, G.J. Russel and J. Woods , "The Effect of Mn on the Positive Temperature Coefficient of Resistance Characteristics of Donor Doped $BaTiO_3$ Ceramics," *J.Appl.Phys.*, **63** [9] 4530-4535 (1988).
9. K. Manfred, "Effect of Heat Treatment on the PTCR Anomaly in Semiconducting Barium Titanate," *Ceramic Bull.*, **50** [8] 676-680 (1971).
10. H.M. Al-Allak, G.J. Russel and J. Woods, "The Effect of Annealing on the Characteristics of Semiconducting $BaTiO_3$ Positive Temperature Coefficient of Resistance Devices," *J.Phys.D:Appl.Phys.*, **20** 1645-1651 (1987).
11. Ho In-chyuan and Fu Shen-Li, "Effect of Reoxidation on the Grain Boundary Acceptor-state Density of Reduced $BaTiO_3$ Ceramics," *J.Am.Ceram.Soc.*, **75** [3] 728-730 (1992).
12. J.R. Macdonald, and W.B. Johnson, "Fundamentals of Impedance Spectrosopy", 12-20, Impedance Spectroscopy, ed. Macdonald J.R., Publ. John Wiley and Sons, New-York, (1987).
13. C.J. Peng and H.Y. Lu, "Compensation Effect in Semiconducting Barium Titanate," *J.Am.Ceram.Soc.*, **71** [1] C44-46 (1988).
14. G.H. Jonker, "Some Aspects of Semiconducting Barium Titanate, Solid State Electron, **7** 895-903 (1964).
15. H. Ihrig, "PTC Effect in $BaTiO_3$ as a Function of Doping with 3d Elements," *J.Am.Ceram.Soc.*, **64** [16] 617-620 (1981).
16. B. Huybrechts, K. Ishizaki, and M. Takata, "Experimental Evaluation of the Acceptor-States Compensation in the PTC-Type Barium Titanate," *J.Am.Ceram.Soc.*, **75** [3] 722-724 (1992).

SURFACE SEGREGATION IN BARIUM TITANATE

Z. Zhang, P. J. Pigram*, R. N. Lamb* and J. Nowotny

Australian Nuclear Science and Technology Organisation, Advanced Materials Program, Lucas Heights, NSW 2234, Australia.
*School of Chemistry, University of New South Wales, Kensington, NSW 2033, Australia.

ABSTRACT

We have investigated thermally-induced surface cation nonstoichiometry in undoped $BaTiO_3$ using X-ray photoelectron spectroscopy (XPS). Ti enrichment is found in oxidized $BaTiO_3$, while the reduced surface shows less Ti segregation. The effect of cooling rate on the cation segregation behaviour is also discussed.

INTRODUCTION

The chemical composition in the surface region is often different from that in the bulk for ceramic materials. The difference is caused by surface segregation of defects resulting in an enrichment of certain lattice elements in the surface region [1, 2]. This difference between the surface and the bulk plays an important role in many applications of ceramic materials. A better understanding of the effect of surface segregation on the properties of ceramics is expected to lead to the preparation of advanced materials with enhanced properties.

$BaTiO_3$-based materials are widely used in industry, and many of them are processed at elevated temperatures [3, 4]. The heat-treatment often induces segregation, leading to the formation of concentration gradients in the boundary layer [2]. In the case of undoped $BaTiO_3$ the segregation may give rise to changes of the Ti/Ba ratio with respect to that in the bulk phase. The purpose of this work is to provide an experimental study on cation segregation in undoped

$BaTiO_3$ using XPS. The effects of the gas phase composition and the cooling rate on thermally induced segregation will be examined.

EXPERIMENTAL

The undoped $BaTiO_3$ ceramic specimens used in this study were cold pressed from powder supplied by Cookson Technology Centre in UK, then sintered in air at 1400°C for 12 h. The bulk impurities in these samples were analyzed by ICPMS (Inductively Coupled Plasma Mass Spectrometry), as summarized in Table I. The samples were mechanically polished and heat-treated in air at 1000°C for 8 h. After this standardization, samples were heated under atmospheric pressure in a processing furnace at 1000°C for 3 h in both reducing (Ar + 1% H_2) and oxidizing (O_2) gas phase compositions (referred to as the reduced and the oxidized $BaTiO_3$ samples, respectively). The heat-treated samples were cooled to room temperature before all measurements (slow cooling: ~ 200 °C/h; rapid cooling: ~ 200 °C/min). During the cooling process the same gas atmosphere as that during the heat-treatment was maintained. Samples were analyzed in ultrahigh vacuum (~ $5x10^{-8}$ Pa) using XPS with a Kratos XSAM 800pci system.

Table I. Impurities in Undoped $BaTiO_3$ (ppm).

Zr	Sr	Zn	Y	Hf	Mg	Ag	Co	Sn	Ta	Nb	Cr	Cd
516	347	43	29	26	25	16	8	7	6	3	2	1

RESULTS AND DISCUSSION

Figures 1(a) and 1(b) illustrate the depth profiles obtained by XPS combined with 5 keV Ar^+-ion-bombardment for both the oxidized and the reduced $BaTiO_3$ specimens which were cooled slowly. The Ti and Ba concentrations were determined by integrating the Ti 2p and Ba 3d peak areas above linear backgrounds. The normalized Ti/Ba concentration ratio was plotted versus the sputtering time, as well as the sputtered depth [the sputtering rate of a standard Ta_2O_5 sample (~ 0.45 nm/sec) was used as a rough guide to calibrate the sputtered depth]. The two figures are similar: after a rapid initial decrease the Ti/Ba ratio assumes a stable value, which is normalized to its stoichiometric bulk value: one. The XPS depth profiles seem to suggest that both the oxidized and the reduced surfaces are enriched in Ti (with respect to Ba), and the degree of Ti enrichment and the thickness of the enriched layer are the same within experimental error. However, the results could also be attributed to preferential sputtering. It is well known that when a multicomponent surface is bombarded

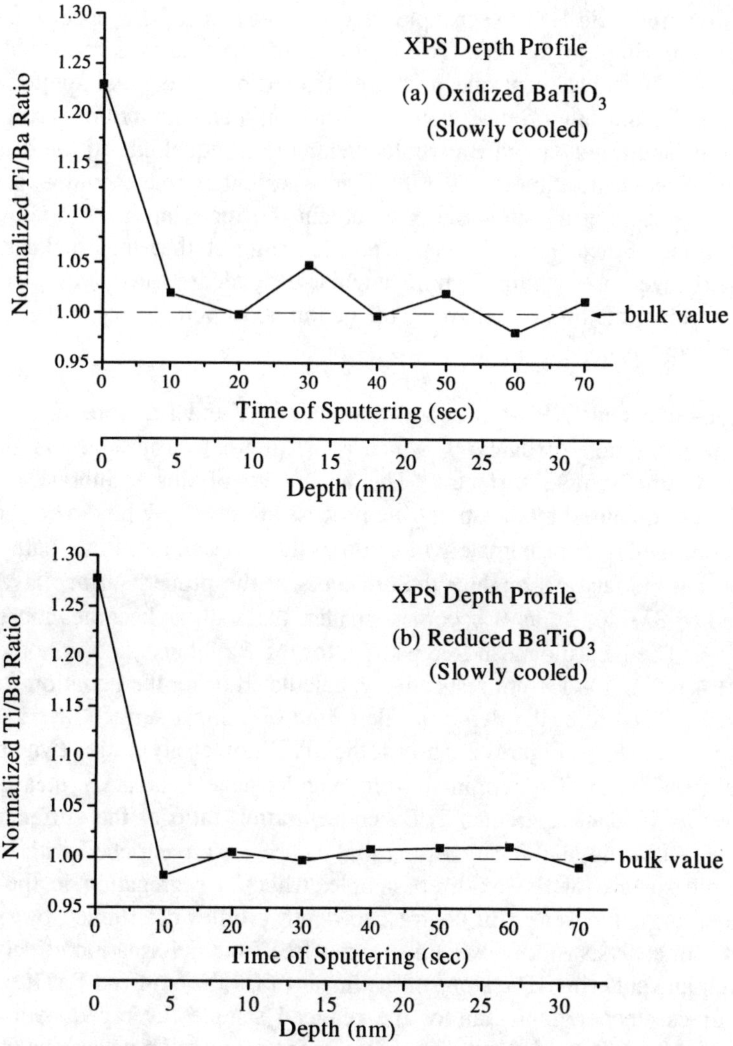

Fig. 1. The normalized Ti/Ba XPS peak area ratio as a function of sputtering time: (a) oxidized $BaTiO_3$, and (b) reduced $BaTiO_3$. The correlation between the sputtering time and the sputtered depth is only indicative, as it is calculated using the sputtering rate of a standard Ta_2O_5 sample.

with an ion beam, different elements are generally not sputtered in proportion to their concentrations on the surface [5]. If Ti is preferentially sputtered with respect to Ba (which is very likely since sputtering of lighter elements should be favoured over heavier ones [5]), the Ti/Ba ratio in XPS would be higher initially for a stoichiometric $BaTiO_3$ sample. Because the initial XPS spectrum is taken before sputtering, so the first Ti/Ba ratio is the true surface ratio and thereby higher than the subsequent ones that are affected by preferential sputtering. The preferential sputtering process would continue until an equilibrium is reached (the amount of sputtered Ti and Ba would eventually be equal due to the decrease of surface Ti concentration). Therefore, it is essential to determine if there is preferential sputtering in order to obtain correct information on surface segregation. Since the XPS depth profiles suggest that the thickness of the enriched layer is within 5 nm, angular-dependent XPS was used as a nondestructive method to determine the cation nonstoichiometry in the outermost surface layers [6].

In angle-dependent XPS the concentrations of Ti and Ba were measured as a function of the take-off angle θ (which is the photoelectron emission angle with respect to the sample surface). The advantage of this technique is that no complications caused by the sputtering process are involved; however, the probed depth is limited to approximately three times the inelastic mean free path (3λ) [7]. By varying the take-off angle θ the thickness of the probed surface layer can be reduced to $3\lambda\sin\theta$, *ie.* as θ becomes smaller, the analysis becomes more surface sensitive. The inelastic mean free paths λ for the Ba 3d and Ti 2p photoelectrons are ~ 1.4 nm and ~ 1.8 nm respectively, calculated using the equation for oxides in Ref. 8. Therefore, the depth profile of the outermost surface layers (~ 5 nm) can be obtained. Figure 2 shows the Ti/Ba intensity ratio (which is not normalized) vs. $\sin\theta$ determined from angular-dependent XPS measurements. The results indicate a greater Ti/Ba concentration ratio at the surface (smaller $\sin\theta$, more grazing angle), suggesting that Ti species are enriched with respect to Ba at the surface of the oxidized sample, while Ti segregation in the reduced specimen is rather weak. In contrast to depth profiles determined by sputtering the present analysis is not destructive and is free from deformation effects such as preferential sputtering. The slope of the linear fit for the oxidized $BaTiO_3$ is about three times steeper than that for the reduced sample (averaged over different runs), which indicates distinct differences in the segregation behaviour of the two specimens. Since the depth profiles in Figs. 1(a) and 1(b) are essentially identical for both oxidized and reduced $BaTiO_3$, this suggests that the initial decrease in the normalized Ti/Ba ratio in Fig. 1 mainly results from preferential sputtering of

Ti which, in consequence, leads to impoverishment of the surface in Ti. This is a good example to illustrate that although XPS is a suitable technique for quantitative depth profiling analysis, preferential sputtering must be considered when interpreting data.

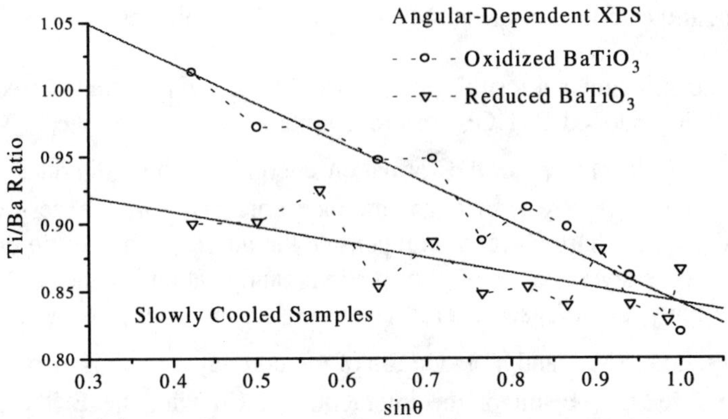

Fig. 2. The Ti/Ba XPS peak area ratio as a function of sinθ (θ: take-off angle) for both the oxidized and the reduced slowly cooled $BaTiO_3$ surfaces (dashed curves are the original data, solid lines are the linear fits).

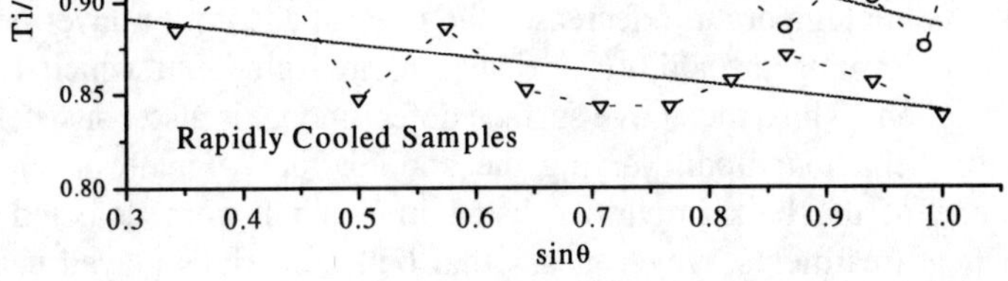

Fig. 3. The Ti/Ba XPS peak area ratio as a function of sinθ (θ: take-off angle) for both the oxidized and the reduced rapidly cooled $BaTiO_3$ surfaces (dashed curves are the original data, solid lines are the linear fits).

Figure 3 presents the angular-dependent XPS results for the oxidized and reduced $BaTiO_3$ samples, which were cooled rapidly to room temperature before the measurements. The results are essentially the same as those obtained from the slowly cooled samples (Fig. 2). The only difference is in the slopes of the linear fits, which are less steep than those for the slowly cooled $BaTiO_3$ samples (the scales in Figs. 2 and 3 are identical for easy comparison). This indicates more Ti enrichment at the surface of the slowly cooled $BaTiO_3$ samples.

Based on the experimental results we propose the following surface defect model. The defects in undoped $BaTiO_3$ are predominantly vacancies: V_{Ti}'''', V_{Ba}'' and $V_O^{\bullet\bullet}$ [3]. The differences in the formation energies of the individual vacancies (G_{Ti}, G_{Ba} and G_O) give rise to an interface core layer with excess electrical charge and a space charge layer to compensate the interface charge. In our model the negative core layer is composed of predominantly cation vacancies (V_{Ti}'''' and V_{Ba}'') and the positive space charge layer consists of mainly oxygen vacancies and electron holes ($V_O^{\bullet\bullet}$and $h^{\bullet}$). On top of the core layer there is also an oxygen adsorption layer as a result of the interaction of O_2 with the $BaTiO_3$ surface [$O_{2(ads)}$, O_2^-, O_2^{2-}, O^- and O^{2-}]. This adsorption layer can be neglected at 1000°C since most of the adsorbed oxygen species are expected to be incorporated into the lattice at this temperature. However, at lower temperatures the role played by the oxygen adsorption layer should also be considered.

It is known that more cation vacancies are created in the core layer when the partial pressure of oxygen in the gas phase is increased. The fact that the concentration ratio of Ti/Ba is greater in an oxidizing environment than in a reducing one, *ie.*, more Ba vacancies than Ti ones, suggests that the free energy formation for V_{Ti}'''' (G_{Ti}) at the interface is greater than that for V_{Ba}'' (G_{Ba}). When the $BaTiO_3$ samples are cooled slowly in the furnace, the surfaces are allowed to interact with O_2 at temperatures between 1000°C and room temperature. As the temperature decreases the oxygen adsorption layer becomes more important, forming an additional electrical field gradient, which leads to more Ti segregation. Thus, the above surface defect model is also consistent with the segregation behaviour upon varying the cooling rate. Finally it should be noted that none of the bulk impurities listed in Table I were detected at the surface after heat treatments, which means that bulk impurities played negligible roles in thermally induced surface segregation.

CONCLUSIONS

A Ti-enriched surface layer was found in the undoped $BaTiO_3$ samples that were heated at 1000°C under atmospheric pressure. This segregation behaviour is promoted by the increase of $p(O_2)$ and the decrease of cooling rate. It has been shown that caution must be exercised when interpreting depth profile data obtained by sputtering. Angular-dependent XPS, on the other hand, is proven to be a very useful nondestructive technique to study surface segregation.

ACKNOWLEDGMENTS

The authors would like to thank M. LaRobina for his help with sample preparations, and the CSIRO Centre for Advanced Analytical Chemistry for the ICPMS results.

REFERENCES

1. P. Wynblatt and R.C. McCune, "Surface Segregation in Metal Oxides" in *Surface and Near-Surface Chemistry of Oxide Materials,* edited by J. Nowotny and L C. Dufour (Elsevier, Amsterdam, 1988), pp. 247-279.
2. J. Nowotny, "Interface Defect Chemistry and its Impact on Properties of Oxide Ceramic Materials" in *The Science of Ceramic Interfaces*, edited by J. Nowotny (Elsevier, Amsterdam, 1991), pp.79-204.
3. J. Nowotny and M. Rekas, "Defect Structure and Electrical Properties of Barium Titanate" in Key Engineering Materials **66** & **67** (1992), pp. 1-44.
4. M. Kuwabara and T. Ide, "CO Gas Sensitivity in Porous Semiconducting Barium Titanate Ceramics", Am. Ceram. Soc. Bull. **66** [9] (1987) 1401-1405.
5. P. Sigmund, "Sputtering by Ion Bombardment: Theoretical Concepts" in *Sputtering by Particle Bombardment I,* Topics in Applied Physics, Vol. 47, edited by R. Behrisch (Springer-Verlag, Berlin, 1981), pp. 9-71; G. Betz and G.K. Wehner, "Sputtering of Multicomponent Materials" in *Sputtering by Particle Bombardment II,* Topics in Applied Physics, Vol. 52, edited by R. Behrisch (Springer-Verlag, Berlin, 1983), pp. 11-90.
6. C.S. Fadley, "Solid State - and Surface - Analysis by Means of Angular-Dependent X-ray Photoelectron Spectroscopy", Prog. Solid State Chem. **11** (1976) 265-343.

7. S. Hoffmann, "Depth Profiling in AES and XPS" Ch. 4 in *Practical Surface Analysis*, Vol. 1, 2nd edn., edited by D. Briggs and M.P. Seah (Wiley, New York, 1990), pp. 143-199.
8. M.P. Seah, "Data Compilations: Their Use to Improve Measurement Certainty in Surface Analysis by AES and XPS", Surf. Interface Anal. **9** (1986) 85-98.

THE CONTROL OF NANO-STRUCTURES FOR SOFT MAGNETIC MATERIALS.

Osamu Inoue, Yasuhiro Sugaya, Nobuya Matsutani* and Koichi Kugimiya
Central Res. Lab., Matsushita Electric Ind. Co., Ltd.
*Corporate Product Development Center, Matsushita Electronic Components Co.,Ltd., 3-4, Hikaridai, Seikacho, Kyoto 619-02, Japan

ABSTRACT

We have studied the structure and loss mechanism of our newly developed metal/insulator NANOCON magnetic material. This material consists of soft magnetic metal grains covered with thin insulating oxide films. The magnetic loss of this material is determined by the eddy current of inner grains, because the eddy current is completely insulated by the oxide films. We have also studied the loss character of poly-crystalline MnZn-ferrites, which have a similar structure as the NANOCON material. It is shown that MnZn-ferrites can be explained by above concept of the NANOCON material. The control of the nano-structure is important to realize excellent magnetic properties at high frequency.

1. INTRODUCTION

Recently, many studies have been conducted on nano-composite materials. These materials have been studied mainly to improve mechanical properties. However, there have been few studies about controlling the nano-structure for electric or magnetic properties in the field of bulk materials.

We have designed a metal/insulator "NANOCON" (Nano-structure Controlled) magnetic material[1-6]. The NANOCON material in this paper consists of Fe-based soft magnetic metal grains, which are completely separated from each other by thin insulating films. This material shows a high electric resistivity because of its structure. Therefore, the NANOCON material is expected to have excellent magnetic properties at high frequency, like thin magnetic films. Furthermore, the concept of the NANOCON material is extended to understand the loss mechanism of magnetic properties of MnZn-ferrites, since poly-crystalline MnZn-ferrites have a similar structure as the NANOCON material[7,8].

In this study, first we report the structure of the NANOCON material, and the magnetic loss at high frequency. Second, we report the effects of nano-structure control on the magnetic loss of low loss MnZn-ferrites.

2. EXPERIMENTAL

The fabrication process of the metal/insulator NANOCON material is shown in Fig.1. The starting material was Fe-based soft magnetic metal powder synthesized by N_2 gas atomizing method. As the first step, this powder was oxidized to form thin insulating films on the surface by heat treatments. The thickness of oxide layers, 5 ~ 50 nm, were controlled by the oxidation conditions. The oxidized powders were then molded and hot-press-sintered at suitable conditions to densify the samples to over 99 % of their theoretical densities.

Poly-crystalline MnZn-ferrites were prepared from $MnCO_3$, ZnO and Fe_2O_3 by using the newly developed ceramic technique, the nano-powder process, to obtain fine and homogeneous powder without impurities[9]. A representative chemical composition was $Mn_{0.74}Zn_{0.2}Fe_{2.12}O_x$. CaO and SiO_2 were used as basic additives.

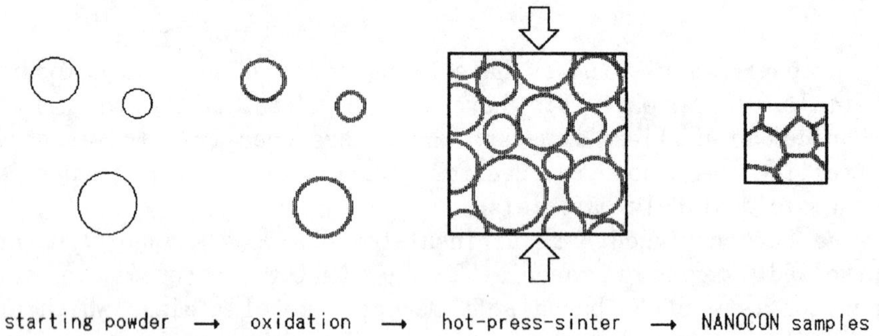

Fig.1 Fabrication process of the metal/insulator NANOCON material.

Electric resistivity of the sintered body was measured by using In-Ga electrodes. The magnetic loss and permeability were measured by a B-H loop analyzer at 1MHz and 50mT(ΔB=100mT).

3. RESULTS AND DISCUSSION

Metal/Insulator NANOCON material

Figure 2 shows a lapped and etched surface of the sintered body. No pore was observed on the lapped surface because of the high relative density of the sintered body. Each grain was separated by thin oxide films, and had a polygonal shape. The spherical particle deformed into a polygonal shape during densification.

Figure 3 shows the relationship between electric resistivity of the sintered body and the thickness of the insulating film. The solid line shows a calculated resistivity based on a simple model. The value of the measured resistivity was in a good agreement with the calculated one at the range of over 30 nm thickness. The measured resistivity suddenly decreased at a thickness of less than 30 nm, but it was still higher than the metal. The resistivity then decreased slowly to the resistivity of the metal particle itself.

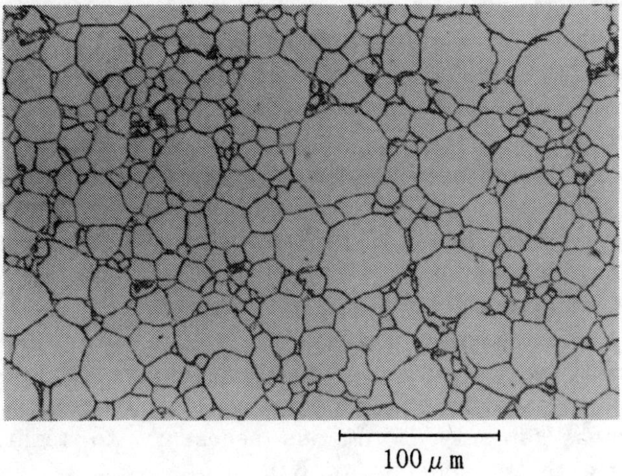

Fig. 2 Lapped and etched surface of the NANOCON material.

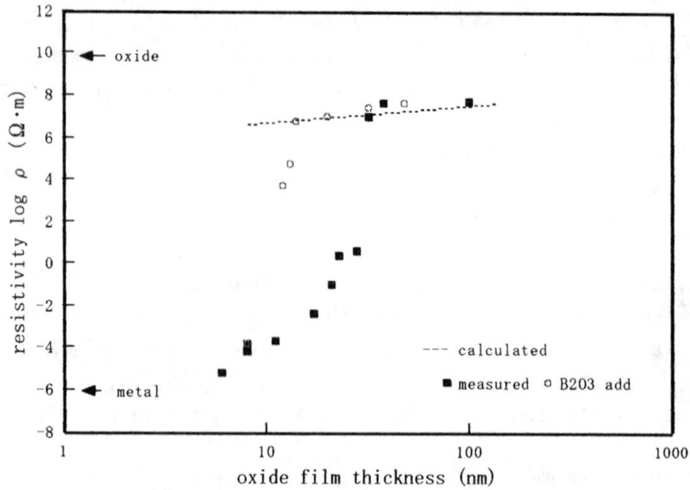

Fig. 3 Relationship between resistivity and oxide film thickness.

The decrease was assumed to be due to the following process. Before the hot-press-sintering, each spherical particle was perfectly coated and insulated by the thin oxide film. As shown in Fig. 1, the spherical particle deformed to a polygonal shape during densification. The surface area became larger than that of the original spherical particle, cracking the insulating oxide film to yield direct contacts of the metal particles. The contacts could be easily made when the film thickness was thinner. The contact naturally caused the breakdown of electric resistivity.

We calculated the resistivity of NANOCON material, which contained small pinholes on the insulating films. Existence of even small pinholes drastically decreased the electric resistivity. For example, a pinhole of 10 nm diameter on each grain of 20 μm in diameter resulted in a resistivity of 1 Ω·m. The calculated direct contact area of metal grains was negligible small (<0.001%). Since the contact area was small, the resistivity breakdown could be prevented by the control of the fabrication process. For example, the effect of B_2O_3 addition is also shown in Fig. 3. The film thickness of only 14 nm was necessary to maintain a high resistivity with the addition of B_2O_3[3].

Figure 4 shows the relationship between grain size and eddy current loss (Pe) of the NANOCON material. Theoretically, the magnetic loss PL consists of the hysteresis loss (Ph) and the eddy current loss (Pe), and is represented by the following equation[10].

$$PL = Ph + Pe = Kh \cdot B^3 \cdot f + \eta \cdot \pi^2 \cdot d^2 \cdot B^2 \cdot f^2 / C \cdot \rho \quad --- (1)$$

where Kh is constant, B magnetic flux density, η anomaly factor, d sample size, C sample shape constant, ρ electric resistivity. Pe becomes dominant at a high frequency because Pe is proportional to the square of frequency.

We assumed that Pe consisted of Pe_{macro} and Pe_{micro} for the NANOCON material because of its structure.

$$Pe_{macro} = \eta \cdot \pi^2 \cdot d_{macro}^2 \cdot B^2 \cdot f^2 / C \cdot \rho_{macro} \quad --- (2)$$

$$Pe_{micro} = \eta \cdot \pi^2 \cdot d_{micro}^2 \cdot B^2 \cdot f^2 / C \cdot \rho_{micro} \quad --- (3)$$

where d_{macro} is the sample size, ρ_{macro} apparent electric resistivity, d_{micro} grain size and ρ_{micro} resistivity of the metal grain itself.

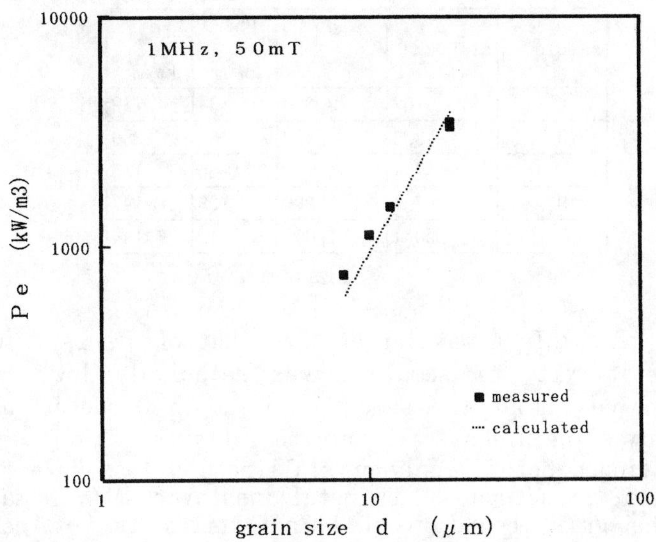

Fig. 4 Relationship between eddy current loss (Pe) and grain size.

However, since the NANOCON material has high resistivity and the eddy current was limited by the oxide film, only the inner grain eddy current loss Pe_{micro} could be effective[6].

The calculated Pe_{micro} by using equation (3) is shown by the dotted line in Fig. 4, asumming $\eta/C = 0.37$. An excellent agreement of measured and calculated Pe was observed. Lower loss samples could thus be expected by using smaller powders.

MnZn-ferrites

The electric resistivity of poly-crystalline MnZn-ferrites is on the order of 10^{-3} Ω·m if there is no additives. It is higher than that of metallic magnetic materials, but not sufficiently high for high frequency use. Therefore, several oxides, such as CaO and SiO_2, are added to poly-crystalline ferrites. These additives segregate at the grain boundaries, and increase the electric resistivity of MnZn-ferrites[7,8]. Table 1 shows several properties of poly-crystalline MnZn-ferrites synthesized by nano-powder process using different amount of additives.

Table 1. Several properties of MnZn-ferrites.

sample	ρ at100kHz (Ωm)	μ at 1MHz	Ph* kW/m³	Pe* kW/m³
A	0.07	3,020	34	1,719
B	3.4	1,403	41	280
C	7.1	1,170	66	135
D	7.7	1,080	108	147
E	20.1	1,018	210	144

*1MHz, 50mT, 60℃

The Pe of sample A was larger than that of others, since the electric resistivity of sample A was relatively low. However, sample E showed almost the same Pe as sample C and D, although sample E showed the highest electric resistivity.

As mentioned before, poly-crystalline low loss MnZn-ferrites have a similar structure as the metal/insulater NANOCON material. However, the grain boundary of MnZn-ferrites is not perfectly shielded by the additives. This is why ρ in Table 1 is not high enough compared with the NANOCON material. Figure 5 shows the distribution map of Ca for the fracture surface of sample B by

SIMS. The samples were easily broken at the grain boundaries. The analysis was performed with prove size of 3 μm at 256 points (48 $\times$ 48 μm area). This figure shows that CaO was imhomogeneosly distributed at the grain boundaries. It seemed that the segregated area of CaO was larger than several tens of μm in diameter. A similar result was also obtained for Si distribution. From this figure, we concluded that the electrically isolated area of poly-crystalline MnZn-ferrites was larger than the real grain size, on the order of 100 μm[8].

Now we can estimate the apparent electric resistivity of MnZn-ferrite to ignore Pe_{macro} similar to the NANOCON material. If the same η/C was assumed, equation (4) is obtained from equation (2) and (3).

$$Pe_{micro} \gg Pe_{macro}$$

$$\rho_{macro} \gg \rho_{micro} \cdot (d_{macro}/d_{micro})^2 \qquad \text{--- (4)}$$

when $\rho_{micro} = 10^{-3}\Omega\cdot m$(resistivity without any additives), $d_{macro} = 10^{-3}$ m(sample size), $d_{micro} = 10^{-4}$ m (size of electrically isolated area) was applied to equation (4), $\rho_{macro} \gg 0.1\ \Omega m$ was necessary to ignore Pe_{macro}. We thus concluded that ρ_{macro} of sample B, C, D and E were sufficiently high to ignore Pe_{macro}, and that Pe was solely dependent on Pe_{micro}, as shown in Table 1.

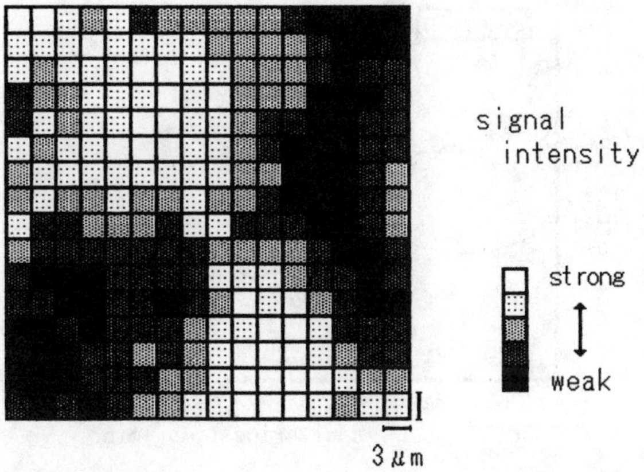

Fig. 5 Distribution map of Ca for fracture surface of sample B.

Samples C and D contained the same additives and showed almost the same electric resistivity. However, these samples showed different permeability and Ph value. Figure 6 shows SIMS depth profiles of fracture surfaces of samples C and D. Since the sputtering rate was about 10 nm/min, it was assumed that additives were segregated at grain boundaries of a few nm thick. The segregation thickness of CaO and SiO_2 were almost the same for two samples, but that of the other additive was different. Therefore, it was considered that the dissolved additive into spinel phase depressed the permeability, and increased the Ph.

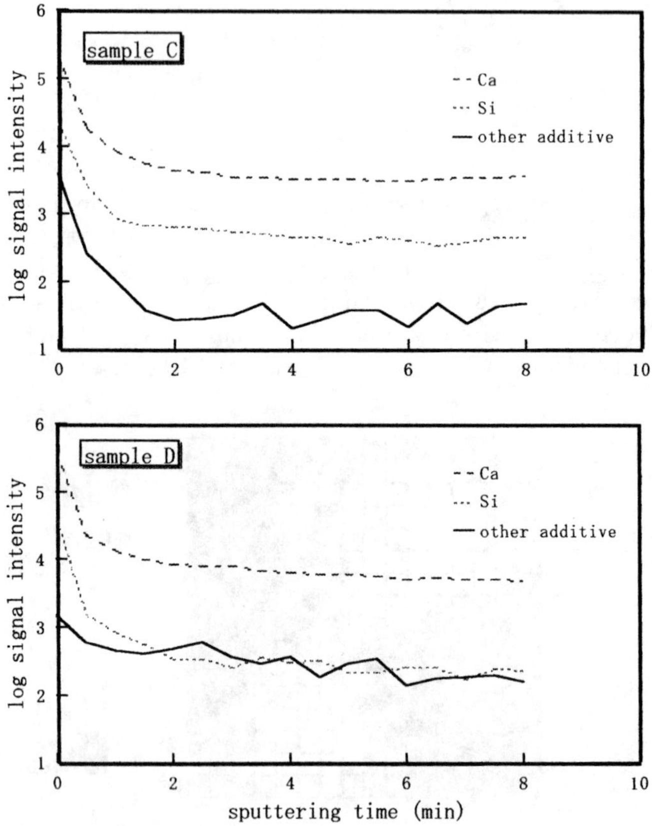

Fig. 6 SIMS depth profiles of samples C and D.

Results indicated that control of the nano-structure was the most important factor to obtain further low loss MnZn-ferrites. A small amount of additives should be homogeneously segregated at grain boundaries of a few nm thick, in order to limit the eddy current and minimize the Ph.

4. CONCLUSIONS

We have studied the structure and loss mechanism of the metal/insulator NANOCON magnetic material. The magnetic loss PL of the NANOCON material was determined by the inner grain eddy current loss Pe_{micro}, because the eddy current is completely limited by the insulating films. Therefore, the size of the eddy current path was more important than the apparent electric resistivity in order to decrease Pe. Our studies also showed that the PL of MnZn-ferrites was similar to the NANOCON material, and that the PL could be efficiently improved by controlling the nano-structure of the grain boundaries.

References

1. K. Kugimiya, K. Hirota, M. Satomi and O. Inoue, "Metal/Dielectric NANOCON Materials," *Journal of the Japan Soc. of Powder and Powder Metallurgy.* vol. 37, p. 333 (1990).
2. O. Inoue, Y. Sugaya, K. Hirota and K. Kugimiya, "Metal/Insulator NANOCON Materials," *1990 American Ceramic Society, Talk Summaries, Electronics Division,* 107-E-90 (1990)
3. Y. Sugaya, O. Inoue, K. Hirota and K. Kugimiya, "Metal/Dielectric NANOCON Materials (Part 2)," *Journal of the Ceramic Society of Japan, International Edition* vol. 96 p. 488 (1991)
4. O. Inoue, Y. Sugaya, K. Hirota and K. Kugimiya, "Metal/Dielectric NANOCON Materials (Part 3)," *Journal of the Japan Soc. of Powder and Powder Metallurgy.* vol. 38, p. 322 (1991).
5. O. Inoue, Y. Sugaya and K. Kugimiya, "Synthesis and Properties of Metal/Insulator "NANOCON" Materials," *Proceedings of the Special Symposium on Advanced Materials-III Nagoya,* p. 150 (1991)
6. Y. Sugaya, O. Inoue and K. Kugimiya, "The Core Loss of Nano-Structure-Controlled Magnetic Materials," *Ferrites, Proceedings of The 6th International Conference on Ferrites (ICF6),* Tokyo and Kyoto Japan, p. 952 (1992)

7. Y. Bando and T. Akashi, "Grain Boundary Structure of Mn-Zn Ferrite," *Fine Ceramics,* Elsevier, p. 122 (1985).
8. O. Inoue, N. Matsutani and K. Kugimiya, "Eddy Current Loss of MnZn-Ferrites," *Ferrites, Proceedings of The 6th International Conference on Ferrites (ICF6),* Tokyo and Kyoto Japan, p. 1155 (1992)
9. M. Nishida, H. Ando and K. Kugimiya, "Preparation of Ultra-Fine Particles by Nanopowder Process," *KONA Powder and Particle.* vol. 9, p. 72 (1991).
10. T. G. W. Stijntjes and J. J. Roelofsma, "Low-Loss Power Ferrites for Frequencies up to 500 kHz," *Advances in Ceramics,* vol. 16, Columbus, p. 493 (1984)

PROCESSING OF COPPER THICK FILM MATERIALS FOR HYBRID CIRCUITS

Shigeki Hayashi, Takahiko Kijima, Hideyoshi Asa, Taishi Hashimoto
and Junichi Date
Research and Development Division, Osaka Sanso Kogyo Ltd.
Sakai-city, Osaka, Japan

ABSTRACT

Recently much attention has been paid to substrates made of copper thick film, which consist of a copper conductor, a dielectric, and a resistor. However, there is not much demand for these copper substrates in spite of the excellent characteristics of copper because the TCR values of the resistors, to be fired in N_2, are insufficient. There are several methods to improve such TCR values. One is to use an air-firable resistor. However in this case the firing process is complicated because different atmospheres are required for all three components. Therefore, we have tried to dope the burn out zone with H_2O and CO_2.

In addition, a profile has been developed in which the burn out zone includes a shoulder-shape and the sintering time and cooling rate are controlled. As a result, TCR's are obtained that fall within 100 ppm/°C, which results in these substrates having great practical value.

INTRODUCTION

As hybrid integrated circuit technology (HIC) contrives to rapidly advance, it is essential to integrate densification of HICs in line with the compactness and high speed of electronic products. In the automobile industry, there has been an increasing demand for ceramic substrates which can be used at high temperatures, in place of plastic substrates now prevailing. In this field, however, Ag/Pd as a conductor is being used overwhelmingly. This is because it is rather difficult to burn out Cu, which is said to be the most suitable material for electronic circuits. In addition to its low price, Cu has a smaller electrical resistance than Ag/Pd and is highly resistant to high frequencies because Cu dose not cause the migration.

We believe that if the efficiency of the resistor is improved by optimizing thefiring conductors, in stead of improving the materials, then the market of

Cu-based ceramic HIC will be expanded considerably.
In this study, we developed the optimum conditions for DuPont's 10 Kohm resistors compatible with Cu conductor. It is also discussed whether these conditions are suitable for other resistors (10 ohm to 1 Mohm).

EXPERIMENTAL METHOD

Experimental sample

The pastes used in the experiment, for use in Cu thick films for firing at 900°C, are as follows:

(1) Conductor pastes : QP153
(2) Resistor pastes : QP601 , QP602 , QP603
QP604 , QP614 , QP615
(3) Dielectric paste : QP445

These pastes are produced by DuPont. The substrates are 96% alumina(1 inch square, 0.635 mm thick) made by KYOCERA.

Processing equipment and evaluation equipment

The processing equipment and evaluation equipment are shown in Table I.
Figure 1 shows the schematic of the firing furnace which is a belt furnace.

Evaluation method

The procedures for measuring the variation in resistance as a function of temperature are shown below.
The test sample was prepared as follows:

(1) The conductor pattern was screen printed onto the substrate.
(2) After printing, the sample was held at room temperature for 10 minutes.
(3) The sample was fired in the belt furnace at the prescribed profile.
(4) Set the printer again with the substrate made at the procedures 1. to 3. above, and printed it.
(5) After printing, the sample was allowed to stand for 10 minutes at room temperature, and then dried at 120 °C for 10 minutes.
(6) After drying made sure that it has been cooled to room temperature. Then measured the thickness.
(7) Burned it in the belt furnace with the prescribed profile set.

Measuring a resistance value and calculating are:

(1) Placed the test sample in the thermostatic chamber, and set to the desired temperature.

(2) Used 4-terminal method to measure resistance value.
(3) Measured the TCR using the thermostatic chamber in the following order:

1. Leaving it for 30 minutes at 25 °C
2. Measuring the initial resistance value for 60 minutes
3. Leaving it for 30 minutes at 125 °C
4. Measuring the resistance value at high temperature for 60 minutes
5. Leaving it for 30 minutes at -55 °C
6. Measuring the resistance value at low temperature for 60 minutes. The temperature was controlled through a temperature controller by using a heater for raising temperature and by using liquid nitrogen source for cooling.

(4) After measuring the resistance at a low temperature, raised the temperature to dry the thermostatic chamber, and then took out the sample.
(5) Calculated the resistance value and thermal coefficients of resistance(TCRs).

$$\text{H.TCR} = \frac{(R(125\ ^\circ C) - R(25\ ^\circ C))}{R(25\ ^\circ C)} \times 10000 \quad \text{ppm} / {}^\circ C$$

$$\text{C.TCR} = \frac{(R(-55\ ^\circ C) - R(25\ ^\circ C))}{R(25\ ^\circ C)} \times 12500 \quad \text{ppm} / {}^\circ C$$

$$\text{SRV} = \frac{R(25\ ^\circ C) \times (\text{mean dried film thickness})}{25\ \mu m}$$

H.TCR is thermal coefficient of resistance (25°C to 125°C)
C.TCR is thermal coefficient of resistance (-55°C to 25°C)
SRV is the specific resistance value, that is the resistance value per unit area. Conversion is based upon dried film thickness, specified by each product. The above examples are the case of 25 μm.

Table I Processing and evaluation equipment

Equipment	Specification
Printer	LS-15GX (New Long Seimitsu)
Screen	325 mesh made with SUS
Drier	DRYOVEN DS-43 (YAMATO)
Belt furnace	47-MT-4841-20AMC-36 (Koyo Lindberg Ltd.)

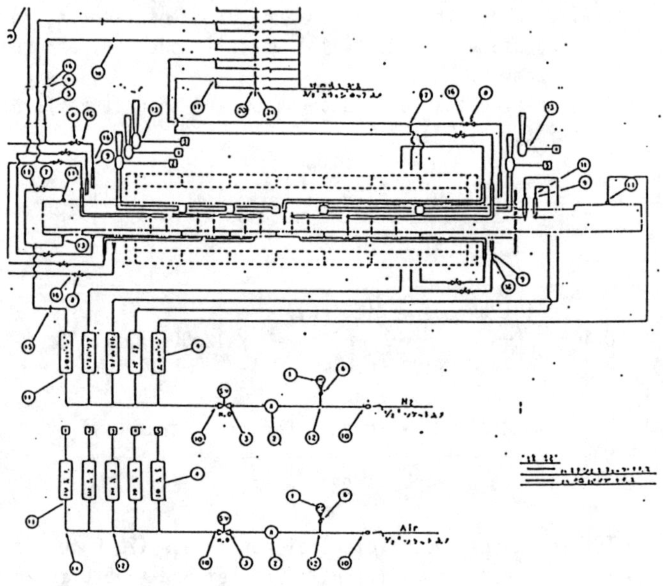

Fig.1 Schematic of Furnace Muffle

RESULTS AND DISCUSSION

Thermal analysis of the pastes

In order to analyze the paste characteristics, we measured, with a TGA, the weight decreasing rate by burnout and the temperature at which weight started to decrease, or starting temperature of decomposition. The measurement was done in nitrogen gas atmosphere with and without 100 ppm oxygen added. First, the paste had been dried at 120 °C in nitrogen atmosphere for 12 hours. The measuring conditions, which were chosen to match set to meet the belt furnace conditions as closely as possible, are as follows:

(1) Temperature raising rate : 50°C/minute
(2) Heating temperature : up to 600°C
(3) Gas atmosphere : N_2 , 100 ppm O_2 / N_2
(4) Sample : QP601 , QP603 , QP604

The results are shown in Figure 2 and Figure 3. The weight decreasing rate proved to be as follows regardless of the atmosphere used.

QP604 > QP603 > QP601

On the other hand, the temperatures at which the weight loss occur are:

(1) First decrease : 100°C
(2) Second decrease : 450°C

These temperatures are the same irrespective of the paste and atmosphere used. According to these results, the first weight loss may be due to the evaporation of organic solvents. The second weight loss may be due to the release of binders. The difference in weight loss depends on the content of solid composition in the paste. The solid composition which differ for each type of pastes mean resistance component(LaB_6) and the glass component in the pastes. Binder releasing temperature is always the same for each paste. It is presumed, therefore, that the chemical composition of the binder is the same temperature at which binder starts to release in O_2-additive N_2, we think that this binder is the decomposable binder. Therefore the binder releasing participate with factor of temperature.

Effect of carbon dioxide

With an increase in CO_2 contents, the resistance value decreased and the TCR is further degraded. However, the TCR value in the case of N_2 with 2% CO_2 added was improved by approx. 20% more than that of H.TCR obtained by burning in only N_2, while C.TCR value is almost unchanged. As a result, we believe that CO_2 at a certain degree of concentration has no negative effect on the resistor QP604. We also confirm that if CO_2 enters the sintering zone, the TCR value becomes only slightly worse). Figure 4 shows these results.

Effect of oxygen

DuPont Japan claims that an atmosphere containing 3-7 ppm O_2 is most suitable for the firing atmosphere in the belt furnace. We attempted to confirm the effect of higher contents of O_2 on TCR value, but were unable to detect any TCR changes because of the instability of the furnace. In other words, the O_2

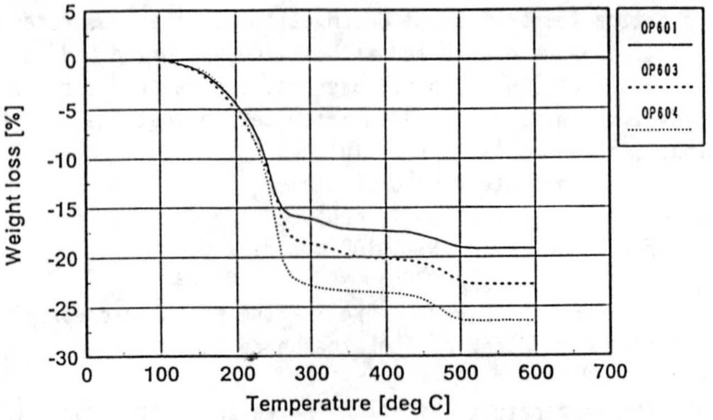

Figure 2 Weight Loss of DuPont QP60 Series (Only N_2)

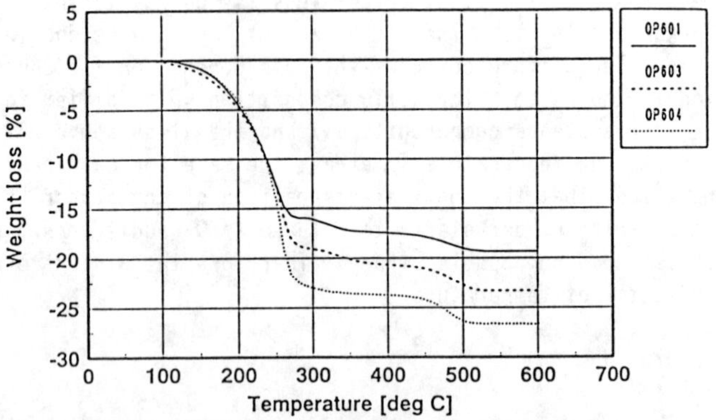

Figure 3 Weight Loss of DuPont QP60 Series (100 ppm O_2 $/N_2$)

concentration changes substantially as a function of temperature. Early in the firing operation, this tendency was pronounced. It was difficult to keep O_2 concentration constant because, at high temperatures, O_2 content measured with a Zr type oxygen analyzer tends to be less than that added to the furnace.

In other words, O_2 is consumed on the chamber of the furnace. When the binder evaporates, however, the change in O_2 content measured with a Zr type oxygen analyzer is clear as shown in the following example. When O_2 is kept at 10 ppm, O_2 concentrations in No.2 zone are monitored, we can clearly find that O_2 is consumed even if one piece of QP604 sample is burnt. (See Figure 5). On the other hand, when H_2 is added to No.3 zone, there is no effect on TCR. Whereas the resistance value showed an increases. Both the resistance and TCR values are not improved in the presence of oxygen and hydrogen.

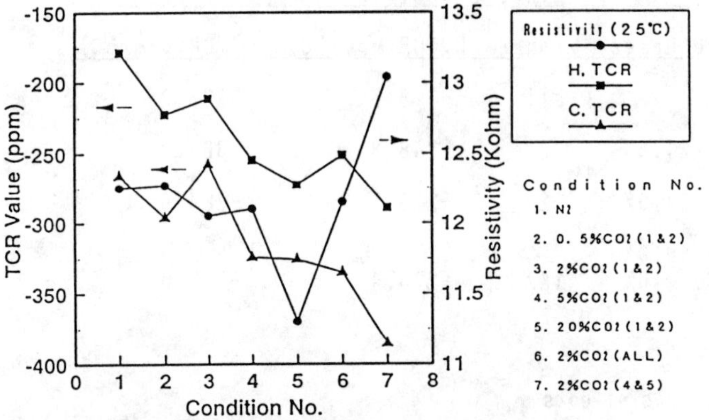

Figure 4 CO_2 effects in Resistivity and TCR

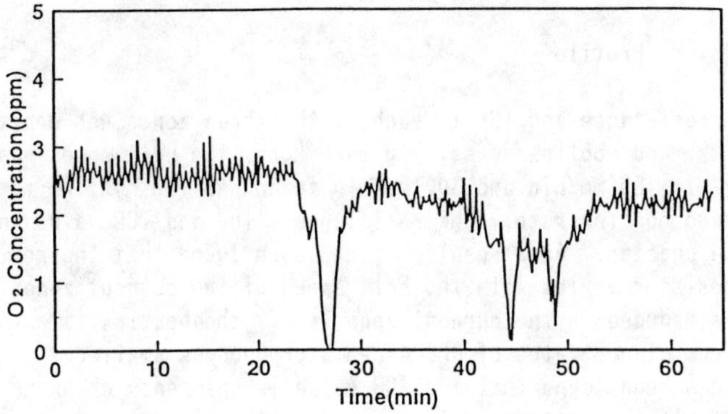

Figure 5 Oxygen Consumption during the Firing of Resistor

Effect of Moisture

As shown in Table II, the addition of H_2O to No.2 zone increase the resistance, and the TCR value becomes better when 13% of H_2O is added, while at other concentrations of H_2O it becomes worse. From these results it is obvious that H_2O has some effect, but this subject is yet to be examined. The presence of hydrogen and oxygen produces similar results.

Table II Moisture effect in H_2O

Dope Gases	Zone	H.TCR(ppm/°C)	C.TCR(ppm/°C)
N2	all	-178.6	-266.2
H2O 21.4%	2	-348.8	-361.2
H2O 13.0%	2	-225.5	-284.0
H2O 10.0% + CO2 10%	1&2	-270.4	-342.0

Consideration of gas effects

Judging from these results, the effect of gas atmospheres in improving resistance and TCR values is not enough. Therefore, we attain to determine the optimum temperature profile.

Effect of Temperature Profile

The variation in resistance and TCR in each of the three zones was monitored: burnout, sintering, and cooling zones. In each zone, the belt speed was changed to 30 mm/min, 85 mm/min and 300 mm/min to change the heating rate, sintering time, and cooling rate. The resistance value and TCR value were measured by these changes. As a result, it has been found that the resistance value of QP604 resistor varies with the belt speed of the burnout zone. If the belt speed is reduced in the burnout zone (i.e., the heating rate is decreased), the resistance value of QP604 resistor becomes smaller. Furthermore, it has been found that the TCR value is changedby changing belt s peed in any zone. If the belt speed is reduced in the burnout zone (i.e. shorter sintering time), and increased in the cooling zone (i.e. lower heating rate), increased in the sintering zone (i.e. higher cooling rate), then the TCR value is better. It has been confirmed that if the heating rate is decreased, sintering time is shortened and the cooling rate is increased, the QP604 resistor has a resistance of near 10 Kohm and a TCR of near 100 ppm/°C. Figure 6 shows these results. Thus, the initial objective has been accomplished for QP604 resistor paste. Subsequently, the resistance and TCR values of the other resistor pastes (QP601, 602, 603, 615 and 616) were measured by burning under the present optimized condition. As a result, as can be seen from Figure 7, the TCR values of QP601 and QP604, when fired under the optimized conditions proved to be far better than fired under those the conditions recommended by DuPont. Though the TCR value is worse for QP603, it still falls within 100 ppm/°C.

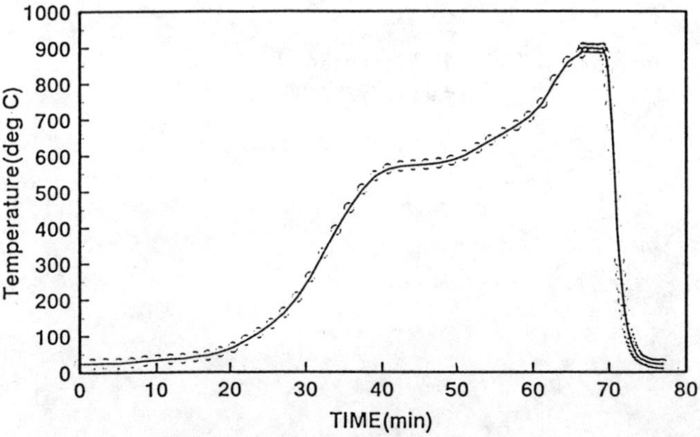

Figure 6 Temperature Profile

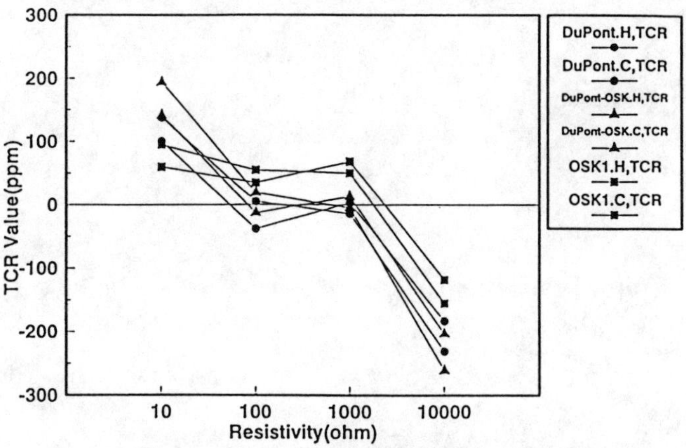

Figure 7 Improvement of TCR in Resisters

Conclusion

The optimum conditions for firing QP604 (10 Kohm) resistor were determined as follows.

(1) The TCR value is substantially improved if the heating rate for burnout is decreased, the sintering time is shortened and the cooling rate is increased.

(2) The CO_2 addition below 2 % did not affect the TCR value.

(3) The H_2O addition has an effect on TCR value; however, the details are yet to be clarified.

Literature Cited

[1] Satish S. Tamahankar, Mark. J. Kirschner
Patent disclosure (KOKAI) number 277282/90

SURFACE AND GRAIN BOUNDARY MODIFICATIONS OF $YBa_2Cu_3O_{7-\delta}$ CERAMICS BY PLASMA-ENHANCED FLUORINATION

C. Magro✧, J.M. Heintz✧, N. Hudakova+, J. Etourneau✧, C. Cardinaud*, G. Turban* and A. Tressaud✧

✧ : Laboratoire de Chimie du Solide du CNRS, Université Bordeaux I, 351 Cours de la Libération, 33405 TALENCE Cedex (FRANCE)

* : Institut des Matériaux de Nantes, 2 rue de la Houssinière, 44072 NANTES cedex 03 (FRANCE)

ABSTRACT

The radiofrequency plasma technique involving mixtures of $CF_4 + O_2$ gases has been applied to the treatment of high T_c superconducting oxides ($YBa_2Cu_3O_{7-\delta}$). The investigation of the various experimental parameters of the process has shown that the improvement of the critical current density J_c mainly depends on the inlet precursor composition $CF_4 + \tau\ \%\ O_2$, on the total pressure, and on the reaction time. The presence of fluorine in the bulk of the ceramics has been observed from electron microprobe analysis, together with an increase of the "Cu^{3+}" content. The plasma enhanced fluorination (PEF) treatment improves the superconducting properties of the materials: both values of the resistivity in the normal state and of the superconducting transition width are reduced and the critical transition temperature is improved of about 1 K. Mechanisms of interaction between the reactive species of the plasma and $YBa_2Cu_3O_{7-\delta}$ ceramics have been proposed through detailed angle resolved X-ray photoelectron spectroscopic analyses. At the surface of the outer grains, the plasma treatment removes $(OH)^-$ and $(CO_3)^{2-}$ species contained in the degradation layer and gives rise to a fluoride-rich layer. In the bulk of the material the occurrence of metal-fluorine bonds in the superconducting phase has to be assumed. Moreover, interactions between atomic fluorine and grain boundaries result in an improvement of intergranular magnetic behavior, according to a.c. susceptibility measurements. An increase of the oxidation state of copper has also been detected, confirming the oxidizing effect of the plasma treatment.

+ Permanent address : Institute of Experimental Physics, Slovak Academy of Sciences, Watsonova 47, 04353 KOSICE (SLOVAKIA)

INTRODUCTION

The discovery of high-temperature superconductors (HTSC), such as $YBa_2Cu_3O_{7-\delta}$, has led to unprecedented activities, mainly because of the possible use of these materials for significant technological advancements. Unfortunately the use of bulk HTSC has been hindered by their relatively low capacity for carrying high densities of electrical current, which is caused by a number of serious material problems such as grain boundary weak links. Several processes have been already proposed to increase the critical current density J_c. Melt-texturing of HTSC ceramics has been especially investigated using either a thermal gradient [1], a magnetic field [2], or a mechanical strain [3]. Other processes have been successfully developed to improve the intergranular properties, for instance Ag-doping or fluorination of $YBa_2Cu_3O_{7-\delta}$ ceramics. It has been previously shown that a fluorination treatment by F_2 gas contributed to remove impurities from grain boundaries [4], and to protect the material against moist air [5]; in addition the critical current density J_c was increased. Unfortunately this process affords little industrial adaptability due to drastic safety precautions.

On the other hand, plasma technologies are widely employed in etching [6] and polymerization [7] processes in microelectronics, especially using fluorocarbon precursors. Therefore an original method, i.e. the radiofrequency plasma-enhanced fluorination (PEF) process can be substituted for the traditional F_2-gas fluorination treatment. This paper is devoted to the investigation of the influence of the various experimental parameters on the superconducting properties of the treated materials and to the investigation of the interaction between the reactive species of a (CF_4 + O_2) plasma and the surface/grain boundaries of the $YBa_2Cu_3O_{7-\delta}$ ceramics through a.c. susceptibility and detailed angle resolved X-ray photoelectron spectroscopic analyses.

EXPERIMENTAL PROCEDURE

Description of Plasma Equipment

The experiments were carried out in a S.E. 80 Barrel-Plasma Technology System. The plasma, obtained from a $CF_4 + \tau\ \%\ O_2$ ($0 \leq \tau \leq 50$) mixture, was excited by a radiofrequency source at 13.56 MHz. The barrel chamber was fitted with a thermostatically controlled heating jacket and a temperature lower than 100°C was maintained during the PEF process. Several experimental parameters could be adjusted from this system and their influence on the superconductive properties are presented in section III.

Preparation of $YBa_2Cu_3O_{7-\delta}$ Ceramics

The starting material was a stoichiometric 1:2:3 powder from Rhône-Poulenc Co. containing few residual carbonated species; the mean grain size was around 2 μm. The powder was pressed to give pellets of 2x2x12 mm^3 in dimension. These were heated up to 920°C for two hours under oxygen flow and slowly cooled under oxygen to achieve the optimum oxidation and to hinder the formation of cracks. From scanning electronic microscopy observations, the $YBa_2Cu_3O_{7-\delta}$ ceramics consisted of homogeneous grains of around 3 μm in diameter and of an interconnected porosity with a compacity of about 0.88. An amorphous layer at the gas/solid interface was observed by transmission electronic microscopy which could be probably ascribed to the presence of $(CO_3)^{2-}$ species [8]. On the other hand, no intergranular phase was observed at the grain boundaries.

Techniques of Characterization

After PEF treatments, the ceramics were characterized by several complementary techniques: (i) Electron microprobe analysis (EMPA) performed on polished fractured surfaces of the sintered pellets, in order to determine the distribution of fluorine in the bulk of the ceramics. (ii) Chemical analysis of copper oxidation state ("Cu^{3+}" content) of the bulk pellet by Mohr-salt titration [9]. (iii) A.c. susceptibility measurement to evaluate the critical current density, $J_{c(a.c.)}$. The temperature dependence of $J_{c(a.c.)}$ was plotted from this measurement, the $J_{c(a.c.)}$ values being deduced from the Bean model [10]. The applied magnetic field amplitudes ranged from 0.1 to 10 Oe and the used frequency was 333 Hz. The temperature dependence of the a.c. susceptibility of a superconducting ceramic generally showed two steps in both real part and imaginary part. The first one, here for $T < 89$ K, corresponds to the intergranular contribution and the other one, for $89 < T < 91$ K, to the intragranular contribution. However in the imaginary part no peak was observed in the higher temperature range, because of a too small size of grains. Therefore, these curves brought us mainly information on the intergranular contribution [11]. (iv) Angle resolved X-ray photoelectron spectroscopy (XPS), using a Leybold AG LH I2 type spectrometer equipped with a monochromator. Two different types of information could be drawn from these experiments. The first one concerned the interaction of the reactive species (F_2 or $F^\bullet$) with the grains at the surface of the pellets. In this case the data were obtained from freshly prepared pellets. The other one dealt with the interaction between the reactive species and the surface of the inner grains of the pellets, the information being obtained from the center of polished fractures. In every case a surface of several mm^2 was analyzed, and the involved depth, d, was $\approx$ 80 Å. During the experiment copper was partly reduced to monovalent species ; this X-ray photoreduction phenomenon has already been observed in Cu-based superconductors [12]. This behavior prevents further interpretation on the oxidation state of copper. Therefore, the XPS study was carried out on the Y 3d, Ba $3d_{5/2}$, Ba 4d, O 1s and F 1s levels respectively.

INFLUENCE OF PEF EXPERIMENTAL PARAMETERS ON THE SUPERCONDUCTING PROPERTIES OF $YBa_2Cu_3O_{7-\delta}$ CERAMICS

Loading Effect

The loading effect is related to the number and the area of samples which are simultaneously treated during one experiment [13]. This loading effect can be explained simply: as the number of simultaneously treated samples increases, the partial pressure of active species (atomic fluorine) per surface unit decreases and so decreases the fluorine content of the ceramics which was evaluated by EMPA. The $J_{c(a.c.)}$ values also decrease when the number of samples which are treated in a same batch increases. Therefore in the following, the plasma treatment will be realized for short duration time (t = 30 min) on only one $YBa_2Cu_3O_{7-\delta}$ pellet with a constant area (1 cm^2) corresponding to a constant loading effect.

Influence of Inlet Precursor Composition

We have been primarily interested in the influence of addition of oxygen to the CF_4 precursor, since these mixtures have been proved to be most efficient in plasma etching processes. PLUMB and RYAN [14] have developed a kinetic model which describes the main chemical reactions occurring in a CF_4 plasma containing different percentages of O_2. According to these authors, when adding O_2 to a CF_4 plasma, additional radicalar reactions occur, which yield the formation of atomic fluorine ($F^\bullet$).

The fluorine content of our treated ceramics reaches a maximum value for an addition of 25 % O_2 to the CF_4 plasma, which corresponds to the maximum amount of atomic fluorine present in the plasma [15,16]. This result can be easily understood considering that the flux of atomic fluorine is proportional to the atomic fluorine density (Knudsen law). On the other hand, although the "Cu^{3+}" content of a sample is increased when treated in a pure CF_4 plasma, the corresponding amount is not as large as that obtained after addition of oxygen. Concerning the influence of the addition of O_2 to a CF_4 plasma on the critical current density, we have observed that the most efficient result also corresponds to a CF_4 + 25 % O_2 mixture (Fig. 1). In the absence of O_2 (i.e. in pure CF_4 plasma), or for higher O_2 rates (CF_4 + 50 % O_2 plasma), the optimal efficiency of the treatment is not reached, which would imply that atomic fluorine is the active species during the PEF process and may be responsible for the oxidation of the $YBa_2Cu_3O_{7-\delta}$ ceramics.

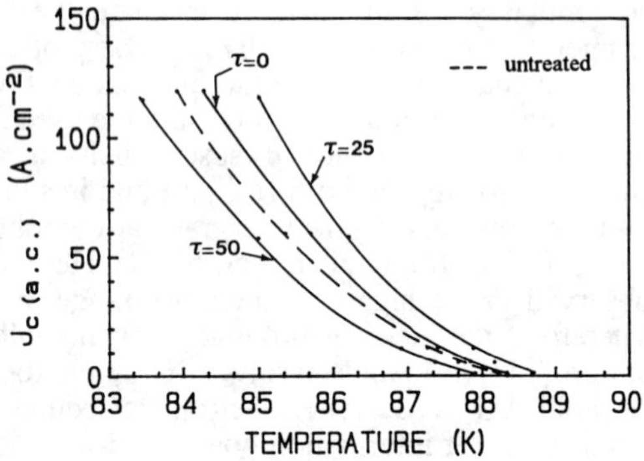

Figure 1 : Influence of addition of O_2 to CF_4 precursor on the temperature dependence of $J_{c(a.c)}$ with Q_{CF_4} =15 $cm^3.s^{-1}$, P=100 W, p=70 mTorr, t=30 min.

Influence of Inlet Gas Flow, Q_{CF_4}

The inlet gas flow, Q_{CF_4}, injected into the reactor may vary over the 15 to 125 $cm^3.s^{-1}$ range. The fluorine content inside the $YBa_2Cu_3O_{7-\delta}$ ceramic is not affected by an increase in the flow rate. This result can be explained by a compensation effect between the increase in the flow rate and the increase of the total pressure inside the reactor [18,19]. Similarly, we have not observed any large change of the temperature dependence of $J_{c(a.c.)}$ when the flow rate is increased. However, a Q_{CF_4} value of 60 $cm^3.s^{-1}$ appears to be an optimum inlet gas flow.

Influence of Total Pressure

The total pressure inside the reactor may vary over the 50 to 400 mTorr range. We have observed that the fluorine content inside the treated $YBa_2Cu_3O_{7-\delta}$

ceramic decreases as the total pressure is increased. This result can be explained by a decrease of the production factor of atomic fluorine mainly due to a decrease in the rate coefficient for dissociation [17]. The subsequent reduction of the flow rate results in a decrease of the fluorine content inside the $YBa_2Cu_3O_{7-\delta}$ ceramics. Simultaneously, the temperature dependence of $J_{c(a.c.)}$ is disadvantaged (Fig. 2).

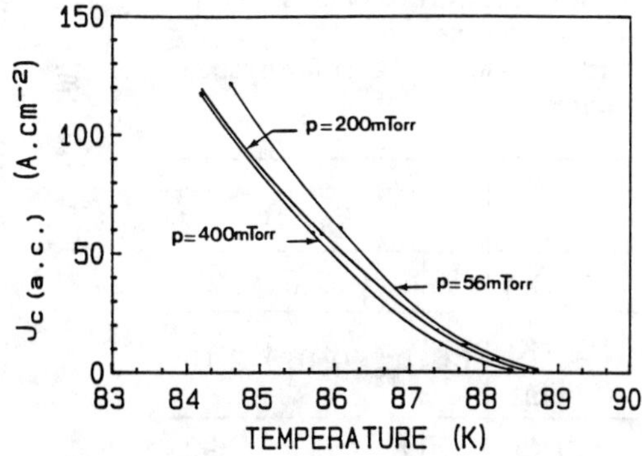

Figure 2 : Influence of the total pressure on the temperature dependence of $J_{c(a.c.)}$ with CF_4+25 % O_2, Q_{CF_4}=15 $cm^3.s^{-1}$, P=100 W, t=30 min.

Influence of Reaction Time

The PEF treatment time also affects the temperature dependence of $J_{c(a.c.)}$ (Fig. 3). For reaction times up to 60 min, $J_{c(a.c.)}$ values are strongly improved by the plasma treatment. On the other hand these values are reduced for reaction times above 60 min and the fluorine content inside the $YBa_2Cu_3O_{7-\delta}$ ceramic decreases simultaneously. It has been shown that the temperature of the sample increases during the experiment (up to about 100°C). It appears that the higher the temperature, the lower the fluorine content inside the $YBa_2Cu_3O_{7-\delta}$ ceramic. This peculiar behavior which corresponds to an apparently negative activation energy has been already observed, since the overall kinetics could be regulated by the adsorption-desorption equilibrium of the active species. This process, which is usually exothermic can become the bottleneck of the reaction at higher temperature [20]. However another assumption can be proposed: when adding O_2 to CF_4, a radicalar reaction occurs which yields the formation of CO_2. The resulting carbonate species could alterate the$YBa_2Cu_3O_{7-\delta}$ ceramic at the surface of the grains and consequently could limit the formation of curing fluorine species.

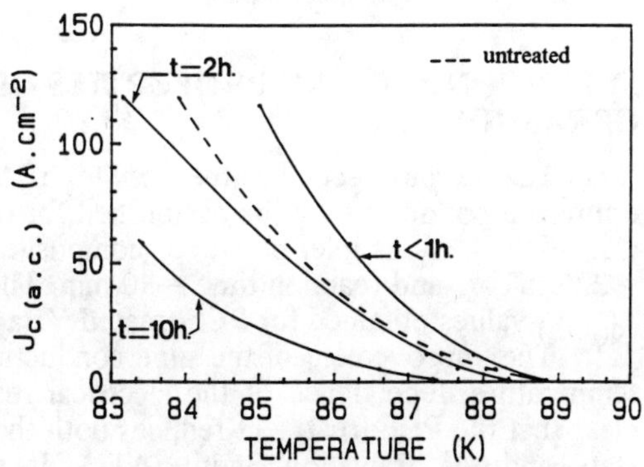

Figure 3 : Influence of the reaction time on the temperature dependence of $J_{c(a.c.)}$ with CF_4+25 % O_2, Q_{CF_4}=15 $cm^3.s^{-1}$, P=100 W, p=56 mTorr).

Influence of Radiofrequency Power

We have not detected any change of the temperature dependence of $J_{c(a.c.)}$ on the power of the generator in the 50 to 300 W range, which would imply that the fluorination is limited by surface reactions.

Table I : Comparison of the superconducting characteristics of $YBa_2Cu_3O_{7-\delta}$ ceramics before and after fluorination treatments.

	as sintered	after F_2 gas treat^t	after PEF treat^t
ΔT_c (K)	3.6	2.6	2.1
$\rho_{300\,K}$ (Ω.cm)	13.10^{-4}	9.10^{-4}	8.10^{-4}
T_c (K)	90.8	91	92
J_c (A.cm^{-2})	375	420	480

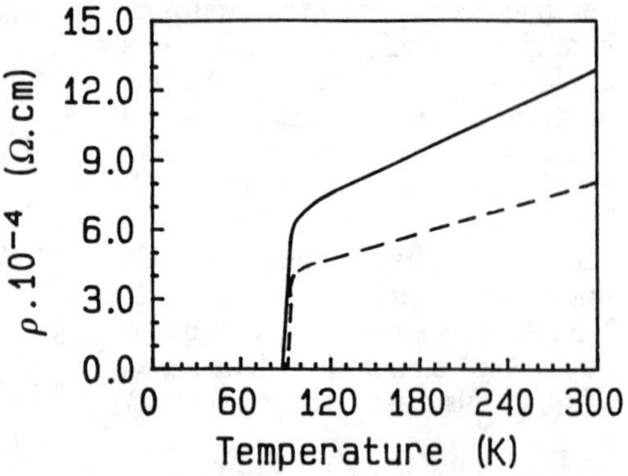

Figure 4 : Temperature dependence of the resistivity of $YBa_2Cu_3O_{7-\delta}$ ceramics before (—) and after (---) PEF treatment under optimized conditions

SUPERCONDUCTING PROPERTIES OF PEF-TREATED $YBa_2Cu_3O_{7-\delta}$ CERAMICS

Taking into account the results of the previous sections, the following optimized conditions can be established for one sample of 1 cm^2 external surface: CF_4 + 25 % O_2 as inlet precursor composition; Q_{CF_4} = 60 cm^3.s^{-1}; total pressure = 225 mTorr and reaction time = 30 min. These conditions correspond to the best $J_{c(a.c.)}$ values obtained for PEF-treated $YBa_2Cu_3O_{7-\delta}$ ceramics.

The improvement of the superconducting properties is also supported by the temperature dependence of the electrical resistivity (Fig. 4). From Table I, it is clear that the PEF treatment reduces both the value of the resistivity in the normal state and the transition width (ΔT_c). It simultaneously improves the critical transition temperature of about 1 K. Finally the improvement of the critical current density of a PEF-treated $YBa_2Cu_3O_{7-\delta}$ ceramic has been confirmed by a direct method involving pulsed currents under 2 μV.cm^{-1}. The J_c value obtained by this method is also compared with that one corresponding to a F_2-gas treatment.

SURFACE AND GRAIN BOUNDARY MODIFICATIONS AFTER PEF TREATMENTS

Distribution of Fluorine within $YBa_2Cu_3O_{7-\delta}$ Ceramics

Fluorine content profiles carried out along 50 µm in the central part of the pellets showed that the PEF treatment leads to an homogeneous distribution of fluorine around 0.5 % in the bulk of the ceramic. This result is in agreement with X-ray maps of fluorine. It can be pointed out that fluorine is much better distributed than in the case of F_2-gas treated samples where the fluorine atomic percentage fluctuated between 0 and 3 % [21].

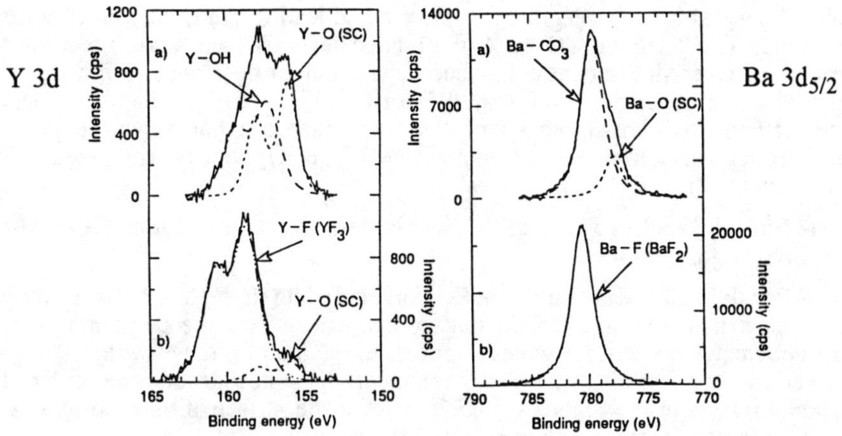

Figure 5 : Y 3d, Ba $3d_{5/2}$ emission spectra obtained at the surface of the outer grains of the $YBa_2Cu_3O_{7-\delta}$ ceramics before (a) and after (b) the CF_4 + 25 % O_2 plasma treatment (the peaks of the superconducting phase are noted S.C.).

Interaction of Reactive Fluorinated Species with Surface of Outer Grains of $YBa_2Cu_3O_{7-\delta}$ Ceramics

Before fluorination, the Y and Ba XPS spectra of $YBa_2Cu_3O_{7-\delta}$ ceramics are currently characterized by two types of contributions (Fig. 5a). For one type, the binding energies (E_b) are strongly shifted towards lower values. These so-called "superconducting peaks" are attributed to metal-oxygen bonds corresponding to the $YBa_2Cu_3O_{7-\delta}$ lattice [22]. The second type of contribution corresponds to the degradation layer (Y-OH and Ba-CO_3 bonds) [23].

After the PEF treatment, for Y and Ba elements, both peaks corresponding to the component of the superconducting phase and to the degradation layer have disappeared (Fig. 5b). New distinct peaks are now observed at 158.9 eV (Y 3d), 780.2 eV (Ba $3d_{5/2}$) and 90.0 eV (Ba 4d). They can be unambiguously ascribed to metal-fluorine bonds (like in the case of a F_2-gas treatment). These observations are supported by a unique F 1s peak at 684.7 eV corresponding to fluorine-metal bonds. However, contrary to a F_2-gas treatment, the PEF process removes the $(OH)^-$ and $(CO_3)^{2-}$ species without getting $(CF)_n$ and HF_2^- species. The interaction

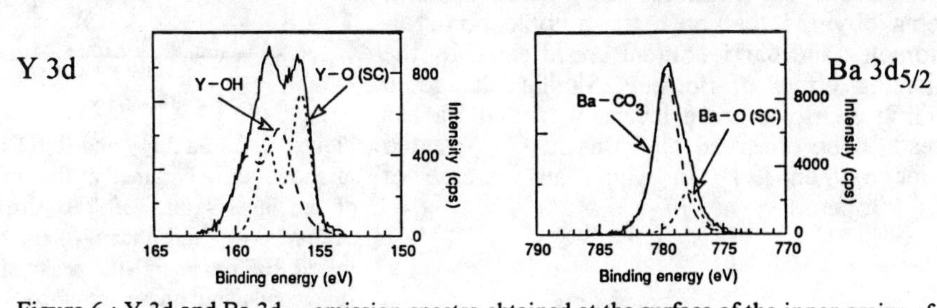

Figure 6 : Y 3d and Ba $3d_{5/2}$ emission spectra obtained at the surface of the inner grains of the $YBa_2Cu_3O_{7-\delta}$ ceramics after the CF_4 + 25 % O_2 plasma treatment .

of the F$^{\bullet}$ species with the degradation layer could lead to the formation of volatile compounds like COF or COF_2. A fluorinated layer is then substituted for the degradation layer. Its thickness has been evaluated to be around 7 nm thick by Auger electron spectroscopy for a PEF-treated $YBa_2Cu_3O_{7-\delta}$ single crystal. It seems actually to consist mostly of fluorides and does not exhibit higher E_b contributions, which have been attributed to M-F-Cu(III) bonds in the case of a F_2-gas treatment [21].

Interaction of Reactive Fluorinated Species with Surface of Inner Grains of $YBa_2Cu_3O_{7-\delta}$ Ceramics

After the PEF treatment the peaks corresponding in the Y 3d, Ba 3d5/2 and Ba 4d spectra to the superconducting component and to the degradation layer remain unchanged, and no new peak appears in the high E_b ranges (Fig. 6). These observations suggest that the PEF treatment does not decompose either the degradation layer or the superconducting phase at the surface of the inner grains.

In addition it can be pointed out that the peaks corresponding to the components of the superconducting phase are slightly shifted towards higher binding energy (+ 0.2 eV) as shown in Figure 7. These displacements cannot be attributed to a charge effect since the peaks corresponding to the degradation layer remain unchanged. Therefore, the following assumption can be launched for explaining this behavior (Fig. 7 a) : metal-fluorine bonds would be present inside the superconducting phase ; the slight shift observed (≈ 0.2 eV) towards higher energies for the peaks Y 3d, Ba $3d_{5/2}$ and Ba 4d, corresponding to the superconducting phase component, is due therefore to an increase of the ionicity of the corresponding bonds.

However the enhancement of the ionicity of these bonds may also result from an increase of the covalency of the competitive Cu-O bonds (Fig. 7 b), due to an increase of the concentration of "Cu^{3+}". This observation is corroborated by the Mohr-salt titrations. From these results it seems obvious that both the anionic non-stoichiometry and Cu^{3+} content are affected by the oxidizing effect of fluorine. Similar change in binding energy of core levels of Y and Ba has already been observed on $YBa_2Cu_3O_{7-\delta}$ treated under oxygen [24], in which an increase of "Cu^{3+}" occurs.

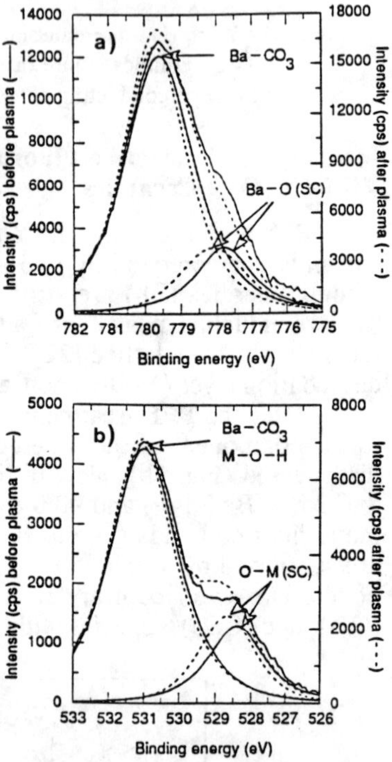

Figure 7 : (a) Ba $3d_{5/2}$ and (b) O 1s emission spectra obtained at the surface of the inner grains of $YBa_2Cu_3O_{7-\delta}$ before (——) and after (---) the optimized PEF treatment (the peaks of the superconducting phase are noted S.C.).

DISCUSSION

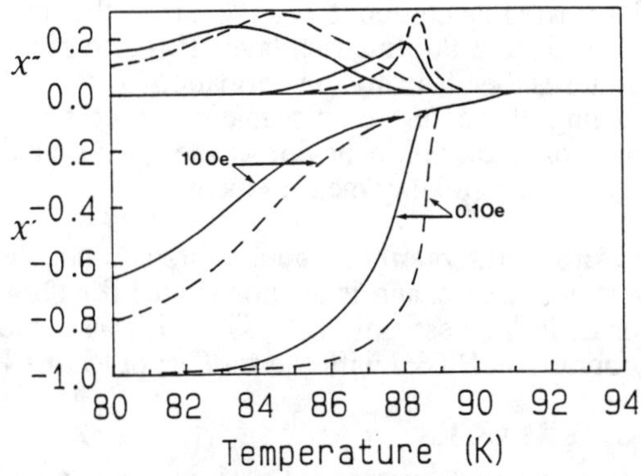

Figure 8 : A.c. susceptibility curves of $YBa_2Cu_3O_{7-\delta}$ before (—) and after (---) the optimized PEF treatment.

HTSC sintered at low temperature such as our $YBa_2Cu_3O_{7-\delta}$ samples are weakly coupled superconductors and they usually present two critical temperatures. One is intrinsic to the phase and the other is characteristic of the coupling between the grains. Resistivity measurements give information on the intrinsic superconducting properties (T_c onset) and especially on the first percolation path and on the surface of the sample. Improvement of T_c, ΔT_c and of the electrical resistivity in the normal state of PEF-treated ceramics reflects the cleaning of the surface and grain boundaries of the outer part of the ceramic as observed by XPS. It could also be related to the increase of the average oxidation state determined by Mohr salt titration. On the other hand, the increase of the J_c value is significant of an improvement of superconducting properties in the bulk of the ceramic. Analysis of the a.c. susceptibility curves (Fig. 8) can help us to understand this improvement in relation with the PEF treatment. Both the width of the transition and the field dependence are reduced for the PEF treated ceramic. It corresponds to a significant improvement of the intergranular junctions. Among many possibilities susceptible to induce coupling phenomena between superconducting grains, the presence of impurities and the departure from stoichiometry at the grain boundaries can be mentioned. Since the presence of fluorine has been established in the bulk of the ceramic, it is likely that the quality of weak link regions has been enhanced by the presence of fluorine in the grain boundary area. It would correspond to the chemical bonding modifications (section V.2.) observed by XPS. The remaining presence of degradation layers in the bulk also suggests that the PEF process actually only improves weak links that were previously carbon free grain boundaries.

CONCLUDING REMARKS

In the present work, we have shown that an improvement of the superconductive properties of $YBa_2Cu_3O_{7-\delta}$ can be obtained using a new technique : the r.f. plasma-enhanced fluorination (PEF) treatment. An average fluorine content of 0.5 at.% has been detected by EMPA, in the bulk of the $YBa_2Cu_3O_{7-\delta}$ ceramics. The atomic fluorine produced inside the fluorinated plasma is likely the responsible species for the oxidation of the material. The effect of PEF would correspond to an increase of the anion rate up to 6.83, which is the content required for optimized superconducting properties (T_c and J_c) [9]. The PEF treatment removes the $(OH)^-$ and $(CO_3)^{2-}$ species located in the degradation layer from the surface of outer grains of the materials and it gives rise to the formation of an homogeneously fluorinated layer, about 7 nm. On the other hand, the PEF treatment does not seem to produce a fluorine rich layer at the surface of inner grains of the materials, but would rather lead to the formation of metal-fluorine bonds inside the superconducting phase due to an oxidation process by $F^{\bullet}$ species. It results in an improvement of weak link behavior of the grain boundaries within the ceramics according to a.c. susceptibility measurements.

Acknowledgements : The authors are indebted to B. Chevalier for valuable suggestions. L. Lozano is acknowledged for fluorination experiments and M. Lahaye for EMPA investigations. This work was carried out as a part of a CNRS research program on HTSC with the support of Rhône-Poulenc Recherche.

REFERENCES

[1] S. Jin, T.M. Tiefel and R.C. Sherwood, *Phys. Rev. B*, **37**, 7850 (1987).

[2] P. de Rango, Ph. D. Thesis, University J. Fourier, Grenoble, France (1992).

[3] S.J. Keating, I. Wei and T.Y. Chien, p.43 in *Ceramic Superconductors II*, edited by M.F. Yan, Am. Ceram. Soc. (1988).

[4] J.M. Dance et al, *Solid State Ionics*, **32-33**, 1188 (1989).

[5] B. Lepine, Ph. D. Thesis, University Bordeaux I, France (1990).

[6] D.W. Hess, p. 459 in *Microelectronic materials and processes*, edited by R.A. Levy, Kluwer Academic-Press (1989).

[7] H. Bui, H. Carchano and D. Sanchez, *Thin Solid Films*, **13**, 207 (1972).

[8] J.M. Heintz et al., *J. Less-Common Metals*, **164-165**, 1377 (1990).

[9] C. Magro, Ph. D. Thesis, University Bordeaux I, France (1992).

[10] C.P. Bean, *Rev. Mod. Phys.*, **36**, 31 (1964).

[11] R.B. Goldfarb et al., in *Magnetic Susceptibility of Superconductors and Other Spin Systems*, edited by R.A. Hein et al., Plenum Press (1992)

[12] W. Herzog et al, *Z. Phys. B*, **71**, 19 (1988).

[13] C.J. Mogab, *J. Electrochem. Soc.*, **124**, 1263 (1977).

[14] J.C. Plumb and K.R. Ryan, *Plasma Chem. Plasma Process.*, **6**, 205 (1986).

[15] Seong-Ju Park, C.P. Sun and J.T. Yeh, p. 65 in *Plasma Processing*, edited by J. Coburn et al., Mat. Res. Soc. Symposium Proceedings, **68** (1986).

[16] R. D'agostino et al., *Appl. Phys.*, 52 (1981) 1259.

[17] A.T. Bell, *J. Macromol. Sci. Chem.*, **A10**, 379 (1976).

[18] G. Turban, Ph. D. Thesis, University Nantes, France (1981).

[19] N. Mutsukura and G. Turban, *Vacum*, **39**, 579 (1989).

[20] R. D'agostino, F. Cramarossa and F. Illuzi, *J. Appl. Phys.*, **61**, 2754 (1987).

[21] C. Magro et al., *J. Mater. Sci.*, to be published (1994)

[22] S. Myhra et al., *Z. Phys. B*, **72**, 413 (1988).

[23] R.P. Vasquez et al., *J. Appl. Phys.*, **66**, 4866 (1989).

[24] P.C. Healy, S. Myhra and A.M. Steward, *Philos. Mag. B*, **58**, 257 (1988).

Composite Materials

TOUGHENING OF Si_3N_4 CERAMICS BY NANO SiC PARTICLES IN GRAIN BOUNDARY PHASE

Tomohiro Yanai and Kozo Ishizaki
School of Mechanical Engineering, Nagaoka Gijutsu-Kagaku Daigaku (Nagaoka University of Technology), Kamitomioka 1603-1, Nagaoka, Niigata 940-21, Japan

ABSTRACT

The purpose of the present work is to toughen the weak grain boundary phase in Si_3N_4 ceramics by forming a nanocomposite grain boundary phase of SiC nano particles. The SiC nano particles are formed by a reaction between coated carbon and an oxide phase on the surface of Si_3N_4 powder. The carbon is coated by thermal decomposition of CH_4 gas. The surface oxygen content is controlled by an oxidation surface treatment. The sintered bodies are obtained by hot pressing at 1800 ℃ for 1 h. With an increase in the carbon content, the fracture toughness and bending strength first increase, then reach maximum values at the same amount of carbon addition, and decrease. The carbon contents at which the fracture toughness and bending strength exhibit maximum values decrease with an increase in the surface oxygen content. The generated SiC particles are found not only in the grain boundary phase but also in the Si_3N_4 grains by TEM observation. The fracture mode changes from inter to transgranular fracture with an increase in the number of SiC particles.

INTRODUCTION

Silicon nitride (Si_3N_4) ceramics have excellent properties, e.g., high strength, wear resistance, high thermal shock resistance and oxidation resistance. Sintered Si_3N_4 ceramics are applied as engineering materials. Highly densified sintered bodies are obtained through liquid phase sintering to overcome the difficulty of sintering due to the strong covalent bonding of Si_3N_4. Sintered Si_3N_4 ceramics contain a weak grain boundary phase, which results from a reaction between sintering additives and the as-received Si_3N_4 powder, especially if there is surface oxygen present. The surface oxygen of as-received powder plays an important role in obtaining good performance of structural Si_3N_4 ceramics.

If the weak grain boundary phase can be made tougher, the bulk mechanical properties are improved significant. A popular toughening method is to disperse a second phase into the matrix of Si_3N_4[1]. An improvement in mechanical properties is achieved when this second phase is nano size[2][3] and is dispersed in the weak grain boundary phase. Toughening mechanisms in such composites result from the Young's modulus mismatch, and the thermal expansion mismatch between the grain boundary and second phases[4]. A residual stress is produced from this mismatch, and thus the fracture energy increases.

The grain boundary phase has poor mechanical properties: has its Debye temperature is about 300 K compared to 1100 K of the Si_3N_4 lattice[5], and the Young's modulus has been to be reported about 80 GPa[6] compared to 300 GPa of Si_3N_4. In this work, nano size SiC particles are formed during sintering. The particles are formed by a reaction between surface oxygen and coated carbon[7][8]. The Young's modulus of SiC is about 440 GPa. Therefore, the high Young's modulus mismatch can be obtained by SiC dispersion in the grain boundary phase. The generated SiC content depends on the surface oxygen and coated carbon contents. The starting Si_3N_4 powder contains surface oxidized impurities, and the surface oxygen content can be controlled by an oxidation or washing treatment. This work reports the influence of oxygen and carbon content on mechanical properties and microstructure in this Si_3N_4–SiC(nano) system.

EXPERIMENTAL

The starting Si_3N_4 powder was a commercial α-phase powder (Ube Industries Co., Ltd., SN-E10 grade, average grain size : 0.3 μm). The surface oxygen amount was increased by a surface oxidation treatment at 1000 ℃ in air for 4 h or 8 h, or decreased by washing in soxhlet with deionized water for 70 h[9]. The carbon was coated by thermal decomposition of CH_4 gas at 800 ℃, and the amount of the carbon coating was controlled by increasing the holding time up to 16 h. The oxygen and carbon content were measured by a combustion type oxygen and carbon analyzer. Assuming that all the oxygen atoms formed SiO_2, the SiO_2 contents were 1.07, 2.68, 3.62 and 4.68 wt% for washed, as-received, oxidized for 4 h and 8 h, respectively. The carbon amount was from 0.06 to 1.50 wt%. The different Si_3N_4 powders were mixed with additives, i.e., 3 mol% each of Y_2O_3 (Shin-etsu Chemical Co., Ltd., average grain size : 0.9 μm) and α-Al_2O_3 (Sumitomo Chemical Co., Ltd., average grain size : 0.4 μm), by ball milling with Si_3N_4 balls using methanol solvent for 24 h in a polyethylene pot. After drying, the mixtures were mixed again by dry ball milling for 12 h. All the mixtures were hot pressed at 1800 ℃ for 1 h in N_2 atmosphere under 30 MPa of uniaxial pressure.

The bulk density was measured by the Archimedes immersion technique in toluene. The bending strength was evaluated by the 3-point bending test. The specimen size was 3 mm × 4 mm × 36 mm, and span length and cross head speed were 30 mm and 0.5 mm/min, respectively. The fracture toughness was estimated by the indentation microstructure method[10] under the following conditions; indenter load 98 N and loading time 15 s. The constitution phase of sintered bodies were analyzed by X-ray diffraction analysis and energy dispersive X-ray spectroscopy (EDX). Microstructure was observed by scanning electron microscopy (SEM) and transmission electron microscopy (TEM).

RESULTS

The β-Si_3N_4 and β-SiC peaks are confirmed by X-ray diffraction analysis[8]. Generated β-SiC particles are found dispersed not only in the grain boundary phase but also in the Si_3N_4 grains by TEM observation, as shown in Fig.1. The SiC particle size is below 30 nm. Figure 2 shows a HREM photograph of nano SiC particles dispersed in grain boundary phase with diffraction patterns and EDX profiles. The diffraction patterns "A" and "B" indicate β-Si_3N_4 and β-SiC crystals, respectively, and the center spot of diffraction patterns "B" and "C" consist of hallow pattern, indicating noncrystalline phases. The EDX profile of (A) and (C) corresponds to grain "A" and "C" in the HREM image, respectively. Si and Al peaks are detected from the β-Si_3N_4 grain, and Y, Al and Si peaks are confirmed from the grain boundary phase. As the light atoms can not be detected by EDX, O and N peaks do not appear. The β-Si_3N_4 and β-SiC peaks are confirmed by X-ray analysis; however, Y, Al and oxide peaks are not found. X-ray diffraction can not analyze

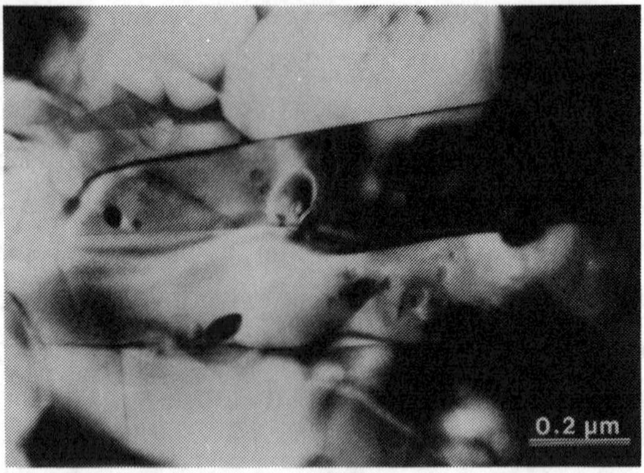

Figure 1 TEM micrograph of Si_3N_4 ceramics prepared from 0.29 wt% carbon coated on as-received (2.68 wt% SiO_2) Si_3N_4 powder. Arrows indicate generated nano size SiC particles.

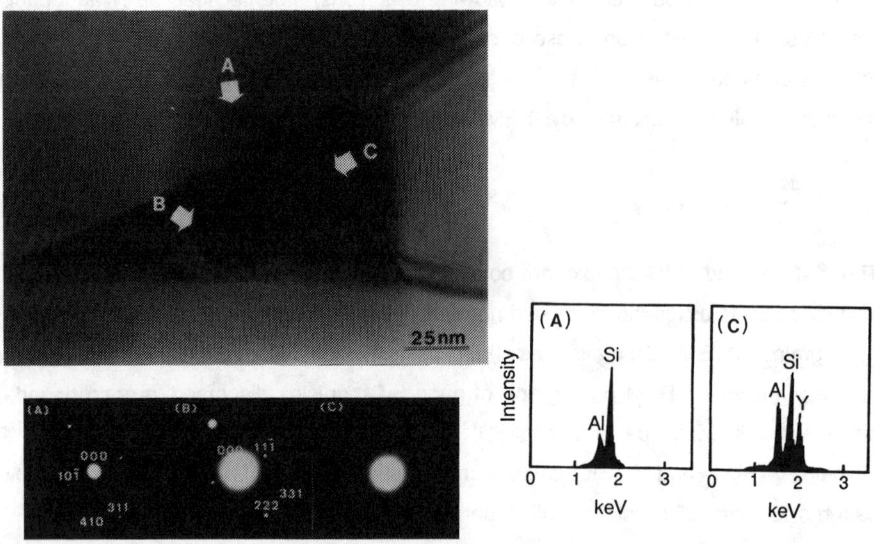

Figure 2 HREM micrograph of nano SiC particles dispersed at Si_3N_4 triple point with the diffraction patterns and EDX profiles. Figures "A", "B" and "C" indicate diffraction patterns and EDX analysis at points in the HREM photograph, respectively. Diffraction patterns "A", "B" and "C" correspond with β-Si_3N_4, β-SiC and amorphus pattern, respectively.

noncrystalline phases. The grain boundary phase is found to consist of Y–Al–Si–O–N and a glassy phase. The EDX profile of the Si_3N_4 grain includes an Al peak, because Al partially forms a solid solution in the β–Si_3N_4 lattice during sintering[11].

Figures 3 and 4 show the fracture toughness and the bending strength as a function of carbon content for raw Si_3N_4 powders with different surface oxygen contents. With an increase in the carbon content, the fracture toughness and bending strength first increase, then reach maximum values, and decrease. The bending strength increases to 1300 MPa at the 0.46 wt% carbon coated surface oxidized Si_3N_4 powder (3.62 wt% SiO_2) from 990 MPa for non–coated one, and fracture toughness shows a similar tendency, i.e., from 5.2 to 6.5 $MN/m^{3/2}$. The maximum peaks of the fracture toughness and the bending strength occur at the same carbon content for the grain amount of oxygen, and shift to higher carbon content with an increase in the surface oxygen content. Figure 5 shows the fraction to the theoretical density as a function of the carbon contents and the surface oxygen contents. The density decreases with an increase in the carbon content, and decreases faster after an inflection point. This point shifts to a higher carbon content with an increase in surface oxygen content. The maximum fracture toughness and the maximum bending strength peaks agree well with the decrease in density.

DISCUSSION

β–Si_3N_4 and β–SiC are detected by X–ray diffraction analysis. In this Si_3N_4 system, the following reaction proceeded during hot press sintering.

$$SiO_2(s) + 3C(s) = SiC(s) + 2CO(g) \qquad (1)$$

In accordance with the HIP phase diagram developed by Ishizaki[12], equation (1) may be explained as follows. As the total gas pressure is only approximately 0.1 MPa or lower under hot-press sintering, the SiC, CO and SiO_2 are stable while C is unstable above 1600 K. Equation (1) will proceed forward to the right hand side.

The generated SiC content depends on the surface oxygen and the coated carbon content on the Si_3N_4 powder surface. Figure 6 (a) shows the carbon content of specimens

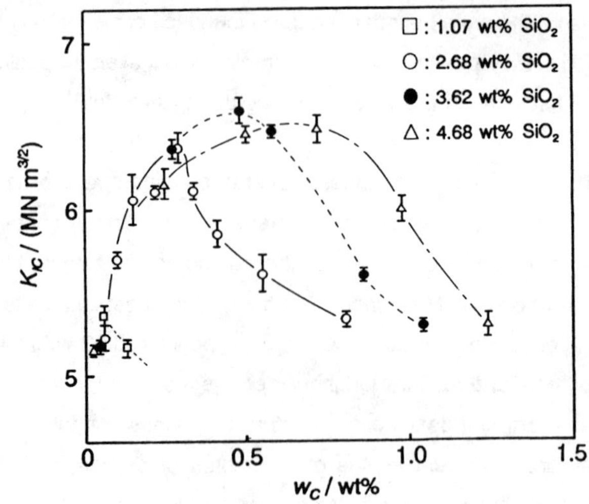

Figure 3 The fracture toughness, K_{IC}, as a function of carbon content, w_C, for raw Si_3N_4 powders with different surface oxygen contents.

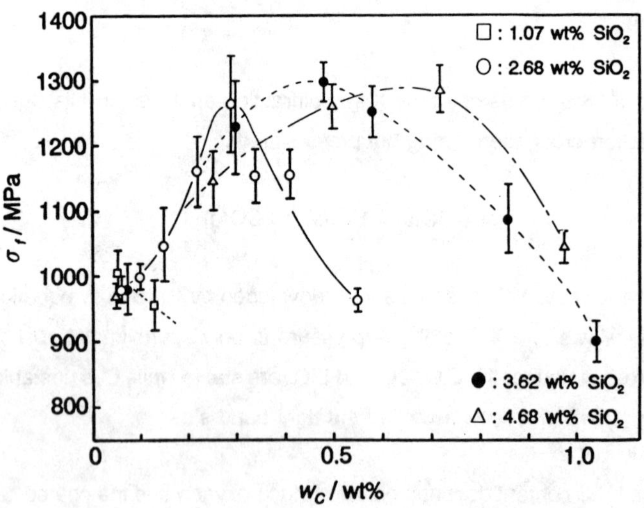

Figure 4 The bending strength, σ_f, for raw Si_3N_4 powders as a function of carbon content, w_C, with different surface oxygen contents.

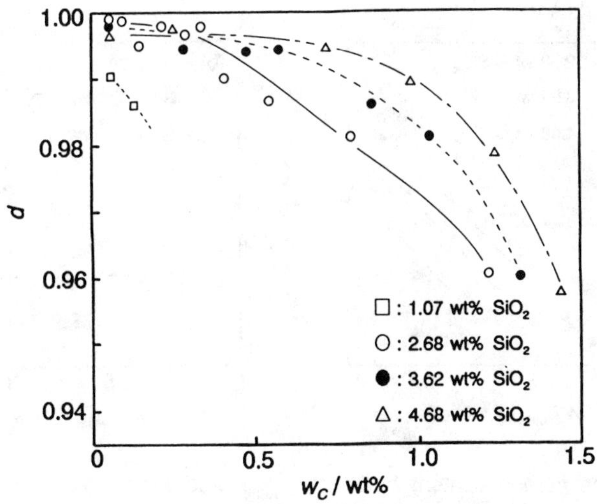

Figure 5 The fraction to the theoretical density, d, as a function of carbon content, w_C, for raw Si_3N_4 powders with different surface oxygen contents.

before and after sintering. The dashed line indicates carbon conditions when the carbon content does not change before and after hot pressing; in other words, SiC particles are not produced this line by equation (1). Figure 6 (b) shows the carbon amount disappeared by sintering due to equation (1). The reacted carbon contents first increases with an increase in carbon content in the raw powders, then reaches a plateau of constant values. The inflection point shifts to a higher carbon content with an increase in oxygen content, which leads to an increase in reacted carbon. When the reacted carbon content increases, the resultant SiC content also increases. To the left side of inflection point, the curves have the same slope. This result suggests that the resultant SiC content is proportional to the surface oxygen content. The inflection points almost agree with the maximum peaks of the bending strength and the fracture toughness, and the inflection points of decreasing density. Therefore, the maximum peaks of the fracture toughness and bending strength, and decreasing inflection points of density occur at higher carbon contents with an increase in surface SiO_2 content. Surface oxidized Si_3N_4 powders are generally regarded as inferior powders; however, these powders can be used to improve mechanical properties by applying the present method. This implies that lower grade Si_3N_4 powders may be used to produce high quality Si_3N_4 ceramics by applying the carbon coating method.

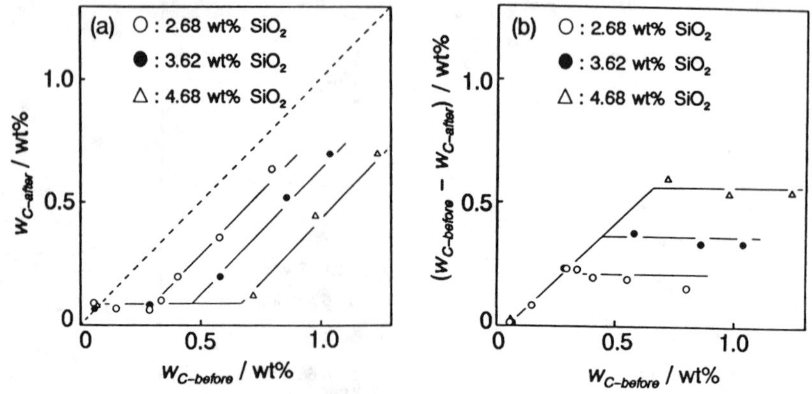

Figure 6 (a) The carbon content before, $w_{C\text{-before}}$, and after, $w_{C\text{-after}}$, sintering. (b) Reacted carbon amount, $w_{C\text{-before}} - w_{C\text{-after}}$, as a function of carbon content of before sintering, $w_{C\text{-before}}$.

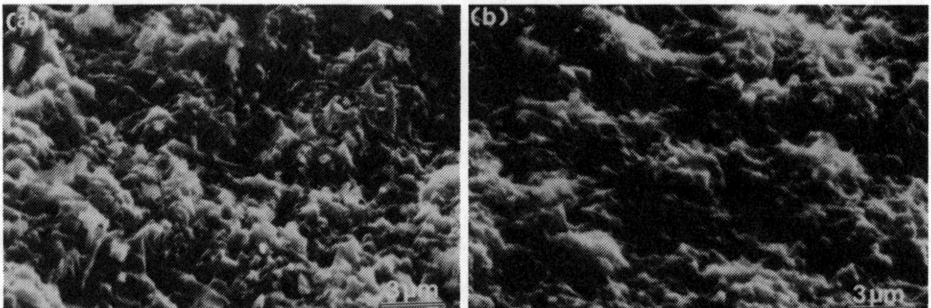

Figure 7 SEM micrographs of fractured surfaces prepared from as-received Si_3N_4 powder (a) non-coated and (b) carbon coated.

Improvement of mechanical properties may be attributed to the residual stress around generated SiC particles due to the difference of Young's modules and the thermal expansion mismatch between the grain boundary phase and SiC, or the matrix Si_3N_4 and SiC. In the case of SiC particles disperses in the grain boundary phase, the effective Young's modulus of the grain boundary is increased by generated nano SiC particles. A residual compressive stress is created around SiC particles dispersed in the grain boundary phase, because the thermal expansion of SiC is smaller than the grain boundary phase. This compressive stress can obstruct the intergranular fracture and crack penetrates the Si_3N_4 grain inside. The week grain boundary phase is toughened by nano SiC particles, and the fracture mode must be change. Figure 7 shows the SEM photographs of fractured surfaces prepared from non-coated and 0.29 wt% carbon coated on the as-received Si_3N_4 powder. The fracture mode is transformed from inter to deflected transgranular fracture by carbon coating. It means that the weak grain boundary phase is toughened by generated nano size SiC particles. While, in the case of SiC particles disperse in Si_3N_4 grain, the residual tensile stress is generated by thermal expansion mismatch between Si_3N_4 and SiC. Because thermal expansion of SiC is larger than Si_3N_4. When crack penetrates inside Si_3N_4 grains and reaches SiC particle, crack is propagated around SiC particle by residual tensile stress. And crack is deflected by SiC particles. Sasaki reported crack propagation of Si_3N_4/SiC nano composite system[3], and showed crack deflection by nano SiC particles in Si_3N_4 grain from TEM and HREM observations. Therefore, because the fracture energy increases for this composite, the fracture toughness increases.

CONCLUSIONS

(1) The β-SiC is confirmed by X-ray diffraction analysis, and this phase is produced by a reaction between the surface oxygen and the coated carbon during hot pressing. This reaction can be explained by the HIP phase diagram.
(2) With an increase in the carbon content, the fracture toughness and bending strength first increase, then reach maximum values at the same carbon content, and decrease. These maximum peaks move to a higher carbon content with an increase in surface oxygen content.
(3) The generated SiC particles are dispersed in both the grain boundary phase and in the

Si_3N_4 grains. As the grain boundary phase is toughened by SiC particles, the fracture mode changes from inter to deflected transgranular fracture.

(4) The result of the change of carbon content before and after hot pressing shows that the reacted carbon content increases with an increase in surface oxygen content. This result suggests that the generated SiC content increases with an increase in SiO_2 content.

(5) Surface oxidized Si_3N_4 powders are generally regarded as inferior powders. The mechanical properties can be improved by generating nano SiC particles, using the present method.

REFERENCES

1) S. T. Buljan, J. G. Baldoni and M. L. Huckabee, "Si_3N_4-SiC Composites", *Am. Ceram. Soc. Bull.*, **66**, 347-52(1987).

2) K. Niihara, K. Izaki and T. Kawakami, "Hot-pressed Si_3N_4 - 32 % SiC Nanocomposite from Amorphous Si-C-N Powder with Improved Strength above 1200 ℃", *J. Mat. Sci. Lett.*, **10**, 112-14,(1990).

3) G. Sasaki, H. Nakase, K. Suganuma, T. Fujita and K. Niihara, "Mechanical Properties and Microstructure of Si_3N_4 Matrix Composite with Nano-Meter Scale SiC Particles", *J. Ceram. Soc. Jpn.*, **100**, 536-40,(1992).

4) K. Niihara, "New Design Concept of Structural Ceramics - Ceramics Nanocomposites -", *J. Ceram. Soc. Jpn.*, **99**, 974-82,(1991).

5) T. Hamasaki and K. Ishizaki, "The Quantitative Evaluation of Grain Boundary Phases by Cryogenic Specific Heat Treatments", in *Grain Boundary Controlled Properties of Fine Ceramics*, ed., K. Ishizaki, K. Niihara, M. Isotani and R. G. Ford, Elsevier Applied Science, London and New York, 1992, pp53-61.

6) K. Oda and Y. Yoshio, "Properties of Y_2O_3-Al_2O_3-SiO_2 Glasses as a Model System of Grain Boundary Phase of Si_3N_4 Ceramics (Part 1)", *J. Ceram. Soc. Jpn.*, **97**, 1493-97,(1987).

7) K. Watari, K. Ishizaki and M. Kawamoto, "Behavior of Carbon in Hot Pressed Silicon Nitride Grain Boundaries and Its Influence on Mechanical Properties", *J. Ceram. Soc. Jpn.*, **96**, 127-33,(1988).

8) T. Yanai and K. Ishizaki, "Mechanical Properties of Si_3N_4 Ceramics Prepared from Carbon Coated Powders", *J. Ceram. Soc. Jpn.*, **101**, 764-68,(1993).

9) M. Kawamoto, K. Ishizaki and C. Ishizaki, in *EURO CERAMICS Volume 2 : Properties of Ceramics*, ed., G. de. With, R. A. Terpstra and R. Merselaar, Elsevier Applied Science, London and New York, 1989, pp120-24.

10) K. Niihara, "A Fracture Mechanics Analysis of Indentation-Induced Palmquvist Crack in Ceramics", *J. Mat. Sci. Lett.*, **2**, 221-23,(1983).

11) A. Tsuge, "The Grain Boundary Engineering on the Strength of Si_3N_4", *Bull. Ceram. Soc. Jpn.*, **18**, 205-10,(1983).

12) K. Ishizaki, "Phase Diagrams under High Total Gas Pressure - Ellingham Diagrams for Hot Isostatic Press Processes", *Acta Met. et Mat.*, **38**, 2059-66,(1990).

EFFECTS OF GRAIN BOUNDARY STRUCTURE ON CREEP BEHAVIOR FOR Si_3N_4/SiC NANOCOMPOSITES

Takeshi Hirano and Koichi Niihara
ISIR, Osaka University, 8-1 Mihogaoka, Ibaraki, Osaka 567, Japan
Kansei Izaki
Mitsubishi Gas Chemical Co., Tsukuba, Ibaragi 300-42, Japan
Fumihiro Wakai
NIRIN, Nagoya, Aichi 462, Japan

ABSTRACT

Effects of grain boundary structure on creep behavior were investigated for hot-pressed Si_3N_4/30vol%SiC nanocomposites. The starting powder for the nanocomposites was prepared using a vapor-phase reaction. The creep tests were conducted in air atmosphere at 1400℃. Comparisons were made among Si_3N_4/30vol%SiC nanocomposites containing 8wt%Y_2O_3 and 2wt%Al_2O_3+ 6wt%Y_2O_3 and monolithic Si_3N_4 containing 8wt%Y_2O_3 and 2wt%Al_2O_3+ 6wt%Y_2O_3. Although Si_3N_4/30vol%SiC nanocomposite containing 8wt% Y_2O_3 was characterized by small grain size, the remarkable creep resistance was observed. The reasons for the high creep resistance will be discussed in relation to the grain boundary structure.

INTRODUCTION

Since Si_3N_4 ceramics show high strength, toughness, and thermal shock resistance, it has become one of the leading ceramic materials with many applications at elevated temperatures. To date Si_3N_4 ceramics have already been used, for example, as turbocharger components in some automobiles. Wider application, however, requires higher toughness, strength, and creep resistance. Because single phase Si_3N_4 is difficult to sinter, Si_3N_4 is sintered with the use of additives. These react with SiO_2 normally present on Si_3N_4 powders to form a glassy phase. This glassy phase enhances liquid-phase sintering. The retained intergranular glassy phase inherent in this process significantly governs the high-temperature creep behavior[1-3].

Recently, fibers and whiskers have been added to Si_3N_4 to improve fracture toughness. This approach gives the potential to not only increase the fracture toughness at room temperature, but also to improve the creep resistance at elevated temperatures. In one study[4], a considerable improvement of creep resistance was observed for SiC whisker reinforced Si_3N_4. However, in other cases the addition of SiC whiskers to the Si_3N_4 sometimes results in little or negative effects on the creep performance[5,6].

Niihara and colleagues have investigated Si_3N_4/SiC nanocomposites, in which the nano-sized SiC particles are dispersed within the matrix grains and/or at grain boundaries[7-11]. These materials were made by hot-pressing amorphous Si-C-N precursor powders prepared by vapor-phase reactions. Though as much as 8wt% of either Al_2O_3 plus Y_2O_3 or Y_2O_3 were used as sintering additives , these nanocomposites exhibited improved fracture strength and hardness at elevated temperatures [7,10-12]. These improved mechanical properties at elevated temperatures for the nanocomposites were attributed to the intergranular fine SiC particles[7,9-12].

The main purpose of this study is to investigate the effects of grain boundary structure on creep behavior of Si_3N_4/SiC nanocomposites. The tensile creep resistance was evaluated at 1400℃ for Si_3N_4/30vol%SiC nanocomposites containing 8wt%Y_2O_3, and the results were compared to Si_3N_4/30vol%SiC nanocomposites containing 2wt%Al_2O_3+6wt%Y_2O_3 and monolithic Si_3N_4 containing 8wt%Y_2O_3 and 2wt% Al_2O_3+6wt% Y_2O_3.

EXPERIMENTAL PROCEDURE

Materials

The starting amorphous Si-C-N precursor powders for the Si_3N_4/SiC nanocomposites were prepared by vapor-phase reaction of $[Si(CH_3)_3]_2NH$- NH_3-N_2 at 1000℃, followed by 4h heat treatment in N_2 at 1350℃. The C/N ratio in the amorphous powder was adjusted so that the SiC content was about 30vol%. The starting powder for the monolithic Si_3N_4 was a commercial α-Si_3N_4 powder made by Toyo Soda Manufacturing Co. Ltd., Tokyo, Japan. The starting powders were mixed with either 8wt%Y_2O_3 or 2wt%Al_2O_3+6wt%Y_2O_3 as sintering aids in a plastic bottle using Si_3N_4 balls and ethanol. The dried mixtures were hot pressed at 1700-1800℃ for 1 to 4hr in N_2 under 34 MPa applied pressure. The hot pressed samples were 60 mm in diameter and 5 mm in thickness. The specimens are labeled as follows (SN: monolithic, SCN: composites) and sintering additives (AY: 2wt%Al_2O_3+6wt%, 8Y: 8wt%Y_2O_3). A Si_3N_4/ 30vol%SiC nanocomposite containing 8wt%Y_2O_3 is, for instance, labeled SCN8Y.

Tensile Creep Testing and microstructural observations

Creep specimens were machined from the hot-pressed samples with a #800 diamond wheel and diamond tip. The specimen has a cylindrical gauge section with a diameter of 3.0mm and a length of 14mm and two target arms at the end of gauge section. The elongation of the distance between them was measured by an optical extensometer. The creep experiments were conducted in air using the tensile

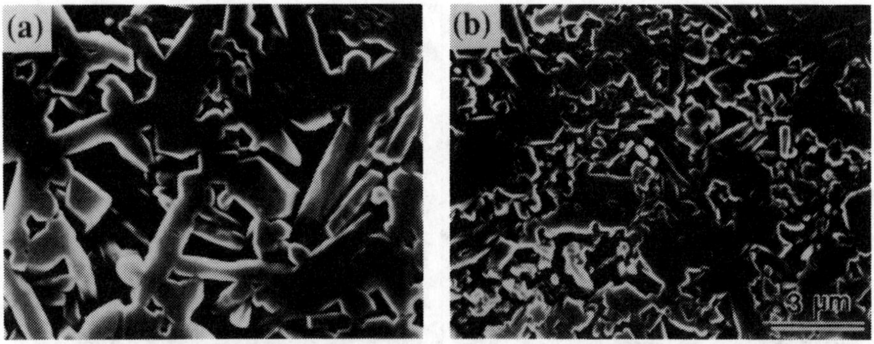

Figure 1. SEM micrographs of SNAY (a) and SCN8Y (b).

creep testing machine equipped with $MoSi_2$ heating elements. To assure thermal stability, equipment and specimen were heated at the test temperature for 20h before testing. During the holding time a pressure of 5.1MPa was applied to the specimen due to the jigs[12].

The microstructure of specimens was observed by scanning electron microscopy (SEM) and transmission electron microscopy (TEM). The microchemical analysis of intergranular phases was attempted using a energy dispersive X-ray (EDX) analyzer. Sections for SEM and TEM were taken from the gauge sections of tensile specimens after high-temperature creep tests and compared with those from as-hot-pressed samples. As intergranular phases strongly affect the creep behavior of Si_3N_4 and Si_3N_4 based composites, XRD analysis was also conducted for all specimens.

RESULTS AND DISCUSSION

Microstructural Analyses of As-Hot-Pressed Samples

Figure 1 shows the representative SEM pictures of as-hot-pressed SNAY and SCN8Y. In the monolithic Si_3N_4, large elongated grains were frequently observed. The morphology of Si_3N_4 grains of SNAY and SN8Y was fibrous, with an aspect ratio of 3 - 8 and a diameter of 0.5 - 1.5μm regardless of the sintering aids. In the Si_3N_4/SiC nanocomposites, the growth of elongated β-Si_3N_4 grains was suppressed by the presence of fine SiC particles, fine equiaxed grains and small rod-like grains were observed. There was little difference in microstructure between SCN8Y and SCNAY.

Figure 2 shows the XRD profiles for the crushed as-hot-pressed specimens. These XRD profiles show that 2wt%Al_2O_3+6wt%Y_2O_3 doped specimens (SNAY and SCNAY) have no clear crystalline grain boundary phases, and that 8wt%Y_2O_3 doped specimens (SN8Y and SCN8Y) contain crystalline phases, such as $Y_5N(SiO_4)_3$, Y_2SiO_5 and $Y_2Si_2O_7$. Smith and Quackenbush [13] reported that the grain boundary phase of sintered Si_3N_4 with 2wt%Al_2O_3+6wt%Y_2O_3 was amorphous, while that of sintered Si_3N_4 with 6wt%Y_2O_3 included crystalline

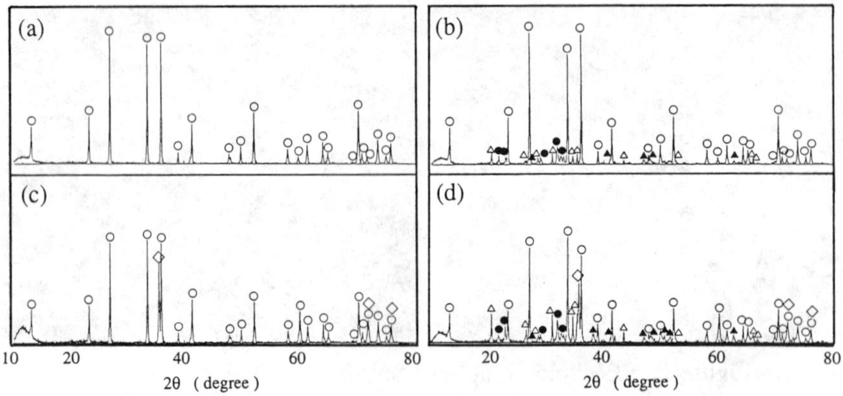

Figure 2. X-ray diffraction patterns of as-hot-pressed samples used in the present study. (a) SNAY, (b) SN8Y, (c) SCNAY and (d) SCN8Y. (○) β-Si_3N_4, (●) $Y_5N(SiO_4)_3$, (△) Y_2SiO_5, (▲) $Y_2Si_2O_7$, (◇) β-SiC.

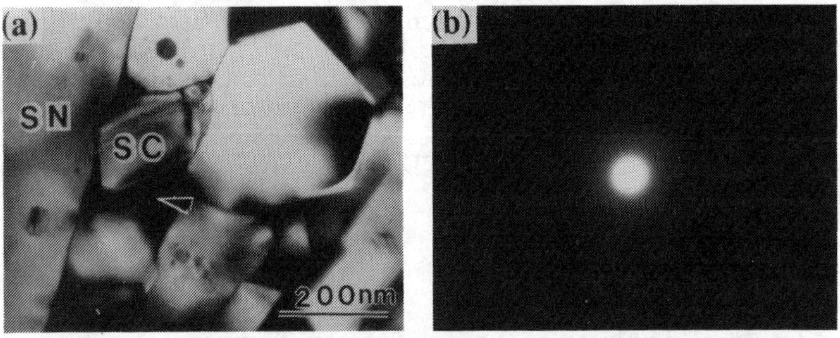

Figure 3. TEM micrograph of SCNAY showing intergranular SiC particle penetrating into β-Si_3N_4 grain (a). Microdiffraction pattern taken from the triple point (indicated by arrow) of SCNAY(b).

phases such as $Y_5N(SiO_4)_3$ and $Y_2Si_2O_7$. Their observation is in good agreement with the present analyses. As the complete crystallization of grain boundary phases are thermodynamically impossible[14], both SN8Y and SCN8Y may contain glassy phases. However, due to crystallization of grain boundary phases, the amounts of glassy phases of SN8Y and SCN8Y are thought to be smaller than those of SNAY and SCNAY.

TEM observations for SCNAY and SCN8Y nanocomposites indicated that nano-sized SiC particles smaller than 100 nm were mainly dispersed within Si_3N_4 grains and fine SiC particle, larger than approximately 100 nm, were dispersed at the grain boundaries. Some intergranular SiC particles are observed to be penetrating

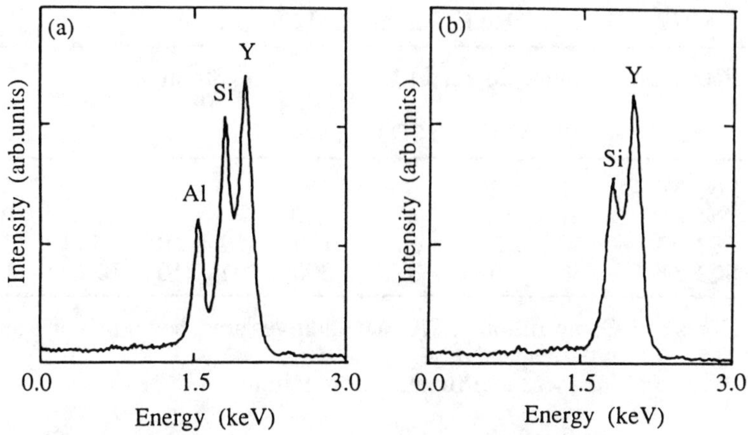

Figure 4. Energy dispersive X-ray spectra from the grain boundary phase at the triple points of SCNAY (a) and SCN8Y (b).

into the β-Si_3N_4grains as shown in Figure 3(a). Niihara et al. [9] reported that these intergranular SiC particles are directly connected with Si_3N_4 grains. The microdiffraction pattern taken from the grain boundary triple points of SCNAY indicated that the grain boundaries are amorphous (Figure 3(b)). The same results were also obtained from the SNAY. These results agree with those from XRD.

As the intergranular films were too narrow to be detected by EDX analysis, EDX spectra were taken only from the grain boundary phase at the triple points. The grain boundary triple point of SCNAY was found to contain Y, Si and Al (Figure 4(a)). The similar results were also obtained from the SNAY. The EDX spectra from the pockets are thought to represent those from most intergranular films, since the pockets are linked with intergranular films without any barriers. Therefore, both triple points and intergranular amorphous films in SNAY and SCNAY are assigned to be the yttrium aluminum silicate glass. On the contrary, the EDX spectra from SCN8Y indicated that the grain boundary phase contains Y and Al (Figure 4(b)) . The same results were also obtained for the SN8Y. However, the chemical compositions of the triple points in SN8Y and SCN8Y may be different from those of intergranular films, since XRD profiles (Figure 2) indicated that both SN8Y and SCN8Y contain crystalline grain boundary phases and such crystallization of grain boundary phases may affect the chemical composition of residual glass [12]. Kuroda et al. [13] reported that the amount of nitrogen in the intergranular films is much smaller than that of the grain boundary crystalline phase of $Y_5N(SiO_4)_3$ for Y_2O_3 doped Si_3N_4.

Creep Behavior

A summary of creep tests is shown in Table 1. The creep rate stated here is either the strain rate taken from an "apparently" straight region of creep curve or the minimum strain rate. During the holding time, the SNAY was deformed at low

Table I. Summary of Creep Tests

Specimen Name	Composition (%) * SiC	Al2O3	Y2O3	Applied Sresss (MPa)	Strain Rate (/h)	Life Time (h)	Total Strain (%)
SNAY	0	2	6	56	-----	0	0.78
SN8Y	0	0	8	200	-----	0.4	0.26
SCNAY	30	2	6	100	2.4×10^{-4}	74	2.48
SCN8Y	30	0	8	200	1.7×10^{-5}	1000**	1.70

Notes : * Composition of SiC and additives are shown by vol% and wt% respectively.
** Test was interrupted before failure.

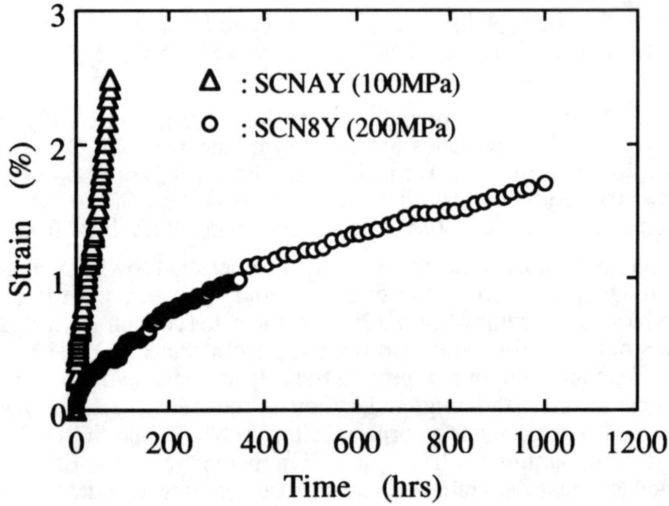

Figure 5. Creep curves of SCNAY and SCN8Y.

applied stress (5.1MPa from the weight of the jigs) and fractured when weights (equivalent to 56MPa) were added. The SN8Y was fractured in 0.4 h after 200 MPa was applied, but did not show the deformation during the holding time. As both SNAY and SN8Y did not show steady state creep, the creep resistance cannot be compared using the steady state creep rate. Since the SN8Y was not deformed at low applied stress during the holding time, the creep resistance of the SN8Y is thought to be superior to that of the SNAY. Pasto et al. also reported the same results that the strain rate of Al_2O_3 and Y_2O_3 doped Si_3N_4 was faster than that of only Y_2O_3 doped Si_3N_4 by 2 orders of magnitude[17]. The difference in the creep resistance between SNAY and SN8Y may be caused by the different grain

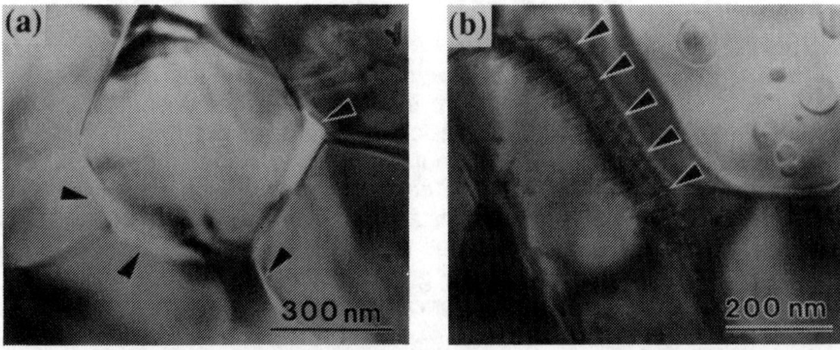

Figure 6. Example of facet-sized cavities in SNAY (a). Nucleation of grain boundary crack introduced by grain boundary sliding in SCNAY (b).

boundary structure and chemical composition as discussed in the previous section. Several researchers have shown that the dominant creep mechanism of Si_3N_4 is grain boundary sliding due to the softening of amorphous phase[1-3]. It is reasonable that the SN8Y which includes intergranular crystalline phases showed the higher creep resistance than the SNAY which has glassy grain boundaries.
In spite of the different applied stress, the tensile creep rate of SCN8Y is much slower than that of SCNAY as shown in Figure 5. This difference may come from the same reason discussed for SNAY and SN8Y, because the grain boundary structure and chemical composition of SCNAY and SCN8Y are similar to those of SNAY and SN8Y, respectively, as described in the previous section. However, the creep resistance of SCN8Y is higher than other 8wt%Y_2O_3 doped hot-pressed Si_3N_4 materials such as Ceradyne HPSN or NCX-34[18]. These are thought to have similar grain boundary phases to SCN8Y. As the creep rate is strongly affected by the grain size, it is not rationalized only by the intergranular crystalline phase that the SCN8Y showed higher creep resistance than other 8wt%Y_2O_3 doped hot-pressed Si_3N_4 which are thought to have larger grains than SCN8Y. The excellent creep resistance of SCN8Y will be due to not only intergranular crystalline phases but also intergranular SiC particles which bond directly on the atomic scale with Si_3N_4 grains. Thus, further research is required to determine the contribution of intragranular SiC particles to the creep behavior of Si_3N_4/SiC nanocomposites.

TEM observations revealed that large cavities were observed in the SNAY in spite of lower applied stress and shorter creep duration. Representative cavities are shown in Figure 6(a). Similar cavities were also found in SN8Y and SCNAY; however, the size and the population of cavities are smaller than those in SNAY. In the SCNAY the nucleation of grain boundary crack introduced by grain boundary sliding was also found, as shown in Figure 6(b). On the contrary, the size and population of cavities in SCN8Y were smaller than those of any other specimens tested in this experiment. These results indicated that grain boundary sliding in the SCN8Y was strongly suppressed by the SiC dispersions.

SUMMARY

Effects of grain boundary structure on creep behavior of hot-pressed Si_3N_4/30vol%SiC nanocomposites were investigated. Comparisons were made among Si_3N_4/30vol%SiC nanocomposites containing 8wt%Y_2O_3, and 2wt%Al_2O_3 +6wt%Y_2O_3 and monolithic Si_3N_4 containing 8wt%Y_2O_3 and 2wt%Al_2O_3+ 6wt%Y_2O_3. Although Si_3N_4/30vol%SiC containing 8wt%Y_2O_3 was characterized by small grain size, remarkable creep resistance was observed. The improvement of high-temperature creep resistance is due to both intergranular crystalline phases and nano-sized SiC particles located at the grain boundaries, both of which strongly suppress the grain boundary sliding and/or cavitation .

Acknowledgement
The authors gratefully acknowledge A.Nakahira and T.Sekino at Osaka University for their discussion and help. A part of this work was supported by the Japan Ministry of Education under a Grant-in-Aid for Science Research No. 03555157 and 04205099, and by the cooperative project "Advanced Materials Creation and Their Limit State Prediction for Environmental Preservation" between ISIR and WRI in Osaka University.

References

1. F. F. Lange, B. I. Davis, and D. R. Clarke, "Compressive Creep of Si_3N_4/MgO Alloys," J. Mater. Sci., 15, 601-18 (1980).
2.T. Chartier, J. -L. Besson, and P. Goursat, "Microstructure, Oxidation and Creep Behavior of a β'-Sialon Ceramics," Int. J. High Tech. Ceram., 2 [1] 33-45 (1986).
3.T. Rouxel, J. -L. Besson, C. Gault, and P. Goursat, "Elastic Properties and Crystallization in SiYAlON Glasses and Ceramics"; pp.351-55 Proceedings of the 1st ECerS (Maastricht, The Netherlands, June 1989). Edited by G.de With, R.A.Terpstra, and R.Metselaar. Elsevier Applied Science, London, U.K., 1989.
4. J. G. Baldoni and S. T. Buljanl, "Creep and Crack Growth Resistance of Silicon Nitride Composites, " ; pp.786-95 in proceeding of Third International Symposium on Ceramic Materials and Components for Engines (Las Vegas, NV, November, 1988). Edited by V.J.Ternnery. American Ceramic Society, Westerville, OH, 1989.
5. B. J. Hockey, S. M. Widerhorn, W. Liu, J. G. Baldoni and S. -T. Buljan, "Tensile Creep of Whisker-Reinforced Silicon Nitride," J. Mater. Sci., 26, 3931-39 (1991).
6.T. Ohji and Y .Yamauchi, "Tensile Creep and Creep Rupture Behavior of Monolithic and SiC-Whisker-Reinforced Silicon Nitride Ceramics," J. Am. Ceram. Soc, 76 [12] 3105-12 (1993).
7. K. Niihara, T. Hirano, A. Nakahira, K. Ojima, K. Izaki and T. Kawakami, "High-Temperature Performance of Si_3N_4-SiC Composites from Fine, Amorphous Si-C-N Powder," pp.107-112 in proceedings of MRS International Meeting on Advanced Materials, vol.5 (Tokyo, Japan, May, 1988). Editted by Y. Hamano, O. Kamigaito, T. Kishi and M. Sakai. Materials Research Society, Pittsuburgh Pa,

1989.
8. K. Niihara, T. Hirano, A. Nakahira, K. Suganuma, K. Izaki and T. Kawakami, "Nanostructure and Thermomechanical Properties of Si_3N_4/SiC Composites Fabricated from Si-C-N Precursor Powders," J. Jpn. Soc. Pow.& Pow. Mett., 36 [2] 243-46 (1989).
9. K. Niihara, K. Suganuma, A. Nakahira and K. Izaki, "Interfaces in Si_3N_4-SiC Nano-Composite," J. Mater. Sci. Lett., 9,598-599 (1990).
10. K. Niihara, K. Izaki and T. Kawakami, "Hot-Pressed Si_3N_4-32% SiC Nano-composite from Amorphous Si-C-N Powder with Improved Strength above 1200℃," J.Mater.Sci.Lett., 10, 112-14 (1990).
11. K. Niihara, T. Hirano, A. Nakahira and K. Izaki, "The Correlation between Interface Structure and Mechanical Properties for Silicon Nitride Based Nanocomposites,"; pp.103-11 in Grain Boundary Controlled Properties of Fine Ceramics. Edited by K. Ishizaki, K. Niihara, M. Isotani and R. G. Ford. Elservier Applied Science, London and New York, 1992.
12. K. Niihara, A. Nakahira, T. Hirano, K. Izaki and F. Wakai "High Temperature Deformation of Si_3N_4 Matrix Nanocomposites," ;pp.42-46 in proceedings of the 11th. Symposium on High-Temperature Materials, (Hakone,Japan), Japan Fine Ceramic Society, Tokyo Japan 1993.
13. J. T. Smith and C. L. Quackenbush, "Phase Effects in Si_3N_4 Containing Y_2O_3 or CeO_2: I Strength," Am. Ceram. Soc. Bull., 59 [5] 529-537 (1980).
14. D. R. Clarke, "On the Equilibrium Thickness of Intergranular Glass Phases in Ceramic Materials," J. Am. Ceram. Soc., 70, 15-22 (1987).
15. M. M. Chadwick and D. S. Wilkinson, "Microstructural Evolution in Annealed and Crept Silicon Nitride," J. Am. Ceram. Soc., 76[2], 376-384 (1993).
16. K. Kuroda, T. Kamino and T. Imura, "Microstructural and Microchemical Characterization of Intergranular Phase in Yttoria-Doped Hot-Pressed Silicon Nitride," J. Electron Microsc.,Vol.35[4], 378-383 (1986).
17. A. E. Pasto, W. C. VanSchalkwyk and F. M. Mahoney, "Creep Behavior of Yttria and Alumina-Doped Silicon Nitride," ; pp.776-785 in proceeding of Third International Symposium on Ceramic Materials and Components for Engines (Las Vegas, NV, November, 1988). Edited by V. J. Ternnery. American Ceramic Society, Westerville, OH, 1989.
18. D. C. Larsen, J. W. Adams, L. R. Johnson, A. P. S. Teotia, and L. G. Hill, "Ceramic Materials for Advanced Heat Engines" ; pp.168-80. Noyes Publications, Park Ridge, NJ.1985.

ANALYSIS OF Ti COMPOUND PARTICLE DISTRIBUTION IN AlN-MATRIX COMPOSITES

Seiji Nakahata, Takahiro Matsuura, Kouichi Sogabe, Akira Yamakawa
Itami Research Laboratory, Sumitomo Electric Industries, Ltd.
1-1-1, Koya-kita, Itami, Hyogo, 664 Japan

ABSTRACT

The Ti compound particle disribution in AlN matrix composites has been studied by X-ray diffraction, scanning electron microscopy, transmission electron microscopy, energy dispersive X-ray, and convergent beam electron diffraction methods. The observation of TiO_2 doped AlN shows Ti compound particles exist inside of AlN crystal grains. These particle diameters are between 200 to 500 nm. Ti compound particles are titanium, nitrogen, aluminum and oxygen ions with rock salt structure. At the interface between the Ti compound particle and the AlN matrix, an AlN crystal lattice distortion is observed, caused by titanium ions diffusing from the Ti compound particles into the AlN matrix.

INTRODUCTION

Aluminum nitride (AlN) has high thermal conductivity (its theoretical value is 319 W/m•K[1]), high electrical resistivity (10^{13} Ω•cm), and high mechanical bending strength (30 kg/mm^2). These excellent properties make it an attractive candidate for packaging materials used in integrated circuits (IC)[2].

Pure AlN crystals with a wurtzite structure (P63mc) are translucent, but Ti compounds such as TiN or TiO_2 doped AlN ceramics blackens AlN, which then becomes opaque. Sogabe[3] reported that the transmittance of infrared rays with wave lengths from 4 to 6 mm is reduced to about one third in 0.5 wt% TiO_2 doped AlN ceramics. However, the structure of Ti compound in AlN ceramics, which are thought to cause AlN ceramics to blacken and become opaque, has not been investigated.

X-ray diffraction (XRD) and electron diffraction in transmission electron microscopy (TEM) were used to analyze the composition and crystal structure of the Ti compound in AlN ceramics. For observation of the small amount of atoms contained in the Ti compound, energy dispersive X-ray (EDX) was used. For observation of the microstructure of Ti compound particles, scanning electron microscopy (SEM) and TEM were used.

To obtain local information on the interface structure between the Ti compound particle and AlN matrix, EDX and convergent beam electron diffraction (CBED) method[4], [5] in TEM was used. The CBED method can detect the lattice parameter in a local area less than ϕ 50 nm by analyzing the Higher Order Laue Zone (HOLZ) lines in bright field diffraction.

In this study, the composition and microstructure of Ti compound particles in AlN ceramics and the interface structure between the Ti compound particle and the AlN matrix was investigated.

EXPERIMENTAL PROCEDURE

AlN powder and 1.0 wt% of Y_2O_3 and 0.5 wt% of TiO_2 were mixed in ethyl-alcohol and molded at 3400 kgf/cm^2 to a pellet-shape. Sintering was performed at a temperature of 1850°C in a nitrogen atmosphere of 1 atm using a graphite furnace. As a comparative sample, an AlN sintered body, which was not doped with TiO_2, was prepared by the same process above mentioned.

The XRD pattern of the sintered sample was obtained using a copper target (Cu Kα) and composition was analyzed in the range of 2θ=20° to 80°. The fracture surface of samples coated with carbon were observed by SEM.

A foil specimen for the TEM observation was prepared by argon ion milling after cutting the sintered body and dimple grinding.

The Transmission Electron Microscope used was JEOL JEM-2000FX equipped with the Tracor Northern TN-2000 EDX system, which is installed at the Research Laboratory of High Voltage Electron Microscope at Kyushu University. The observations were performed at room temperature and at a operating voltage of 200 kV. The specimen was oriented near to the [3 -3 0 1] zone axis to observe the CBED pattern. The diameter of the electron probe was less than about 20 nm for EDX analysis and the CBED method in TEM.

Density and thermal diffusivity of the samples were measured at room temperature by the Archimedes method and laser flash method, respectively.

RESULTS and DISCUSSION

The Properties of AlN Ceramics

The color of TiO_2 undoped AlN sintered body was milky-white and the doped one appeared black. The densities of doped and undoped sintered bodies were 3.28 g/cm^3 and 3.27 g/cm^3, respectively, and exceeded 3.26 g/cm^3 of the theoretical density of AlN, because of the higher density of Y_2O_3 and TiO_2. The thermal conductivity of these samples at room temperature was calculated from the equation $\lambda = C \bullet \alpha \bullet \varrho$, where α is the thermal diffusivity, ϱ is the density and C is the heat capacity, using a density of 3.26 g/cm^3 and heat capacity of 0.736 J/g• K, respectively. Thermal conductivity of 260 W/m• K and 240 W/m• K were obtained for TiO_2 doped and undoped samples, respectively. It is considered that by increasing the second phase (such as Y_2O_3 and TiO_2 which have low thermal conductivity), the thermal conductivity of the AlN sintered body becomes lower. In this study, however, the addition of 0.5 wt% TiO_2 does not adversely affect the thermal conductivity of the AlN sintered body.

Estimation of Ti Compound Particles

<u>Analysis by X-ray diffraction</u> : The composition of the Ti compound in the AlN sintered body was measured by XRD. Figure 1 shows the XRD pattern obtained from the TiO_2 doped sample. Three peaks indicated by arrows in Figure 1 do not correspond to the AlN crystal or $Y_2O_3 \bullet Al_2O_3$ crystal. The three peaks agree with the TiO or TiN crystal with rock salt structure (Fm3m) and the lattice parameter is a=0.4261 nm. As the literature value of TiO crystal [6] and the TiN crystal [7] lattice parameters is 0.41766 nm 0.4235 nm respectively, it is suggested that the composition of the Ti compound in AlN ceramics is similar to that of the TiN compound.

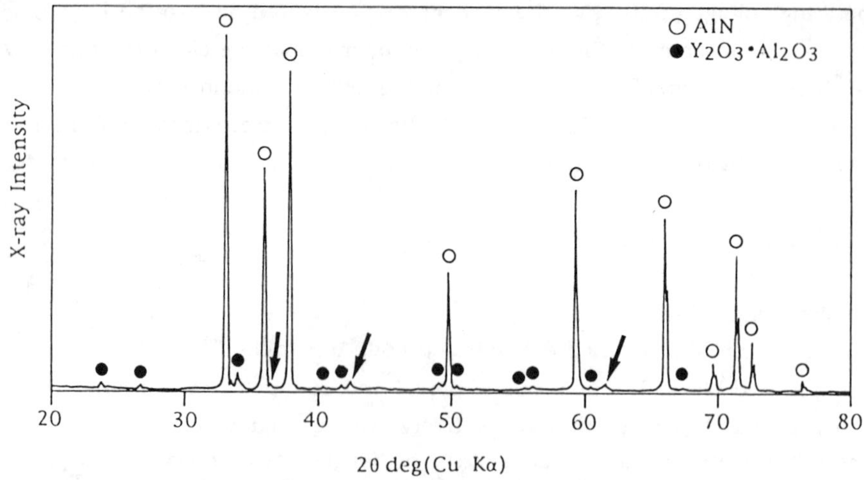

Figure 1 X-ray diffraction pattern obtained from AlN doped TiO_2.

Figure 2 SEM images (a) and (b) obtained from TiO_2 doped AlN and undoped AlN, respectively.

Observation of Microstructure by SEM and EDX: The differences of microstructure of the Ti doped sample and the undoped sample were observed using SEM and the composition of the Ti compound was analyzed using EDX. Figure 2 (a) and (b) show SEM images of fractured surfaces of a doped TiO_2 sample and an undoped sample, respectively. In both samples, the AlN crystal grain size is about 5 to 8 μm, and no differences of AlN crystal grain size can be observed. In Figure 2 (a), however, particles exist inside of the AlN crystal grain, as indicated by the arrow, which is not seen in Figure 2 (b). Figure 3 shows an EDX spectrum obtained from the particle indicated by the arrow in Figure 2 (a). This spectrum shows the particles consisted of titanium and nitrogen ions, as well as small amounts of aluminum and oxygen ions. Here, observed carbon ions were coated on the sample surface.

From the results of XRD and EDX analysis, it is suggested that the compound of this particle is TiN with a rock salt structure containing small amounts of aluminum and oxygen ions. It is believed that the difference of the lattice parameter from the literature value of the TiN crystal was caused by aluminum and oxygen ions dissolved into the TiN crystal.

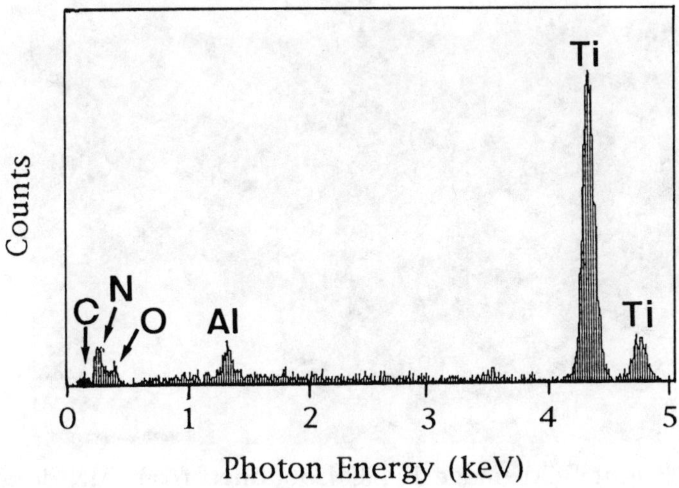

Figure 3 EDX analysis spectrum in SEM obtained from the particle indicated by arrow in Figure 2 (a).

Observation by TEM and EDX: TEM observation and EDX analysis were carried out to more closely observe the microscopic structure of Ti compound particles. Figure 4 shows the bright field images of TEM obtained from a TiO_2 doped sample. The particles with 200 nm to 500 nm diameter included in the AlN crystal grains are observed in Figure 4 indicated by the arrows. Figure 5 (a) and (b) show a bright field image and EDX spectrum obtained from the particle, respectively. Titanium and aluminum ions are observed, but nitrogen and oxygen ions cannot be detected since the beryllium window in the EDX system was used. Electron diffraction was observed to estimate the crystal structure of the particle. Figure 6 shows the electron diffraction pattern at [001] zone axis obtained from the particle. The symmetry of this pattern agrees with the one obtained from crystal with the FCC structure. The lattice parameter estimated from this pattern is a=0.42 nm and agrees with the one measured by XRD in the range of measurement error. Diffused reflections indicated by the arrow in Figure 6 are observed and are believed to be caused by impurities, such as aluminum ions, contained in the TiN particle.

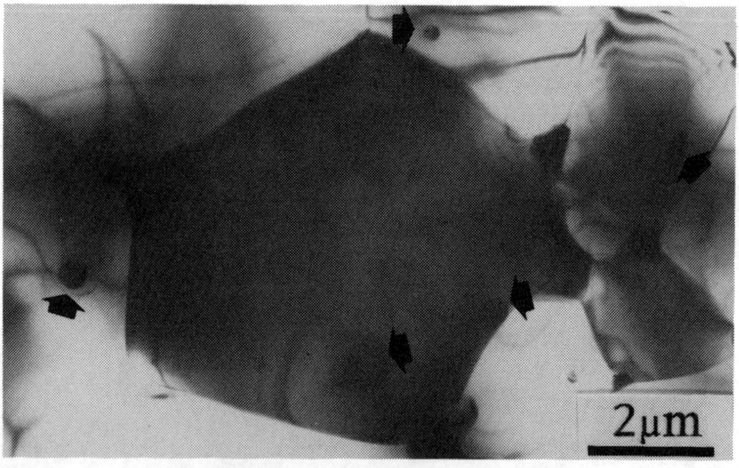

Figure 4 Bright field image in TEM obtained from AlN doped TiO_2.

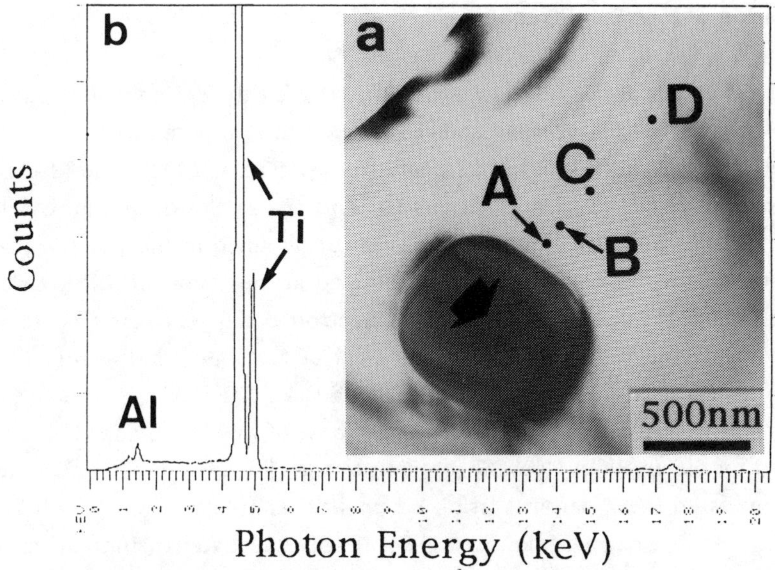

Figure 5 Bright field image (a) obtained from TiO_2 doped AlN and EDX analysis spectrum (b) obtained from the particle indicated by arrow in (a).

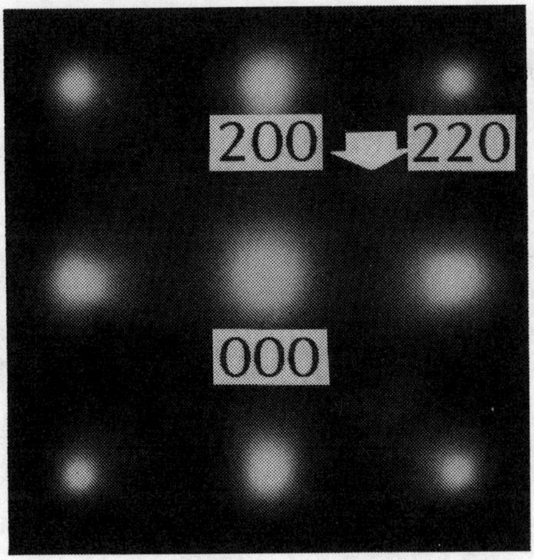

Figure 6 Diffraction pattern at [0 0 1] zone axis obtained from the particle indicated by arrow in Figure 5 (a).

Observation of Interface Structure

For investigation of the interface structure between the Ti compound particle and AlN matrix, the CBED method and EDX analysis in TEM were used.

Figure 7 shows the HOLZ patterns (a) and (b) observed at 200 kV and [3 -3 0 1] zone axis and obtained from positions (D) and (A) as shown in Figure 5 (a), respectively. Broader HOLZ lines in Figure 7 (b) obtained in the vicinity of the TiN particle than those in Figure 7 (a) obtained at the center of AlN grain are observed. This result indicates that the electron beam was scattered in the sample because of the disorderly arrangement of aluminum and nitrogen ions in the AlN crystal in the vicinity of the TiN particle. The cross positions of the HOLZ line of (0 3 -3 10) and (-3 0 3 10), and those of (3 0 -3 -8) and (0 -3 3 -8), indicated by the arrow in Figure 7 (b), shift to the [1 1 -2 0] direction. Figure 8 shows simulated HOLZ patterns using kinematical approximation[8] at the 200 kV and [3 -3 0 1] zone axis. Figure 8 (a) was calculated using the lattice parameter a=b=0.3111 nm and c=0.4980 nm; on the other hand, (b) was calculated using a=0.3145 nm, b=0.3111 nm, and c=0.4980 nm. This result suggests that the shift of the cross position of HOLZ lines is caused by the difference of lattice parameters of a and b axis.

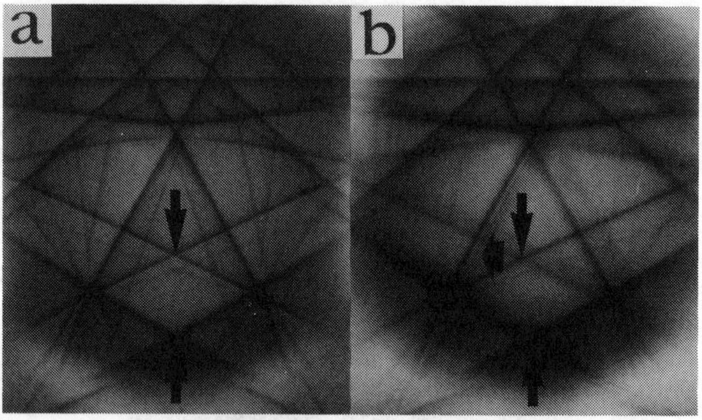

Figure 7 [3 -3 0 1] HOLZ patterns (a) and (b) obtained from positions (D) and (A) indicated in Figure 5 (a).
The direction of the arrow indicated in (a) show [-1 -1 2 0] axis.
Note the change in position of crossing.

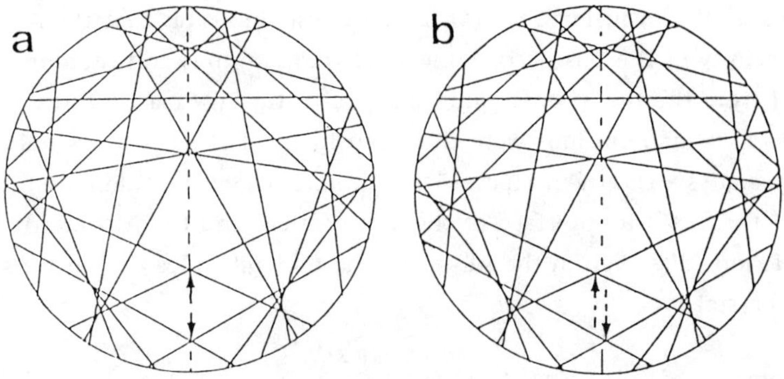

Figure 8 Simulated HOLZ patterns at the [3 -3 0 1] zone axis by kinematical approximation. These HOLZ patterns (a) and (b) were simulated using lattice constants a=b=0.3111 nm, c=0.4980 nm and a=0.3145 nm, b=0.3111 nm, c=0.4980 nm, respectively.

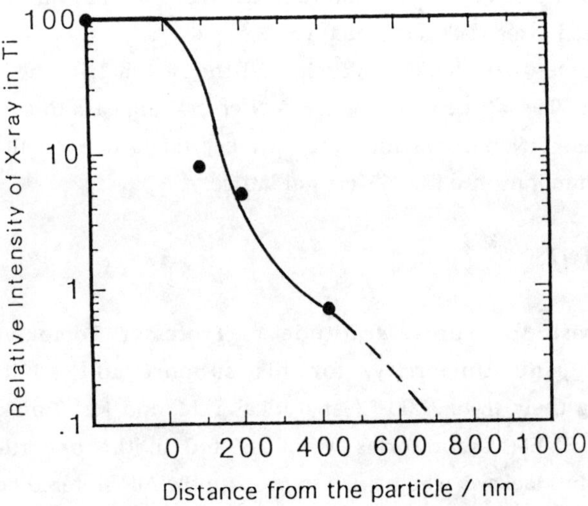

Figure 9 Concentration profile for Ti ion in AlN crystal grain in the vicinity of the TiN particle.

For analysis of the composition of the interface between the TiN particle and the AlN matrix, EDX analysis was performed. Figure 9 shows the concentration profile for Ti ions analyzed from EDX spectrums obtained from various positions as shown in Figure 5 (a). EDX analysis indicates that the AlN crystal in the vicinity of the TiN particle included titanium ions and titanium ions are diffused from the the Ti compound particle to the AlN matrix in the range of 400 nm. As aluminum ions were dissolved in the TiN particle, as indicated in the EDX analysis shown in Figure 5 (b), titanium ions and aluminum ions are diffused through the interface of the TiN particle and the AlN matrix. In this study, Ti ions dissolved in the AlN crystal is thought to lead to lattice strain of the AlN crystal.

CONCLUSION

TiO_2 doped AlN has been sintered and the morphology of Ti compound particles described. Ti compound particles exist inside of the AlN crystal grains. The diameters of the Ti compound particles are from 200 nm to 500 nm, and their compositions consist of titanium, nitrogen, aluminum and oxygen and these particles have a rock salt structure.

In the interface between the TiN particle and the AlN matrix, the AlN crystal lattice is strained. The lattice strain of the AlN crystal appears to titanium ions diffused from the TiN particle into the AlN crystal lattice result from the diffusion of titanium ions into the AlN crystal lattice.

ACKNOWLEDGMENTS

The authors wish to express gratitude to Professor Tomokiyo of HVEM Laboratories, Kyushu University, for his support and advice during observation using their JEOL 2000-FX analytical TEM, and Mr. Tanaka and Mr. Manabe also of HVEM Laboratories, who assisted in the operation of the microscope and Mr. Maekawa who helped to prepare the AlN sintered body.

REFERENCE

[1] G. A. Slack, R. A. Tanzilli, R. O. Pohl, and J. W. VanderSande, "The Intrinsic thermal Conductivity of AlN" J. Phys. Chem. Solids, **48** [7] 641 (1987)

[2] N. Kuramoto and H. Taniguchi, "Transparent AlN Ceramics," J. Mater. Sci., Lett., **3**, 471 (1984)

[3] K. Sogabe, A. Yamakawa and M. Miyake, "Properties of AlN Nanocomposite" p 21 in Proceedings of the Ceramics Society of Japan, 1991.

[4] J. A. Eades, ed, Special issues of the Journal of Electron Microscopy Technique on Convergent Beam Electron Diffraction, vol 13. Nos.1, 2 (1989).

[5] J. W. Steeds, p 387 in Introduction to analytical Electron Microscopy. Edited by J. J. Hren, J. J. Goldstein, and D. C. Joy. Plenum, New York, 1979.

[6] M. E. Straumnis and H. W. LI, Z. Anorg, "Gitterkonstanten, Ausdehnumgskoeffizienten, Dichten, Fehlordnung und Aufbau der Phase Titan(II)-Oxyd" Allegem. Chem., **305**, 143 (1960)

[7] Francis S. Galasso, International Series of Monographs in Solid State Physics Volume 7, Structure and Properties of Inorganic Solids, Pergamon Press Ltd., Oxford.

[8] M. Tanaka, M. Terauchi, T. Kaneyama, Convergent Beam Electron Diffraction, Tokyo. JEOL-Maruzen (1988).

Characterization of V/Al_2O_3 Interface

Y.Ikuhara*, P.Pirouz**, A.H.Heuer**, C.P.Flynn***
and Y.Kubo*
Japan Fine Ceramics Center*, 2-4-1 Mutsuno,
Atsuta-ku Nagoya, 456,Japan
Case Western Reserve University**, Cleveland OH
44106, USA
University of Illinois at Urbana-Champaign***,
Urbana, IL 61801, USA

ABSTRACT

Vanadium films were deposited on both the (0001) basal plane and $(1\bar{1}02)$R-plane of sapphire by molecular beam epitaxy (MBE). The interface structures between vanadium and sapphire were studied by conventional and cross-sectional high resolution electron microscopy (HREM) to clarify the detailed atomic configuration at the interface. The HREM observations were performed from three directions for both samples; two cross-sectional views and a plan view. SADPs (Selected Area Diffraction Patterns) indicated that vanadium and sapphire have an unique three dimensional orientation relationship irrespective of the substrate orientation. Plan-view analysis showed moire fringes which is related to geometrical misfit dislocations at the interface. Cross-sectional HREM observations showed the atomistic configuration of the interface, and the existence of periodic arrays of geometrical misfit dislocations. Computer-simulations showed that sapphire is aluminum-terminated for the basal plane and oxygen-terminated for the R-plane at the interface with vanadium.

INTRODUCTION

The study of the atomic structure of metal/ceramics interfaces is finding increasing importance because of the influence of interfaces on the mechanical and electronic properties of metal/ceramics composites. Consequently, the structural characterization of such interfaces play an important role for understanding such properties.

Transmission electron microscopy (TEM), in particular, high resolution electron microscopy (HREM), is one of the most useful techniques to characterize the interface. HREM has been successfully applied to various metal/ceramics and ceramics/ ceramics interfaces [1,2].

The Nb/Al_2O_3 interfaces has been extensively investigated as a model system, and it has been found that the atomic structure of Nb/Al_2O_3 interface depends not only on substrate orientation, but also on the technique employed in manufacturing the composites [3-5]. The V/Al_2O_3 system, like the Nb/Al_2O_3 system, in many ways provides a model system for understanding the structure of a metal/ceramic interface. Thus, for instance, the thermal expansion mismatch between V and Al_2O_3 (as well as between Nb and Al_2O_3) is relatively negligible and hence thermal mismatch does not play an important role in determining the interface structure. In addition, a comparison of the two systems Nb/Al_2O_3 and V/Al_2O_3 is in fact interesting because the misfit parameter between V and Al_2O_3 is much larger than that of Nb/ Al_2O_3.

The objectives of this study are to clarify the orientation relationship between V and Al_2O_3, the atomic structure of the

V/Al_2O_3 interface, and the nature of misfit dislocations accommodating the large lattice mismatch between the two components by selected area diffraction pattern (SADP), conventional TEM and HREM.

EXPERIMENTAL PROCEDURE

Vanadium films with a thickness of about 50 nm were deposited by molecular beam epitaxy (MBE) at a rate of 0.03 nm/min on the (0001) basal plane and $(1\bar{1}02)$R-plane of sapphire at a substrate temperature of 920 K in a vacuum of 10^{-10} torr.

TEM specimens were prepared using standard techniques involving mechanical grinding to a thickness of 0.1 mm, dimpling to a thickness of 20 μm and ion beam milling to electron transparency at 4 kV in a cold stage (liquid nitrogen) in order to minimize specimen damage [6]. The samples were subsequently examined in a top-entry JEOL 4000EX high resolution electron microscope with a point-to-point resolution of better than 0.18 nm from three directions: two cross-sectional views and plan view .

RESULTS AND DISCUSSION

Orientation relationship

Figure 1 shows SADPs obtained from the interfaces for (a)the basal plane substrate and(b) the R-plane substrate. These SADPs were taken from cross-sectional view parallel to $[10\bar{1}]_V//[1\bar{2}10]_{Al_2O_3}$ and $[110]_V//[\bar{1}\bar{1}20]_{Al_2O_3}$, respectively, from which the following orientation relationships were obtained for the basal plane substrate and the R-plane substrate.

$$(111)_V//(0001)_{Al_2O_3}$$
$$[10\bar{1}]_V//[1\bar{2}10]_{Al_2O_3} \qquad (1)$$

$$(001)_V//(1\bar{1}02)_{Al_2O_3}$$
$$[110]_V//[\bar{1}\bar{1}20]_{Al_2O_3} \qquad (2)$$

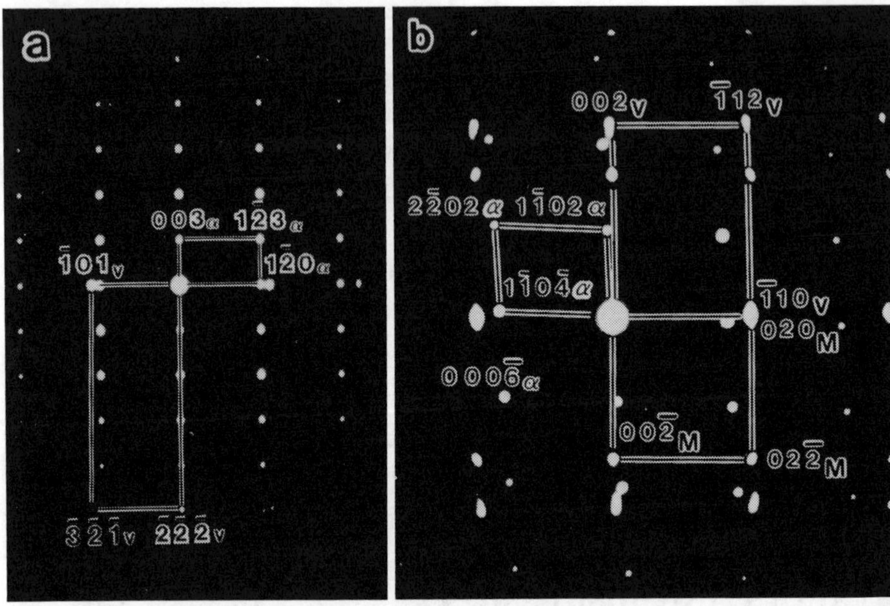

Fig. 1. Selected area diffraction patterns along (a) $[10\bar{1}]_V//[1\bar{2}10]_{Al_2O_3}$ cross-sectional direction for basal plane substrate, and (b) $[110]_V//[\bar{1}\bar{1}20]_{Al_2O_3}$ cross-sectional direction for R-plane substrate.

Equations (1) and (2) are equivalent in crystallography, indicating that there is an unique (three-dimensional) orientation relationship between V and Al_2O_3. These orientation relationships are the same as the one obtained for the Nb/Al_2O_3 system [4,5].

Interfacial dislocations

Figure 2 (a) is a cross-sectional HREM micrograph of the interface for the basal plane substrate obtained from the $[10\bar{1}0]_{Al_2O_3}$// $[1\bar{2}1]_V$ direction. In this micrograph, the basal planes of alumina are parallel to the interface, and $(10\bar{1})_V$ planes of vanadium are parallel to the $(1\bar{2}10)_{Al_2O_3}$ planes of sapphire and normal to the interface. An array of misfit dislocations can be seen, which occur periodically at every eleven (on the average) $(10\bar{1})_V$ planes, corresponding to about ten $(1\bar{2}10)_{Al_2O_3}$ planes. The large misfit of 10.6% between $(10\bar{1})_V$ and $(1\bar{2}10)_{Al_2O_3}$ planes is accommodated by these misfit dislocations. The spacing between these misfit dislocations is 2.2 nm.

Figure 2(b) shows a cross-sectional HREM micrograph of the interface for the R-plane substrate observed along the $[0\bar{2}2\bar{1}]_{Al_2O_3}$//$[010]_V$ direction. The interface is quite smooth, sharp and free of any interfacial phases. The interface plane is almost parallel to $(001)_V$ and $(1\bar{1}02)_{Al_2O_3}$, but there is a misorientation about 3° with respect to $(1\bar{1}02)$ plane of Al_2O_3. This crystal tilt is originated from the tendency to maintain the three-dimensional orientation relationship of V with respect to Al_2O_3 [7]. At the interface, there is one extra half-plane (arrowed) at every ten $(\bar{1}01)_V$ planes on average (i.e. a separation of 3.1nm) even though the misfit parameter is as large as 8.4% between the projection of $(\bar{1}01)_V$ and $(2\bar{1}\bar{1}0)_{Al_2O_3}$ planes [6].

According to Bollmann [8], a moire pattern is closely related to the location of misfit dislocations. In fact, a moire pattern from two overlapping mismatched lattices ideally represents the position of these misfit dislocations and forms a Wigner-Seitz cell. The moire fringe obtained from plan-view direction for basal plane substrate, i.e. the [111] zone-axis of V showed that moire fringes lie parallel to <121> directions of vanadium and form a hexagonal

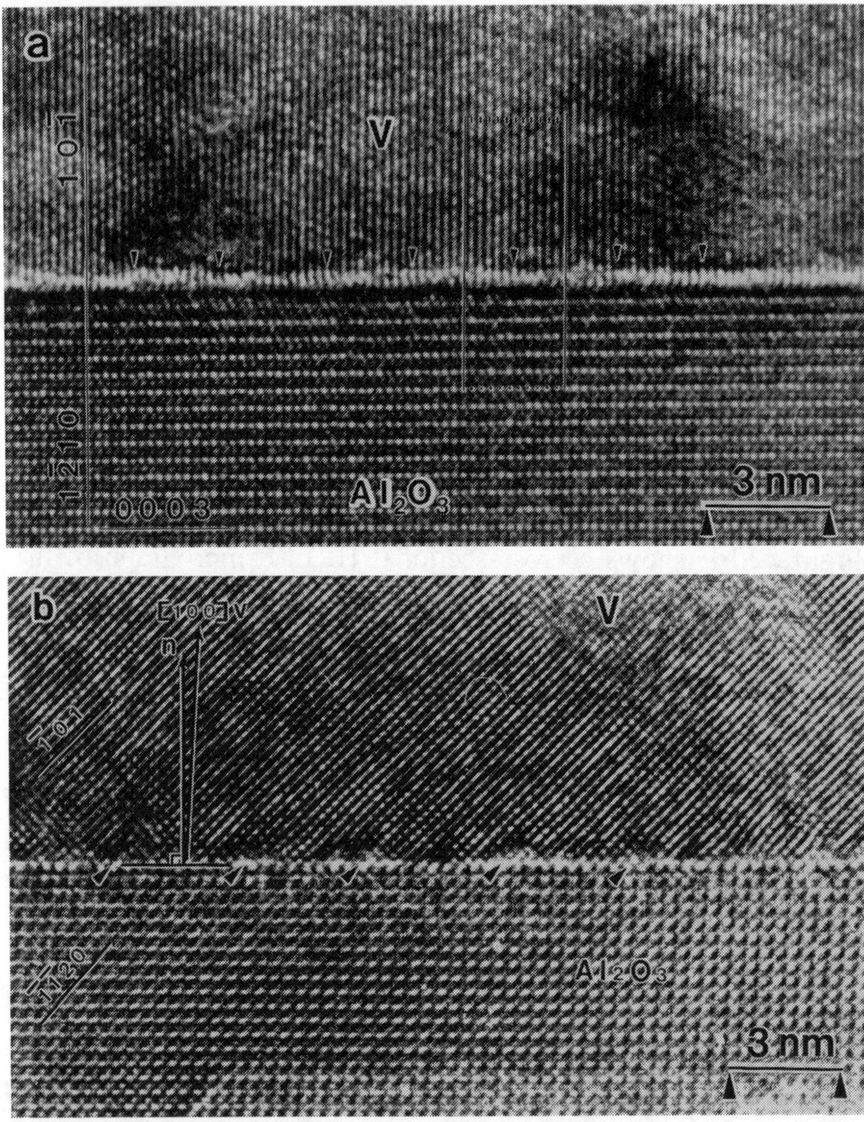

Fig. 2. Cross-sectional HREM micrographs of the V/Al_2O_3 interface along (a) the $[10\bar{1}0]_{Al_2O_3}$// $[1\bar{2}1]_V$ for basal plane substrate, and (b) along the $[0\bar{2}2\bar{1}]_{Al_2O_3}$//$[010]_V$ direction for the R-plane substrate. Note that both micrographs show a periodic array of geometrical misfit dislocations as indicated by arrows.

set of fringes, and the spacing of moire fringes is ~2.2 *nm*., which is consistent with the value determined from cross-sectional HREM observations of the interface in Fig.2 (a). On the other hand, the orthogonal sets of moire fringes were observed for the R-plane substrate [9] . The moire fringe spacing along $[1\bar{1}0]$ is 1.3nm, which is also consistent with the value from cross-sectional HREM observations with the incident beam parallel to $[\overline{11}20]_{Al_2O_3}//[110]_V$. Thus, the spacing of moire fringes are equal to the location of present misfit dislocations.

In general, when the mismatch between two crystals is large, the interface between them may be incoherent with a complete lack of, or very weak, interfacial bonding. But, the present study showed that the dislocation-like contrast was observed along the interface, and the spacing of this contrast is related to that of moire fringes. In addition, cross-sectional CTEM (Conventional Transmission Electron Microscopy) observations also showed that a periodic array of features exhibiting strain contrast along the interface [6].

The fact that there is an unique orientation relationship in the present system, V/Al_2O_3, implies that there is moderate interfacial bonding between vanadium and sapphire and that the (metal) atoms in the immediate vicinity of the interface will be displaced slightly from their lattice positions to accommodate the bonding configuration, which minimizes the energy of the system. This displacement of (metal) atoms at the interface in turn implies a very localized strain at the interface. This localized strain is considered to correspond to the dislocation-like contrast in the CTEM images; however, this has not a proper Burgers vector. We call this localized strain geometrical misfit dislocations, because the spacing between them is fixed geometrically. In a system with a large lattice mismatch but weak interfacial bonding, a set of geometrical misfit dislocations are present. These dislocations are obtained from the overlap of two dissimilar lattice spacings just like a set of moire fringes. Geometrical misfit dislocations across other sets of parallel planes with different spacings can be seen at the interface from any direction, as long as the parallel planes in the two lattices are parallel to the incident beam and the interplanar spacings are within the resolution of the electron microscope (~0.18 nm in the present case).

Atomic Structure of the V/Al_2O_3 Interface

To determine the location of atoms at the V/Al_2O_3 interface, computer simulations were performed on the various structure models. The simulations were actually carried out over a range of thicknesses from 4 to 25 nm, and range of defoci from -10 nm to -80 nm for various models. The best fit interface models and simulated results for(a) basal plane and (b) R-plane are shown in Figure 3, in which the simulated images are obtained at thicknesses of 8 nm, 8 nm and a defocus values Δf= -65 nm, -56 nm, respectively.

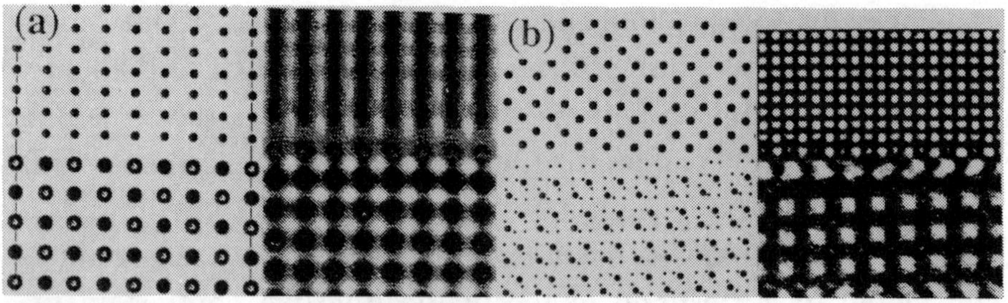

Fig. 3. Models and computer-simulated images of the V/Al_2O_3 interface on (a) basal plane and (b) R-plane of sapphire. Upper and lower crystals correspond to vanadium and Al_2O_3, respectively. The smallest circles represent the oxygen atoms in Al_2O_3.

The results indicate that the vanadium deposited on the basal plane of sapphire bonds to an aluminum layer at the interface, and that vanadium atoms are bonded to oxygen atoms which form an extra monolayer on the original O-terminated R-plane. Recently, Guo, Ellis and Lam [10] have carried out detailed calculations on the electronic structure and energetics of the basal plane and R-plane of sapphire surface and concluded that the lowest cleavage energy for these surface corresponds to the surfaces which was terminated by an aluminum plane and oxygen plane, respectively. These results also support the present results.

REFERENCES

1. M. Rühle, A. G. Evans, M. F. Ashby and J. P. Hirth, Eds., *Metal-Ceramics Interfaces* , Pergamon Press, New York, (1990).
2. M. Rühle, A. H. Heuer, A. G. Evans and M. F. Ashby, Eds., Proc. of the International Symposium on Metal-Ceramic Interfaces, *Acta Metall. Mater.*, **40** (1992).
3. W. Mader and M.Rühle, " Electron Microscopy Studies of Defects at Diffusion-Bonded Nb/Al_2O_3 Interface" *Acta Metall. Mater.*, **37** , 853-866 (1989)
4. D.X.Li,P.Pirouz, A.H.Heuer, S.Yadavalli and C.P.Flynn, " The characterization of Nb-Al_2O_3 and Nb-MgO interfaces in MBE grown Nb-MgO-Nb-Al_2O_3 multilayers", *Acta Metall.Mater.*, **40**, S237-248 (1992)
5. J.Mayer, C.P.Flynn and M.Rühle,"High-Resolution Electron Microscopy Studies of Nb/Al_2O_3 Interfaces", *Ultramicroscopy*, **33**, 51-61 (1990)
6. Y. Ikuhara, P. Pirouz, A. H. Heuer, S. Yadavalli and C.P.Flynn, "Structure of V-Al_2O_3 interfaces grown by molecular beam epitaxy" *Philos.Mag. A.*, **69** (1994)
7. Y.Ikuhara and P.Pirouz, "A high-resolution electron microscopy study of vanadium deposited on the basal plane of sapphire", *Ultramicroscopy*, **52**, 421-428 (1993)
8. W. Bollmann, *Crystal Defects and Crystalline Interfaces*, Berlin, Springer-Verlag (1970).
9. Y. Ikuhara, P. Pirouz, A. H. Heuer, C. P. Flynn and S. Yadavalli, " A High-Resolution Electron Microscopy Study of V/Al_2O_3 Interface", *Proc.50th EMSA Meeting, Boston,* San Francisco Press, 146-147 (1992)
10. J.Guo, D.E.Ellis and D.J.Lam, "Electronic structure and energetics of sapphire (0001) and ($1\bar{1}02$) surfaces", *Physical Rev.B* **45,** 647-656 (1992)

INTERFACE STRUCTURE OF CERAMIC/METAL NANOCOMPOSITES

Tohru SEKINO, Toshio NAKAJIMA, Shyuzo MIHARA, Satoru UEDA and Koichi NIIHARA
The Institute of Scientific and Industrial Research, Osaka University
8-1 Mihogaoka, Ibaraki, Osaka 567, Japan

ABSTRACT

Sintering processes, microstructure and mechanical properties have been investigated for the Al_2O_3-based composites containing nanometer-sized metal dispersoids such as W, Ni, and Ti. The Al_2O_3/metal nanocomposites were fabricated by controlled reduction and hot-pressing of fine Al_2O_3 and metal-oxide powder mixtures in hydrogen gas. During these processes, the metal oxides were reduced to metal particles. TEM observation revealed that the nanometer-sized W and Ni particles were dispersed in the matrix with various sizes and dispersion states depending on the metal phase and also the preparation processes. High-resolution TEM observation revealed that no impurity phases were found at the interfaces between the W, Ti and Ni metal particles and Al_2O_3 matrix. However, the interface structures were disturbed in the Al_2O_3/Ni and Ti systems. The mechanical properties at room and high-temperatures of these composites were improved by dispersing small amounts of nanometer-sized metal particles.

INTRODUCTION

Recent research on the nanocompositing of ceramic materials has shown the high possibility for improving mechanical properties of ceramic-based composites [1-8]. On the basis of this concept, we succeeded in developing new types of ceramic/metal nanocomposites, Al_2O_3/W [9,10] and Al_2O_3/Mo [11], by using a conventional sintering technique. The ceramic/metal nanocomposites are characterized by those typical microstructures, with nanometer-sized metal particles dispersed throughout the ceramic matrix. The nanometer-sized second dispersion has a strong effect on the physical properties of matrix grains, and the resultant mechanical properties of these nanocomposites are significantly enhanced[10,11]. However, the relationships between mechanical properties and nano-structure, as well as correlation between mechanical properties and interface structure, are not well-understood for the ceramic/metal nanocomposites.

In this study, the fabrication processes and mechanical properties will be described for Al_2O_3/W, Al_2O_3/Ni and Al_2O_3/Ti composites. Observation of characteristic nano- and interface structures by the transmission electron microscopy will be also discussed for these composite systems.
℃

EXPERIMENTAL PROCEDURE

Materials and Fabrication Processes

The Al_2O_3/W nanocomposites were prepared by the reduction and following hot pressing technique (R/H method) [10]. In this method, mixtures of α-Al_2O_3(Sumitomo Chem. Co., AKP-53, particle size less than 0.2 μm) and WO_3 powder (Nakarai Chem. Co., 350 mesh) corresponding to 2.5 to 10 vol.% of W metal, were obtained by ball milling. They were reduced by H_2 gas in a die with two steps of heating at 900℃ and 1100℃. The hot pressing was consecutively carried out at 1400℃ under an applied pressure of 30 MPa for 1 hr in Ar atmosphere. Another process for fabricating the Al_2O_3/W nanocomposites was the powder-metallurgical (PM) method [9]: The 2.5 to 30 vol.% of fine W powder (Nippon Tungsten Co., particle size less than 0.6μm) was mixed with α-Al_2O_3. The mixture was hot-pressed at 1400 to 1600℃ under an applied pressure of 30 MPa for 1 hr in Ar atmosphere.

For the Al_2O_3/Ni composites, both nickel oxide (NiO) and nickel nitrate ($Ni(NO_3)_2 \cdot 6H_2O$) were used as starting materials of the Ni phase. These materials, corresponding to the 5 to 20 vol.% of Ni in the final sintered bodies, were mixed with α-Al_2O_3 by ball milling. The dried mixtures were calcinated at 400℃ in air and ball milled again. Hot press sintering was carried out at 1450℃ using the same procedure as described above.

Reducing the titanium oxide (TiO_2 or Ti_2O_3) to metallic Ti is so difficult because of its high stability that we applied another procedure for the Al_2O_3/Ti composite. In this system, titanium dihydride (TiH_2) was selected as a source material. TiH_2 was mixed with α-Al_2O_3 by wet ball milling. Dried mixtures, which contained 5 vol.% of Ti, were hot pressed at 1500℃ for 1 hr under a vacuum condition. During this process, TiH_2 was decomposed to metallic Ti.

Characterization of Al_2O_3/Metal Nanocomposites

Sintered bodies were machined and polished to observe mechanical properties. Fracture strength was measured by the 3-point bending test (span; 30 mm, cross-head speed; 0.5 mm/min) up to 1300℃ in Ar atmosphere. Fracture toughness was estimated by the indentation fracture (IF) method with 49N-load and the indentation strength in bending (ISB) method [12,13]. Hardness was estimated from the Vickers indentation with 49N-load at room-temperature and with 9.8N-load up to 1400℃ under vacuum.

Transmission electron microscopy (TEM; HITACHI, H-8100) was used to observe the micro-, nano- and interface structures of composites. The W distribution in the composite was estimated from the TEM photographs by using an image analysis technique.

RESULTS AND DISCUSSION

Microstructural Observation

The Al_2O_3/W and Al_2O_3/Ni composites were composed of only α-Al_2O_3 and the corresponding metals, and no reaction phase or remained oxide was found by the XRD analysis. In the Al_2O_3/Ti composite, however, a small amount of TiO was found. This may result from the oxidation of TiH_2 or the reduction of Ti during sintering.

Typical TEM images of the Al_2O_3/W composites fabricated by the PM method and R/H method are shown in Figures 1(A) and 1(B), respectively. The W particles in the specimen by the PM method were located both within the Al_2O_3 matrix grains and at the grain boundaries. Contrary to the PM method, much finer W particles were dispersed within the matrix grains in the composite fabricated by the R/H method, although large W particles of over 1 μm were still located at the Al_2O_3 matrix grain boundaries (Figure 1(B)).

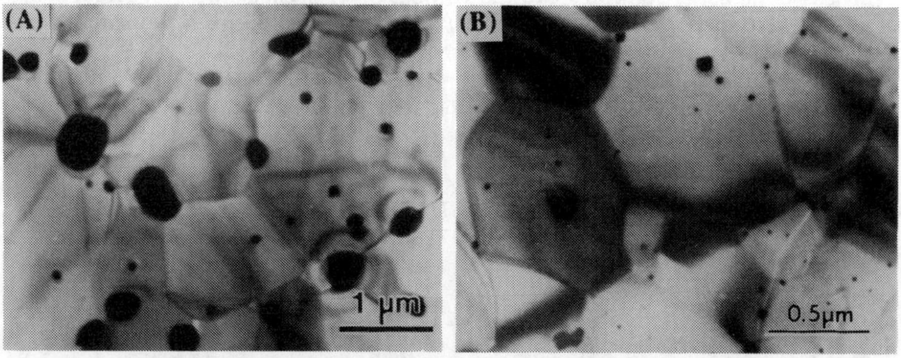

Figure 1 TEM images of the Al_2O_3/5vol.%W nanocomposites prepared from W powder(A) and WO_3 powder (B).

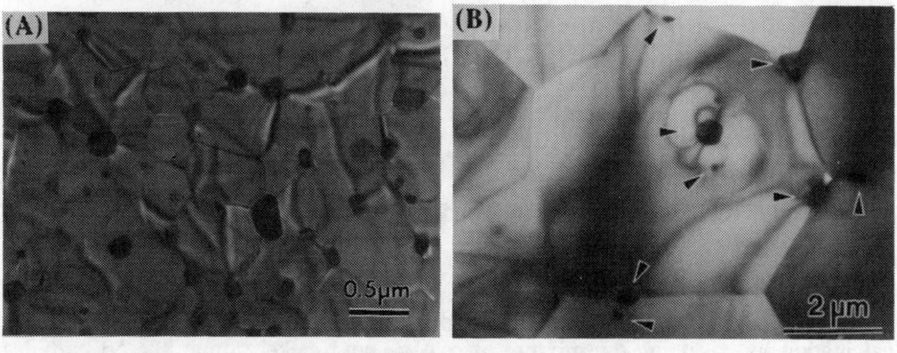

Figure 2 TEM image of the Al_2O_3/5vol.%Ni nanocomposite (A) and Al_2O_3/5vol.%Ti composite prepared from TiH_2 (B).

A homogeneous dispersion of a fine-grained metal phase was also observed for the Al_2O_3/Ni composites as shown in Figure 2(A). Most of the Ni particles, however, were dispersed at the grain boundary or at the triple point in the specimen prepared from nitrate. In the Al_2O_3/Ti system, most of the Ti particles were about 1 μm in diameter and located at the grain boundaries. Some small particles, however, were dispersed within the Al_2O_3 matrix grains as show in Figure 2(B). Compared to the Al_2O_3/W and Al_2O_3/Ni systems, many stress contrasts were found as seen in Figure 2(B), which may indicate the existence of a large strain in the specimen.

The particle-size distributions of W in the Al_2O_3/5vol.%W composites prepared by PM and R/H methods are shown in Figure 3. The W distribution in the composite by the PM method was found to reflect that of the raw W powder. However, that of the W in the R/H specimen was very narrow, and almost all W particles were less than 80 nm. The average particle size and the inter-/ intragranular dispersion ratio in number of the metal dispersoid were evaluated for the Al_2O_3/5vol.%W and Al_2O_3/5vol.%Ni composites and summarized in Table I. In case of the Al_2O_3/5vol.%W composite, about 65% of the number of W particles with an average size of 41 nm were dispersed within the matrix grains for the specimen prepared by the R/H method. In the specimen by the PM method, however, only 17% of the intragranular W particles was observed with the average size of 137 nm. This observation indicates that the R/H method produces an intragranular dispersion of much finer W particles.

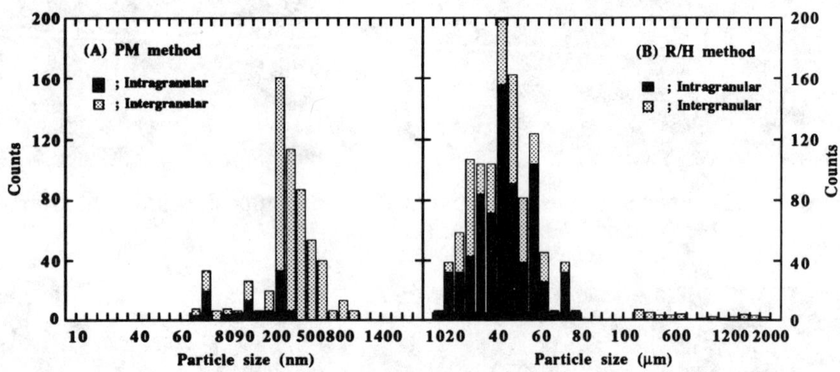

Figure 3 Particle size distribution of the Al_2O_3/W nanocomposites prepared using W powder (A) and WO_3 powder (B).

In the Al_2O_3/5vol.%Ni prepared from nickel nitrate, the average Ni particle size in the specimen was 96 nm. However, the intragranular dispersion was only 18%, which was much smaller than the Al_2O_3/W composites by the R/H method. Thus, the Al_2O_3/Ni composite is regarded as an intergranular type of nanocomposites. From TEM observation for the composite powder, which was obtained by calcinating and reducing the mixture of α-Al_2O_3 and nickel nitrate, the Ni particles of about 20 nm were found to precipitate homogeneously around the α-Al_2O_3

particles. In addition, the sintering temperature of Al_2O_3/Ni composites in this experiment was 1450℃, which was almost the same as the melting point of Ni. The correlation of sintering temperature and melting point of Ni as well as characteristic structure of the composite powder might influence the typical intergranular type of nano-structure in the sintered body for the Al_2O_3/Ni system.

Table I Fraction and average particle size of the intra-/intergranular metal particles in the Al_2O_3/5vol.%W and Al_2O_3/5vol.%Ni nanocomposites.

Starting mixture	Dispersed metal in composites: Intragranular fraction(%)/size(nm)	Intergranular fraction(%)/size(nm)	Total average size(nm)	Al_2O_3: Grain size (μm)
Al_2O_3/W	16.6/137	83.4/340	306	0.96
Al_2O_3/WO_3	65.3/41	34.7/107	64	1.17
Al_2O_3/NiO	33.7/101	66.3/224	185	0.96
Al_2O_3/$Ni(NO_3)_2$	17.9/49	82.1/130	96	0.64

Note ; fraction of the metal dispersion is in particle number.

The interface between Al_2O_3 and the metal dispersion is thought to have an effect on the mechanical and physical properties of the composites. To reveal the Al_2O_3/metal interface structure in the composites, high resolution TEM observations were carried out. Figure 4 shows the interface between the Al_2O_3 matrix and the intragranular W nanoparticle in the Al_2O_3/5vol.%W composite fabricated by the R/H method. As is clearly seen from this photograph, the nano-sized W particle was bonded directly with Al_2O_3 grains without any impurity phase, and facet planes were generated at the interface. A ledge-structured interface in the atomic-layered scale was also observed. No dislocation, however, was observed within the intra-

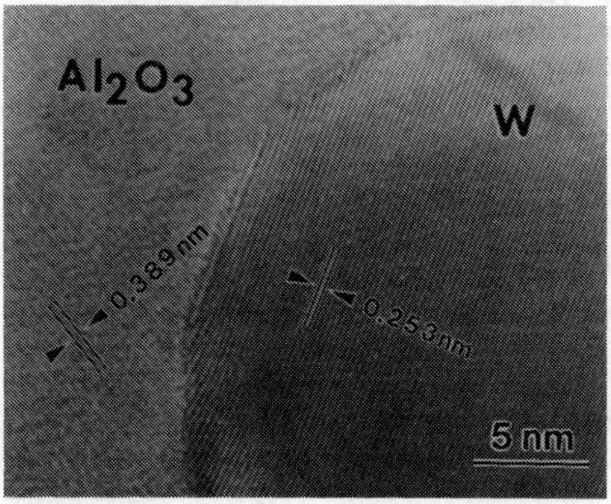

Figure 4 HR-TEM image of the Al_2O_3/intragranular W interface.

granular W particle. These characteristic structures not only at the interface but also in the W particle will play an important role to relax the localized interface strain and to stabilize the intragranular type of nanostructure.

Figure 5 shows the high resolution images of the Al_2O_3/Ni interface prepared by using NiO(A), and Al_2O_3/Ti interface(B). In both systems, the distinct inter-mediate phase was not observed at the interfaces although the metal phases were large (micrometer-sized metal particles) and located at the grain boundaries. However, the structure of both interfaces were strongly disturbed as clearly seen from the pictures. Further observation on the interface structure, particularly between the Al_2O_3 matrix and nano-sized Ni or Ti particle, is under investigation.

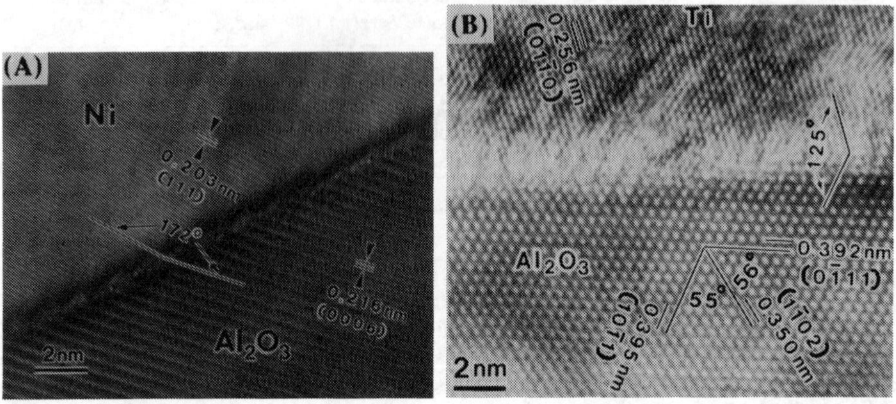

Figure 5 HR-TEM images of the interfaces between Al_2O_3 and Ni interface (A) and between Al_2O_3 and Ti interface (B). In both pictures, metal particles are large and existing between the matrix grains.

Mechanical Properties of Al_2O_3/Metal Composites

Mechanical properties of the Al_2O_3/metal nanocomposites observed at room-temperature are summarized in Table II. The Al_2O_3/W composites which contained finer nano-sized W dispersion indicated higher strength and hardness as expected from the microstructural observation results. The fracture strength of over 1.1 GPa was achieved for the Al_2O_3/5vol.%W composites. It must be noted that many intragranular W particles are dispersed in this composite as mentioned before. From these observations, it is reasonable to think that the strengthening effect is accelerated when a finer and larger number of nano-sized particles are incorporated within the matrix grains. The strengthening by the nano-sized W dispersion may be attributed to the existence of a localized residual stress by the thermal expansion mismatch between Al_2O_3 matrix and W dispersion. The clean Al_2O_3-W interface structure will be another important factor for the hardening and strengthening of the matrix grains, since the strong bonding between the matrix and metal particles are expected.

The Al_2O_3/Ni composite fabricated from nickel nitrate also indicated a high strength of about 1.1GPa. However, the strengthening mechanism may be different from those discussed for the Al_2O_3/W system. In this system, the correlation of thermal expansion coefficient between matrix Al_2O_3 and Ni dispersion can not induce the effective stress field around the dispersion. The strengthening of the composites is, therefore, attributed to the refinement of matrix grains by the nano-sized Ni dispersion at the grain boundaries. The fracture toughness of the Al_2O_3/Ni composites was enhanced when a larger amount of dispersion was incorporated. This is explained by the increment of plastic deformation of soft Ni dispersion. On the contrary to the Al_2O_3/W and Al_2O_3/Ni systems, the fracture strength and toughness of the Al_2O_3/Ti composite were not so much enhanced because most of the Ti dispersion was fairly large. This composite was thus characterized as a microcomposite. In addition, the interface between Al_2O_3 and Ti was disturbed, which may affect the mechanical properties of this composite system.

Table II Mechanical properties for Al_2O_3/metal nanocomposites.

System (vol.%)	Starting material of metal	σ_f (MPa)	Hv (GPa)	K_{1C} ($MPam^{1/2}$)
Monolithic Al_2O_3	-	528	18.0	3.2
Al_2O_3/10%W	W	831	18.0	4.1
Al_2O_3/5%W	WO_3	1105	18.7	3.8
Al_2O_3/20%Ni	NiO	720	13.0	4.6
Al_2O_3/5%Ni	$Ni(NO_3)_2$	1090	18.1	3.4
Al_2O_3/5%Ti	TiH_2	820	17.8	3.2

Table III High-temperature strength for the Al_2O_3/W nanocomposites prepared by the PM method.

Composition	Test temperature (℃)			
(vol.%)	R.T.	1000	1100	1200
Monolithic Al_2O_3	528.1	260.5	-	198.6
Al_2O_3/15vol.%W	781.0	653.9	518.2	419.1

High-temperature fracture strengths for the Al_2O_3/W composite is listed in table III. The strength of monolithic Al_2O_3 decreased with increasing test temperature and reached about 200 MPa at 1200℃. The strength of the Al_2O_3/15vol.%W composite was higher than those of monolithic Al_2O_3 and was maintained over 500 MPa up to 1100℃. A sudden decrease in strength, however, was found over 1200℃.These results suggest that the nano-sized metal dispersion may inhibit the plastic deformation of the Al_2O_3 matrix below 1100℃, and the high strength is consequently maintained. The high-temperature Vickers hardness of the Al_2O_3/W nanocomposites was also improved, especially above 1000℃, and was about 3 times higher than that of the monolithic Al_2O_3 at 1400℃. This result is also rationalized by the characteristic nanostructure of the composites. It can be attributed to the hardening of Al_2O_3 matrix at high-temperatures by the residual stresses and the dislocation pinning effect of the fine intragranular W particles.

CONCLUSIONS

Fine-metal particulate-dispersed Al_2O_3 nanocomposites can be fabricated by using controlled sintering techniques. The incorporation of nano-sized W dispersoid into the Al_2O_3 grains significantly improved both room and high-temperature mechanical properties. These results can be attributed to the characteristic dispersion state of the W particles, the interface structure between the ceramic and metal nano-particles. Dispersing Ni particles of nanometer-size could also improve the fracture strength of Al_2O_3. These enhancements were mainly attributed to the grain growth inhibition of the matrix by the nano-sized dispersion at the grain boundaries. The characteristic nanostructure of the Al_2O_3/Ni composite was related to the morphology of its staring mixture and sintering condition.

Interfaces between metals and Al_2O_3 matrix are basically clean, even though a disturbed interface structure was observed in the Al_2O_3/Ni and Al_2O_3/Ti composites. The lattice adjustment of intragranular W with Al_2O_3 is excellent, and the ledge structure was generated, which should extract the significantly improved mechanical properties.

ACKNOWLEDGEMENTS

The authors acknowledge the financial supports from the Ministry of Education, Science and Culture under a Grant-in-Aid for Scientific Research, Nos. 03555157, 04205099 and 05750749, and also from the cooperative project "Advanced Materials Creation on their Limit State Prediction" between ISIR and WRI in Osaka University, for the achievement of this investigation.

References

1. K. Niihara, A. Nakahira, G. Sasaki and M. Hirabayashi, "Development of Strong Al_2O_3/SiC Composites"; pp. 129-134 in *Proc.of MRS Int'l.Mtg on Adv. Mater*. Plenum Press,Tokyo, 1988.
2. K. Niihara and A. Nakahira, "Particulate Strengthened Oxide Ceramics-Nanocomposites"; pp. 637-644 in *Advanced Structural Inorganic Composites*. Edited by P. Vincenzini. Elsevier Science Publishers, Trieste, 1990 .
3. E. Breval, G. Dodds, and C.G. Pantano, "Properties and Microstructure of Ni-Alumina Composite Materials Prepared by the Sol/Gel Method," Mater. Res. Bull., **20**[10] 1464-1468 (1985).
4. K.Niihara, "New Design Concept of Structural Ceramics -Nanocomposites-," J.Ceram.Soc.of Jpn., **99**[10] 974-982 (1991).
5. A. Nakahira and K. Niihara, "Sintering Behaviors and Consolidation Process for Al_2O_3/SiC Nanocomposites," J. Ceram. Soc. Japn., **100**[4] 448-453

(1992).
6. A. Nakahira and K. Niihara, "Microstructure and Fracture Behaviors at High Temperatures for Al_2O_3-SiC Nanocomposites"; pp.165-178 in *Fracture Mechanics of Ceramics, Vol.9*. Edited by M. Sasaki, R.C. Bradt, D.P.H. Hasselman and D. Munz. Plenum Press, New York, 1992.
7. M. Ishitsuka and M. Tamai, "Fabrication and Mechanical Properties of Pressureless Sintered Al_2O_3/SiC Nanocomposites," J. Jpn. Soc. of Powd. and Powd. Metal., **39**[12] 1109-1112 (1992).
8. E.Yasuda, Q.Bao, Y.Tanabe and K.Niihara, "High Temperature Creep of SiC/MgO Nanocomposite"; pp.431-436 in *Proc. 1st Int. Symp. Sci. of Eng. Ceram.* Edited by S. Kimura and K. Niihara. The Ceramic Society of Japan, Tokyo, 1991 .
9. T.Sekino, A.Nakahira, M.Nawa and K. Niihara, "Fabrication of Al_2O_3/W Nanocomposites," J. Jpan.Powd. and Powd. Metal., **38**[3] 326-330 (1991).
10. T.Sekino, A.Nakahira, M.Nawa and K. Niihara, "Fabrication and Mechanical Properties of Tungsten Metal Dispersed Al_2O_3 Based Nanocomposites"; pp.745-750 in *Proc. Austceram '92*. Edited by M.J. Bannister. CSIRO Pub., Melbourne, 1992.
11. M. Nawa, T. Sekino and K. Niihara, "Microstructure and Mechanical Properties of Al_2O_3/Mo Nanocomposites," J. Japan. Powd. and Powd. Metal., **39**[12] 1104-1108 (1992).
12. P.Chantikul, G.R. Anstis, B.R. Lawn and D.B. Marshall, "A Critical Evaluation of Indentation Techniques for Measuring Fracture Toughness: II, Strength Method," J. Am. Ceram. Soc., **64**,[9] 539-543 (1981).
13. M.V. Swain and N. Claussen, "Comparison of K_{IC} Values for Al_2O_3-ZrO_2 Composites Obtained from Notched-Beam and Indentation Strength Techniques," J. Am. Ceram. Soc., **66**[2] C-27 (1983).

THERMODYNAMICS AND KINETICS OF PASSIVE OXIDATION OF SI_3N_4-BONDED SIC CERAMICS

Benyan Pei*, G. Q. (Max) Lu+ and G. X. Wang#
* Division of Process Metallurgy, Lulea University of Technology, 97187 Lulea, Sweden
+ Division of Thermal Engineering, School of Mechanical and Production Engineering, Nanyang Technological University, Singapore 2263
# Department of Metallurgy, Wuhan Iron and Steel University, Wuhan P. R. of China

The oxidation reactions of Si_3N_4-bonded SiC ceramics are more complicated than a single component like SiC or Si_3N_4. The thermodynamics and kinetics of air oxidation of such ceramics in the temperature range of 900-1300°C are studied in this work. The active to passive transition of oxidation was analyzed by thermodynamic calculations using the JANAF table . The weight gain was measured using a TGA technique to study the kinetics. It is shown that all components contribute to the formation of a silica film. The kinetic results indicate that the oxidation reactions of Si_3N_4-bonded SiC ceramics are mixed controlled by both an interface reaction and diffusion through the product layer. The activation energy obtained for the specimen studied is found to be comparable with that for SiC oxidation as reported in the literature.

INTRODUCTION

Silicon-based ceramics, such as SiC and Si_3N_4, are used for many high temperature applications due to their low oxygen permeability at very high temperatures. They are thermally stable and form a protecting silica film upon oxidation (passive) in high oxygen atmosphere. Si_3N_4-bonded SiC ceramics with a small amount of silicon offer more advantages, such as increased oxidation resistance and wear resistance. For instance, Si_3N_4-bonded SiC ceramics, are widely used as linings in blast furnaces, and were first introduced for this application in the later sixties [1]. Compared to traditional oxide refractories, SiC bricks possess low porosity, high mechanical strength, and better alkali and thermal shock resistance [2]. Si_3N_4-bonded SiC bricks containing approximately 20 ~ 25 wt.% Si_3N_4 and small amounts of silicon are the major types of silicon carbides used in blast furnaces. Oxidation by air and other oxidants is a major wear factor in service. To understand the oxidation behavior is of importance to achieve longer service life for SiC bricks and to improve product quality. There have been many studies [3- 14] related to oxidation of synthesized pure SiC or Si_3N_4. However, the oxidation of Si_3N_4-bonded SiC bricks is much more complicated than a single component like Si, SiC or Si_3N_4. So far, there is little understanding of oxidation of Si_3N_4-bonded SiC bricks in terms of thermodynamics and kinetics.

The present work intends to study the air oxidation of SiC bricks both thermodynamically and kinetically. The stability of such bricks in air will be discussed with respect to active and passive oxidation. The transition between active and passive oxidation are also analyzed. A kinetic model of unreacted core will be used to discriminate the rate mechanisms by fitting to the weight gain versus time data.

THERMODYNAMIC ANALYSIS

Gibbs energy data for α-SiC and α-Si_3N_4 from JANAF table [15] are used in the following analysis. The oxidation product SiO_2 may exist in the amorphous or crystalline state. However, the differences between the Gibbs energy data for different forms of SiO_2 are negligibly small (within ± 4 kJ/mol). The Gibbs energy data of SiO(g), CO(g) are also taken from JANAF table [15].

A number of thermodynamic analyses of stability upon oxidation have been made for Si [4, 8, 9], SiC [4 - 6, 8, 9] and Si_3N_4 [5 - 8, 9]. According to these studies, direct vaporization of Si, SiC and Si_3N_4 can be neglected, especially in environments where both carbon and nitrogen potentials are relatively high. The stabilities of Si, SiC and Si_3N_4 in an oxidizing atmosphere depend mainly on the stability of the oxidation product, SiO_2. In other words, the stability of Si_3N_4-bonded SiC bricks depends essentially on the protection layer, SiO_2 film. This is because Si-based materials may be oxidized either passively or actively at high temperatures, depending on oxidant potentials. Thermodynamic calculations were first carried out using the SOLGASMIX program [16] to determine the predominant gaseous species at the solid-gas interface.

(1) if the oxygen potential is low, active oxidation with formation of SiO(g)

occurs according to the following reactions in the range of 900-1300°C:

$$2\ Si + O_2\ (g) = 2\ SiO(g)$$
$$\Delta G^0 = -196.733 + 0.031T\log T - 0.270T \quad (kJ) \qquad (1)$$

$$SiC + O_2\ (g) = SiO(g) + CO(g)$$
$$\Delta G^0 = -138.837 + 0.015T\log T - 0.230T \quad (kJ) \qquad (2)$$

$$Si_3N_4 + 1.5\ O_2\ (g) = 3\ SiO(g) + 2\ N_2$$
$$\Delta G^0 = 450.790 + 0.046T\log T - 0.736T \quad (kJ) \qquad (3)$$

Under active oxidation, the surface of Si, SiC and Si_3N_4 will be exposed and these materials will continue to be deteriorated due to the formation of gaseous silicon monoxide, SiO(g).

In environments with high oxygen potential, passive oxidation will result:

$$Si + O_2\ (g) = SiO_2$$
$$\Delta G^0 = -904.760 + 0.174T \quad (kJ) \qquad (4)$$

$$SiC + 1.5\ O_2\ (g) = SiO_2 + CO(g)$$
$$\Delta G^0 = -945.230 + 0.078T \quad (kJ) \qquad (5)$$

$$Si_3N_4 + 3\ O_2\ (g) = 3\ SiO_2 + 2\ N_2$$
$$\Delta G^0 = -1968.39 + 0.190T \quad (kJ) \qquad (6)$$

Carbon monoxide CO(g) may be further oxidized to CO_2 (g), depending on the oxygen potential. Passive oxidation is characterized by weight gain due to the formation of a solid SiO_2 product scale. This will prevent further oxidation of Si, SiC, and Si_3N_4.

The transition from passive to active oxidation can occur by consumption of SiO_2 reacting with substrate Si, SiC and Si_3N_4, according to the following reactions:

$$Si + SiO_2 = 2\ SiO(g)$$
$$\Delta G^0 = 708.027 + 0.031T\log T - 0.444T \quad (kJ) \qquad (7)$$

$$SiC + 2\ SiO_2 = 3\ SiO(g) + CO(g)$$
$$\Delta G^0 = 1473.950 + 0.046T\log T - 0.848T \quad (kJ) \qquad (8)$$

$$Si_3N_4 + 3\ SiO_2 = 6\ SiO(g) + 2\ N_2\,(g)$$
$$\Delta G^0 = 2869.970 + 0.092T\log T - 1.662T\ (kJ) \tag{9}$$

The transition between active and passive oxidation has been studied both thermodynamically [4, 8 - 11] and kinetically [3, 7, 10, 11]. The kinetic studies take into account the mass-transfer constraints and determine the transition point in terms of oxygen potentials in the bulk gas for given temperatures. Wagner [3] proposed a model for transition between active and passive oxidation of Si in oxygen-inert gas mixtures in the viscous gas flow condition. According to this model, the active-passive transition oxygen partial pressures, $P_{O2,tr}$, for oxidation of Si is:

$$P_{O_2,tr} = 0.5\left(\frac{D_{SiO}}{D_{O_2}}\right)^{0.5} P_{SiO,eq} = 0.34P_{SiO,eq} \tag{10}$$

Where D_{SiO}/D_{O2} is the ratio between the diffusion coefficients of $O_2(g)$ and SiO(g) molecules in inert N_2 gas . It was estimated as 0.47 at 1500 K in the present work by the Chapman-Enskog theory [17]. $P_{SiO,\ eq}$. is determined from the reaction (7).

The same equation as (10) applies to the case for Si_3N_4, but $P_{SiO,\ eq}$ for this component is determined by the reaction (9). The problem for the active-passive transition of SiC oxidation is more complex than that for Si and Si_3N_4. In the oxidation of SiC, both SiO(g) and CO(g) carry oxygen atom and diffuse away from the surface. Balat *et al.* [11] extended the Wagner's model to SiC and formulated the following equation:

$$P_{O_2,tr} = \left(\frac{D_{SiO}}{D_{O_2}}\right)^{3/8}\left(\frac{D_{CO}}{D_{O_2}}\right)^{1/8} P^{3/4}{}_{SiO,eq}P^{1/4}{}_{CO,eq} = 0.75P^{3/4}{}_{SiO,eq}P^{1/4}{}_{CO,eq} \tag{11}$$

where $P_{SiO,eq}$ and $P_{CO,eq}$ are determined by the reaction (8). D_{CO}/D_{O2} was estimated to be 0.99 at 1500 K.

As an approximation, Wagner's model (Eqn. 10) is used to calculate the active-passive transition pressures for Si and Si_3N_4, and Eqn. (11) for SiC in this work.

In Figure 1, the transition oxygen pressures are shown for Si, SiC and Si_3N_4 as a function of temperature.

OXIDATION EXPERIMENTS

Si_3N_4-bonded SiC firebrick samples were cut into small specimens (10x10x20 mm). The bricks contain about 73 wt.% SiC, 22 wt.% Si_3N_4, and 1 wt.% Si (on average)[18]. Apparent porosity is about 18% and sintered density of 2981 kg/m^3. In order to reduce the experimental error caused by inhomogeneity of the brick, two specimens were used at the same time for each oxidation experiment.

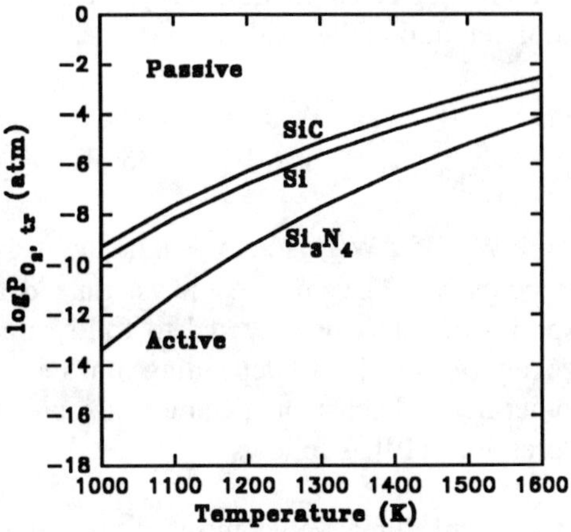

Figure 1. Transition oxygen pressure for Si, SiC and Si_3N_4 (P_{N2} =1 atm)

Oxidation was carried out in an electrical resistant furnace. The furnace was heated at a rate of 8 °K/min to the desired temperature, with Ar gas used as a purging gas. When the desired temperature (900 - 1300 °C) was reached, the Ar gas was stopped and air was introduced. After a certain time of oxidation the specimen was discharged from the furnace upon cooling to room temperature. Each specimen was then weighed using an analytical balance (TG328A) with an accuracy of ± 10^{-8} kg. The average value of the weight gains of the two specimens were calculated.

KINETICS OF OXIDATION

Since the oxidation of the firebrick specimen was conducted in natural air flow the oxidation reactions are expected to be passive, with a protecting silica scale formed on the surface of the specimen. The weight gain is therefore attributed to the oxidation of all three components. Kinetically the specimen can be treated as a spherical particle because of its dimension. Assuming that three components are oxidized stoichiometrically in air, it is not difficult to demonstrate from Eqs. (4-6) that the final weight of a specimen upon complete oxidation would be

$$W_\infty = 60W_0\left(\frac{0.71}{40} + 3\times\frac{0.22}{140} + \frac{0.01}{28}\right) = 1.369W_0 \qquad (12)$$

where W_0 is the initial weight of the specimen. The amount of weight losses due to emissions of C and N from SiC and Si_3N_4 are found to be insignificant and therefore neglected in Eqn. (12). The fractional conversion in terms of SiO_2 formation during the oxidation process, X, at any time *t* is defined as follows

$$X = \frac{W_t - W_0}{W_\infty - W_0} = \frac{\Delta W(t)}{0.369 W_0} \tag{13}$$

where W_t and $\Delta W(t)$ are the weight and weight gain of the specimen at any time *t*, respectively. The conversion vs. time data were obtained from the oxidation experiments and are expected to follow one of the kinetic models according to Szekely *et al.* [19], depending on the rate-controlling mechanisms. For isothermal and constant specimen size processes, there are generally three kinetic expressions [19] as follows:

If the chemical reaction controls at the surface of a shrinking core:

$$1 - (1-X)^{1/n} = k_c\, t \tag{14}$$

where n= 1, 2, 3 for long slab, cylinder and sphere. n = 3 for in this study with the spherical particle assumption.

For diffusion control through the product layer,

$$1 - 3(1-X)^{2/3} + 2(1-X) = k_d\, t \tag{15}$$

For mixed control of reaction at the interface and diffusion through the product layer,

$$1 - (1-X)^{1/3} + \varphi^2 \left[1 - 3(1-X)^{2/3} + 2(1-X) \right] = k_m\, t \tag{16}$$

where k_c , k_d , and k_m are the rate constants respectively for the three control mechanisms. φ^2 is called the Thiele Modulus representing the ratio of chemical reaction rate to the diffusion rate. A large value of this ratio (>> 1) means that the oxidation is diffusion-controlled with a relatively fast interface reaction.

In this study, the appropriate rate expression is determined by fitting the conversion-time data to each of the above equations (14-16) using a standard curve fitter (Marquardt-Levenberg algorithm). The results are summarized in Table I.

Table I. Comparison of deviations of X by different rate expressions

Sum of absolute deviations of X						
Eqn. No	900 °C	1000 °C	1100 °C	1200 °C	1300 °C	All temp.
14	3.50e-3	9.05e-3	2.16e-2	8.48e-2	3.40e-2	0.153
15	2.50e-3	1.80e-3	1.20e-2	4.30e-2	1.80e-2	0.077
16	1.02e-5	4.83e-5	3.60e-4	2.50e-3	5.80e-4	0.0035

RESULTS AND DISCUSSION

It is seen from the above table that equation (16) gives the best fit to the experimental data available. It shows that in the temperature range of 900-1300°C the oxidation of Si_3N_4-bonded SiC brick occurs by mixed control of interface reaction and diffusion through the product layer. This is clearly illustrated in Figures 2, 3 and 4 where the conversion vs. time data are respectively fitted to the three equations.

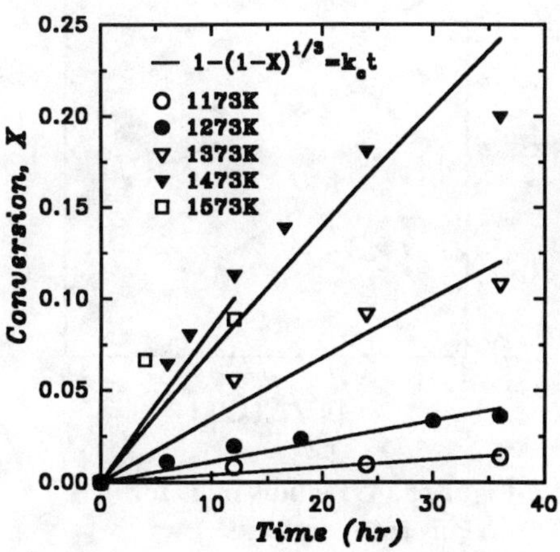

Figure 2. Conversion-time profiles fitted to reaction control equation

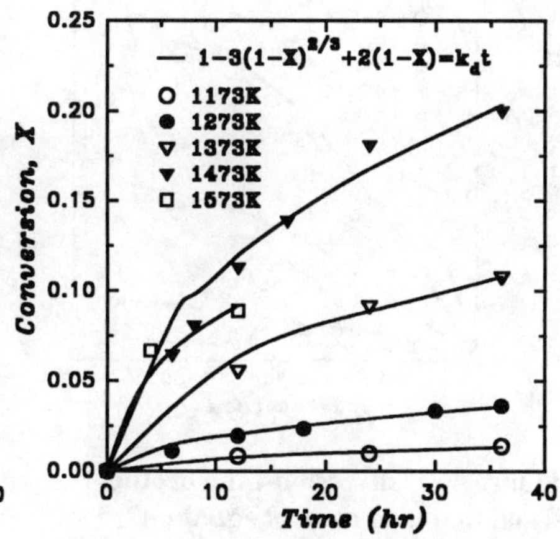

Figure 3. Conversion-time profiles fitted to diffusion control equation

It is seen that all equations fit the low temperature data reasonably well because at such temperatures the reaction rates are slow and the conversion is small. At higher temperatures, the mixed-controlling rate expression fits better than the diffusion-controlling one, which in turn fits better than the reaction control equation. It shows that the oxidation process does not really follow the parabolic law nor the linear reaction-controlling rate expression alone. The mixed control of both mechanisms (a complex power law) represents the overall rate process the best. Mixed rate controlling mechanisms were also observed by Luthra [12] for the oxidation of SiC and Si_3N_4.

In the process of curve fitting, the rate constants are varied for various temperatures to obtain the optimal fit. It is expected that the rate constant follows the Arrenhius law.

$$k = A_0 \exp(-E/RT) \qquad (17)$$

Therefore, one can plot the logarithm of the fitted values of the rate constants versus the reciprocal of the temperature to obtain the kinetic parameters such as activation energy, E and pre-exponential factor, A_0. Figure 5 shows these plots for different rate constants, k_c, k_d, and k_m. The solid lines are the linear regression for the data points.

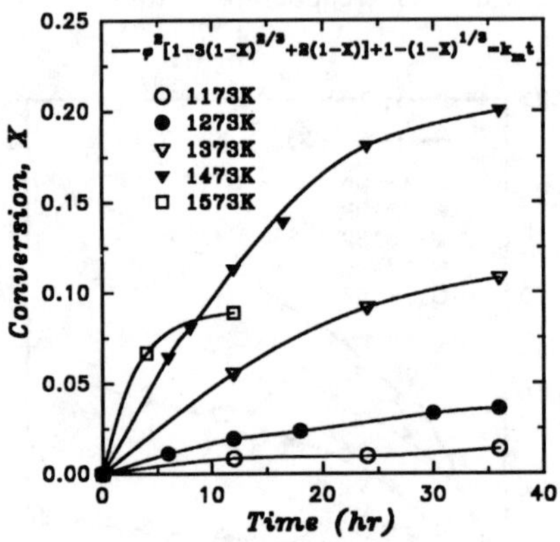

Figure 4. Conversion-time profiles fitted to mixed control equation

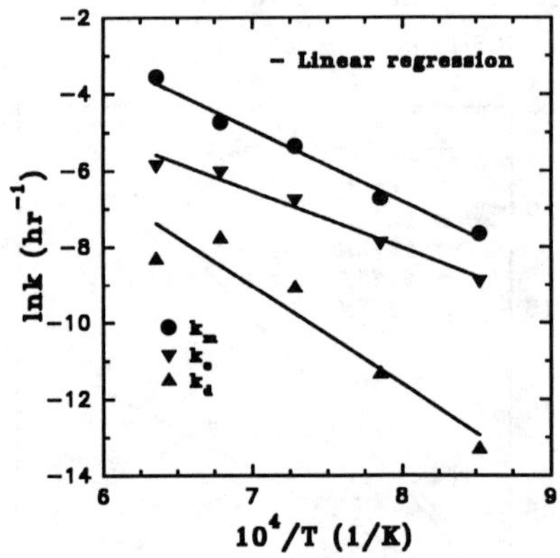

Figure 5. Arrenhius plots for various rate constants

The intercepts and the slopes of these lines determine the pre-exponential factors and activation energies, respectively. Table II lists these kinetic parameters derived from different rate expressions.

It is noted that with the mixed controlling rate expression the temperature dependency of the rate constants are more linear. The activation energy obtained

Table II. Kinetic parameters derived from different rate expressions

Rate constant	Activation energy, E (kJ/mol)	A_0 (hr^{-1})	Regression coefficient
k_c	123.63	6.74×10^{-3}	0.987
k_d	213.29	1.50×10^{-3}	0.952
k_m	155.99	0.050	0.994

from k_m is larger than that for Si oxidation (119kJ/mol), comparable with that for SiC (134 -498kJ/mol), and smaller than that for pure Si_3N_4 oxidation (300-400 kJ/mol) [13, 14].

It is also noted from Figure 5 that for a temperature of 1300 °C, the extracted values of rate constants are much lower than those for the low temperatures. This might be due to the possible nucleation and crystallization of SiO_2 at this temperature. Thus, the rate is neither controlled by surface reaction (linear law) nor by the parabolic diffusion law. However, the rate constant, k_m, at this temperature as extracted from the mixed rate expression is much higher. This is possibly due to the fact that the curve fitting optimized both parameters, i.e. φ^2 and k_m. This does not necessarily give the true information at this temperature. Besides, the lack of more data points at 1300 °C could well constitute large errors in the rate constants.

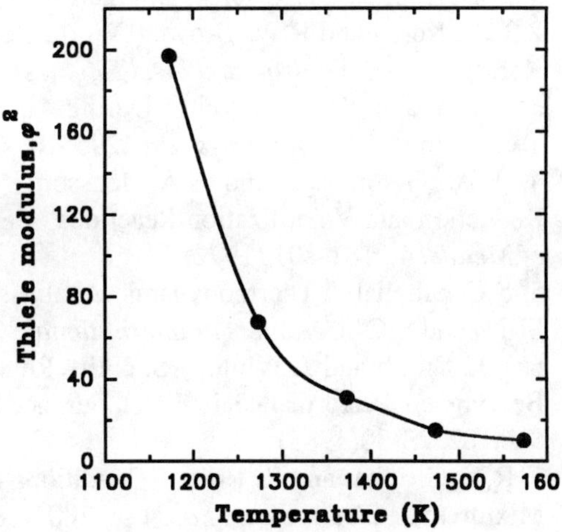

Figure 6. Thiele Modulus as a function of temperature

The fitting of conversion-time data using equation (16) gives various values of the Thiele Modulus for different temperatures. Figure 6 is a plot of φ^2 as a function of temperature. It shows that as the temperature increases the Thiele Modulus actually decreases, which means the increase in reaction rate is slower than the decrease in diffusional resistance when the temperature increases. This could be attributed to the crystallization at higher temperatures slowing down the intrinsic reaction.

CONCLUSIONS

In this paper, thermodynamic analysis on the active and passive oxidation of Si_3N_4-bonded SiC bricks is presented. The transition between the two types of oxidation is also studied. Three kinetic rate controlling expressions are used to fit the experimental data of the weight gain versus time (conversion vs. time). The best fit determines that the mixed controlling mechanism of both reaction and diffusion is observed for the oxidation of this ceramic in air. Neither the parabolic law nor the linear rate law is suitable for the overall oxidation rate. The activation energy obtained from the Arrenhius plot for oxidation of Si_3N_4-bonded SiC brick is comparable with literature data, smaller than that for pure Si_3N_4, and larger than that for single-crystal Si.

REFERENCES:

1. A. Fichel and J. Rramss, " Silicon Carbide Lining in the Blast Furnace - Theory/Application/Results" *Interceram Special Issue* (1983)
2. C.L. Keran and R.W. Brown, "World Report on Silicon Carbide Blast Furnace Refactories -1987" *Iron and Steel Engineer*, 12, 35-42 (1987)
3. C. Wagner, " Passivity During the Oxidation of Silicon at Elevated Temperature " *J. Appl. Phys.* 29, 1295 - 1297(1958)
4. E.A. Gulbransen and S.A. Jansson, " The High Temperature Oxidation, Reduction and Volatilization Reactions of Silicon and Silicon Carbide" *Oxidation of Metals*, 4, 181-201 (1972)
5. S.C. Singhal, " Thermodynamic Analysis of the High-Temperature Stability of Si_3N_4 and SiC" *Ceramurgia International,* 2, 123-130 (1976)
6. P.L. Smith and J. White, " Stability Relationships of SiC and Si_3N_4 and Their Bearing on Performance in Blast Furnace Environments" *Trans. J. Br. Ceram. Soc.* 82, 23-30(1983)
7. R.V. Giridhar and K. Rose, " Conditions for Thermal Nitridation of Si in N_2-O_2 Mixtures" *J. Electrochem. Soc.* 135, 2803-2807 (1988)
8. A. Yamaguchi, "Oxidation Behavior of Si and SiC" *Refractories*, 39, 44-49 (1987) (in Japanese).
9. A.H. Heuer and V.L.K. Lou, " Volatility Diagrams for Si, Si_3N_4 and SiC" *J. Am. Ceram. Soc.* 73, 2785-3128 (1990)
10. T. Narushima, T. Goto, Y. Iguchi and T. Hirai, " High Temperature Active Oxidation of Chemically Vapor-Deposited Silicon Carbide in an Ar-O_2 Atmopshere" *J. Am. Ceram. Soc.* 74, 2583-2586 (1991)
11. M. Balat, G. Flamant, G. Male and G. Pichelin, " Active to Passive Transition in the Oxidation of SiC at High Temperature and Low Pressure in Molecular and Atomic Oxygen" *J. of Materials Science*, 27, 697-703(1992)

12. K. L.Luthra," Some New Perspectives on Oxidation of Silicon Carbide and Silicon Nitride", *J. Am. Ceram. Soc.*, 74, 1095-1103(1991)
13 J. A. Costello and E. Tressler," Oxidation Kinetics of Hot-Pressed and Sintered α-SiC", *J. Am. Ceram. Soc.,* 64, 327-331(1981)
14. D. J. Choi, D. B. Fischback, W. D. Scott, "Oxidation of Chemically-Vapor-Deposited Silicon Nitide and Sinngle-Crystal Silicon", *J Am. Ceram. Soc.*, 72, 1118-23(1989)
15. M. Chase, C. Davies, J. Downey, D. Frurip, R. McDonald and A. Syverud, *JANAF Thermodynamic Tables*, 3rd ed., ACS and AIP, New York (1986.)
16. G. Eriksson, " Thermodynamic Studies of High Temperature Equilibria XIII: SOLGASMIX" *Chem Scripta* 8, 100-103 (1975)
17. R.B. Bird, W.E. Stewart and E.N. Lightfood, *Transport Phenomena*. Wiley, New York, 508-513 (1960)
18. D.Y. Peng, J.H. Li, S.Z. Liu, Y. Yu, Y.L. Wang and Z.T. Wang, " Properties and Application of Si_3N_4-Bonded SiC Bricks for Blast furnaces" *UNITECR'89, Proceedings*. Vol. 1, L.J. Trostel ed. American Ceramic Society, 101-108(1989)
19. J. Szekely, J. W. Evans, and H.Y. Sohn, *Gas-Solid Reactions,* Academic press, New York, 76, (1976)

IMPROVEMENT OF MECHANICAL PROPERTIES BY THE FORMATION OF A GRAIN BOUNDARY PHASE IN ZrO_2 - 3 MOL% Y_2O_3 - Al_2O_3 CERAMICS

H.Tsubakino, T.Fujiwara, A.Yamamoto and S.Ioku

Department of Materials Science and Engineering,
Faculty of Engineering, Himeji Institute of Technology
2167 Shosha, Himeji, Hyogo 671-22, Japan

ABSTRACT

The effect of alumina addition on bending strength and fracture toughness in ZrO_2-3mol% Y_2O_3 (PSZ) was studied by a combination of three point bending tests, an indentation technique, dilatometry, and transmission electron microscopy. The addition of alumina to PSZ improves these mechanical properties. This improvement is not related to the variation in density and grain size caused by the addition of alumina, but is a result of the formation of an amorphous phase located at the grain boundaries. This phase contains alumina, silica and yttria.

INTRODUCTION

Yttria-partially stabilized zirconia (PSZ) ceramics have remarkable advantages in mechanical properties, such as strength and toughness. This improvement of the mechanical properties is believed to be related to the stress-induced martensitic transformation from the metastable tetragonal to stable monoclinic phases, the transformation-toughening mechanism [1]. On the other hand, these properties are greatly decreased by thermal cycling or ageing at relatively low temperatures particularly in water vapor environments, i.e., cracks are formed due to partial transformation that occurs during the cycling or ageing [2-7].

Previous studies [8-10] have shown that the addition of alumina improves mechanical properties and retards the transformation. Other studies [2,11,12] also reported that the addition of alumina improved thermal shock and mechanical properties.

In this report, the fundamental effect of alumina addition to PSZ has been studied by using the three-point bending test, an indentation method, dilatometry, and transmission electron microscopy (TEM) combined with energy dispersive spectroscopic (EDS) analysis.

EXPERIMENTAL PROCEDURE

The starting powders were ZrO_2-3mol% Y_2O_3 (PSZ) and ZrO_2-3mol% Y_2O_3-5mass% Al_2O_3 (5A) produced by a co-precipitation method. The chemical composition of these powders is shown in Table 1. These powders were pressed unidirectionally at 100 MPa and sintered at 1673∼1873 K (T_s) at several different times (t_s) ranging from 3.6 to 28.8 ks in air, and then air cooled to avoid cracks caused by the martensitic transformation. X-ray diffraction pattern has indicated that there is no monoclinic phase in the as-sintered specimens.

Table 1 Chemical composition of co-precipitated powders (mass%).

	Y_2O_3	Al_2O_3	CaO	MgO	SiO_2	Fe_2O_3	TiO_2
PSZ	5.63	0.02	0.09	0.10	0.10	0.01	0.15
5A	5.27	5.12	0.14	0.23	0.13	0.02	0.10

Bending strength and fracture toughness were obtained from as-sintered specimens. Three-point bending test was performed using specimens of size 32x8x3 mm. The cross head speed was 8.3 μm/s and the span length was 20 mm. The fracture toughness was evaluated using the indentation microfracture method [13]. Disc specimens of 10 mm in diameter were used.

SEM and TEM observations, and EDS analysis were also performed by using FE-SEM and high resolution TEM, respectively. The foils for TEM observations were prepared by an ion-thinning technique.

Thermal expansion was measured during thermal cycling, at temperatures ranging from room temperature to 1073K. The heating and cooling rate of the thermal cycling was 0.022 K/s.

RESULTS AND DISCUSSION

1. Mechanical Properties

The typical cracks produced by the indentation method (load=98 N) are shown in Fig. 1. The crack length in 5A is obviously shorter than that in PSZ. Many of these cracks propagate through the grain boundaries. Therefore, it is predicted that some changes will occur at the grain boundaries in PSZ by the addition of alumina. The critical stress intensity factor evaluated by using the value of Young's modulus, 207 GPa [14], is shown in Fig. 2. The K_{IC} in 5A is much higher than that in PSZ. The bending strength also increases with the addition of alumina, as shown in Fig. 3. The fracture toughness and bending strength increase at higher sintering temperatures (T_s).

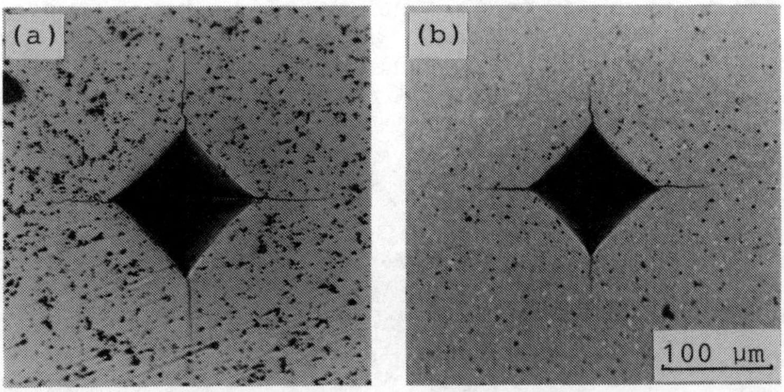

Figure 1 Typical optical micrographs, showing cracks produced by indentation test in PSZ (a) and 5A (b), (T_s=1773 K, t_s=18 ks)

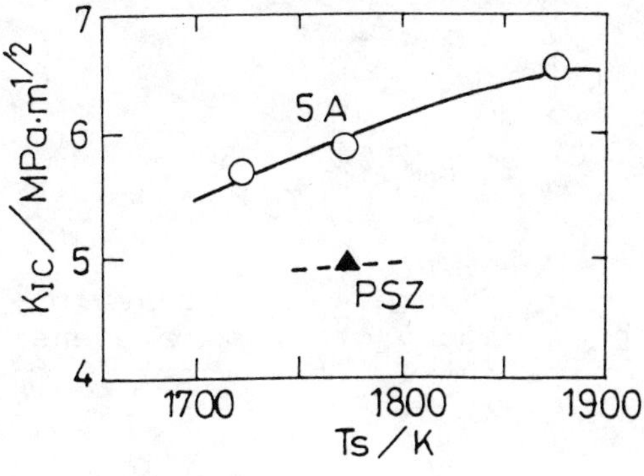

Figure 2 Relationship between critical stress intensity factor, K_{IC}, and sintering temperature, T_s (t_s=18 ks).

The increase of mechanical properties by the addition of alumina has been explained by several factors [1,11, 12] such as grain size and density, i.e., specimens with smaller grain size and higher density show higher mechanical properties. However, it has been clearly shown from the grain size (D) and the density (ρ) data obtained in this study by an intercept method and the Archimedes method, respectively (Fig. 4) that the increase in mechanical properties (shown in Figs. 2 and 3) are not related to these factors.

The fracture toughness was evaluated from 5A specimens sintered for various times (t_s) at constant T_s, 1873 K. The K_{IC} increases with t_s up to 15 ks and remains constant for t_s over 20 ks (Fig. 5). However, the grain size increases with an increase in t_s whereas the density is almost constant, as shown in Fig. 6. Therefore, this data confirms that the improvement of mechanical properties is not related to the factors of grain size and density.

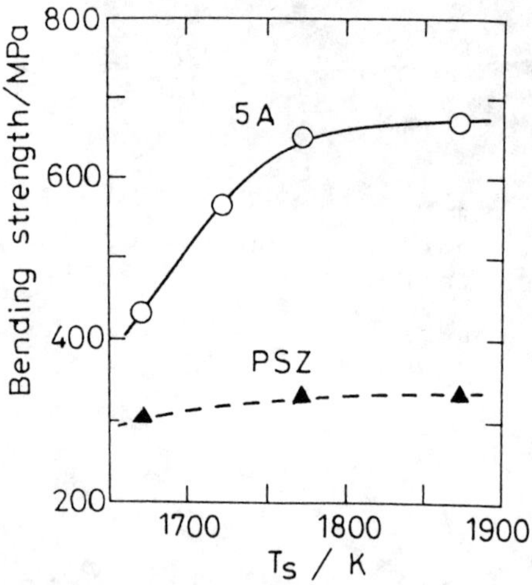

Figure 3 Relationship between bending strength and T_s (t_s=18 ks).

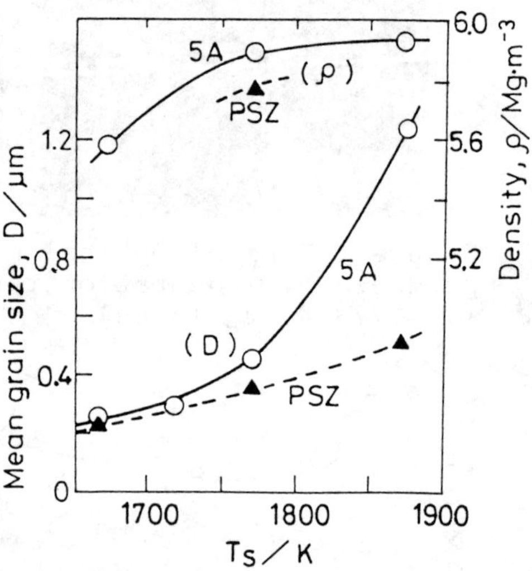

Figure 4 Relationship among grain size, density and T_s (t_s=18 ks).

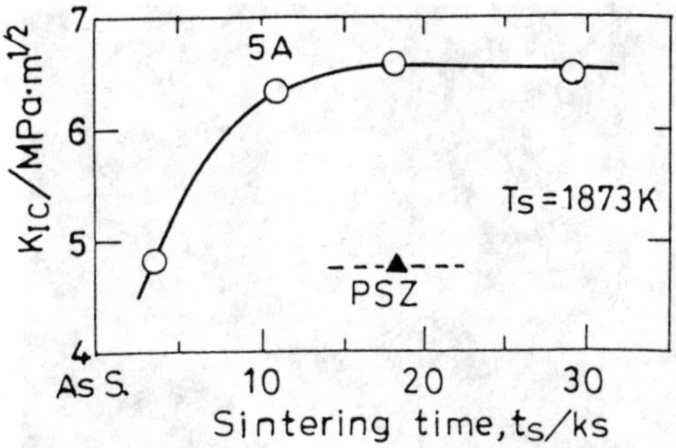

Figure 5 Relationship between K_{IC} and sintering time, t_s.

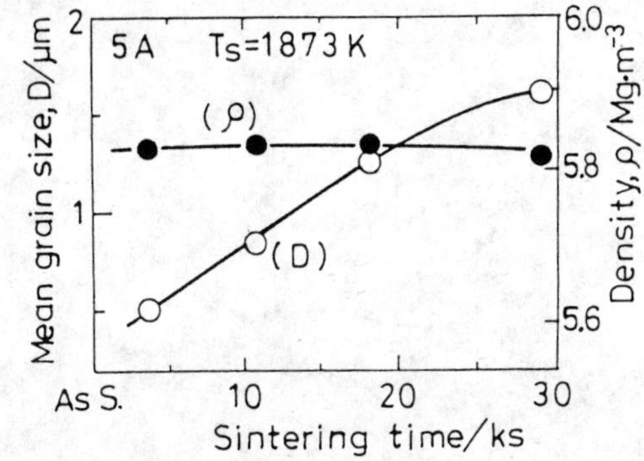

Figure 6 Relationship among grain size, density and t_s.

2. TEM observations

Typical results of TEM observations of 5A are shown in Fig. 7. A grain boundary phase can be seen, particularly at the triple junctions of grains. A similar grain boundary phase was also observed in PSZ, though the micrograph of PSZ is not shown here. However, this phase in 5A is much thicker than that in PSZ. Convergent beam diffraction patterns of each region of the grain boundary phase and grain center in 5A are shown in Fig. 8, in which the incident electron beam is concentrated within 10 nm in diameter. The diffraction pattern at the grain boundary phase shows hollow one; thus, it can be concluded that the grain boundary phase consists of an amorphous phase. A similar grain boundary phase was observed in commercial TZP by Rühle et al [15], though Niel et al [16] did not find evidence of this grain boundary phase in TZP.

Typical EDS spectra obtained at the grain center and grain boundary phase formed in 5A are shown in Fig. 9. The EDS spectrum at the grain boundary phase was obtained from the triple junctions, because the width of the phase at the grain surface is narrow. Though

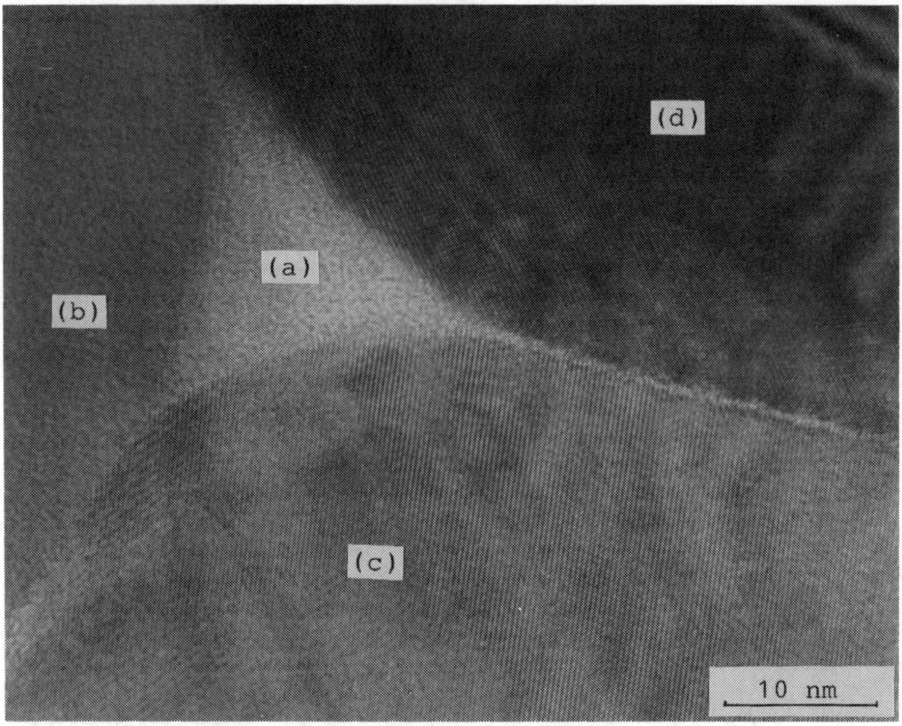

Figure 7 Typical TEM micrograph in 5A, showing grain boundary phase (T_s=1773 K, t_s=18 ks).

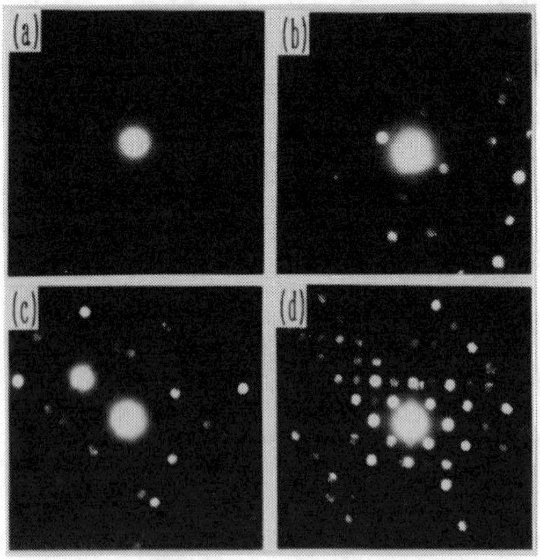

Figure 8 Convergent beam diffraction patterns of the regions in Fig.7.

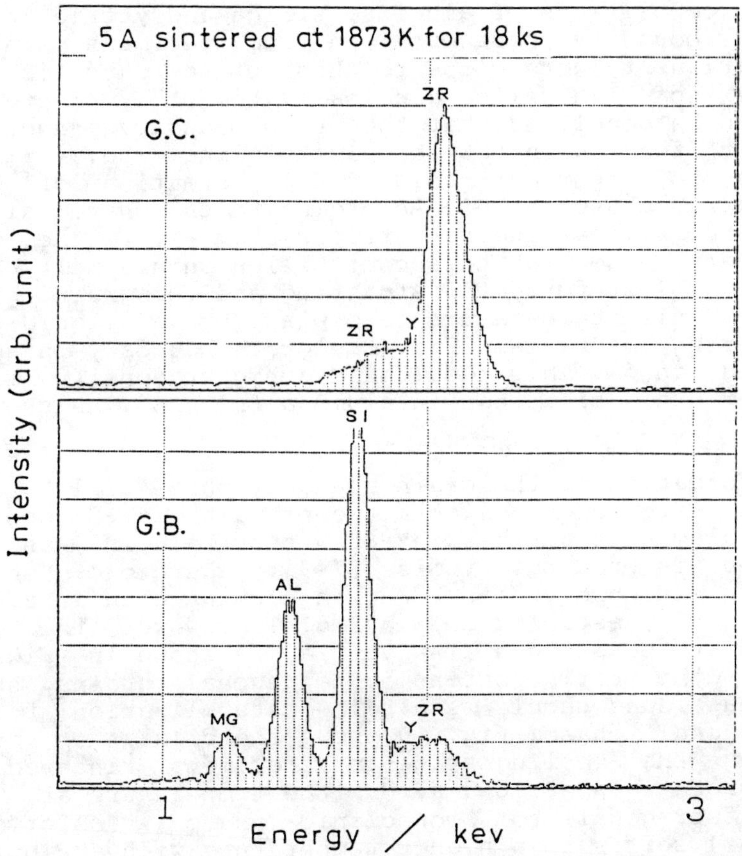

Figure 9 Typical EDS spectrum at grain center and grain boundary regions in 5A.

oxides such as silica, alumina, magnesia and yttria are usually concentrated at the grain boundary phase, they are not detected except for yttria at the grain center regions. In addition, the content of such oxides are very few except for yttria at grain boundary phase in alumina-free PSZ. The oxides of silica and magnesia originate from impurities contained in the original powder. The formation of the grain boundary phase enriched with these oxides has been suggested in previous studies [8-10]. In this study, this theory was confirmed experimentally.

The compositions of alumina, silica and yttria in the grain boundary phase obtained from specimens sintered at various t_s were close to those of the eutectic line in silica-rich silica-alumina system [17]. Therefore, it can be concluded that the grain boundary phase more easily forms when PSZ contains alumina; i.e., alumina plays an important role in the formation of grain boundary phase in PSZ. According to the phase diagram of the silica-alumina-yttria system [17], the grain boundary phase with this composition should melt above 1800 K. Therefore, the sintering and the grain growth above this temperature take place by a liquid-phase sintering mechanism after the grain boundary phase is formed in 5A. This results in a higher density and a larger grain size than in alumina-free PSZ, as shown in Figs. 4 and 6.

The formation of the grain boundary phase will produce a much stronger cohesion between grains in PSZ containing alumina than that in PSZ without alumina, as suggested in previous studies [7-9]. Therefore, shorter cracks are observed in PSZ with alumina than in alumina-free PSZ, as indicated in Fig. 1. However, the grain boundary phase contains yttria, as shown in Fig. 9. Then, the yttria content in tetragonal grains should decrease and should result in destabilization to the tetragonal phase. Fig. 10 shows the dilatation curves during the cooling stage in specimens sintered for various times at 1873 K. M_s (the temperature at which the tetragonal to monoclinic phase transformation begins) shifts to a higher temperature with an increase

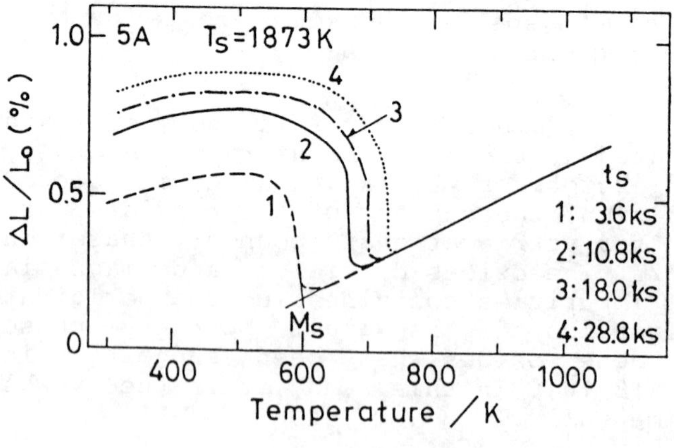

Figure 10 Dilatation curves during cooling stage in 5A sintered for various time (T_s=1873 K).

in sintering time. This large shift of M_s results from the decrease in yttria content in tetragonal grains. This decrease is caused by the formation of the grain boundary phase and by the increase in grain size accompanied with sintering time.

CONCLUSIONS

1. The bending strength and fracture strength in PSZ containing alumina are higher than those in alumina-free PSZ.
2. Both the grain size and density increase with the addition of alumina. Therefore, the improvement of mechanical properties due to this addition is not related to the effect of alumina addition on grain size and density.
3. In PSZ containing alumina, an amorphous phase enriched alumina, silica, yttria and magnesia formes at the grain boundaries. The formation of this grain boundary phase will produce a much stronger cohesion between grains, resulting in the improvement of mechanical properties in PSZ containing alumina.
4. Liquid phase sintering will occur at higher temperatures above 1800 K in PSZ containing alumina.

REFERENCES

1. D.J.Green, R.H.J.Hannink and M.V.Swain, "Transformation Toughening of Ceramics", CRC Press, Florida (1989).
2. T.Sato and M.Shimada, "Crystalline Phase Change in Yttria-Partially Stabilized Zirconia by Low-Temperature Annealing", *J.Am.Ceram.Soc.*, 67, C-212(1984).
3. M.Yoshimura, T.Noma, K.Kawabata and S.Somiya, "Role of H_2O on the Degradation of Y-TZP", *J.Mater.Sci. Lett.*, 6, 465(1987).
4. H.-Y.Lu and S.-Y.Chen, "Low-Temperature Aging of t-ZrO_2 Polycrystals with 3 mol% Y_2O_3", *J.Am.Ceram. Soc.*, 70, 537(1987).
5. F.F.Lange, G.L.Dunlop and B.I.Davis, "Degradation During Aging of Transformation-Toughened ZrO_2-Y_2O_3 Materials at 250°C", *ibid.*, 69, 237(1986).
6. S.Sato, S.Ohtani, T.Endo and M.Shimada, "Transformation of Yttria-Doped Tetragonal ZrO_2 Polycrystals by Annealing under Controlled Humidity

Conditions", *ibid.*, 68, C-320(1985).
7. H.Tsubakino, M.Hamamoto and R.Nozato, "Tetragonal-to-Monoclinic Phase Transformation during Thermal Cycling and Isothermal Aging in Yttria-Partially Stabilized Zirconia", *J.Mater.Sci.*, 26, 5521(1991).
8. H.Tsubakino, R.Nozato and M.Hamamoto, "Effect of Alumina Addition on the Tetragonal-to-Monoclinic Phase Transformation in Zirconia-3 mol% Yttria", *J.Am.Ceram.Soc.*, 74, 440(1991).
9. H.Tsubakino, M.Hamamoto, K.Sonoda and R.Nozato, "Effect of Alumina Addition on Bending Fracture Behavior in Y-PSZ", *J.Soc.Mater.Sci.Jpn.*, 41, 631(1992).
10. H.Tsubakino, K.Sonoda and R.Nozato, "Martensite Transformation Behaviour during Isothermal Ageing in Partially Stabilized Zirconia with and without Alumina Addition", *J.Mater.Sci.Lett.*, 12, 631 (1992).
11. H.Watanabe and M.Chigasaki, "The Effect of Al_2O_3 or SiO_2 Addition on the Thermal Shock Resistance of Y_2O_3-Stabilized ZrO_2", *Yogyo Kyokaishi*, 94, 255 (1986).
12. M.Kihara, T.Ogata, K.Nakamura and K.Kobayashi, "Effect of Al_2O_3 Addition on Mechanical Properties and Microstructures of Y-TZP", *J.Jpn.Ceram.Soc.*, 96, 646(1988).
13. K.Niihara, "A Fracture Mechanics Analysis of Indentation-Induced Palmqvist Crack in Ceramics", *J.Mater.Sci.Lett.*, 2, 221(1983).
14. F.F.Lange, "Transformation Toughening, Part 2 Contribution to Fracture Toughness", *J.Mater.Sci.*, 17, 235(1982).
15. M.Ruhle, N.Claussen and A.H.Heuer, "Microstructural Studies of Y_2O_3-Containing Tetragonal ZrO_2 Polycrystals (Y-TZP)"; p.352 in Advances in Ceramics, Vol.12, *Science and Technology of Zirconia II*. Edited by N.Claussen, M.Ruhle and A.H.Heuer. Amer.Ceram.Soc., Columbus, OH,(1984).
16. T.G.Nieh, D.C.Yaney and J.Wadsworth, "Analysis of Grain boundaries in a Fine-Grained, Superplastic, Yttria-Containing, Tetragonal Zirconia", *Scripta Metall*, 23, 2007(1989).
17. E.M.Levin, C.R.Robbins and H.F.McMurdie, *Phase Diagrams for Ceramists*, Edited by M.K.Reser. Am.Ceram.Soc., Columbus, OH, 165(1969).

EFFECTS OF DIFFERENT STATES OF SiC ON THE MICROSTRUCTURAL DEVELOPMENT OF AL_2O_3-SiC NANOCOMPOSITE

Yueping Xu, Atsushi Nakahira and Koichi Niihara
The Institute of Scientific and Industrial Research, Osaka University
8-1 Mihogahoka, Ibaraki-shi, Osaka 567, Japan

ABSTRACT

The sol-gel technique has been used to make advanced Al_2O_3-SiC nanocomposites. Effects of the particle size and agglomerate state of SiC on the configuration of the grain boundary, microstructural development, as well as sinterability of Al_2O_3-SiC nanocomposites under different sintering conditions, were studied by TEM, SEM and X-ray diffraction. The reasons why SiC nano-particles are entrapped within the Al_2O_3 grains or at the grain boundaries were discussed on the basis of the movement of grain boundaries during the sintering processs. Finally, the good dispersion of nanosized SiC particles within the Al_2O_3 matrix of Al_2O_3-SiC nanocomposite was obtained by sol-gel powder processing and by controlling the sintering conditions.

INTRODUCTION

Recently, significant improvement of mechanical properties was achieved by transforming from a microcomposite to a nanocomposite structure in the Al_2O_3-SiC system. Nanocomposites are a new class of materials in which the matrix is reinforced by using ultrafine second particles, which has become more and more attractive [1-3]. Many investigations have addressed the effect of second phase particles on the densification of ceramic powders both theoretically and experimentally, in order to improve the properties of particulate ceramic matrix composites [4-6]. However, in most of these preparations, the particle size of second phases is in the micrometer range.

In this paper, sol-gel processing is used to prepare Al_2O_3-SiC nanocomposites, allowing nanometer range particles of SiC to be used as the second phase. The influence of different particle sizes of SiC on the sinterability, on the movement of the grain boundary, as well as on the microstructure of the Al_2O_3 matrix was investigated.

EXPERIMENTAL PROCEDURES

Materials of the Al_2O_3-SiC system containing 2.5, 5, 7.5, 10 vol% SiC were chosen in this experiment. Commercially available high purity boehmite with a BET surface area of 240 m^2/g (Jap. Abrasive Co.) was used as the source of Al_2O_3 sol. A small amount of α- Al_2O_3 (2wt%, 99.99% Sumitomo AKP-53, Japan) was used to enhance the transformation and densification of alumina matrix. The β-SiC particles were from Ibiden Co. Two kinds were used; one was obtained directly from the company, the average particle size of which was 0.65 μm (SG-1), and the other had an average diameter of 0.17 μm after the hard agglomerates had been removed (SG-2). The details of how the composite is processed are reported in reference [7].

In order to characterize the agglomerate state of powders, the pore size distribution of green compacts was measured by the mercury intrusion porosimetry. The green compacts were obtained by pressing at 150 MPa, followed by pressing isostatically at 300 MPa. The measurement of pore size was performed at a contact angle of 130° and a surface tension of 4.8 $x10^7$ N/M. Powders were hot pressed in a graphite die. The hot pressing was performed in the temperature range from 1400°C to 1800°C in N_2 atmosphere at an applied pressure of 30MPa. The densities of sintered composites were measured using the Archimedes method in toluene. The matrix grain size was determined by the line intercept method after the polished surface was thermally etched at 1400°C for 40 minutes in vacuum. X-Ray diffraction analyses of Al_2O_3-SiC nanocomposite powders were made using a standard diffractometer after the calcination at different temperatures from 400°C to 1200°C. The distribution state of SiC and the configuation of grain boundaries were observed by scanning electron microscopy (SEM) and transmission electron microscopy (TEM). The thinned specimens for TEM were prepared using the Ar ion sputtering technique after mechanical grinding to 30 μm thickness.

RESULTS AND DISCUSSION

Figure 1 shows the characteristics of Al_2O_3-5 vol% SiC (0.17 μm) nanocomposite powders before sintering. Most SiC particles were surrounded by the Al_2O_3 matrix powder in the Al_2O_3-SiC nanocomposite powder. This is because the very fine SiC particles of nanoscale size could be mixed with transparent Al_2O_3 sols after choosing the optimum conditions [7].

The pore size distribution of powder compacts is an excellent indicator of the state of the powder. The pore size distributions of Al_2O_3-5 vol% SiC (0.65 μm) and Al_2O_3-5 vol% SiC (0.17 μm) are shown in Figure 2. It is evident that the pore size distribution of Al_2O_3-5 vol% SiC (0.17 μm) is narrower and smaller than that of Al_2O_3-5 vol% SiC (0.65 μm). This is because the Al_2O_3-5 vol% SiC (0.65 μm) contains agglomerates of SiC and the large pores are formed by the packing of agglomerates or large SiC particles.

During the sintering process, the transformation from γ-AlOOH to α- Al_2O_3 will occur. In general, with increasing temperature, the phase changes from γ-AlOOH, through transition alumina, to α- Al_2O_3. For the Al_2O_3-SiC nanocomposite, the experimental results of X-ray analysis indicate that it first transforms from boehmite to γ-alumina at 400°, then to transition alumina and at 1100°C, it changes into α- Al_2O_3. It has

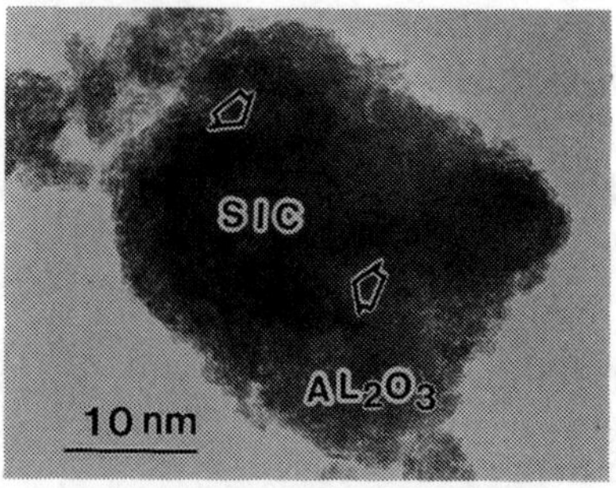

Fig. 1. Characteristics of Al2O3-5vol% SiC (0.17μm) nanocomposite powder

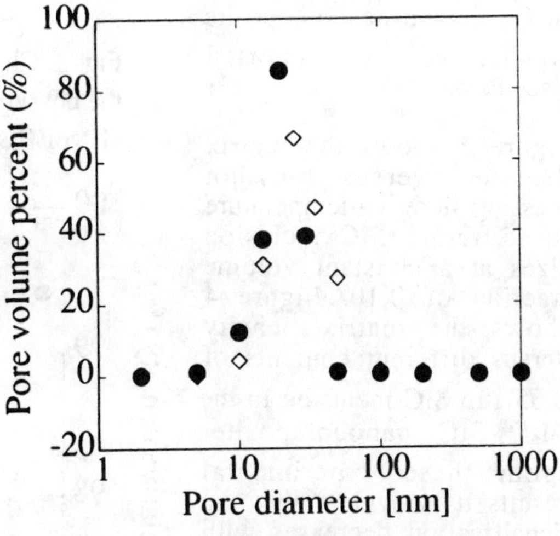

Fig. 2. Pore size distribution for Al2O3-5 vol%SiC powder compacts with different size of SiC 0.65μm and 0.17μm

◇: 0.65μm SiC inclusion
●: 0.17μm SiC inclusion

been suggested that the phase transformation from γ-into α-phase is caused by the boundary migration between γ/α interfaces. The rapid grain growth of α- Al_2O_3 during the γ--α phase transformation is explained by a mechanism involving interface boundary migration and lattice epitaxy [8]. More detailed studies on how the SiC particles influence the phase transformation of Al_2O_3 in the nanocomposite powder will be reported elsewhere.

Figure 3 shows the matrix densities versus the hot pressing sintering temperature for different SiC inclusion sizes at a constant volume fraction of 0.10. Figure 4 shows the matrix density versus different contents of 0.65 μm SiC inclusion in the Al_2O_3-SiC nanocomposite. From these experimental results it is evident that the densification decreases with decreasing particle size of SiC and with increasing volume of SiC inclusion in the initial period of sintering of Al_2O_3-SiC nanocomposites. The reason is that in ceramic composites the second phase particles have a strong effect on the sinterability of the

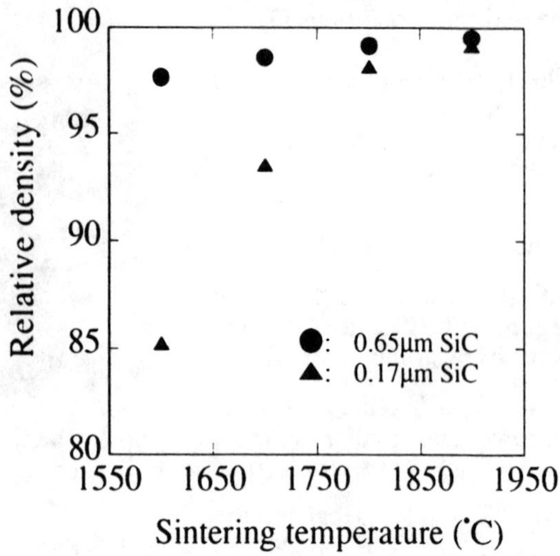

Fig. 3. Change of the relative density with the hot pressing temperature in the Al2O3 -10vol% SiC nanocomposite

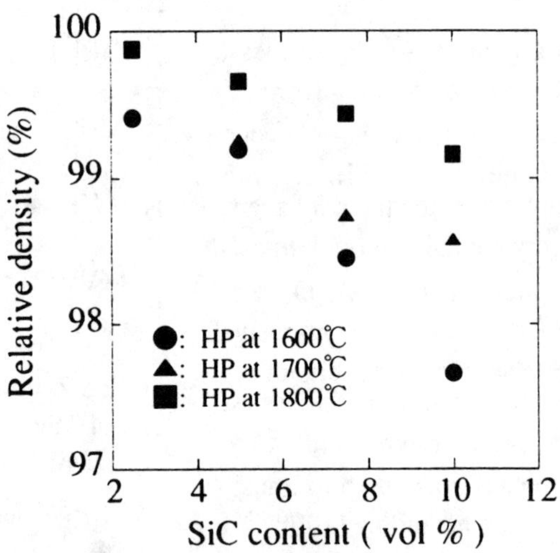

Fig. 4. Variation of the relative density with the different content of SiC in Al2O3-SiC (0.65μm) nanocomposite

matrix. The mismatch in shrinkage rates between the powder matrix of Al_2O_3 and the inert dispersed phase of SiC leads to an effective hydrostatic backstress which opposes densification. Such hydrostatic stress can be expressed as [9]:

$$\sigma = -\varepsilon_m f / (1/4G_m + f/3nK_m)$$

where K_m and G_m are the bulk and shear moduli, respectively. ε_m is the shrinkage strain due to the sintering of matrix, f is the volume fraction of the second inclusion, and n is the number of individual particles within the volume fraction f. By this theoretical hypothesis the experimental results can be clearly explained. When the second inclusion particle size decreases under the constant volume fraction of the second inclusion, the number of second phase particles increases. This in turn causes an increase in hydrostatic stress that reduces the sintering rate of the matrix. By TEM analysis (Figure 5), the porous region surrounding the SiC inclusions can be observed. This indicates that the interaction between the matrix and SiC inclusion particles causes reduction of the sinterability of the

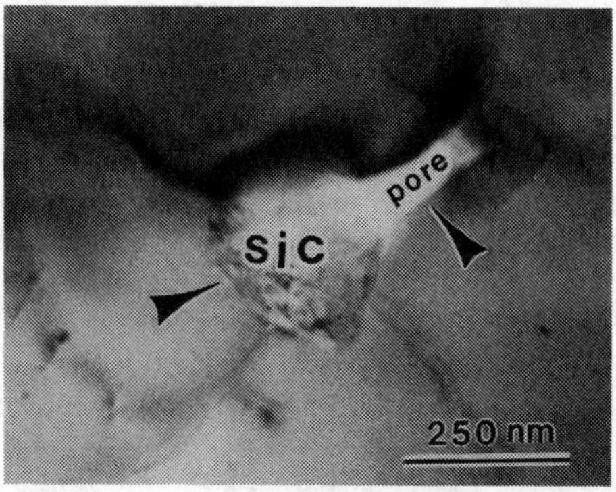

Fig. 5. TEM micrograph of Al2O3-5vol% SiC (0.17μm) nanocomposite after hot pressed at 1550°C for 1 hour

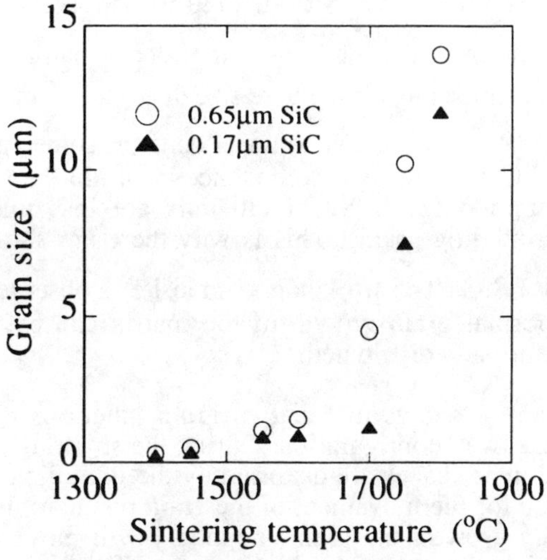

Fig. 6. Change of the grain size with the hot pressing temperature in Al2O3-5vol%SiC nanocomposite

nanocomposites.

Figure 6 shows the change of matrix grain size with increasing hot pressing temperature for different SiC inclusion sizes at a constant volume fraction of 0.05. A decrease in the grain size is observed with decreasing second phase particle size. This can be explained by the second phase particles residing at the grain boundary producing a drag force that inhibits the movement of the grain boundary. The interaction of a grain boundary with a spherical inclusion has been considered by Zener [10]. By analogy with surface tension, the maxium force F exerted by particles would be $F = \pi d\gamma$, where d is the size of the second particle, and γ is the surface energy per unit boundary area. If a plane boundary is inserted at random into N particles per unit volume, each with the particle size d, it will touch Nd of them per unit area. The particles will exert a pinning force per unit area grain boundary equal to $\pi d^2\gamma N$. Comparing Al_2O_3-5 vol% SiC (0.65 μm) and Al_2O_3-5 vol% SiC (0.17 μm) nanocomposites, the smaller is the SiC particle size, the larger the number of particles in the same volume fraction. This is why the nano-sized SiC in the Al_2O_3-5 vol% SiC (0.17 μm) nanocomposite has a stronger effect on the inhibition of the grain growth process than that of Al_2O_3-5 vol% SiC (0.65 μm) nanocomposite. As the pore size distribution analysis showed that the Al_2O_3-5 vol% SiC (0.65 μm) green compact contains some agglomerates of SiC, the distribution of SiC particles is inhomogeneous within the matrix. Therefore, some grains that contained fewer SiC inclusions are less inhibited from growing relative to surrounding grains. This is why there are some abnormal grains in the Al_2O_3-5 vol% SiC (0.65 μm) nanocomposite as observed by SEM. It also suggests that the abnormal grain growth of the matrix can be inhibited when the inclusions are uniformally distributed.

Figure 7 shows that at two grain junctions the SiC grains are in the drag and breakaway configuration. During the sintering process, the grain boundaries move and their shapes are distorted by the inhibition of SiC particles. When the driving force for the movement of the grain boundary is greater than the force barrier for grain growth, the grain boundary will move over the SiC particle and the SiC particle enters into the Al_2O_3 grain. The previous theoretical treatment showed that such barrier will decrease with decreasing particle size. This is why compared to SG-1 (10 vol% SiC) nanocomposite, most of the nanometer SiC particles in SG-2 (10 vol% SiC) nanocomposites are located within the Al_2O_3 grains as shown by SEM analysis (Figure 8).

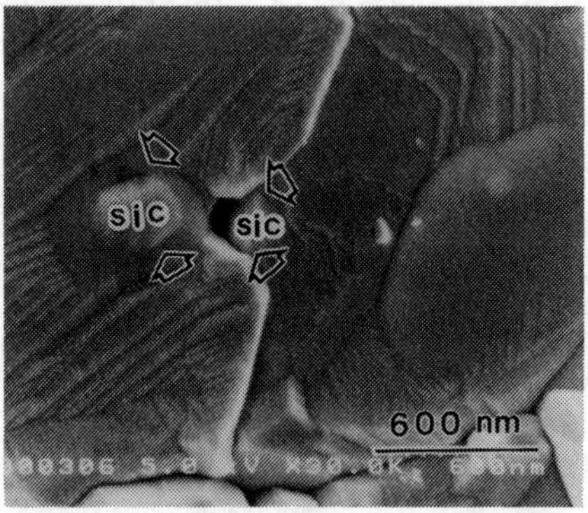

Fig. 7. Drag and breakaway configuration of SiC at 2-grain junction in Al_2O_3-5vol% SiC (0.17 μm) after hot pressed at 1650°C

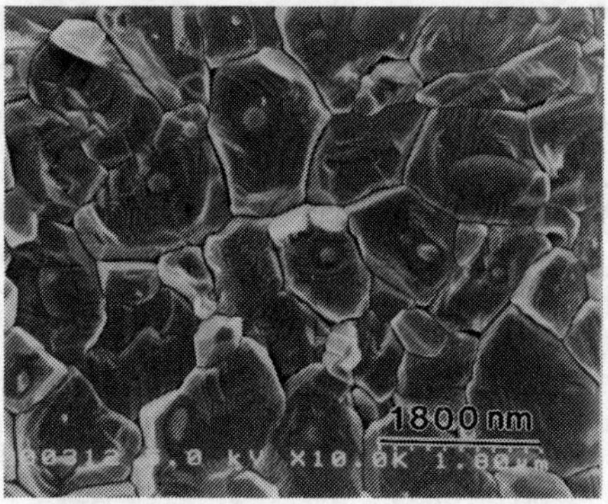

Fig. 8. SEM micrograph of thermal etched surface for Al_2O_3-10 vol% SiC (0.17 μm) after hot pressed at 1700°C

CONCLUSION

1. In Al_2O_3-SiC nanocomposites, the density and grain growth of the alumina matrix decrease with decreasing SiC particle size and increasing SiC partial volume fraction. The presence of SiC particles has a strong effect on the sinterability of the Al_2O_3-SiC nanocomposites.

2. From SEM and TEM analyses it is concluded that the nano-sized SiC particles are easily entrapped into Al_2O_3 grains. This is because the decrease of second SiC inclusion particle size results in a decrease in the grain boundaries pinning force.

3. By using sol-gel processing and controlling the sintering conditions, Al_2O_3-SiC nanocomposites can be obtained with homogeneous distribution of SiC particles into the Al_2O_3 matrix.

ACKNOWLEDGEMENT

The authors gratefully acknowledge T. Sekino at Osaka University and Prof. Guo Jingkun at Shanghai Institute of Ceramics, Chinese Academy of Sciences for their discussion and help. A part of this work was supported by the Japan Ministry of Education under a Grant-in-aid for Science Research No. 03555157 and 04205099, and by the cooperative project " Advanced Materials Creation and Their Limit State Predication for Envirmental Preservation" between ISIR and WRI in Osaka University.

REFERENCES

1. K. Niihara, "NEW Design Concept of Structural Ceramics-Ceramic Nanocomposites" J. Ceram. Soc. Japan, 99[10]974-982 (1991)
2. K. Niihara, A. Nakahira, "Strengthening of Oxide Ceramics by SiC and Si_3N_4 Dispersion", in Ceramic Materials and Components for Engines. Edited by V. J. Tennery, American Ceramic Society, Westerville, OH, p.919-926, 1989.
3. K. Niihara, A. Nakahira, "Particulate Strengthened Oxide Ceramics Nano-composite" in Advanced Structural Inorganic Composites, ed. P. Vincenzini, Elsevier Science Publishers, London, p637-664, 1990
4. A. G. Evans, "Considerations of Inhomogeneity Effects in Sintering" J. Am Ceram. Soc., 65[10] 497-501(1982)
5. F. F. Lange, M. H. Hirlinger, "Hindrance of Grain Growth in Al_2O_3 by ZrO_2 Inclusions" J. Am. Ceram. Soc., 67[3] 164-168(1984)
6. R. Raj, R. K. Bordia, "Sinterin Behavior of Bi-Modal Powder Compacts" Acta Metall., 32[7] 1003-1019(1984)

7. Yueping Xu, A. Nakahira, K. Niihara "Characteristics of Al_2O_3-SiC Nanocomposites Prepared by Sol-gel Processing" J. Ceram. Soc. Jpn, to be published.
8. T. C. Chou, T. G. Nieh., "Nucleation and concurrent Anomalous Grain Growth of α- Al_2O_3 During a Phase Transformation" J. Am. Ceram. Soc., 74[9] 2270-2279(1991)
9. Chun-Hway Hsueh, " Sintering Behavior of Powder Compacts with Multiheterogeneities" J. Mater. Sci.,21 (1986) 2067-2072
10. C. Zener, private communication to C.. S. Smith, Trans. Met. Soc. AIME, 175,15,1949

Synthesis and Sintering

THE SINTERING REACTION OF SIALON COMPOSITES

Yoshio Ukyo and Shigetaka Wada
Toyota Central Research and Development Labs. Inc.
Nagakute-cho, Aichi-gun, Aichi-ken 480-11, Japan

ABSTRACT

The sintering reaction for sialon composites composed of Y-α'-sialon and β'-sialon was investigated in the system Si_3N_4-Y_2O_3-AlN. It is assumed that Y-α'-sialon with higher solubility is formed by the reaction between AlN and a liquid phase at the initial stage of sintering, and that Y-α'-sialon with lower solubility is formed by the reaction between Y-α'-sialon with higher solubility and the remaining Si_3N_4 at the final stage of sintering. Y_2SiO_5 and YAM were formed as intermediate compounds.

INTRODUCTION

We have previously reported that a sialon composite composed of Y-α'- sialon and β'-sialon is obtained by hot-pressing a mixture of Si_3N_4,Y_2O_3 and AlN(1-2). The mechanical properties of sialon composites are dependent on the amount of Y-α'-sialon. This means that the sialon composite with excellent mechanical properties can be obtained by controlling the amount of Y-α'-sialon. For example, we have obtained a sialon composite with high strength, about 1.3 GPa at room temperature and 1.0 GPa at 1400℃, by hot-pressing the mixture of Si_3N_4,Y_2O_3 and AlN(3). This excellent strength is closely related to the microstructure, in which very fine Y-α'-sialon grains exist between fine prismatic β'-sialon grains and a very small amount of glassy phase is formed at the grain boundaries. This is because most of the Y_2O_3 and AlN used as additives were dissolved in sialon grains during sintering.

The crystal structure and the properties of Y-α'-sialon have been often reported(4-11). However, only a few works have been reported on the sintering reaction in the system Si_3N_4-Y_2O_3-AlN (12-14). Nagel et al.(12), Mitomo et al.(13), and Cao et al.(14) have reported the shrinkage behavior and the reaction sequence during gas pressure sintering in this system. These reports, however, focus only on the region where a large amount of Y_2O_3 and AlN (single Y-α'-sialon phase) exists.

We have also previously reported that crystalline phases obtained by hot-pressing mixtures of Si_3N_4-Y_2O_3-AlN powders consisted of Y-α'-sialon with various solid solubilities and β'-sialon with a nearly constant solid solubility. These results did not agree with those obtained by Mitomo et al.(5-6). Moreover, the compounds including SiO_2, such as $10Y_2O_3\ 9SiO_2\ Si_3N_4$ and $4Y_2O_3\ SiO_2\ Si_3N_4$, were formed in the region of small amount of additives(15). The formation of these compounds cannot be expected from the phase (behavior) diagram reported by Mitomo et al.(4) and Huang et al.(16). These disagreements are thought to lie in the difference in sintering conditions and the impurities, especially oxygen, in the raw powders used. The oxygen contained in Si_3N_4 powder seems to play a very important role, especially in the region where a small amount of Y_2O_3 and AlN, because the compounds including SiO_2 are formed.

The sintering in this system is promoted by Y_2O_3 and AlN as sintering additives, which dissolve into the Si_3N_4 grains to form Y-α'-sialon and β'-sialon. As a result of this dissolution, the amount of glassy phase at the grain boundary is reduced and the excellent strength of the composite at high temperatures can be obtained. Therefore, it is very important to investigate the behavior of Y_2O_3 and AlN and to understand the sintering behavior and the formation mechanisms of Y-α'-sialon and β'-sialon. In this report, the sintering behavior of the system

Si_3N_4 -Y_2O_3 - AlN in the region where Y- α '- and β '-sialon are formed were investigated by hot-pressing.

EXPERIMENTAL PROCEDURE

A powder mixture of Si_3N_4 (UBE-SN-E10) -2.3mol% Y_2O_3 -20.7 mol% AlN was used in this experiment. The ratio of AlN to Y_2O_3 was 1 to 9 because all of the Y_2O_3 and AlN can be dissolved ideally into Si_3N_4 grains according to the following equation ;

$$\text{(4-3a) } Si_3N_4 + a\,(Y_2O_3 + 9AlN) = Y_{2a}\,(Si_{12-9a},\ Al_{2a})\,(O_{3a},\ N_{16-3a})$$

The powder mixture was sintered by hot-pressing at 1850℃ for 1 hr under 20 MPa pressure. The heating rate was 20℃/min. Sialon composite composed of about 55% Y- α '-sialon and 45% β '-sialon was obtained. The shrinkage was determined by measuring the movement of die. The reaction sequence was investigated by XRD, and the micostructure was observed by SEM.

EXPERIMENTAL RESULTS

Shrinkage Behavior

Figure 1 shows the shrinkage curve and the derived shrinkage rate. The shrinkage of the specimen sintered at 1850℃ for 1 hr was defined as the standard (100%). The density of the specimen sintered at 1850℃ for 1 hr was 3.26 g/cm^3 and almost full density. Shrinkage started at about 1350℃ and proceeded rapidly above 1550 ℃. Shrinkage almost finished when the temperature was kept at 1850℃ for about 10 min. The shrinkage rate showed maxima at about 1600℃ and 1750℃.

Reaction Sequence

Figure 2 shows the reaction sequence determined by X-ray diffraction method.

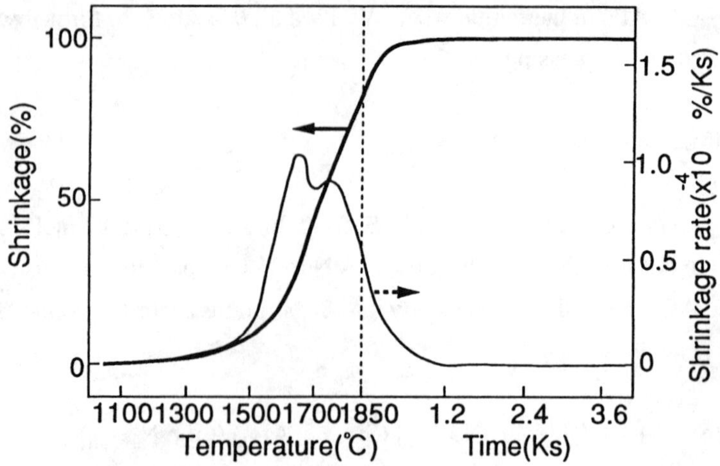

Figure 1 Shrinkage and shrinkage rate curves for sialon composite composed of Y-α'-sialon and β'-sialon at a constant heating rate of 0.34 ℃/S.

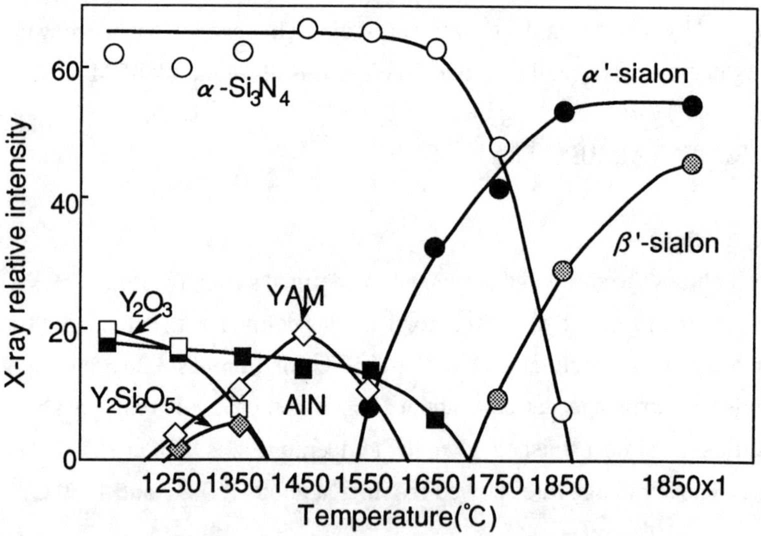

Figure 2 Phase changes during the formation of sialon composite composed of a Y- α'-sialon and β'-sialon.

Y_2O_3 disappeared between 1350 and 1450℃. AlN remained up to higher temperatures than Y_2O_3 and disappeared between 1650 and 1750℃. This result agreed well with those reported by Nagel et al (12). and Mitomo et al.(13). Y_2SiO_5 and YAM ($2Y_2O_3$ Al_2O_3) were formed as intermediate phases between 1250 and 1550℃. The temperature at which AlN decreased rapidly corresponded to that at which Y- α '-sialon increased rapidly. This indicates that AlN played an important role in the initial formation of Y- α '-sialon. The formation of β '-sialon started above 1750℃. Finally the sialon composite containing about 55% Y- α '-sialon and about 45% β '-sialon was obtained by hot-pressing at 1850℃ for 1hr.

Observation of Microstructure

Figure 3 shows the change in microstructures observed by SEM during sintering. As shown in this figure, the surface of Si_3N_4 particles seems to be covered with amorphous phases between 1250 and 1350℃. This means that liquid phases have already been formed at this temperature. The amount of liquid phase increased with increasing the temperature and showed the maximum between 1650 and 1750 ℃. This temperature range coincided with the temperature at which the shrinkage rate showed the maximum. Moreover, this temperature range coincided well with that above which Y- α '-sialon increased rapidly. This suggests that the liquid phase formed is closely related to the initial formation of Y- α '-sialon. The amount of the liquid phase decreased due to the dissolution of Y, Al and O into Si_3N_4 grains and the formation of Y- α '-sialon above this temperature.

Change in Lattice Constant (solubility) of Y- α '-sialon During Sintering

Figure 4 shows the change in diffraction angle of (031) plane of Y- α '-sialon formed at various temperatures. As shown in this figure, the diffraction angle of (031) plane moved toward a high angle with an increase in the temperature. This means that the solubility of Y- α '-sialon decreased with an increase the temperature. The same phenomenon was observed by Walls et al.(18) in the formation of Y- α '- sialon from a starting mixture consisting of Si_3N_4, AlN, Al_2O_3 and Y_2O_3.

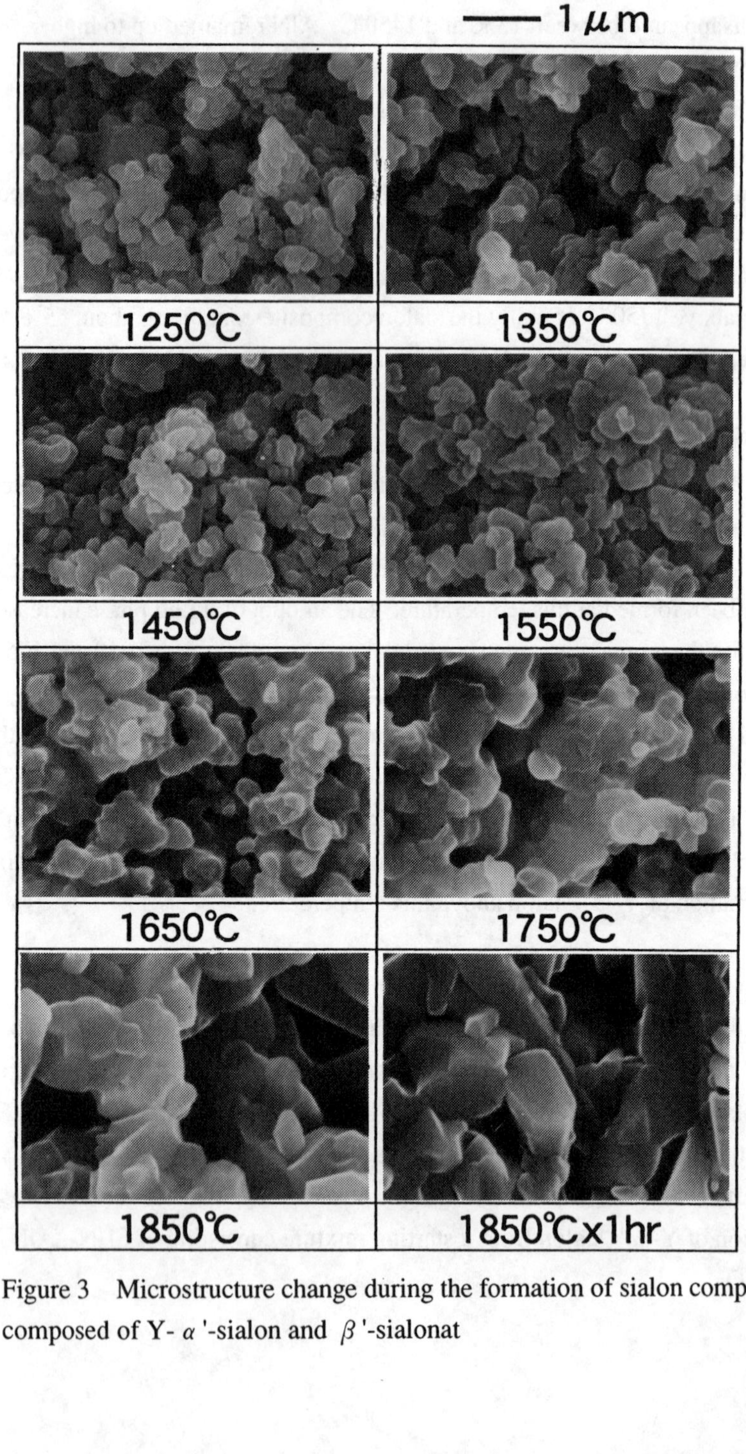

Figure 3 Microstructure change during the formation of sialon composite composed of Y- α '-sialon and β '-sialonat

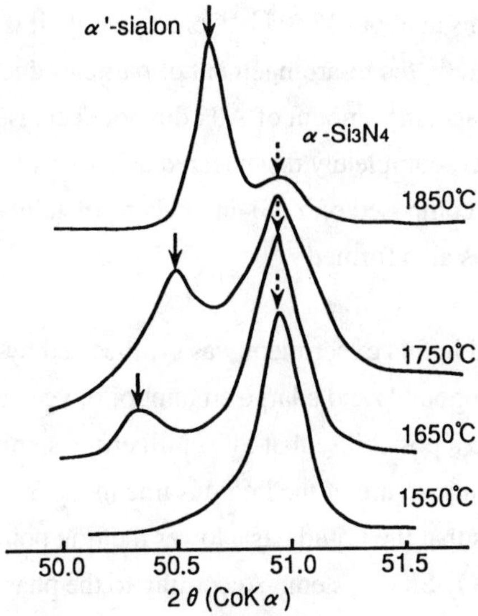

Fig.4 Change in X-ray diffraction angles of (031) plane for Y- α '-sialon and α -Si_3N_4 during the formation of sialon composite composed of Y- α '-sialon and β '-sialon

DISCUSSION

Sintering Reaction

The reaction sequence was devided into the following three stages; initial, middle and final stages. The initial stage is up to 1550℃, in which the shrinkage is small. Y_2O_3 disappears during this stage. The crystalline phases formed are intermediate compounds such as $Y_2Si_2O_5$ and YAM. The Middle stage is between 1550 and 1750℃. In this stage, a large amount of liquid phase is formed, Intermediate compounds disappear, the shrinkage proceeds rapidly, and the formation of Y- α '-sialon starts. The final stage is over 1750℃, in which the amount of liquid phase decreases and β '-sialon is formed. Crystalline phases in this stage are only Y- α ' -sialon and β '- sialon .

Initial Stage

As described above, shrinkage started at about 1300℃ and proceeded rapidly at about 1550℃. As shown in Figure 3, the surface of Si_3N_4 particle was already covered with liquid films at about 1350℃. Consequently, it is thought that shrinkage was promoted by the re-arrangement of particles due to the increase in the fluidity. At this first stage, the amount of AlN did not decrease, while that of Y_2O_3 already decreased. Y_2O_3 completely disappeared at above 1350℃. From this result, the liquid phase composed of Y, Si and O is formed by the reaction of Y_2O_3 and SiO_2. $Y_2Si_2O_5$ was also formed.

The Si_3N_4 powder used in this experiment was synthesized by the thermal - decomposed imido-compounds and a large amount of oxygen exists at the surface (18). Therefore, there is a possibility that SiO_2 influences significantly in the first stage. The minimum temperature of the liquidus line in the Y_2O_3-SiO_2 system is 1660 ℃. This suggests that the liquid has a lower melting point than the minimum melting point in the Y_2O_3-SiO_2 system. According to the phase diagram of the Y_2O_3-SiO_2-Al_2O_3 system, the minimum liquidus temperature is lower than 1400℃. This suggests that the Al_2O_3 existing at the surface of AlN powder takes part in the formation of liquid phase, and therefore, the liquid phase is composed of Y-Al-Si-O. Some part of Y_2O_3 which decreased between 1250 and 1350℃ forms $2Y_2O_3$ Al_2O_3 (YAM). The amount of YAM increased with an increase the temperature and shows a maximum at 1450℃. YAM is thought to be formed not only by the reaction between Y_2O_3 and Al_2O_3 existing at the surface of AlN but also by the reaction between Y_2O_3, AlN and SiO_2 shown in the following equation:

$$4AlN + 3SiO_2 + 4Y_2O_3 = Si_3N_4 + 2\,(2Y_2O_3\ Al_2O_3)$$

In this equation, the formation of β'-sialon is expected. However, β'-sialon was not detected by XRD because the temperature is too low.

Middle Stage (1550-1750℃)

The shrinkage proceeded rapidly and the shrinkage rate shows maxima due to the increase in the amount of liquid phase during this stage. This increase in liquid phase is attributed to the disappearance of Y_2SiO_5 and YAM at this stage. Figure 2 shows that at the same temperature (about 1550℃) the formation of Y- α '- sialon started whereas the amount of AlN was reduced. The amount of Si_3N_4 did not change at 1650℃ although a large amount of Y- α '-sialon was formed. Therefore, it is assumed that, at 1550-1650℃, the Y- α '-sialon was formed by the reaction between the liquid phases, which were composed of Y, Si, Al and O, and AlN. This means Si_3N_4 particle did not take part in the initial formation of Y- α '-sialon. The solubility of the Y- α '-sialon formed initially is higher than that of the Y- α '-sialon formed later.

Final Stage (1750-1850℃)

The formation of β '-sialon started, and the amount of Si_3N_4 was reduced rapidly at 1750℃. The amount of liquid phase decreased above 1750℃. This decreae in liquid phase is attributed to the dissolution of some elements from the liquid phase into Si_3N_4 grains and the formation of Y- α '-sialon and β '-sialon. AlN disappeared completely at this final stage. When the temperature was raised to 1850℃, the shrinkage was about 80%. The amount of Y- α '-sialon became nearly constant when the temperature was raised to 1850℃, while that of β '-sialon increased slightly. It is estimated that β ' -sialon was formed by the precipitation from the remaining liquid phase. As described later, the formation mechanism of Y- α '-sialon in the final stage is thought to be different from that of Y- α '-sialon formed in the middle stage. That is, it is assumed that Y- α '-sialon at the final stage is formed by the reaction between the Y- α '-sialon formed initially in the middle stage and the remaining Si_3N_4.

Formation Reaction of Y- α '-sialon

As shown in Figure 2, the formation of Y- α '-sialon starts at about 1550℃ under our experimental condition and the solubility of Y- α '-sialon formed initially is higher than that of Y- α '-sialon formed later. Nagel et al.(12) and Mitomo et al.(13) assumed in their papers that the formation of Y- α '-sialon occured by a solution-precipitation mechanism simultaneously with dissolution of Si_3N_4 and AlN. Since they did not refer to the solubility of Y- α '-sialon during sintering, it is impossible to discuss the formation mechanisms.

Walls et al.(7) have also reported that in the formation of Y- α '-sialon from the mixture consisting of Si_3N_4, Y_2O_3, AlN and Al_2O_3, the Y- α '-sialon with large lattice constants (higher solubility) was formed at initial stage of sintering. They explained that the change in the lattice constants (solubility) might be due to different Y- α '-sialon liquid tie line at different temperatures, but is more likely to be due to the more reactive oxide constituents in the starting mixture promoting an oxygen rich Y- α '-sialon at lower temperatures which at higher temperatures reacts with the remaining Si_3N_4 to give the final equilibrium composition.

Chen et al.(19) have also investigated the sintering reaction of the system Si_3N_4-Y_2O_3-AlN-Al_2O_3. They reported that Al, N rich melt and β '-sialon were formed due to the raction between AlN and the oxide melts formed at lower temperatures. They also showed that AlN trapped the liquid phase because the wettability between AlN particles and the liquid phase is higher than that between Si_3N_4 and the liquid phase. Moreover, they observed that essentially all sialon grains must grow by a heterogeneous nucleation mechanism started by some kind of identifiable seed.

As shown in Figure 2, the relative amount of Si_3N_4 did not change although a large amount of Y- α '-sialon formed at 1550-1650℃ and the amount of AlN decreased. The large amount of liquid phase is formed above 1550℃ at which the formation of Y- α '-sialon starts. These facts suggest that the initial formation of Y- α '-sialon is caused by the reaction between the liquid phase and AlN. Therefore, Si_3N_4 used as the starting powder did not take part in the formation of Y- α '-sialon between 1550 and 1650℃. The Y- α '-sialon with lower solubility was formed above 1650℃ by the reaction between Si_3N_4 and Y- α '- sialon with higher solubility formed at lower temperatures. This assumption is supported by the solubility change of Y- α '-sialon during sintering as shown in Figure 4.

CONCLUSIONS

The sialon composite composed of Y- α '- sialon and β '-sialon is obtained by hot-pressing the mixture of Si_3N_4, Y_2O_3 and AlN powders. The sintering reaction was studied, and the following results have been obtained.

(1) YAM ($2Y_2O_3$ Al_2O_3) and Y_2SiO_5(Y_2O_3 SiO_2) were formed between 1250 and 1550℃ as intermediate compounds.

(2) Above 1550℃, shrinkage proceeded rapidly due to the formtion of a large amount of liquid phase.

(3) The formation of Y- α '-sialon started at about 1550℃ and Y- α '-sialon formed initially has higher solubility than that formed later.

(4) The initial formation of Y- α '-sialon is believed to be due to the reaction between the liquid phase and AlN. It is also assumed that the Y- α '-sialon with lower solubility was formed at the final stage of sintering by the reaction between Si_3N_4 and Y- α '-sialon having a higher solubility .

References

(1) Y. Ukyo and S. Wada, "Formation and Stability of Y- α '-sialon co-existing with β '-sialon" pp 567-571 in *Euro-Ceramics.* Edited by G. de With, R. A. Terpstra and R. Metaselaar. Elsevier Applied Science, 1989.

(2) S. Wada and Y. Ukyo, " Microstructure and Properties of α '/ β ' Sialon Composite" pp29-33 in Proc. of the 34th Japan Congress on Materials Research (Kyoto, Japan, 1991). Published by The Society of Materials Science, Kyoto, Japan, 1991.

(3) Y. Ukyo and S. Wada, " High Strength Si_3N_4 Ceramics " Nippon Seramikkusu Kyokai Gakujutu Ronbunshi, 97[8] 872-874 (1989)

(4) M. Mitomo, F. Izumi, Y. Bando and Y. Sekikawa, "Characterization of

α '-SiAlON Ceramics" pp377-386 in Proc. Int. Symp. on Ceram. Component for Engines,. Edited by Somiya et al. KTK Science Publishers, Tokyo, Japan, 1983

(5) M. Mitomo and O. Fukunaga, "The stability of α -sialon at High Temperature" Yogyo-kyokai-shi, 89[11] 631-633 (1982)

(6) F. Izumi, M. Mitomo and Y. Bando, "Refinements for Cacium and Yttrium Containing α '-SiAlON," J. Mater. Sci., 19, 3115-3120 (1984)

(7) P. A. Walls and D. P. Thompson, " Reaction Mechanisms in the Formation of Calcium and Yttria α '- β ' Sialon Composites " Br. Ceram. Soc., No.37, 35-50 (1986)

(8) S. Slassor and D. P. Thompson, "Preparation and Characterization of Yttrium α '-SiAlON" pp223-230 in Proc. Int. Conf. on Non-Oxide Technical and Engineering Ceramics, Edited by S.Hampshire, Elsevier Applied Science, New York, 1986.

(9) T. Ekstrom, "Effect of Composition, PhaseContent and Microstructure on the Performance of Yttrium Sialon Ceramics," Mater. Sci. Eng., A109, 341 (1989)

(10) C. Chatfield, T. Ekstrom and M. Mikus, "Microstructural Investigation of Alpha-Beta Yttrium SiAlON Materials," J. Mater. Sci., 21, 2297-2298 (1979)

(11) G. Grand, J. Demit, J. Ruste and J. P. Torre, "Composition and Stability of Y-Si-AlO-N Solid Solutions Based on α -Si_3N_4" J. Mater. Sci.Lett., 14, 1749 - 1751 (1979)

(12) A. Nagel, P. Greil and G. Petzow, "Reaction Sintering of Yttrium Containing α -Silicon Nitride Solid Solution" Rev. Chim. Minerale, 22, 437-448 (1985)

(13) K. Ishizawa, N. Ayuzawa, A. Shiranita, M. Takai, N. Uchida and M. Mitomo, Proc. Int. Symp. on Ceramic Component for Engines, (1986) pp511

(14) G. Z. Cao, R. Metselaar and G. Ziegler, "Gas Pressure Sintering of α '-Sialon Ceramics " pp346-350 in *Euro-Ceramics.* Edited by G. de With, R.

A. Terpstra and R. Metaselaar. Elsevier Applied Science, 1989.

(15) N. Sugiyama, Y. Ukyo and S. Wada, " Phase Relations between α '- and β '-sialons in $Si_3N_4(SiO_2)$-AlN-Y_2O_3 System"J. Ceram. Soc. Jpn., 101 [7] 729-732 (1993)

(16) Z-K. Huang, P. Greil and G. Petzow, "Formation of α -Si3N4 Solid Solutions in the System Si_3N_4-AlN-Y_2O_3," J. Am. Ceram. Soc., 66 [6] c96-97 (1983)

(17) P. Greil, R. Nitzsche, H. Friedrich and W. Hermel, "Evaluation of Oxygen Content on Silicon Nitride Powder Surface from the Measurement of the Isoelectric Point" J. European Ceram. Soc., 7 [7] 353-359 (1991)

(18) M. Peukert and P. Greil, "Oxygen Distribution in Silicon Nitridde Powders" J. Mater. Sci., 22[10] 3717-3722 (1987)

(19) I-W. Chen and S-L. Hwang, "Superplastic SiAlON- A Bird's Eye View of Silicon Nitride Ceramics" pp209-222 in Mat. Res. Soc. Symp. Proc., Vol.287.Edit by I-WeiChen, P. F. Becher, M. Mitomo, G. Petzow and T. S. Yen, Materials Research Society, Pittburg, Pennsylvania, 1993.

DENSIFICATION OF MIXED α'/β' SIALON CERAMICS AND EVALUATION OF THEIR GRAIN-BOUNDARY PHASE.

Koji Watari, Masaki Yasuoka, Shuji Sakaguchi and Shuzo Kanzaki
Department of Ceramics Science, National Industrial Research Institute of Nagoya, Hirate-cho, Kita-ku, Nagoya, 462 Japan.

ABSTRACT

This paper reports on sintering behavior of mixed α'/β' sialon ceramics, fabricated from α-Si_3N_4, AlN and Ln_2O_3 (Y_2O_3, Er_2O_3 and Yb_2O_3), and discusses the densification process by considering the chemical reactions to form α'- and β'- sialon and the kinetical analysis of shrinkage. The quantity and character of the intergranular phase in mixed α'/β' sialon ceramics are also evaluated by internal friction measurements.

INTRODUCTION

One effective approach to solve the degradation of mechanical strength of Si_3N_4 ceramics at elevated temperatures is by the formation of a transient liquid phase during sintering. Sialon ceramics, which are solid solutions of Si_3N_4, are obtained by such transient liquid-phase sintering, with the ingredients of the liquid incorporated into the Si_3N_4 structure during the middle and final stages of sintering [1].

The solid solution in β-Si_3N_4, which is called β'-sialon, is formed by the simultaneous equivalent substitution of aluminum and oxygen for silicon and nitrogen. The reaction is commonly described by the following equations:

$$(4-z)Si_3N_4+2zAlN+zSiO_2 \rightarrow 2Si_{6-z}Al_zO_zN_{8-z} \quad (1)$$
$$(6-z)Si_3N_4+zAlN+zAl_2O_3 \rightarrow 3Si_{6-z}Al_zO_zN_{8-z} \quad (2)$$

In the equations, z takes values between 0 and 4.2 [2,3]. On the other hand, α'-sialon, which is the solid solution of α-Si_3N_4, is formed by a substitution of similar elements as for β'-sialon, with some metals cations accommodated into the two large interstices of the unit cell. This gives a general formula of the type $M_x(Si,Al)_{12}(O,N)_{16}$, where M is Ca, Li, Mg, Ca or lanthanide metals. When x is $0.3 < x < 2.0$, single-phase α'-sialons are fabricated. The reaction for the formation of single-phase α'-sialon containing Y_2O_3 proceeds as follows [4]:

$$(4-1.5x)Si_3N_4+0.5x(Y_2O_3+9AlN) \rightarrow Y_xSi_{12-4.5x}Al_{4.5x}O_{1.5x}N_{16-1.5x} \quad (3)$$

With x values less than 0.3, mixtures of α'-sialon and β'-sialon are obtained [5]. Single-phase sialon materials were expected to be the ideal structural ceramics with high strength at temperatures higher than 1200℃, however, several investigators have demonstrated that flexural strength and fracture toughness of these materials are much lower than those of β-Si_3N_4 ceramics containing an oxide addition. Poor mechanical properties of single-phase sialons are attributed to their microstructure which consists of grains with low aspect ratio, and weak pullout and absence of crack-branching effect due to the strong bonding between the grain interfaces [7].

Recently, mixed α'/β' sialon ceramics have received a great deal of attention, because of their high mechanical strength at room and elevated temperatures, and high fracture toughness [8]. These properties are a result of the microstructure, which consists of prismatic β-grains and roundish fine α-grains. Furthermore, their microstructure and mechanical properties, as well as their thermal properties, can be controlled by the phase composition. It is also possible to widely alter the quantity and type of elements that can be accommodated into a Si_3N_4 structure [7]. However, the densification mechanism, which links directly to control of microstructural development and phase composition, has not yet been clarified.

In this work, we investigate the sintering behavior of mixed α'/β' sialon ceramics with AlN and lanthanide oxides (Ln_2O_3) such as Y_2O_3, Er_2O_3 and Yb_2O_3 by high temperature dilatometry. We then discuss the densification process in combination with the change of reaction sequence during sintering. The evaluation of quantity and character of intergranular phase in mixed α'/β' sialon ceramics is also obtained by internal friction measurements.

EXPERIMENTAL PROCEDURE

α-Si_3N_4 (Denki Kagaku Kogyo, K.K., Japan), AlN (Tokuyama Soda. Co. Ltd., Japan), Y_2O_3, Er_2O_3 and Yb_2O_3 (Hokko Chemicals Industry Co. Ltd., Japan) powders were used as the starting materials. The raw Si_3N_4 powder had an oxygen content of 1.3 wt.%, and a specific surface area of 11.3 m^2/g. This powder was composed of 94% α- and 6% β-Si_3N_4. The starting AlN powder had an oxygen content of 1.05 wt.% and a specific surface area of 3.2 m^2/g. Nominal composition of powder mixtures was 83.10 for Si_3N_4, 1.69 for AlN and 15.21 for Ln_2O_3 in mol%, and the atomic proportion in the overall composition is represented by the following formula: $Ln_{0.15}Si_{11.325}Al_{0.675}O_{0.225}N_{15.775}$.

These powders were mixed using a Si_3N_4 pot and balls. After drying at 800℃ in a N_2 gas with 10% H_2, rectangular bars with dimensions of 10 x 10 x 45mm and discs with 14mm diameter x 8mm height were made from mixed powders using a home-made die, followed by cold-isostatic-pressing (CIP) at 490 MPa for 60 s. The green density of all the specimens was 1950 ± 20 kg/m^3.

Thermal shrinkage of bars was measured at temperatures from 1000 to 1850 ℃ at a heating rate of 10 ℃/min. under a N_2 gas pressure of 1MPa by using a high-temperature dilatometer [9], in order to investigate the sintering phenomena. Simultaneously, discs were put within a mixture of Si_3N_4/AlN/Ln_2O_3/BN into a graphite crucible, and fired in a graphite furnace at different temperatures starting from 1300 up to 1850℃ for 15 min. and 4 h at a heating rate of 10 ℃/min. under a N_2 gas pressure of 1MPa, followed by quenching at a cooling rate of above 100 ℃/min. until near 1000℃.

X-ray diffractometry of the crushed powders was carried out using CuK α radiation at 40kV and 100mA to identify crystalline phases and to obtain the phase composition. The α-Si_3N_4, α'-sialon, and β-Si_3N_4/β'-sialon contents in the sintered bodies were determined as follows: after XRD measurements, the X-ray peak areas of α-Si_3N_4 and α'-sialon for {102} and {201} planes, and of β-Si_3N_4/β'-sialon for {101} and {201} planes, were calculated using the Lorents-function, and the relative amount of these phases was obtained by comparing their peak areas. Lattice parameters were measured with Si as an internal standard, and the solid solubility of AlN and oxides into a Si_3N_4 structure was determined by referring to the published data.

The internal friction measurement was carried out by the torsion pendulum with a frequency of about 8Hz at room temperature to 1400℃ in the atmosphere [10].

3. RESULTS AND DISCUSSION

3.1. Sintering process

The temperature dependence of the shrinkage rate is shown in Figure 1. The shrinkage of all the compacts starts around 1150℃, and the type of lanthanide oxide produces a different sintering behavior. In the case of the Y_2O_3 addition, the shrinkage rate increases gradually at temperatures higher than 1350℃, and reaches a maximum at around 1570℃. Subsequently, the shrinkage rate shows a slight decrement and again an increment at temperatures between 1570 and 1620℃, decreasing rapidly at temperatures higher than 1620℃. For Er_2O_3 and Yb_2O_3 additions, the shrinkage rate shows an increase till 1510℃ for the Er_2O_3 addition, and 1480℃ for the Yb_2O_3 addition, decreasing thereafter. At temperatures over 1580℃, the shrinkage rate shows a rapid increase, with a maximum at 1630℃ for the Er_2O_3 addition and at 1650℃ for the Yb_2O_3 addition, decreasing thereafter with increasing sintering temperature.

Figure 2 reveals the phase change during the formation of α'- and β'-sialons at temperatures between 1300 and 1850℃. The lattice parameters of α'-sialon are shown in Figure 3. α' -sialon with large unit cell dimensions is initially formed at temperatures between 1400 and 1500℃, and its content increases by raising the sintering temperature, but decreases above 1700℃. The amount of α'-sialon varies with the type of oxides. Lattice parameters of α'-sialon attain a maximum at 1530℃, but thereafter decrease as the temperature is raised. The influence of the type of lanthanide oxide on the lattice parameters has not been identified in the present work. The x value in $Y_x(Si, Al)_{12}(O,N)_{16}$ is 0.4 for the specimen obtained at 1500℃ and 0.25 for the one at 1850℃.

β'-sialon is formed at temperatures higher than 1550℃ for Y_2O_3 and Er_2O_3 additions, and at temperatures higher than 1600℃ for the Yb_2O_3 addition. A remarkable increment in the amount of β'-sialon is observed at

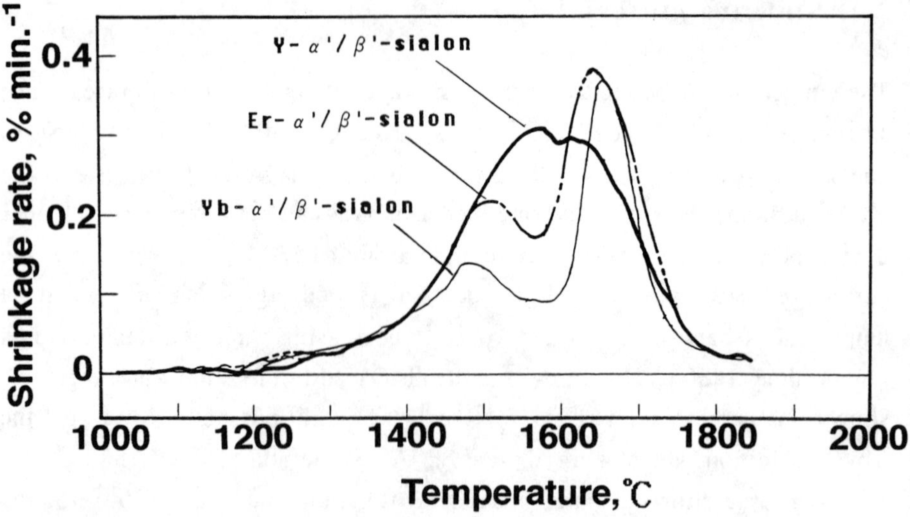

Figure 1. Temperature dependence of shrinkage rate of Ln_2O_3 doped α'/β' sialon.

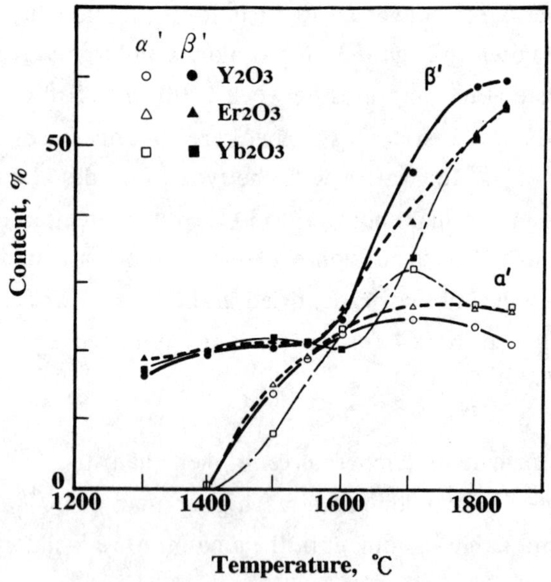

Figure 2. Phase change during the formation of α'/β'sialon ceramics at temperatures between 1300 and 1850℃.

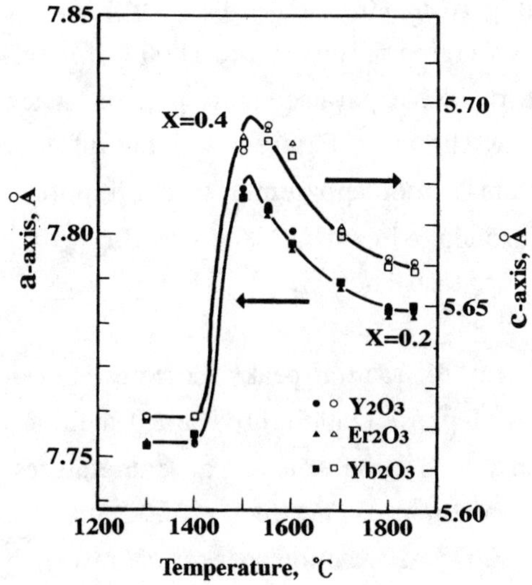

Figure 3. Change of lattice parameters of α'- sialon.

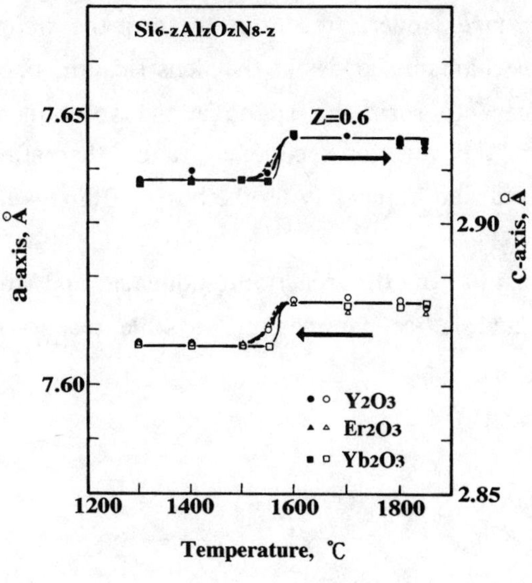

Figure 4. Change of lattice parameters of β'- sialon.

temperatures ranging from 1600 to 1700℃ for Y_2O_3 and Er_2O_3 additions, and at temperatures ranging from 1700 to 1800℃ for the Yb_2O_3 addition. A significant change in lattice parameters is present at temperatures between 1550 and 1600℃, as shown in Figure 4, but the influence of the oxide type on lattice parameters is not apparent. According to published data, the value of z in the formula, $Si_{6-z}Al_zO_zN_{8-z}$, was 0.6 for the specimen sintered at temperatures higher than 1600℃.

Identification of X-ray diffraction peaks indicates, besides the formation of α' and β'-sialons, J-phase ($Ln_4Si_2O_7N_2$) and melilite ($Si_3N_4 \cdot Ln_2O_3$) in specimens with Ln_2O_3 addition sintered at temperatures between 1300 and 1400℃, and at temperatures between 1500 and 1700℃, respectively. Disappearance of X-ray diffraction peaks of raw AlN and Ln_2O_3 was recognized above 1500 and 1300℃, respectively.

It is generally accepted that densification of Si_3N_4 ceramics with oxide additives and Si_3N_4 solid solutions proceeds via liquid-phase sintering, which consists of three stages: (I) particle rearrangement, (II) solution-precipitation and (III) coalescence of grains [11]. At the solution-precipitation stage, several reactions to form the sialons take place, and a number of mechanisms to assist the densification occur. In this work, a reaction process to form the sialon at the solution-precipitation stage is derived from the reaction sequence, and, thereafter, the densification process is discussed kinetically on the basis of the results in Figure 1.

From the results of the reaction sequence and the change in phase composition and lattice parameters, possible reactions are summarized in Table I.

Table I. Possible reactions to form the α' and β'-sialons.

α' -sialon:	
$SiO_2+Al_2O_3+Ln_2O_3+Si_3N_4/AlN$	
$\rightarrow$ Nitrogen glass + J-phase ($Ln_4Si_2O_7N_2$)	(a)
J-Phase+N-glass+Si_3N_4/AlN	
$\rightarrow \alpha$'-sialon^H +Melilite ($Si_3N_4 \cdot Ln_2O_3$)+N-glass	(b)
Si_3N_4+9AlN:$Ln_2O_3 \rightarrow \alpha$'-sialon^H	(c)
Si_3N_4+α'-sialon$^H \rightarrow \alpha$'-sialon^L	(d)
β'-sialon:	
Si_3N_4+Melilite+N-glass$\rightarrow \beta$'-sialon	(e)
Si_3N_4+α'-sialon^L+N-glass$\rightarrow \beta$'-sialon	(f)

H and L indicate the α'-sialon materials with high and low solid solubility, respectively.

In liquid-phase sintering of mixed sialons, nitrides and oxide additives as well as oxide impurities react to form a liquid and to subsequently precipitate a partial crystalline phase at the grain boundaries. The phenomenon is described by reaction (a). α'-sialon is compatible with melilite. Furthermore, α'-sialon with high solid solubility is found at the initial stage. The reactions in precipitation of α'-sialon are expressed as equations (b) and (c). But, as the content of α'-sialon increased and lattice parameters decreased with raising the temperature, the reaction (d) should be also considered. In the present work, we could not, however, confirm the presence of α'-sialon with various solid solubilities, because the lattice parameters measured strike an average one of diverse solid solubilities.

Melilite disappears due to the precipitation of β'-sialon, and the chemical reaction in the formation of β'-sialon is expressed as equation (e). It is also considered that precipitation of β-sialon could proceed through reaction (f). Han et al. [12] investigated the phase transformation from α' to β'-sialon by liquid infiltration in the Y-Si-Al-N-O system. They reported that during heating, the glass matrix penetrated into the sintered α'-sialon specimen to induce the transformation of α'- into β'-sialon, and that no trace of α'-sialon existed in β'-sialon grains. Sugiyama et al. [13] reported recently that α'-sialons with low solid solubility disappeared during the heat-treatment, while the ones with high solid solubility were stable. Such phenomena are anticipated to occur during precipitation of β'-sialon in the present work, since the increase in the amount of α'-sialon is slowed by the formation of β'-sialon, and thereafter the amount of α'-sialon decreases slightly, as do, simultaneously, the lattice parameters.

Reactions in formation of α'- and β'-sialon greatly influence the sintering behavior. In the case of Er_2O_3 and Yb_2O_3 additions, the formation of α'-sialon yields to a reduction of shrinkage rate at temperatures between 1480 and 1580℃, because the ingredients in the liquid, i.e., rare-earth and aluminum, are removed from it, causing the quantity and the viscosity of the liquid-phase to decrease and to increase, respectively. This hinders the dissolution of residual α-Si_3N_4 into the liquid. Furthermore, as melilite formation consumes rare-earth cations, the amount of transient liquid is reduced. Therefore, reduction of the shrinkage rate at temperatures from 1480 to 1580℃ is due to the formation of α'-sialon and melilite.

At temperatures higher than 1600℃, precipitation of β'-sialon and a rapid increase of shrinkage rate occurred concurrently, when Er_2O_3 and Yb_2O_3 were added. This is attributed to reactions (e) and (f). The progress of reactions (e) and (f) results in an increase in the amount and a decrease in viscosity of the liquid. Therefore, the solution-precipitation process is considered to be accelerated, because the precipitation of β'-sialon progresses, generally, when the liquid volume becomes larger and its viscosity becomes lower.

Figure 5 presents the isothermal shrinkage of compacts containing Yb_2O_3 addition. The kinetic parameter obtained takes the values between 0.76 and 0.70, which are much larger than values predicted for the solution-precipitation process in the classical liquid-phase sintering model proposed by Kingery et al. [11]. It is not possible here to clarify the sintering process through a mathematical analysis.

A number of additional solution-precipitation mechanisms have been identified, such as pore-elimination due to liquid flow and cooperative flow of particle/liquid mixture, Ostwald ripening with shape accommodation and particle disintegration [14]. It is suggested that sintering of mixed α'/β'-sialon is a concurrent process involving these mechanisms. The pore-elimination process due to the cooperative flow of particle/liquid mixture causes the rapid shrinkage observed at 1580 to 1700℃ for Er_2O_3 and Yb_2O_3 additions, since the rate of particle migration toward the interstice between particles is very fast. Furthermore, it is expected from the results of Han et al. that in the present sintering, disintegration of α'-sialon grains occurs at temperatures higher than 1600℃. In this case, it is easy for the α'-sialon grains to migrate toward the interstice between particles, resulting in rapid shrinkage. Such an interpretation, however, may be too simplistic, and further work such as microstructural observation, is required to understand the densification process exactly.

3.2. Evaluation of grain-boundary phase

The temperature dependence of internal friction for mixed α'/β' sialon is shown in Figure 6. The results of internal friction for Si_3N_4 ceramics containing Y_2O_3 and Al_2O_3 addition [15], annealed Si_3N_4 ceramics with Y_2O_3 and Al_2O_3 additives [15], and Si_3N_4 ceramics without sintering aids [16] are also plotted in the same figure. The internal friction of mixed α'/β' sialon ceramics increases gradually up to 1000℃. The result is similar

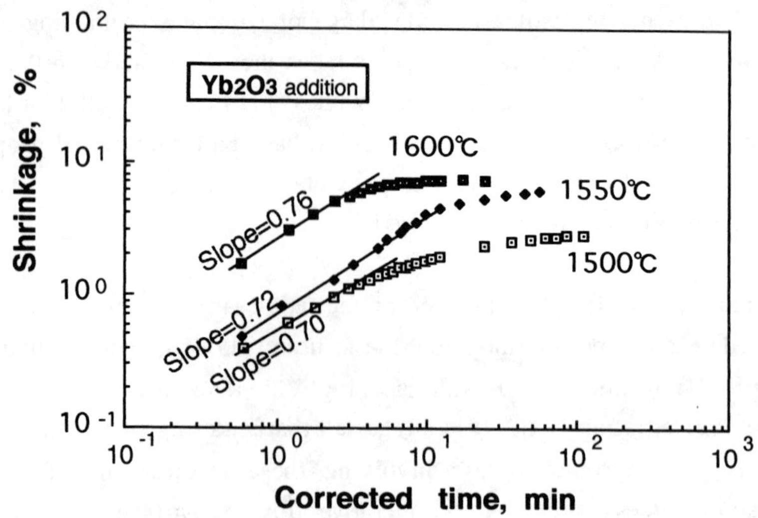

Figure 5. Isothermal shrinkage of compacts with Yb_2O_3 addition.

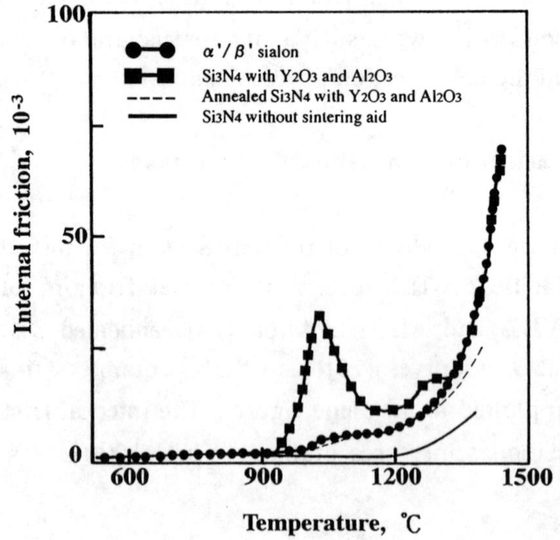

Figure 6. Temperature dependence of internal friction.

to that of annealed Si_3N_4 ceramics and Si_3N_4 ceramics without sintering aids, but Si_3N_4 ceramics containing Y_2O_3 and Al_2O_3 addition have an internal friction peak around 1080℃. Sakaguchi [17] and Awazu [15] have mentioned that an internal friction peak at 1080℃ relates closely to the presence of a glassy phase. Moreover, in the case of Si_3N_4 ceramics containing Y_2O_3 and Al_2O_3 additions, it was clarified from low-temperature specific heat measurements that the combination of these additives causes the formation of a large amount of glassy phase after sintering [18]. Referring to results of diverse Si_3N_4 ceramics, it is considered that the amount of the intergranular glassy phase in the present sialon materials is very small, and it is postulated that the type of grain boundary of mixed α'/β' ceramics resembles that of annealed Si_3N_4 ceramics and Si_3N_4 ceramics without sintering aids. Consequently, it can be predicted that the materials obtained possess excellent strength at elevated temperatures.

CONCLUDING REMARKS

The sintering process of mixed α'/β' sialon ceramics, fabricated from α-Si_3N_4, AlN and Ln_2O_3 (Y_2O_3, Er_2O_3 and Yb_2O_3), was studied. Though the shrinkage rate increased with increasing the temperature, the formation of α'-sialon caused the retardation of densification rate temporally, whereas that during the formation of β'-sialon the densification proceeded remarkably. The shrinkage accompanied by the precipitation of β'-sialon is probably ascribed to the dissolution of residual α-Si_3N_4 and α'-sialon into a liquid and to the phase transformation from α' to β'-sialon by liquid infiltration, which result in an increase in liquid volume and in a decrease of liquid viscosity. These phenomena are considered to induce the solution-precipitation process, cooperative flow process of the particle/liquid mixture and particle disintegration, yielding to rapid shrinkage. The type of oxide greatly influenced the sintering behavior and phase composition, and the densification at the initial and middle stages and the reactions from α' and α to β' accelerated in order of $Y_2O_3 > Er_2O_3 > Yb_2O_3$ addition. The internal friction of mixed α'/β' sialon showed only a relaxation peak, which increase exponentially by increasing the temperature, suggesting the existence of a small amount of intergranular glassy phase in sintered materials.

ACKNOWLEDGEMENT

The authors would like to thank Mrs. Maria Cecilia Valecillos for her cooperation.

REFERENCES

[1] A.Nagel, P.Greil and G.Petzow, "Reaction Sintering of Yttrium Containing α-Silicon Nitride Solid Solution," ***Revue de Chemie Minerale*, 22**, 437-47 (1985).

[2] Y.Oyama and O.Kamigaito, "Solid Solubility of Some Oxides in Si_3N_4," ***Japan J. Appl. Phys.*, 10**, 1637-42 (1972).

[3] K.H.Jack and W.I.Wilson, "Ceramics Based on the Si-Al-O-N and Related System," ***Nature*, 238**, 28-9 (1977).

[4] S.Hampshire, K.H.Park, D.P.Thompson and K.H.Jack, "α-Sialon Ceramics," ***Nature*, 274**, 880-82 (1978).

[5] K.Ishizawa, N.Ayuzawa, A.Shiratani, M.Takai, N.Uchida and M.Mitomo, "Some Properties of α-Sialon Ceramics," in *Ceramic Materials and Components for Engines*, Edited by W.Bunk and H.Hausner, German Ceramic Society, Bad Honnef. FRG (1986) pp.511-18.

[6] S.Umebayashi, K.Kishi, E.Tani and K.Kobayashi, "The Strength of Hot-Pressed β-Sialons with Z=1 in $Si_{6-z}Al_zO_zN_{8-z}$," ***Yogyo-Kyokai-Shi*, 92**, 35-51 (1984).

[7] T.Ekstrom and M.Nygren, "Sialon Ceramics," ***J. Am. Ceram. Soc.*, 75**, 259-76 (1992).

[8] Y.Ukyo and S.Wada, "High Strength Si_3N_4 Ceramics," ***J. Ceram. Soc. Japan*, 97**, 872-74 (1989).

[9] O.Abe, "Sintering Process of Y_2O_3-Added Si_3N_4," ***J. Mater. Sci.*, 25**, 3641-48 (1990).

[10] S.Sakaguchi, T.Kodama, N.Murayama and F.Wakai, ***J. Alloy Compound***, *in press*.

[11] W.D.Kingery, "Densification During Sintering in the Presence of a Liquid Phase: I," ***J. Appl. Phys.***, **30**, 301-6 (1959).

[12] S.M.Han, S-J.Kang and Y-T.Lee, "Phase Transformation from α' to β'-Sialon by liquid infiltration in Y-Si-Al-O-N System," ***J. Eur. Ceram. Soc.***, **12**, 431-34 (1993).

[13] N.Sugiyama, Y.Ukyo and S.Wada, "Thermal Stability of Phases in Si3N4 (-SiO2)-AlN-Y2O3 System," ***J. Ceram. Soc. Japan***, **102**, 145-48 (1994).

[14] G.Petzow and W.A.Kaysser. "Basic Mechanisms of Liquid-Phase Sintering" in Sintering Key Papers. Edited by S.Somiya and Y.Moriyoshi. Elsevier Applied Science, London and New York, 1990, pp.595-614.

[15] K.Awazu, O.Komura and M.Miyake, "High Temperature Properties of Sintered Silicon Nitride," ***Funtai oyobi Funmatsuyakin*** (J.Japan Soc. Powder and Powder Metall.), **39**, 499-503.

[16] I.Tanaka, G.Pezzotti, K.Matsushita, Y.Miyamoto and T.Okamoto, "Impurity-Enhanced Intergranular Cavity Formation in Silicon Nitride at High Temperature," ***J. Am. Ceram. Soc.***, **74**, 752-59 (1991).

[17] S.Sakaguchi, N.Murayama and F.Wakai, "Internal Friction of Si3N4 at Elevated Temperatures," ***Yogyo-Kyokai-Shi*, 95**, 1219-22 (1987).

[18] K.Watari, K.Ishizaki and K.Mori, "Evaluation of Glassy Phase in Sintered Silicon Nitride by Low-temperature Specific Heat Measurement," ***J. Am. Ceram. Soc.*, 74**, 244-46 (1991).

INFLUENCE OF OXYGEN CONTENT ON Si_3N_4 GRAIN BOUNDARY

Toyohiro Hamasaki, Tomohiro Yanai and Kozo Ishizaki
Nagaoka Gijutsu-Kagaku Daigaku (Nagaoka University of Technology)
Kamitomioka 1603-1, Nagaoka, Niigata 940-21, JAPAN

ABSTRACT

Three types of Si_3N_4 with Y_2O_3 and Al_2O_3 are hot pressed at 1800℃ under 30 MPa for 1 h. The first one is sintered from a commercial Si_3N_4 powder. The second powder is modified from the same powder by aqueous Soxhlet washing. This modified powder has less surface oxides. The third one is made from the same commercial powder and is oxidized by a heat treatment. The surface oxygen content of the Si_3N_4 powder is controlled by washing or oxidation surface modification treatment. The quantity and types of grain boundary phases of sintered Si_3N_4 are evaluated by cryogenic specific heat measurements. The bending strength at high temperature increases with a decrease in the amount of grain boundary glassy phase.

INTRODUCTION

Si_3N_4 ceramics have increasingly become a promising material for heat engine components because of their excellent performance in high temperature environments. However, Si_3N_4 is difficult to sinter. Metal oxides are used as sintering aids to provide densification. During densification these additives react with Si_3N_4 surface oxides to form a liquid phase[1]. Upon cooling the liquid phase remains at the grain boundary phases and forms a glassy phase or crystalline phase. The presence of a grain boundary glassy phase in sintered Si_3N_4 has a deleterious effect on mechanical properties, particularly at high temperatures[2, 3]. Therefore, the quantitative evaluation of grain boundary phases is of great importance. The quantitative evaluation of grain boundary phases is usually determined by a destructive metallographic method, on chemically and thermally etched sample surfaces[4-8]. The authors have reported a new quantitative evaluation method of grain boundary phases by cryogenic specific heat measurements[9, 10].

In this study, the results of a quantitative evaluation of grain boundary phases of Si_3N_4 by cryogenic specific heat measurements have been correlated to bending strengths at high temperatures.

EXPERIMENTAL PROCEDURE

Three types of hot pressed Si_3N_4 ceramics were investigated in this study. The first one (sample A) was sintered from a commercial Si_3N_4 powder (Ube Industry Co., Ltd., UBE-SN-E10, specific surface area 11.1 m^2/g, average particle size 330 nm). The second one (sample B) was made from the same powder with modified surfaces by aqueous Soxhlet washing. This modified powder has less surface oxides[11, 12]. The third one (sample C) was made from the same powder as A and was oxidizing in air at 1000℃ for 12 h. Powders of Y_2O_3 (Shin-Etsu Chemical Co., Ltd., 1.3 m^2/g, 900 nm) and Al_2O_3 (Onoda Cement Co., Ltd., 1.7 m^2/g, 900 nm) were added as sintering aids (3 mol% each) to all samples.

The powders were mixed for 24 h in ethanol using by a ball milling machine with Si_3N_4 balls in a plastic vessel. The powder slurries were dried in a vacuum furnace at 50°C for 3 h and crushed. The resultant powders were hot pressed at 1800℃ under 30 MPa for 1 h in nitrogen atmosphere. Density measurements by the Archimedean method showed full densification (above 99.9%). The sintered bodies were machined to rectangular bars of $3 \times 4 \times 36$ mm. Three points bending strengths were measured from room temperature to 1500℃. The crosshead speed was 0.5 mm/min, and the span was 30 mm. The cryogenic specific heat was measured at temperatures from 10 to 40 K using a laboratory designed adiabatic calorimeter[9].

RESULTS AND DISCUSSION

Figure 1 shows the temperature dependence of the bending strength of samples A, B and C. The bending strength degradation of all samples began above 800℃. The bending strength of sample B is the highest in these samples at 800℃. The reason may be due to the fact that sample B has less grain boundary glassy phase than samples A and C.

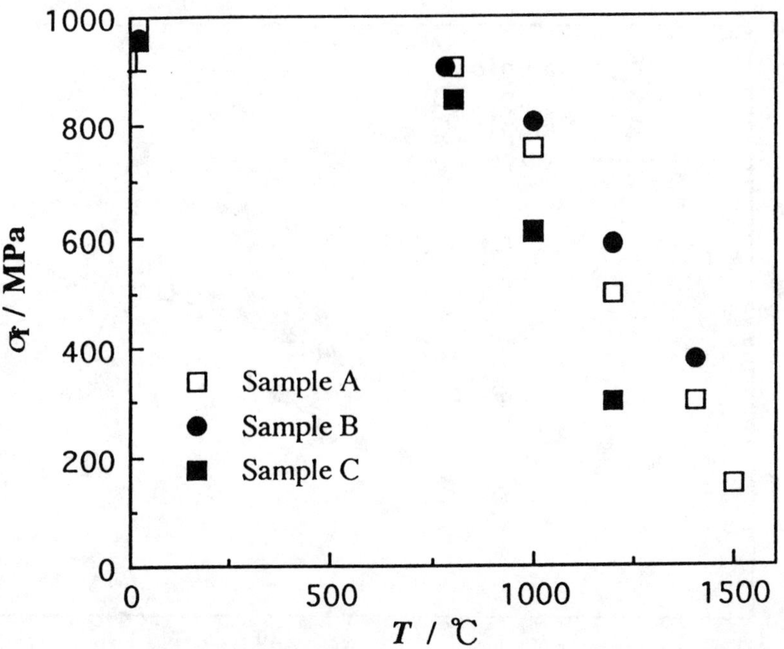

Figure 1 The temperature dependence of bending strength of samples A, B and C. Open squares (sample A) and solid circles represent the bending strength of hot pressed Si_3N_4 made from commercial powder and from the same powder after 100 h aqueous washing(sample B), respectively. Solid squares represent the bending strength of sample C made from the same powder as A with oxidized in air at 1000℃ for 12 h.

The temperature dependence of cryogenic specific heats of samples A, B and C is shown in Fig. 2. The solid line is calculated by Debye theory using the Debye temperature of 1100 K[13]. The isochoric specific heat of an electrically insulating material with a perfect crystalline phase at low temperatures decreases in accordance with the Debye T^3 law[14]. At cryogenic temperatures the difference between the isochoric specific heat and the isobaric one becomes negligible (less than 0.03% at 50 K for Si_3N_4). The transition to a glassy state is associated with a departure from the Debye T^3 law[15]. The difference between the specific heat of glassy and crystalline phases has been referred to as an abnormal specific heat. The abnormal specific heat is caused by extra phonon modes characteristic for the glassy phase[15]. In the case of Si_3N_4, the difference which exists between the measured cryogenic specific heats of the samples and the calculated value gives evidence of the existence of a grain boundary glassy phase in sintered Si_3N_4.

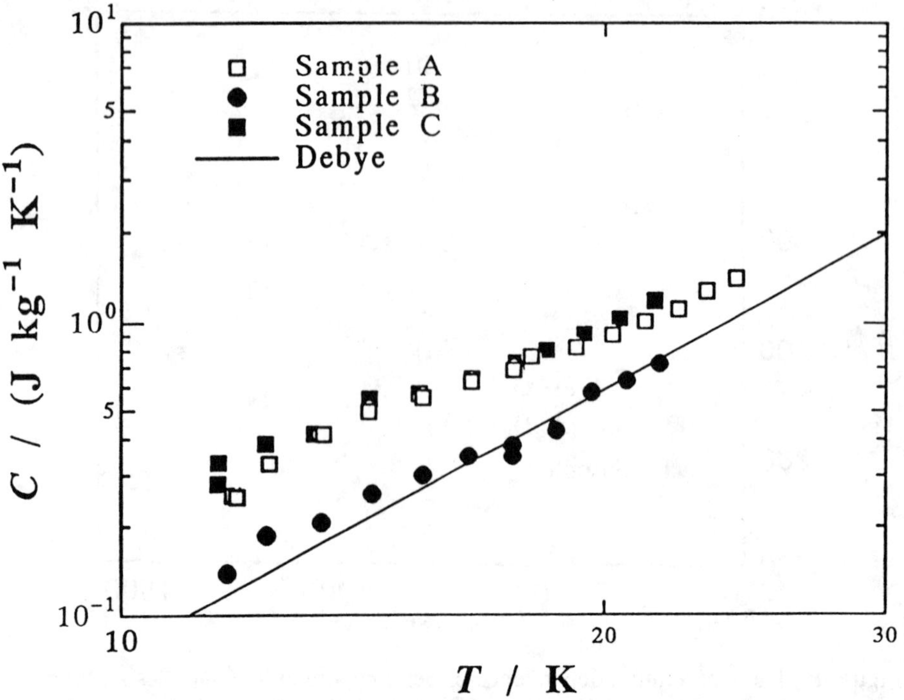

Figure 2 The temperature dependence of cryogenic specific heats of samples A, B and C. The symbols for samples are the same as in Fig.1. The solid line corresponds to the calculated specific heat by Debye theory using Debye temperature of 1100 K[13].

The specific heat of a sintered body can be represented by:

$$C = xC_1 + yC_2 + zC_3 \tag{1}$$

where x, y, and z are respectively the amounts of grain lattice, grain boundary crystalline phase, and grain boundary glassy phase, and C, C_1, C_2, and C_3 are respectively the total specific heat of sample, the lattice specific heat of Si_3N_4, the specific heat of the crystallized grain boundary phase, and the specific heat of the grain boundary glassy phase. C_1 and C_2 can be estimated by the Debye theory using the Debye temperature of 1100 K[13] and by the Debye theory using the Debye temperature of 350 K that was calculated by the cryogenic specific heats of a simulated partial crystalline grain boundary phase, respectively. C_3 is obtained by the cryogenic specific heats of a simulated glassy grain boundary phase[10]. The composition of the simulated glass ($8.0Si_3N_4 45.9Y_2O_3 20.7Al_2O_3 25.4SiO_2$ in wt%) is fixed by the X-ray determination of the maximum solubility of nitrogen, under the same heat treatment. This assumes that the oxygen of Si_3N_4 reacts with sintering additives to form a grain boundary glassy phase[1, 16-18]. The composition of the partially crystallized grain boundary phase was $Y_5N(SiO_4)_3$ ($4.8Si_3N_4 18.4SiO_2 76.8Y_2O_3$ in wt%), which was determined from the X-ray diffraction analysis of the heat treated sample A[10]. The amount of each phase can be evaluated to solve eq. (1).

Table I. The Amount of Lattice Si_3N_4 Phase in Weight Fraction, x, Grain Boundary Recrystallized Phase, y, and Grain Boundary Glassy Phase, z.

	Sample A	Sample B	Sample C
x	0.935±0.0011	0.974±0.0004	0.916±0.0012
y	0.010±0.0001	0.020±0.0002	0.015±0.0008
z	0.054±0.0011	0.006±0.0003	0.068±0.0012

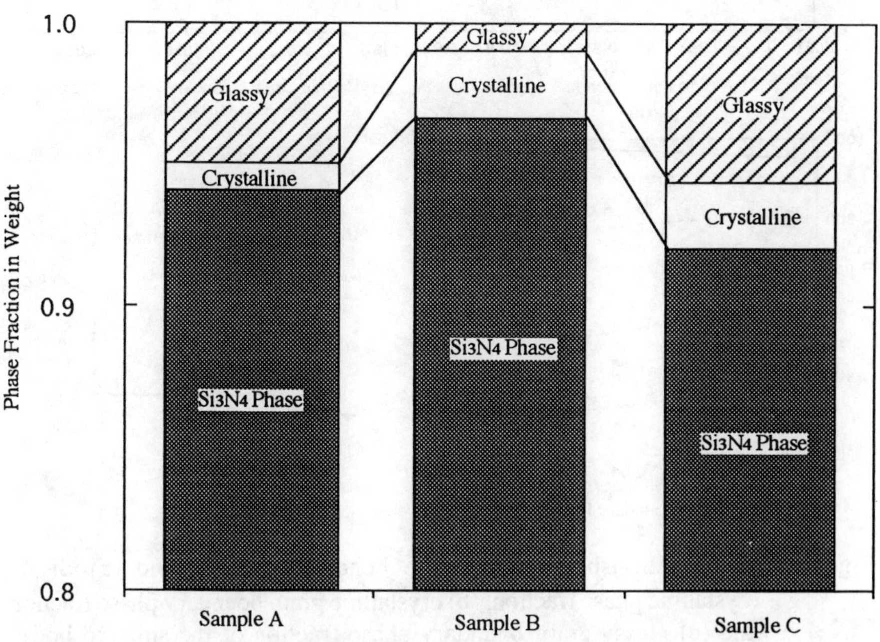

Figure 3 The results of the quantitative evaluation of constituent phases for hot pressed Si_3N_4.

Figure 3 and Table I show the result of the quantitative evaluation of the grain boundary phase. Sample B has the most lattice Si_3N_4 phase. For the surface modified powder that has the smallest amount of surface oxides, the smallest amount of liquid phase was formed during the sintering process and, consequently, the smallest amount of grain boundary phases is found in sample B. Sample C has the most grain boundary phases. The powder of sample C has the highest amount of surface oxides. The highest amount of grain boundary phases is found in sample C. The surface oxygen content of the starting powder influences the amount of grain boundary phases in sintered Si_3N_4 ceramics.

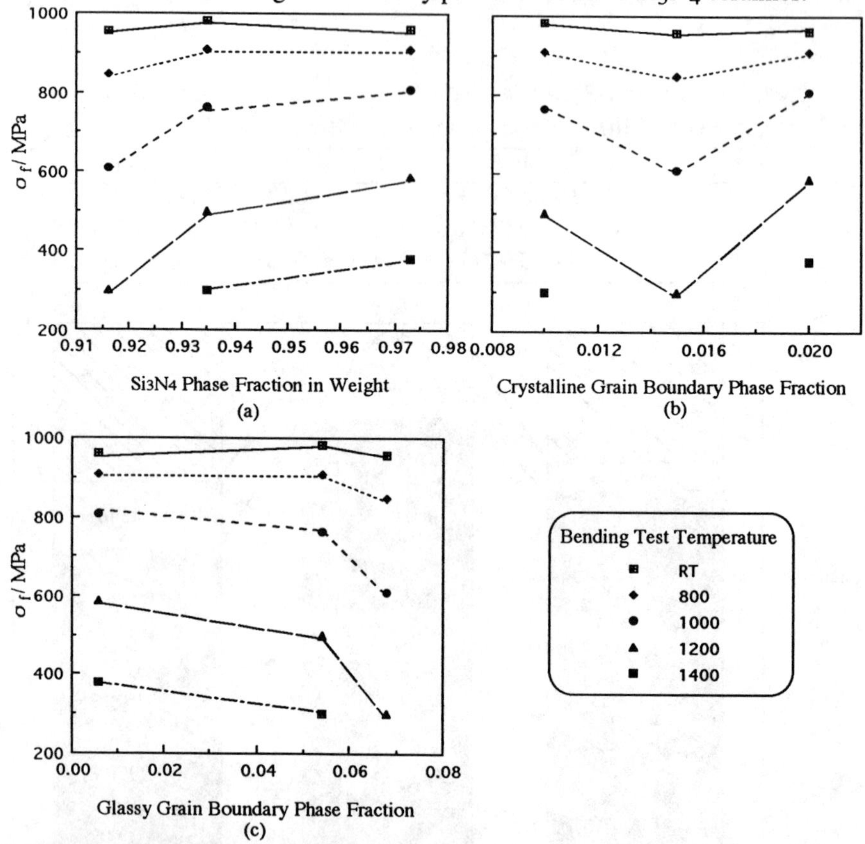

Figure 4 The relationship between the bending strength and (a) Si_3N_4 crystalline phase fraction, (b) crystalline grain boundary phase fraction and (c) glassy grain boundary phase fraction of the sintered body. The marks indicate different test temperatures at which the bending strength was measured.

Figures 4 (a), (b) and (c) show the relationship between (a) Si_3N_4 crystalline phase fraction, (b) crystalline grain boundary phase fraction and (c) glassy grain boundary phase fraction of sintered body and the bending strength tested at different temperatures. The bending strength increased with an increase in the Si_3N_4 phase fraction. This tendency can be observed by raising the bending test temperature. The bending strength decreases with an increase in the amount of glassy grain boundary phase. Therefore, the high temperature bending strength depends on the amount of the glassy grain boundary phase in the system Si_3N_4-Y_2O_3-Al_2O_3.

CONCLUSION

The quantity of grain boundary crystalline and glassy phases of sintered Si_3N_4 can be evaluated by cryogenic specific heat measurements. The amount of grain boundary phases depends on the content of surface oxygen of the starting powder. The high temperature bending strength decreases with an increase in the amount of glassy grain boundary phase.

REFERENCES

[1] R. E. Loehman, "Oxynitride Glasses," *J. Non-Cryst. Solids*, **42**, 433-446(1980).

[2] N. Hirosaki, A. Okada and M. Mitomo, "Effect of Oxide Addition on the Sintering and High-Temperature Strength of Si_3N_4 Containing Y_2O_3," *J. Mater. Sci.*, **25**[3], 1872-1876(1990).

[3] J. Zeng, O. Yamada, I. Tanaka and Y. Miyamoto, "Hot Isostatic Pressing and High Temperature Strength of Silicon Nitride-Silica Ceramics," *J. Am. Ceram. Soc.*, **73**[4], 1095-1097(1990).

[4] W. E. Lee and G. E. Hilmas, "Microstructural Changes in β-Silicon Nitride Grains Upon Crystallizing the Grain-Boundary Glass," *J. Am. Ceram. Soc.*, **72**[10], 1931-1937(1989).

[5] M. K. Cinibulk and G. Thomas, "Grain-Boundary-Phase Crystallization and Strength of Silicon Nitride Sintered with a YSiAlON Glass," *J. Am. Ceram. Soc.*, **73**[6], 1606-1612(1990).

[6] G. N. Babini, A. Bellosi and P. Vincenzini, "Factors Influencing Structural Evaluation in the Oxide of Hot-Pressed Si_3N_4-Y_2O_3-SiO_2 Materials," *J. Mater. Sci.*, **19**[11], 3487-3497(1984).

[7] K. Yabuta, H. Nishino and K. Uematsu, "Quantitative Analysis of o'-Phase in β'-o' Sialon Ceramics by Image-Analysis Technique," *Nippon Seramikkusu*

Kyokai Gakujutsu Ronbunnshi, **99**[2],173-177(1991).
[8] A. Yamakawa, M. Miyake and K. Ishizaki, "Changes of YAP-Phase Content and Some Properties by the Annealing of Si_3N_4-Al_2O_3-Y_2O_3-AlN Ceramics," *J. Ceram. Soc. Jpn.*, **101**[9], 996-1000(1993).
[9] T. Hamasaki and K. Ishizaki, "The Quantitative Evaluation of Grain Boundary Phase by Cryogenic Specific Heat Measurements," in *Grain Boundary Controlled Properties of fine Ceramics,* ed., K. Ishizaki, K. Niihara, M. Isotani and R. G. Ford, Elsevier Applied Science, London and New York, 1992, 53-60.
[10] T. Hamasaki and K. Ishizaki, "Qualitative and Quantitative Evaluation of Si_3N_4 Grain Boundary Phase by Cryogenic Specific Heat Measurements," *J. Am. Ceram. Soc.*, accepted.
[11] M. Kawamoto, C. Ishizaki, and K. Ishizaki,"Fluidity-Increasing Behavior of Silicon Nitride Powder by Aqueous Washing," *J. Mater. Sci. Letters*, **10**[5]279-81(1991).
[12] K. Ishizaki, M. Kawamoto, and C. Ishizaki, "Improving Flow Behavior of Silicon Nitride Powder by Surface Modification," in *Surface Modification Technology IV*, ed., T. S. Sudarshan, D. G. Bhat and M. Jeandin, The Minerals, Metals and Materials Society, Pennsylvania, USA, 1991, pp.655-63.
[13] K. Watari, Y. Seki and K. Ishizaki, "Temperature Dependence of Thermal Coefficients for HIPped Silicon Nitride," *J. Ceram. Soc. Jpn.*, **97**[2]174-81(1989), and *J. Ceram. Soc. Jpn. International*, **97**[2]170-78(1989).
[14] H. M. Rosenberg, *Low Temperature Solid State Physics*, Oxford Univ. Press, London, U.K., 1963, pp.7.
[15] R. C. Zeller and R. O. Pohl, "Thermal Conductivity and Specific Heat of Noncrystalline Solids," *Phys. Rev. B*, **4**[6]2029-41(1971).
[16] K. H. Jack, "Sialons and Related Nitrogen Ceramics", *J. Mater. Sci.*, **11**[6]1135-58(1976).
[17] R. E. Loehman, "Preparation and Properties of Yttrium-Silicon-Aluminum Oxynitride Glasses", *J. Am. Ceram. Soc.*, **62**[9-10]491-4(1979).
[18] K. Oda and T. Yoshio, "Properties of Y_2O_3-Al_2O_3-SiO_2 Glasses as Model System of Grain Boundary Phase of Si_3N_4 Ceramics (Part 1)," *J. Ceram. Soc. Jpn.*, **97**[12]1493-97(1989).

The Effect of Amount of Halogen Impurities on High Temperature Mechanical Properties of Sintered Si_3N_4

M.Asayama, K.Nishida
Toshiba Corporation, New Materials Engineering Lab.
2-4 Suehiro-cho, Tsurumi-ku, Yokohama, Japan
A.Tsuge
Toshiba Corporation, Research and Development Center
1 Toshiba-cho, Komukai, Saiwai-ku, Kawasaki, Japan
S.Kanzaki, T.Oji, K.Watari
Goverment Industrial Research Institute, Nagoya
1 Hirate-cho, Kita-ku, Nagoya, Japan
K.Isozaki, H.Hirotsuru
Denki Kagaku Kogyo Co.Ltd
1 Shinkai-Cho, Ohmuta, Japan

ABSTRACT

The relation between the amount of halogens in Si_3N_4 raw powder and high temperature mechanical properties of sintered Si_3N_4 was investigated. The halogen impurities had the marked effect on the decrease in strength of Si_3N_4 at 1250℃. From X-ray diffraction analysis, the halogens were found to prevent the grain-boundary phase crystalization in sintered bodies. The creep behavior of the sample containing less halogen was drastically improved by reducing the amount of additives.

INTRODUCTION

Many engineering ceramic components made of silicon nitride have been developed because of their excellent physical and chemical properties. However, there is a growing requirement for silicon nitride ceramics with improved mechanical properties. Attainment of these properties would lead to the use of silicon nitride ceramics in more severe application fields. Improvement of high-temperature strength is one of the major concerns regarding sintered Si_3N_4, which has great advantages in terms of production cost and manufacturing flexibility. In order to densify Si_3N_4, however, a relatively large amount of sintering additives are required, which tend to form grain-boundary phases of low refractoriness [1]. In many cases, the additives exist as glassy phases in grain boundaries and the glassy phases deteriorate mechanical properties at elevated temperatures [2]. In order to obtain a more refractory boundary, it is necessary to control grain-boundary phases with respect to various factors. Impurity is one of the most important factors [3] to be controlled, and reducing the amount of total additives is also thought to be an another effective factor. However, the effect of the total amount of additives and impurities on the formation of glassy phases have not yet been clarified.

In this study, the effects of impurities(F,Cl) and the amount of additives on the high-temperature mechanical properties are investigated for sintered silicon nitride.

EXPERIMENTAL

Materials

The results of chemical analysis of Si_3N_4 raw powders are listed in Table I. Total halogen contents range from 50 ppm to 1040 ppm. Each powder was mixed with $5Y_2O_3$-$2HfO_2$-5AlN(wt%) additives by ball milling and sintered at 1750℃ in N_2 atmosphere. The sintered bodies were HIP'ed at 1700℃ at 1000 N_2 pressure.

Table I. Chemical Analysis of Si_3N_4 Powder

Impurities	Powder Number					
	No.1	No.2	No.3	No.4	No.5	No.6
F (ppm)	670	440	70	50	130	10
Cl (ppm)	10	10	70	240	910	40
F+Cl (ppm)	680	450	140	290	1040	50
Fe (ppm)	50	40	20	20	10	10
Al (ppm)	60	30	80	80	60	10
Ca (ppm)	190	210	250	170	180	110
Mg (ppm)	30	50	50	40	30	20
O (%)	1.33	1.35	1.28	1.23	1.23	1.13
C (%)	0.12	0.11	0.07	0.05	0.14	0.09

Characterization

The 3-point bending strength was measured at 1250 under cross head speeds of 0.5 and 0.02 mm/min.

The grain boundary crystalline phase of the sintered bodies were determined by X-ray difraction. The microstructures of sintered bodies were investigated using scanning electron microscopy (SEM) and transmission electron microscopy (TEM).

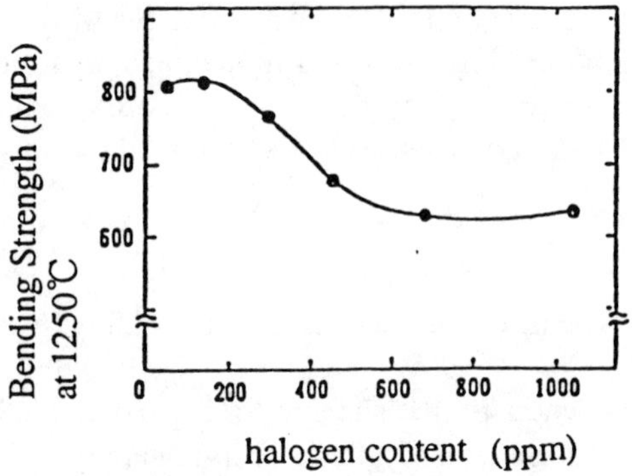

Figure 1. Bending strength (1250℃) as a function of halogen content

RESULTS AND DISCUSSION

The Effect of the Halogen Impurities

The fracture strength at 1250℃ for the samples containing various total halogen contents is shown in Fig.1. The high-temperature strength tends to decrease with the increase of the total halogen content. It is found that the strength of the sample with the total halogen content of 680 ppm is 20% lower than that of the sample with the halogen content of 140 ppm. The X-ray diffraction patterns for the samples with various total halogen contents are shown in Fig.2. The diffraction patterns, in the case of halogen contents of 290 and 140 ppm, show the distinct crystalline phase of n-merillite (X). On the other hand, the sample containing a large amount of total halogen (680 ppm) shows no crystalline phase. The halogens are thought to prevent the crystallization of grain-boundary

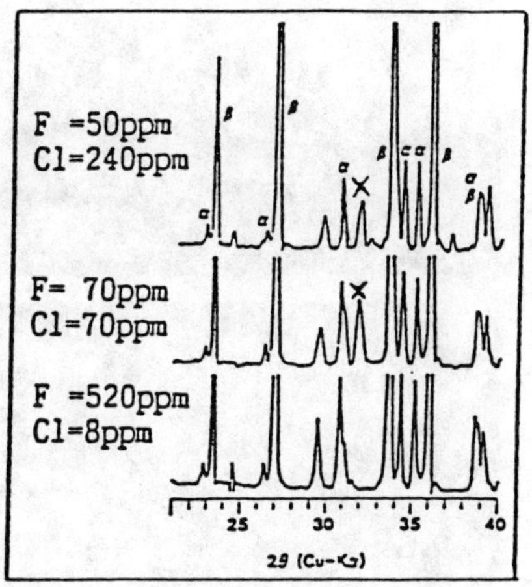

Figure 2. X-ray diffraction patterns of sintered Si_3N_4 with various halogen contents

phases.

Observation of microstructure by SEM and TEM reveals that no change in grain morphology occurred in samples regardless of the different amounts of halogen impurities (Fig. 3). On the other hand, the amount of the grain-boundary amorphous phase clearly increased with an increase in halogen impurities (Fig. 4).

The Effect of Total Amount of Additives

Generally, the sintered Si_3N_4 requires a large amount of additives, which results in formation of a thick grain-boundary layer.

Softening of this thick grain-boundary layer at high-temperatures brings a decrease in high-temperature strength, but it is not clear whether the total amount of additives or that of halogen impurities has the greater effect on high-temperature strength.

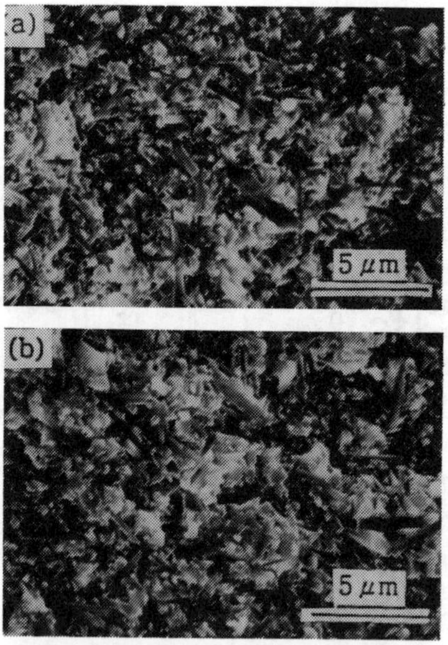

Figure 3. SEM photographs of fracture surface of Si_3N_4 prepared from powders with (a) 140 ppm and (b) 680 ppm halogen contents.

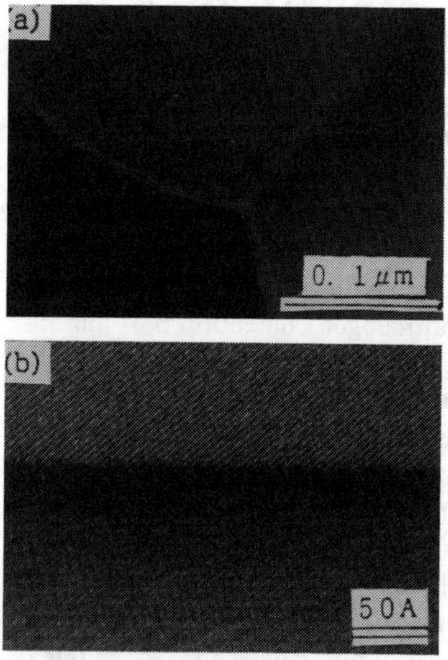

Figure 4. TEM photographs of grain boundary (a) dark field image (680 ppm halogen) (b) lattice image (140 ppm halogen).

Figure 5 shows the effect of the total amount of additives in the No.3 powder on the cross head speed (CHS) dependance of the high-temperature strength. It is found that the high-temperature strength does not depend on the total amount of additives at 0.5 mm/min CHS. However, strength degradation accompanied by stress-strain curve deformation occurs at 0.02 mm/min CHS in the case of a larger amount of additives. It is also observed that the high-temperature strength is lower from about 400 MPa to 600 MPa in the case of Si_3N_4 derived from other raw powders with larger amounts of halogens. In this figure, the stress-strain curve of 60% additives shows almost no decrease of high-temperature strength. The deformation behavior at high-temperatures is clearly improved by reducing the amount of additives. Namely, the stress-strain curve for the sample with 12 wt% additives (100%) shows large deformation, but that for the sample with 7.2 wt% additives (60%) shows almost no deformation.

This excellent deformation behavior at 60% suggests that the material is superb with respect to creep behavior.

Figure 6 shows the tensile creep curves of the samples with 100% and 60% additives at 1250℃ under the stress of 250 MPa. The creep behavior is drastically improved by reducing the amount of additives.

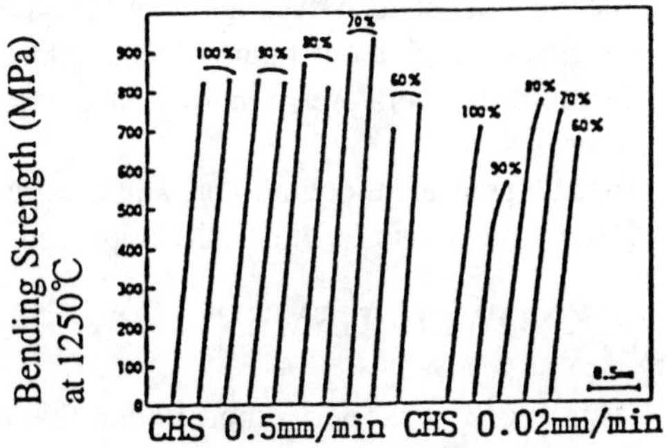

Figure 5. Effect of total amount of additives (shown as percetages of inintial total amount of 12%) on CHS dependance of high-temperature strength.

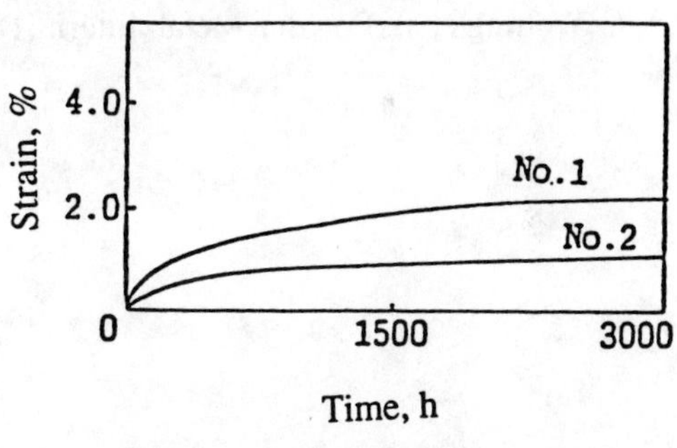

Figure 6. Creep curves at 1250 ℃ under 250MPa No.1:100% additives. No.2:60% additives

CONCLUSIONS

The relation between the amount of halogens in raw Si_3N_4 powder and mechanical properties at high temperature of sintered Si_3N_4 bodies was investigated.

The high temperature strength is decreased about 20% by the 700 ppm halogen impurities.

The halogens are thought to prevent the grain-boundary phase crystallization.

The creep behavior is drastically improved by reducing the amount of additives.

ACKNOWLEDGEMENT

This work was performed under the management of Engineering Research Association for High Performance Ceramics as a part of The Fine Ceramics R&D Project of Basic Technology for Future Industries supported by New Energy and Industrial Technology Development Organization.

REFERENCES

1. M.Mitomo, J.Mater.Sci.,11,(1976),1103
2. A.J.Kiehle et al.,J.Am.Ceram.Soc.,58,(1975),17
3. G.Wotting et al.,Powder Metal. Intern.,18, (1986),25

Mechanical Properties

MACHINING PERFORMANCE OF ZIRCONIA CUTTING TOOLS

S.B. Senthil Kumar R.Krishnamurthy M.V. Gopalakrishnan
C.V. Gokularathnam

Indian Institute of Technology
Madras 600 036. India
E-mail: mctools@iitm.ernet.in

ABSTRACT

Grain boundary phases can control the properties and performance of the engineering ceramics. It is well documented that the presence of the grain boundary phases in zirconia plays an important role in microstructural development as well as influencing the electrical and mechanical properties. The present paper deals with the techniques of fabricating and evaluating the performances of newly developed hybrid zirconia cutting tool containing ceria stabilized zirconia and yttria stabilized zirconia.

INTRODUCTION

Zirconia (ZrO_2) is a remarkable material, which has attracted a great deal of attention from scientists, technologists and users. The progress in our understanding of this material and in exploiting it has been substantial[1]. The phenomenon of phase transformation of zirconia is utilized in the development of tougher grades of alumina-zirconia cutting tools[2]. The increased toughness is due to the stress-induced transformation of zirconia particles at the vicinity of a propagating crack (by absorbing the energy at the crack front), or due to the volume change from phase transformation. Ceria and Yttria are the two materials most frequently used in partially stabilized zirconia (PSZ). Recent machining trials on yttria-stabilized zirconia cutting tools and ceria-stabilized zirconia cutting tool have evaluated the performance characteristics[2,3]. However such materials have certain basic drawbacks.

The disadvantage of Y-PSZ is that it loses its strength (degrades) when annealed in the temperature range of 200 - 300°C in humid air or at lower temperatures in water[4]. Such problems are not present in Ce-PSZ, because it has a very good resistance to transformation during low temperature annealing[5]. One noted disadvantage of Ce-PSZ is its lower hardness. This property has led to the concept of adding Ce-PSZ to Y-PSZ in an alumina matrix[6]. Thus, the thermal stability of Y-PSZ could be improved without losing its mechanical properties, by homogeneous mixing with Ce-PSZ. The present paper deals with the techniques of fabricating and evaluating the performance of a newly developed hybrid zirconia cutting tool containing ceria stabilized zirconia and yttria stabilized zirconia.

FABRICATION

The main criteria for transformation toughening to occur is the submicron size of the tetragonal phase particles. Transformation toughening can be operative only, if the tetragonal phase particle is smaller than a particular critical size diameter. For 3Y-TZP the critical grain size is about 0.3μm[7], whereas, for 12 Ce-TZP, the critical grain size is about 3 μm, about ten times larger[8] .

Carefully weighed quantities of 3Y-PSZ and 12Ce-PSZ powders were mixed with alumina powder. Thus, hybrid agglomerates will facilitate grain boundary characteristics and consequently phase stability.The fabrication details is given in Table 1.

The nose radius of the insert was made by grinding. From the knowledge gained from the previous research[9], conditions optimal for inducing completer cyclic (t-m-t) transformation were used to grind the nose radius in an optical profile grinding machine.

Table I Fabrication Details

Ceramic-Composite (wt %)	85 % Al_2O_3 10 % 12Ce-PSZ 5 % 3Y -PSZ
Binder used	Poly Vinyl Alcohol
Compacting Pressure	200 MPa
Sintering Temperature	1500° C
Sintering Time	2 hours
HIPing Temperature	1470° C
HIPing Pressure	200 MPa
HIPing Time	2 hours
Insert Designation	ISO 120416
Insert Hardness Hv10	1350 GPa

For evaluating the phases present after grinding, the ground inserts of ceramic-composites was subjected to X-ray Diffraction. Typical XRD profiles for ceramic-composites for as sintered, HIPed and ground conditions are presented in Figure 1.

Referring to Figure 1, it can be seen that HIPing and grinding, have resulted in stress-induced transformation of tetragonal to monoclinic phase. This will facilitate cyclic transformation of t-m-t during subsequent service of the ceramic inserts as a cutting tool.

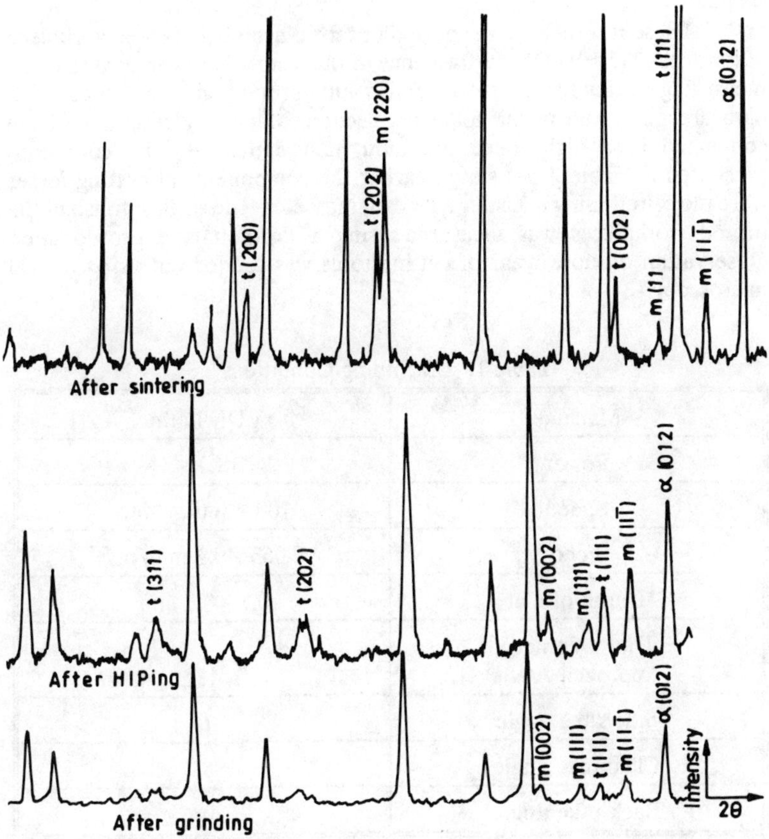

FIGURE 1 XRD Profiles

From the fig 1 it can be calculated that the ceramic inserts contains 66 vol% m phase and 34 vol% t phase. From the Ref[10] it has been proved that for 15 wt % ZrO_2-Al_2O_3 an optimum resistance to fracture during high-speed turning of steel is obtained. Most likely, all three mechanisms of toughening contribute to the toughening in this case, namely transformation to increase volume phase, absorbing of crack tip energy by phase transformation and diffusion of crack tip by deflection [11].

EXPERIMENTAL PROCEDURE

These inserts have been tried out for machining of steel workpiece of hardness 275 Hv. The performance of the inserts has been evaluated by measuring cutting forces, surface finish and observation on tool-wear. The data are presented in the following section. Machining tests have been conducted in a high speed precision VDF lathe with the conditions presented in Table 2. As stated earlier the components of cutting forces were measured using a Kistler type dynamometer. The surface finish of the turned workpieces was measured using a Perthon type profilometer. Observation on nose wear of cutting tools was carried out using optical microscopy.

Table II Machining Conditions

Machine	VDF Lathe
Power	18 kw
Speed	100 - 300 m/min
Feed	0.063 - 0.1 mm/rev
Depth of Cut	0.1 - 0.5 mm
Tool Geometry Approach Angle	45°
Side rake angle	-6°
Clearance angle	6°
Back rake angle	0°

RESULTS AND DISCUSSIONS

Cutting Forces

Typical variations of cutting forces as a function of cutting speed are presented in Figure 2.

It is seen that with an increase in cutting speed there is a reduction in cutting force components. Referring to Figure 2 one can visualize two distinct zones in the trend of variations of cutting forces with cutting speed.

There is a zone of higher order of cutting forces with lower cutting speed. This is attributed to the ploughing of the cutting tool on the work surface. As the speed increases the cutting force component decreases, indicating steady cutting. This is over a region of 150 - 200 m/min. Beyond 200 m.min the nature of variation further changes. In this region the slope changes, where cutting force component tends to increase. This can be attributed to an increase in the nose radius/or deformation of nose associated with phase transformation of t - m phase.

This observations clearly indicates the occurrence of certain critical velocity for nose deformation, which is around 200 - 250 m/min. The reduction in the cutting force with increasing speed can also be due to the softening of work material at temperatures associated with high cutting speeds and consequent reduction in shear strength. This facilitates easy chip removal without much chip strain and consequent reduction in cutting force.

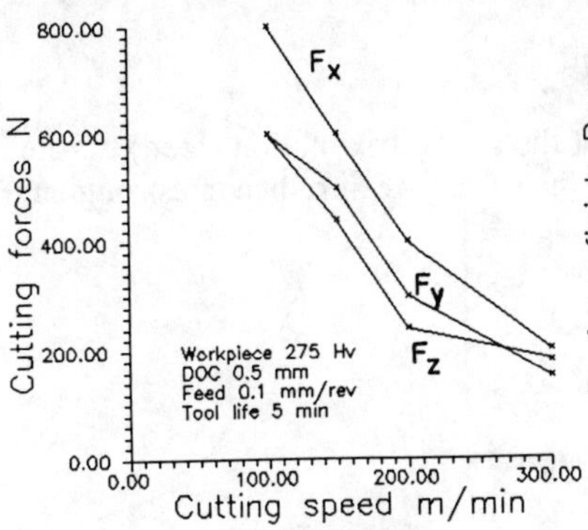

FIGURE 2 Cutting forces vs Cutting speed

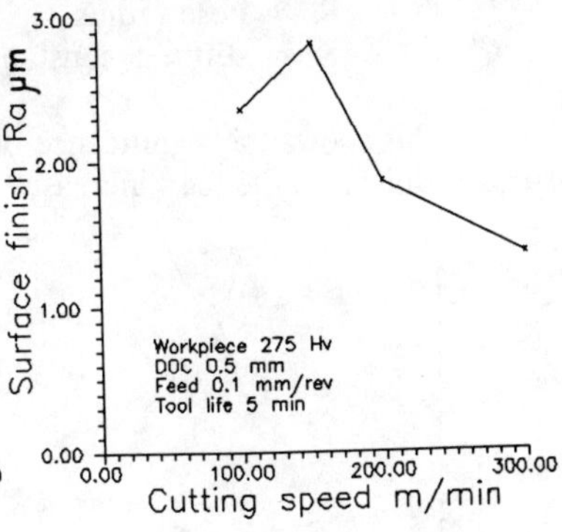

FIGURE 3 Surface finish Ra vs Cutting speed

Surface finish

The influence of cutting speed on surface finish is illustrated in Figure 3.

It is seen that in the lower cutting speed range the machined surface was relatively rougher which is mostly attributed to the ploughing trend. As the speed increases the cutting becomes steady with easy chip removal, resulting in an improvement of surface finish in the range of 150 - 200 m/min. An increase in cutting speed produces an increase in temperature in the primary deformation zone, which softens the workpiece and the increase in the real area of contact at the tool-workpiece interface. Relative motion between the tool edge and workpiece produces plastic deformation within the surface, resulting in a better surface finish[12].

Beyond the critical velocity of 200 m/min the nose experiences a deformation with consequent roughening of surface. The variation of surface finish with cutting feed is presented in Figure 4. In turning the surface texture is obtained by replication of the cutting nose on the work surface. The accuracy of the replication is largely dependent on the form stability of the cutting tool and the feed rate. It is known that surface finish Rt is simply related to the above parameters by the relationship.[13]

$$Rt = K * (s^2/8R) \qquad (1)$$

where

s = feed rate
R = nose radius
K = stiffness constant

This shows the significance of the generatrix motion (feed) on the surface finish. As the feed increases the surface texture becomes rougher.

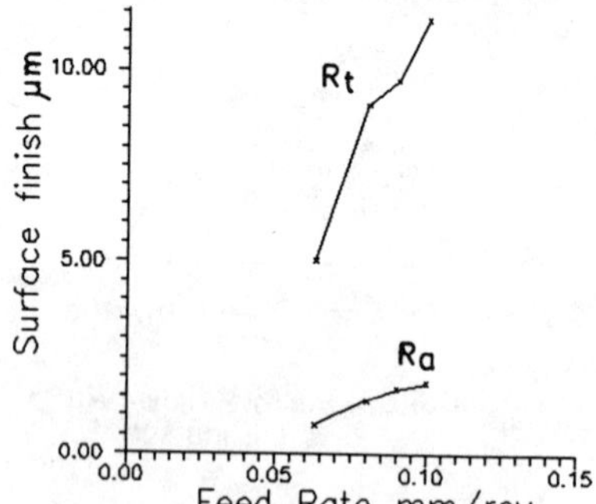

FIGURE 4 SURFACE FINISH vs Feed rate

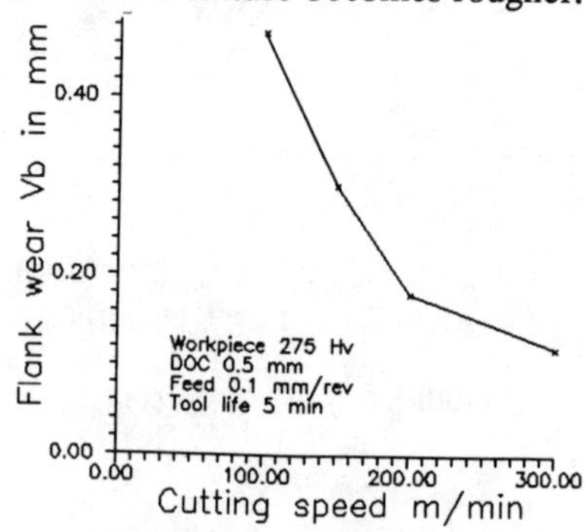

FIGURE 5 FLANK WEAR vs Cutting speed

Flank Wear

The performance of cutting tools in turning is evaluated by measuring the stability to maintain dimensional tolerances, as well as the surface finish. The dimensional tolerance in turning is largely influenced by the flank wear of the cutting tool. During machining the just machined work surface slides past the flank surface of the cutting tool establishing an increase in adhesion contact wear. During this period of adhesion the material on the flank surfaces may undergo smoother flank wear due to deformation at the contacting asperities or exhibit abrasion wear due to rubbing of the strain hardened work surface.

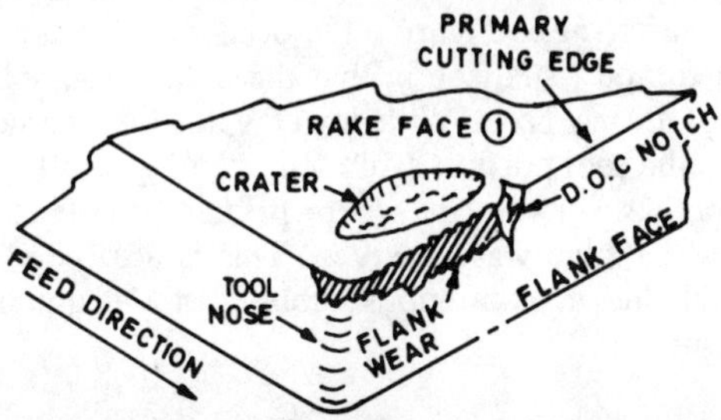

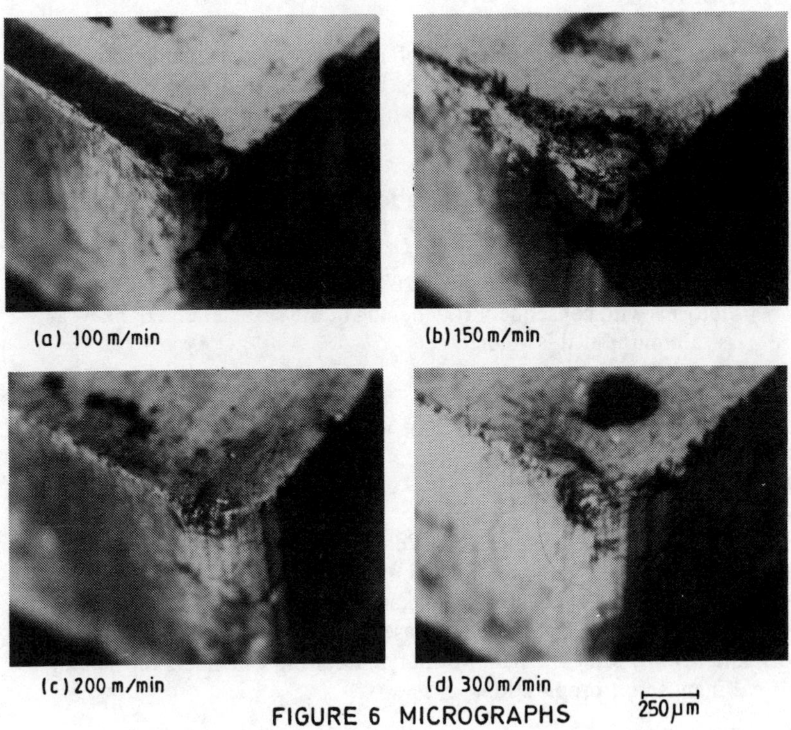

FIGURE 6 MICROGRAPHS

All this results in removal of material from the flank resulting in occurrence of flank wear.The variation of flank wear with cutting speed is illustrated in Figure 5. It is seen that, as the cutting speed increases the flank wear decreases indicating steady cutting. Beyond the critical velocity of 200 m/min there is a small increase in flank wear. During machining the material over the flank surface will exhibit monoclinic to tetragonal transformation due to heating and subsequent tetragonal to monoclinic transformation due to contact stress. The occurrence of the cyclic tetragonal to monoclinic transformation results in softening of the tool material and a small increase in flank wear.

The optical micrographs of the worn portion of the cutting tool are presented in Figure 6. Referring to Figure 6 the occurrence of the critical velocity of 200 m/min can be visualized in that the cutting tool exhibited minimum wear on the cutting nose. With lower velocities, namely 100 m/min and 150 m/min the tool portion exhibited chipping and triangular nose wear. With an increase in velocity severe primary grooving on the Depth of Cut Line (DCL) zone was observed. This is attributed to the cutting of tool material due to continuous rubbing of the unmachined diameter of the work surface.

CONCLUSIONS

The machining performance of newly-developed hybrid ceramic composite cutting tools containing ceria stabilized zirconia and yttria stabilized zirconia has been evaluated. The following conclusion were made.

1. There is a reduction in cutting forces with an increase in cutting speed in the range of 100 - 300 m/min.

2. There is a critical velocity of 200 m/min beyond which the nose deforms with consequent roughening of the surface. The flank wear is minimum at this velocity.

3. Beyond the critical velocity the increase in flank wear is attributed to the cyclic transformation of tetragonal to monoclinic phase.

Acknowledgements

The authors wish to thank Mr. Peter Donnelly, M/S TOSOH Corp., Japan for giving free ceramic powder samples and Mr. R. Sreenivasan, Technical Officer, Mr. C.N. Karunakaran, Supervisor, Mr. G. Paneer Selvam, Junior Mechanic, IIT, Madras for helping in the experimental work and Mr. M. Ranganathan, Mr. Surya, Drafting Engineers for helping in final manuscript preparation.

REFERENCES

1. E.C. Subbarao, "Zirconia - An Overview", Advances in Ceramics, Volume 3, Science and Technology of Zirconia, Edited by A.H. Heuer and L.W. Hobbs, Published by The American Ceramic Society, Inc. Columbus, Ohio. 1981, Page 1.
2. A.K. Chatopadhyay and A.B. Chatopadhyay, "Wear Characteristics of Ceramic Cutting Tools in Machining Steel", Wear, Vol 93 n3 (1984) pp 347-359.
3. Y. Yamane, H. Usuki, B. Yan and N. Narutaki, "The Formation of Protective Oxide Layer in Machining Re-sulphurized Free Cutting Steels and Cast Irons", Wear, Vol 139 n6 (1990) pp 195-208.
4. Tatsuo Noma, Mashiro Yoshimura and Shigeyuk Somiya, "Degradation Behaviour of Rapidly Quenched Y-TZP", Int. J. Ceram. Soc. Japan, Vol 96 (1988) pp1090-2.
5. Koji, Tsukuma, "Mechanical Properties and Thermal Stability of CeO_2 Containing Tetragonal Zirconia Polycrystals", Am. Ceram. Soc. Bull., Vol 65 issue 7 (1986) pp1386-9.
6. Sato, T., Ohtaki, S., Fukushima, T., Endo, T. & Simada, M., "Mechanical Properties and Thermal Stability of Yttria Doped Tetragonal Zirconia Polycrystals with Diffused Ceria in Surface", In Advanced Structural Ceramics, Vol 78, edited by P.F. Becher, M.V. Swain & S. Somiya. Materials Research Society, Pittsburgh, 1987, pp 147-54.
7. Gupta, T.K., Lange, F.F. & Bechold, J.H., "Effect of Stress Induced Phase Transformation on the Properties of Polycrystalline Zirconia containing Metastable Tetragonal Phase", J. Mater. Sci., Vol 13 n8 (1978), pp 1464-70.
8. Tsukuma, K. & Takahata, T., "Mechanical Properties and Microstructure of TZP and TZP/AL_2O_3 Composites", in Advanced Structural Ceramics, Vol 78, edited by P.F. Becher, M.V. Swain & S. Somiya. Materials Research Society, Pittsburgh, 1987, pp 1234-135.
9. Swain, M.V. & Hannink, R.H.J., "Metastability of the Martenstic Transformation in a 12 mol% Ceria-Zirconia Alloy: II, Grinding Studies". J. Amer. Ceram. Soc., Vol 72 n4 (1989) pp 1358-64.
10. Nils Claussen and Manfred Ruhle, "Design of Transformation Toughened Ceramics", Advances in Ceramics, Volume 3, Science and Technology of Zirconia, Edited by A.H. Heuer and L.W. Hobbs, Published by The American Ceramic Society, Inc. Columbus, Ohio. 1981, Page 137-163.
11. V. Sivasankaran, Performance of Ceramic Cutting Tools in Finish Machining of Alloyed Centrifugally Cast Cast-Iron, Ph.D Thesis, IIT, Madras. 1986

12. S.K. Bhattacharyya, E.O. Ezugwu and A. Jawaid, "The Performance of Ceramic Tool Materials for the Machining of Cast Iron", Wear, Vol 135 n9 (1989) pp147-159.
13. M.K. Balakrishnan Nayar, Evaluation and Enhancement of The Performance of Powder Metallurgy High Speed Steel Tools in Machining, Ph.D Thesis, IIT, Madras. 1986

ROLE OF DIAMOND IN PRECISION MACHINING OF PLASMA SPRAYED FINE CERAMICS

S. Gowri
Mechanical Engineering Department
K.R.E.College, Srinivasnagar-574 157

K. Narayanasamy
R. Krishnamurthy
Mechanical Engineering Department
Indian Institute of Technology , Madras-600 036
India

ABSTRACT

Plasma sprayed fine ceramics such as Al_2O_3 , Al_2O_3-TiO_2 and partially stabilized ZrO_2 were ground and lapped with diamond abrasives both in the bonded and free form. The influence of diamond at different cutting conditions in these precision machining methods was analyzed and discussed.

INTRODUCTION

Since ceramics are superior to metallic or organic materials in many ways, they promise a nearly unlimited range of applications. Hence ceramic coated surfaces also have many applications and it is not an exaggeration to say that they are the "solutions in search of problems". Advances in space technology, automotive technology and electronics technology have led to increasingly stringent requirements for ceramic coated parts with a high degree of precision. But the sprayed ceramic coatings cannot satisfy the requirements for accuracy without precision machining. Development of the new ceramics has mainly concentrated on improvement of the characteristics of ceramic materials; minimal attention has been given to improving their machinability (1). Being hard and brittle, ceramics present many obstacles to precision machining. Figure 1 shows the relationship between the hardness of the material to be machined and the hardness of the tool material. Precision machining of ceramics or ceramic coatings can be done with harder cutting abrasives like diamond, cubic boron nitride (CBN) and green grit silicon carbide. The abrasives can be of a bonded type, as in the case of grinding, or loosely fed as in the case of lapping.

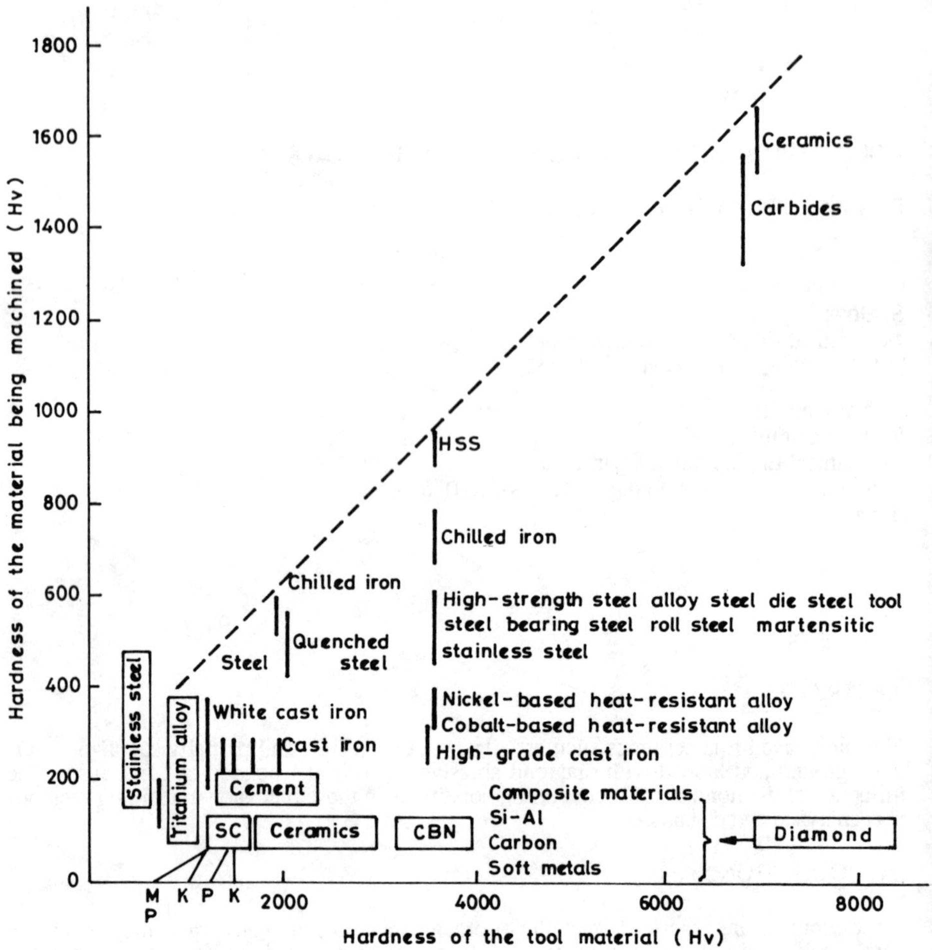

Fig. 1 Relationship between hardness of tool material and work material (1)

PRECISION MACHINING OF FINE CERAMICS

Grinding

Ceramics are usually machined by abrasive processes, such as, grinding, superfinishing techniques and other abrasive process. Ceramic components therefore often have been diamond ground or lapped (2). Caveney and Thiel (3) have conducted an experimental investigation on the grinding of high density Al_2O_3 with diamond abrasive grinding wheels. The variation in grinding efficiency, machining cost and surface damage were determined based on the grades of Al_2O_3 ceramics. Inasaki, I.(4) in his study, reported that the efficiency of speed-stroke grinding, was featured by high work speed and a small depth of cut. Studies conducted by Nakagawa, T. et al. (5), suggest that, in the final stage of manufacturing of ceramic mechanical parts, grinding by diamond wheel is usually adopted in order to provide configurational and dimensional accuracy to the parts. Ceramic components have often been diamond ground or lapped (2). Studies conducted by Pai, D.M., et al. (6) on grinding swarf, highlight some of the points. The important difference between the chips from ductile to brittle work piece material is that long ribbon like chips tend to form with ductile materials, whereas small powder like particles form with hard brittle materials such as ceramics.

Ceramics/ ceramic coatings are known for the presence of isolated strength field over their matrix; hence, during grinding of hard brittle ceramic materials, depending upon the intensity of stress and strength fields, material removal would be associated with either deformation (ductile) induced shear fracture or rupture of the material (7). The occurrence of deformation or rupture could be visualized by analogizing grinding to an indentation process. The abrasive grain entering the ceramic surface, the ceramic material would be deformed and hence flow of material around the indenting abrasive takes place with possible surface cracking. The limitation of depth of grinding based on crack formation can also be understood from Fig. 2. Brinksmeier and Roth (8) have evaluated the influence of microstructure on precision grinding of fine ceramics and reported that the measured increase in grinding energy reflected larger amount of rubbing and ploughing of grinding abrasives against ceramics. Zhon et al (9) have established that generally by reducing the depth of grinding, the grinding force reduced and the surface finish was poor because of dominant ploughing of cutting edges. Hence the study on grinding force components is considered to be very important to produce 'crack-free' ground surfaces in case grinding of ceramics.

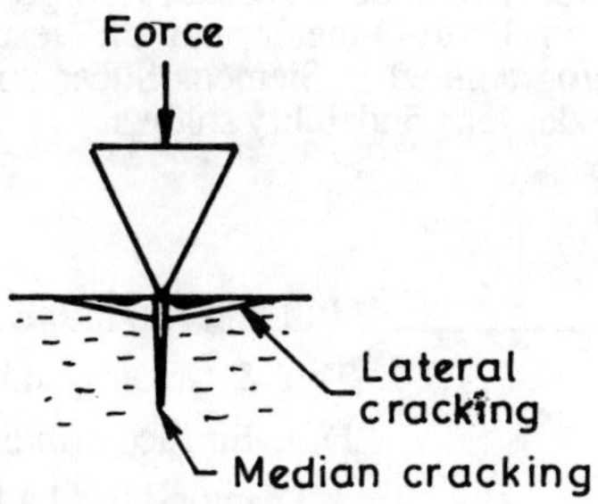

$$\sigma_{all} \propto \frac{(k_1 c)^{n_1}}{(d)^{n_2}}$$

c - Fracture rupture strength

d - Depth of grinding

k_1 - Constant

n_1, n_2 - Material exponents

Fig. 2 Possible crack formation during grinding of brittle materials (14)

Lapping

Grinding with a diamond wheel is the only method which has been practically applied to machining of advanced ceramics (10). It is, however, quite difficult to completely avoid the initiation and propagation of cracks during the grinding process (11). These cracks result in considerable decrease in the strength of ground ceramic parts as mentioned earlier. Hence some more finishing is required to remove the impaired surface layer. Davis C.E.,(12) has made a study of the influence of diamond as a lapping medium on ceramic materials such as Al_2O_3, WC and synthetic sapphire with respect to lapping plate material, lapping time, lapping pressure and abrasive size, distribution and concentration. It was observed that an increase in mean-size of diamond abrasive produced an increase in stock removal rate. For each set of abrasive size, it was found that surface finish could not be improved beyond an optimum value. Lapping, the conventional free abrasive machining process, has been well investigated and practically applied to glass finishing (13). Chen et al.(10), have conducted studies on lapping of advanced ceramics; however, lapping of plasma sprayed ceramic coatings has not been a subject of study so far. The present work, hence, aims at obtaining data regarding the basic features of lapping of plasma sprayed and ground ceramic coatings.

EXPERIMENTATION

Ceramics such as Al_2O_3 , Al_2O_3-TiO_2, Y_2O_3 stabilized ZrO_2 (PSZ) were plasma sprayed on bright drawn steel substrate using Miller's Plasma Dyne, USA. Pure Ar was used as plasma gas. Sophisticated 6-axis microprocessor controlled, programmable robot, Motoman, Japan was used as the spray unit. The sprayed specimens were kept in a desiccator. Precision machining methods such as grinding and lapping were carried out with diamond as cutting abrasive under the conditions given in Table I and II. The surface finish was measured with Perth-O-meter and the results have been discussed. The experimental data was programmed in Siemens Super computer using SPSS software to develop mathematical model for grindability studies.

Table I	Grinding conditions
Machine	Tool & Cutter grinder
Operation	Dry, Surface grinding
Wheel	a. Diamond BZ 1A1-100-6-1-6 grit 185
	b. CBN-B120 RR 100D-150-6-grit 120
Wheel speed	5.7, 11.5, 15.7 & 31.5 m/s
Depth of grinding	10, 20, 30 & 40 μm
Work feed	0.5 mm/s

Table II	Lapping conditions
Machine	Wentzky, Germany
Operation	Flat surface lapping
Lapping medium & abrasive	Hifin diamond compound
Diamond size	0.25-2, 6-8, 10-14 μm
Lapping speed	0.30 m/s
Lapping pressure	52 kPa
Lapping time duration	2-30 min.

RESULTS AND DISCUSSION

Parametric Influence on Grinding Force

Typical variations of normal force with velocity for diamond grinding of Al_2O_3, Al_2O_3-TiO_2 and PSZ are presented in Fig. 3. It is seen that the grinding force variations with velocity exhibit three different zones. In the case of Al_2O_3, with increasing grinding velocity up to 10 m/s, the force was higher but mostly steady; in the range of grinding velocity of 10-15.7 m/s there is a drop in the normal force; beyond 15.7 m/s of grinding velocity a mild rise in the normal force component was observed. As stated earlier, a ceramic material is known to posses isolated pockets of strength field. With smaller grinding velocity, the diamond will tend to plow on to the ceramic surface, rather than cutting the surface. This can result in the observed higher order force component. However, as the velocity increases, the grinding becomes steady and the diamond is able to cut with reduced force component. Between 15.7-31.5 m/s of grinding velocity, the normal force component tends to increase; this may be due to possible reduction in the grinding performance of diamond at higher temperature such as that associated with higher grinding velocity (diamond becomes unstable above 850° C in atmospheric conditions). Referring to Fig. 3, it is seen that as the depth of grinding increases, the grinding force (normal component) increases. Unlike the case of ductile materials, one has to limit the depth of grinding in case of ceramics to avoid the possible crack formation on the surface. Around 30 μm depth of grinding is seen to be the critical depth, beyond which there is a rise in the normal force of grinding.

In the case of PSZ, there is a change in the trend of variations of normal force with grinding velocity. In the case of PSZ, there is a rising trend in the normal force with grinding velocity. The normal force exhibits three distinct zones of variations (Z_A, Z_B and Z_C). In the lower velocity range, the normal force increases with velocity (up to 11.5 m/s), in the median range of velocity (10-15 m/s) there is a drop in the normal force and with the higher order velocity (15 m/s) the normal force components increase. During grinding, PSZ ceramic experienced mechanical stress and rise in temperature. Depending upon the temperature or stress magnitude, it was possible that the PSZ surface during grinding could undergo stress-induced tetragonal (t) to monoclinic (m) transformation or with higher order temperature, it may experience t—>m—>t transformation. In the low speed range (up to 10 m/s), there would be greater tendency to plowing, with the mechanical stress being dominant. This might be resulting in t—>m transformation promoting transformation toughening of the surface material; this along with the plowing tendency, has resulted in an increase in the normal force component. However with further increase in velocity (10 m/s), the grinding capability of diamond increased, resulting in a reduction in force.

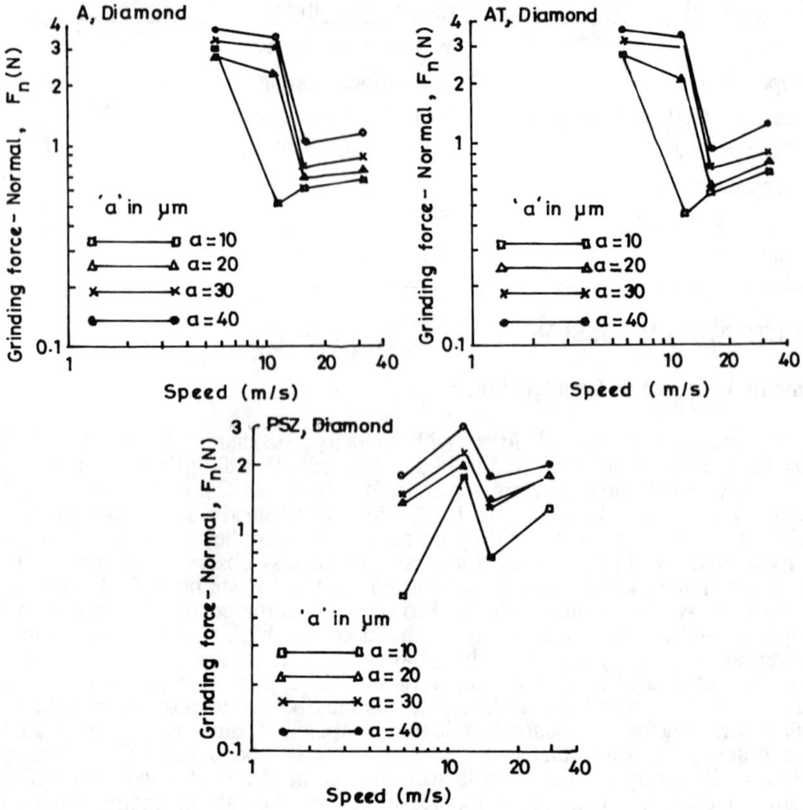

Fig. 3 Influence of grinding speed on normal grinding force
(A - Al_2O_3, AT - Al_2O_3-TiO_2, PSZ - Y_2O_3-ZrO_2)

Further, the increase in temperature with higher order grinding velocity may result in t—>m—>t transformation maintaining a hard surface. This can also reduce the normal grinding force component (absence of side flow of material around the abrasive as in the case of a toughened surface). For higher order range of grinding velocity (15-20 m/s), the grinding performance of diamond becomes questionable due to its thermal instability at higher order temperature. Figure 4 presents typical variations of tangential components of grinding force with grinding speed. In the case of diamond grinding, three distinct zones of variation on tangential components of grinding force can be seen. The low velocity grinding, where the diamond has to plastically deform and shear the surface asperities (plowing), exhibits increasing tangential force; with increasing velocity of grinding the process becomes steady, results in rupturing the materials (grinding stress overcoming the resistance or the strength in the ceramic material): with the zone of higher order of grinding velocity, the possible thermal degradation of diamond, can result by the observed rise in the tangential component of grinding force.

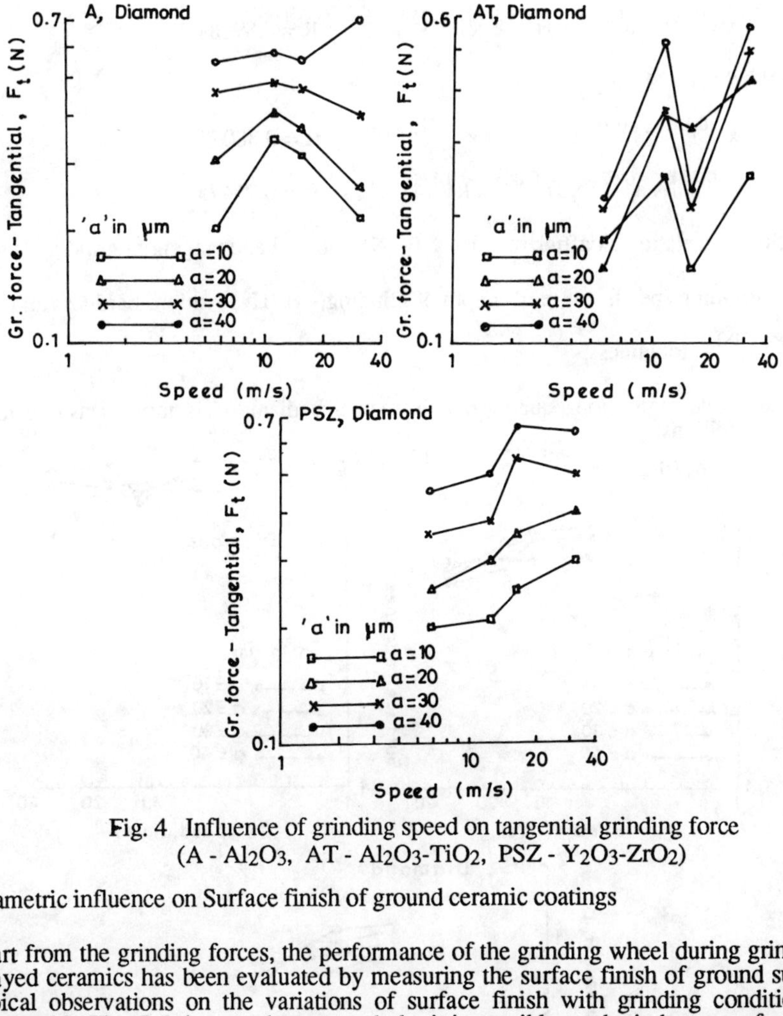

Fig. 4 Influence of grinding speed on tangential grinding force
(A - Al_2O_3, AT - Al_2O_3-TiO_2, PSZ - Y_2O_3-ZrO_2)

Parametric influence on Surface finish of ground ceramic coatings

Apart from the grinding forces, the performance of the grinding wheel during grinding of sprayed ceramics has been evaluated by measuring the surface finish of ground surfaces. Typical observations on the variations of surface finish with grinding conditions are presented in Fig. 5. It is seen that, as a whole, it is possible to obtain better surface finish with diamond for all the three ceramic coatings. With the relatively sharp cutting edge of the diamond, it is possible to obtain better surface finish. With the increase in grinding speed for all the three ceramics there is an improvement in the surface finish.

Mathematical modelling

For Diamond Wheel

$F_n = e^{3.66} \times v^{-0.128} \times a^{0.923} \times H^{-0.88} \times k^{0.172}$ R = 0.82369

$F_t = e^{-12.52} \times v^{0.299} \times a^{0.587} \times H^{1.369} \times k^{-0.254}$ $\qquad R = 0.89981$

For CBN Wheel (14)

$F_n = e^{0.478} \times v^{0.145} \times a^{0.14} \times H^{0.047} \times k^{0.072}$, $\qquad R = 0.88077$

$F_t = e^{-2.41} \times v^{0.284} \times a^{0.527} \times H^{-0.068} \times k^{0.171}$ $\qquad R = 0.95475$

R- multiple correlation c oefficient, F_n & F_t (N)- normal and tangential grinding forces

v (m/s) - grinding speed, a (mm)- depth of grinding, H (Hv) - micro hardness and

k (MPa. m $^{1\backslash 2}$)- toughness.

From the above regression equations, it is seen that diamond is more sensitive to the grinding conditions.

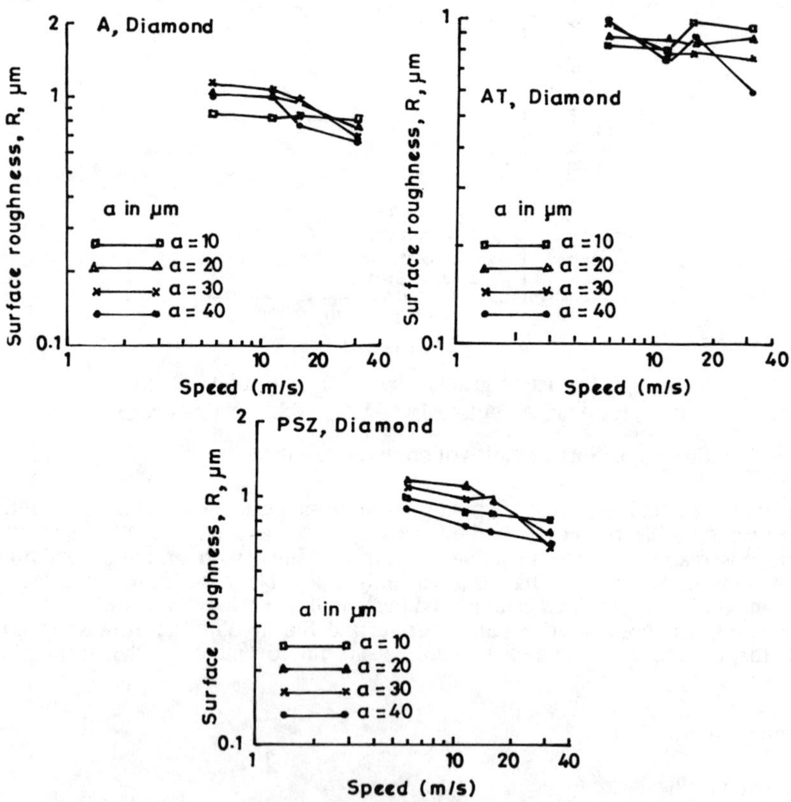

Fig. 5 Influence of grinding speed on surface roughness (R_a)
(A - Al_2O_3, AT - Al_2O_3-TiO_2, PSZ - Y_2O_3-ZrO_2)

Parametric Influence on Lapped surface quality

As stated earlier, lapping trials were conducted on three types of ceramic coated surfaces for different abrasive (diamond) size and duration of lapping. The observed variation in R_t value with duration of lapping is presented in Fig. 6.

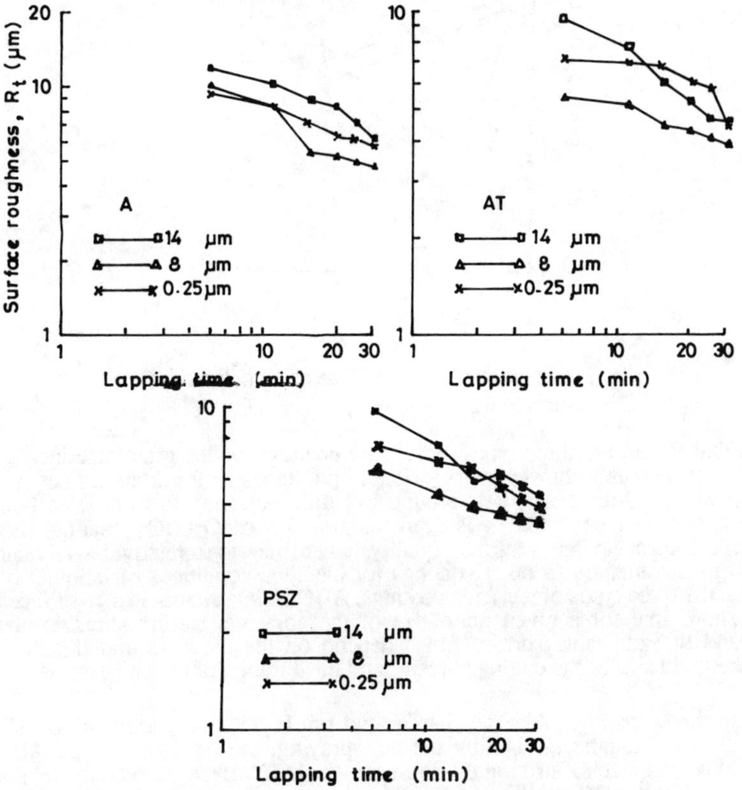

Fig. 6 Influence of lapping time on surface roughness (R_t)
(A - Al_2O_3, AT - Al_2O_3-TiO_2, PSZ - Y_2O_3-ZrO_2)

It is seen that for all the three ceramic coatings, there is an improvement in surface finish with a tendency to saturate after around 15-20 minutes of lapping. With increasing lapping duration, the diamond abrasive will cut/shear both the active and dummy asperities resulting in improved surface quality. However, for larger duration, it is possible that the surface becomes glazed, exhibiting a tendency to saturation. If lapped further, the surface quality may deteriorate due to pull-out of localized, glazed material. Referring to Fig. 6, it is seen that both Al_2O_3-TiO_2 and PSZ have yielded better results compared to Al_2O_3 coating. Further for both Al_2O_3 and Al_2O_3-TiO_2 coatings, there exists a critical abrasive size of 6 - 8 µm for achieving better lapped textures. However in the case of PSZ, with the increase in grain size, the surface quality deteriorates. Generally , increase in R_t means the presence of deep scratches on the surface. It may be that pre-lapping ground surface texture of Al_2O_3 or Al_2O_3-TiO_2 with medium size of diamond abrasives could correct larger amount of surface irregularities (grinding marks) by medium size of abrasives during lapping. The influence of abrasive size and stock removal on surface finish is presented in Fig. 7.

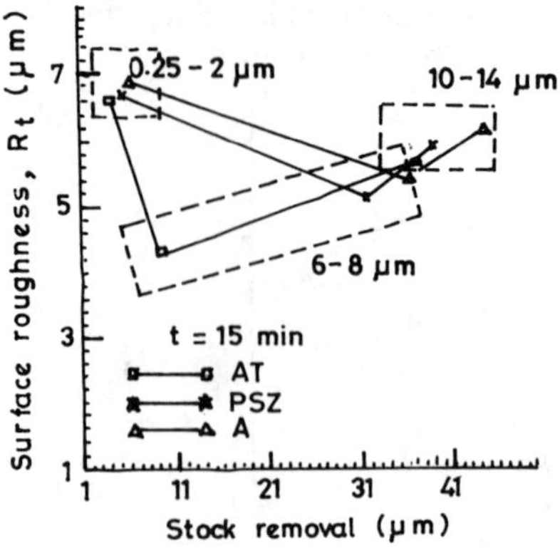

Fig. 7 Influence of abrasive size and stock removal on surface roughness (R_t)

It is seen that for all the three types of ceramic coatings, as the grain size increases, the stock removal increases; however, there is an optimum size for attaining better surface finish. For all the three cases, it is around 6-8 μm. Referring to Fig. 7, an interesting observation can also be made; it is seen that for the Al_2O_3-TiO_2 coating, there is a considerable improvement in surface quality, when the stock removal was maintained around 10 μm, compared to other coatings for identical conditions of lapping. It means that among the three types of ceramic coatings, Al_2O_3-TiO_2 attains a finer surface texture within a short time for a given grain size of the abrasive. During abrasion, the stock removal and the attainable surface finish depend on the hardness and ductility of the coating, the grain size of the cutting abrasive and the duration of the process.

Among the three coatings, Al_2O_3 is harder and brittle with poor toughness. PSZ is the softest with good toughness. During plasma spraying, the hard and brittle Al_2O_3 will fragment into finer size on striking the substrate, the PSZ will form an elongate, thin layer on striking due to transformation toughening characteristics. However, Al_2O_3-TiO_2 with its high hardness and limited toughness will form a layer of material (with medium size) on striking the substrate. This can explain the observed relationship between surface finish and stock removal rate. With the very small grain size of the diamond (0.25 -2 μm), there is only a marginal difference in the finish and stock removal rate values among the coatings. As the grain size increases to 6-8 μm, the brittle (fragmental) Al_2O_3 coating and relatively softer PSZ exhibit higher stock removal rate with rougher surface. In the case of Al_2O_3-TiO_2 with 6-8μm grain size, the stock removal rate of the surface layer is minimum with considerable improvement in surface finish. With a further increase in stock removal rate, the surface finish deteriorates. The hard Al_2O_3-TiO_2 coating, with limited toughness (better than Al_2O_3) facilitates good surface finish with controlled material removal rate. These observations can be attributed to the hardness, toughness and size of the grain in the sprayed layer. On extrapolation, it may be possible to attain a relatively better surface finish and reduced stock removal rate in the case of PSZ and Al_2O_3 by using abrasive of size in the range of 2-6μm.

CONCLUSIONS

Ceramic coated surfaces could be successfully precision machined with diamond abrasives in both bonded and free form conditions. The major conclusions drawn are as follows :

Grinding with bonded diamond abrasives gave better surface finish. ie., R_a= 1.15, 0.82 and 0.98 μm for Al_2O_3, Al_2O_3-TiO_2 and PSZ ceramic coatings (for a = 30 μm). At higher depth of grinding and speed, diamond became thermally unstable. Hence grinding velocity, ranging from 11.5 to 16 m/s and depth of grinding of 30 μm were evaluated to be critical.

Lapping with free diamond abrasives improved the surface finish of the ceramic coatings with lapping time attaining saturation after about 15 minutes for Al_2O_3-TiO_2, 18 minutes for PSZ and about 25 minutes for Al_2O_3. Diamond grains of size 6-8 μm were found to be optimal giving better surface finish of lapped ceramic coatings.

REFERENCES

1. Akiro Kobayashi," Precision Machining Methods for Ceramics", *Advanced Tech. Ceramics*, (edt.) S. Somiya, Acad. Press In., Japan, 261-282, (1989).

2. Ida, J., " Grinding Characteristics of Ceramics for Circuit components with Metal Bonded Diamond Wheels", *Annals of CIRP*, **17**,1, 232-234, (1969).

3. Caveney, R.J. and Thiel," Grinding Alumina with Diamond Abrasive", NBS Special Publication, USA, (1972).

4. Inasaki, I.," Speed stroke Grinding of Advanced Ceramics", *Annals of CIRP*, **37**, 2, 299-302, (1988).

5. Nakawaga, T. and Suzuki, K.," High Coefficient of Grinding of Ceramics and Hard Metals on Grinding Centre", *Annals of CIRP*, **35**,1, 205-210, (1986).

6. Pai, D.M., Ratterman, E. and Shaw, M.C.," Grinding Swarf ", *Jl. of Wear*, **131**, 1,329-339, (1989).

7. Namba, Y., Yamada, Y., Tsuboi, A., Unno, K. and Nakao, H.," Structure of Mn-Zn Ferrite Single Crystals Ground by an Ultraprecision Surface Grinder with Various Diamond Wheels", *Annals of CIRP*, **41**, 1, 347-351, (1992).

8. Brinskmeler, E. and Roth, P., " Influence of Microstructure in Precision Grinding of Fine Ceramics", *Proc. of ASPE Annual Meeting*, Orlando, USA, 73-78, (1992).

9. Zhon, L., Soyoji, K. and Kuriagawa, T.," Ceramic Mirror Grinding", *Proc. of ASPE Annual Meeting*, Orlando, USA, 79-83, (1992).

10. Chen, C., Saki, S. and Inasaki, I.," Lapping of Advanced Ceramics", *Jl. of Materials and Manufacturing Processes*, **6**, 2, 211-226, (1991).

11. Spurr, G. and Tio, T.H.," Surface Layer Damage in the Grinding of Advanced Engineering Ceramics", *Proc. of SME 16th North American Manufacturing Research Conference*, USA, 224, (1988).

12. Davis, C.E.," A Study of the Influence of Flat Lapping with Diamond Micro Abrasives", *Diamond Information Series 1.32*, De Beers Industrial Division, De Beers Publication, London, UK., (1984).

13. Kasczmarek, J., *Principles of Machining by Cutting, Abrasion and Erosion*, (edt.) Kasczmarek, Peter Pergins Ltd., NY., USA., 413, (1976).

14. Gowri, S., Ph.D Thesis report on " Characterization and Precision Machining of Plasma Sprayed Ceramic Coatings", IIT., Madras, (1993).

SOL- GEL DERIVED THICK ZrO_2 COATINGS BY LIQUID THERMAL SPRAYING

A. R. Di Giampaolo, P. Capellucci *, H. Vesteghem, J. Kriegesmann* and J. Lira**

Universidad Simon Bolivar, Surface and Interfaces Research Group, Apartado 89000, Caracas 1080-A, Venezuela
*Fachhochschule des Landes Rheinland-Pfalz, Abteilung Koblenz, Fachbereich Keramik, Höhr-Grenzhausen, Germany
**Laboratoire Materiaux Céramiques et Traitements de Surface, URA320, CNRS, ENSCI, 47 Av. A. Thomas, 87065 Limoges, France

ABSTRACT

Thick monoclinic ZrO_2 coatings (> 5 μm) were obtained by spraying a sol-gel derived nanoscale suspension on stainless steel substrates. The ceramic deposition was carried out spraying the sol with a commercial flame-spray equipment. This new process is compared with a more traditional technique such as the wet spraying of the sol at room temperature followed by thermal treatments between 500 and 900 °C.

Both, wet and thermal sprayed coatings, were analyzed by scanning microscopy and X-ray diffraction. Coatings by wet-spray depositions showed a suitable densification behavior at low temperature however the high shrinkage of the film induced cracks. The thermal spraying process yield crack-free homogeneous layers with good adhesion and low porosity.

1. INTRODUCTION

A recent report on coating's technology [1] showing predictions of the world coating business by the year 2000, indicates that in the advanced ceramic market, ceramic coatings are the fastest growing group. In fact ceramic coatings have proved to be a viable solution to improve the oxidation resistance, the wear performance and the thermal insulation of metal parts.

There is a wide range of processes used for ceramic coating deposition. For thin films, sol-gel methods have gained major interest, due to low densification temperatures of nanoscale materials, and the simple deposition procedure[2]. More over, thermal spray methods, with plasma or flame torch, have permitted the deposition of millimetrically thick films, allowing for a better chemical and thermal protection of metal substrates[3].

Among a large number of ceramic coating materials, ZrO_2 finds practical application as thermal barrier coatings (TBC), in gas sensor and as substrate buffer layers [4,5]. TBC have to be multi-functional, answering for heat insulation, wear resistance and corrosion resistance. Dense and thick coatings, applicable by simple, reproducible and inexpensive methods are thus required.

In a previous work a new hybrid process for coating metals was presented, combining the sol-gel and the thermal spraying methods. Taking advantage of the high sinterability of nanoscale materials, an amorphous aluminosilicate thick coat was obtained by flame spraying of a xerogel powder [6]. However in that process, the drying step induced the formation of agglomerates and the gel partially lost its favorable particle characteristics (nanometric size and narrow particle size distribution). This motivated searching for a more efficient process that could make use of the advantages of thermal spraying and sol-gel methods and lead to the innovation presented in this work, that is the thermal spraying of the sol. In this report we compare the new technique with a more conventional process such as the wet spraying of the sol using compressed air, at room temperature.

2. REVIEW OF THE LITERATURE

2.1. ZrO_2 synthesis and coatings via sol-gel

Several reports [7-14] concerning the synthesis of ZrO_2 via sol-gel show the wide range of material characteristics attainable by this method. Stabilized and non-stabilized ZrO_2 coatings on stainless steels have been produced by sol-gel using the dip-coating or spinning methods. These authors have investigated the improvement of properties of stainless steel, in particular the improvement of oxidation and corrosion resistance.

Using different precursors, Izumi et al [12] elaborated pure ZrO_2 coatings. Unstable solutions obtained by alkoxide precursors tended to produce discontinuous films, while sols from Zr-tetraoctylate and Zr-acetylacetonate were more stable and formed transparent continuos films.

Atik et al [13] using Zr-isopropoxide prepared transparent sols by sonocatalysis and obtained tetragonal ZrO_2 coatings with thickness less than 50 nm. Samples were dip-coated, dried and then sintered at different temperatures up to 900°C. Tested in H_2SO_4 at 81 °C they showed weight loss less than 14 mg cm^{-2} after 28h .

Yttria (8%) stabilized zirconia coatings were obtained by M. Shane et al.[2] by spinning a sol elaborated from Zr-tetrabutoxide and yttrium acetate. Cracked coatings were observed after heat treatments at low temperatures (<750°C), however at 1050°C the cracks self-healed during crystallization and grain growth. Qualitative corrosion resistance tests showed little or no hydrogen evolution on coated samples sintered at high temperatures (> 850°C). Excellent adhesion due to significant coatings-substrate interfacial reactions were evidenced.

Unsupported ZrO_2 films have been obtained by Vesteghem et al.[14] from a colloidal zirconia suspension, using Zr-oxyacetate. Hydrolysis of Zr-oxyacetate was incomplete and remaining acetate groups, while covering the colloidal ZrO_2 particles, protected them against coarsening. No occurrences of aggregation or coalescence of particles were observed during a period of three years. Monoclinic ZrO_2 with particle size of about 24 nm and crystallites of about 11 nm were obtained. Thermal treatment. of an unsupported film showed that densification began below 1000°C and was complete when the martensitic transformation occurs. Due to the sol stability as well as the crystallization and sinterization behavior of the ZrO_2 film, this sol-gel method was chosen for the coating of steel by our proposed sol flame spraying process.

2.2. Zirconia Coatings by Thermal-Spraying

Thermal-spraying is one of the most versatile coating technique and many thermal-spray systems are commercially available today such as powder flame, wire flame, wire arc, conventional plasma, detonation gun, high-energy plasma and high-velocity oxyfuel (HVOF) systems [1,3]. Thermal sprayed coatings have typical thickness between 0.1 and 10 mm and consist of bonded particles that span a range of size, shape and degrees of melting [3]. Despite tthat hermal spraying is a widely used technique, substantial improvements are required for energy and material savings. In fact, only 3 to 5 % of the available thermal energy for plasma spraying is used in melting the coating material. Furthermore, due to decomposition or evaporation the efficiency of deposition of most coating materials is around 50 % [3,15].

The ceramic materials mostly used for thermal-spray coatings are Al_2O_3, Cr_2O_3, $MgZrO_3$, TiO_2, ZrO_2 and mixtures of these oxides. Usually an intermediate bonding layer of MCrAlY (M: Ni, Co or Fe) is applied to improve the adhesion of the ceramic particles to the substrate, specially for flat surfaces. Commercial thermal-spray ceramic powders, usually consist of grit or hollow spherical particles having typical particle size in the 10 to 100 μm range [15,16]. Flame spraying yields porous ceramic coatings. For this reason, plasma spray, that produces a more dense coating is preferred for ceramics.

Many ZrO_2 based thermal sprayed coatings have been used in industrial applications, the most common being doped ZrO_2 for thermal barrier coatings. Numerous authors [1,4,5,15,18] have shown that the most popular ZrO_2-TBC, because of its high thermo-mechanical resistance, consist of plasma sprayed Y_2O_3 (6-9 % wt) partially stabilized zirconia (Y-PSZ), with a MCrAlY bonding interlayer. TBC usually fail either from spalling of the Y-PSZ coating or due the oxidation of the MCrAlY interlayer [4,18].

In the range of composition between 6 and 9 wt % Y_2O_3 a non equilibrium tetragonal phase forms when Y-PSZ is plasma sprayed. This phase with high yttria content is called *t'* to distinguish it from the low yttria content tetragonal *t* phase which is the equilibrium phase. The *t'* phase forms by diffusionless transformation of the cubic high temperature phase, due to the fast cooling rate of the plasma spray process. The *t'* phase is non transformable into the monoclinic phase during rapid cooling to room temperature. However, on further heating in the high temperature range used in TBC applications, it transforms to an equilibrium mixture of cubic and low-yttria *t* phase. The last may further transform to monoclinic phase on cooling to room temperature causing failure of the ceramic coatings by delamination and spalling [4,5,17,18]. Thus the non-equilibrium behavior of plasma sprayed ZrO_2 play a deleterious role on long term durability of the coating.

Although ZrO_2 based TBC are presently being used successfully under relatively clean high temperature environments, such as those deriving from natural gas and jet aircraft fuel. Lost of adherence have been reported when the coatings were subjected to combustion environments of industrial quality fuel, containing impurities such as potassium, sodium, vanadium, phosphorous, sulfur and lead compounds [1,5,19].

Two major failure modes have been reported: a) saturation of coatings' porosity by molten salts, resulting in damage by mechanical effects and b) hot corrosion by molten salts, which react with the most commonly used ZrO_2 stabilizers (CaO, Y_2O_3 and MgO). Research destined to achieve ZrO_2 coatings for applications in industrial fuel environments, is focused on reducing the porosity of the thermal sprayed coatings, either by hot isostatic pressing (HIP) or by surface densification treatments (laser, ion beam, electron bean), also to trying new corrosion-resistant stabilizers (i.e., In_2O_3) [1,5].

New zirconia TBC sources are also being investigated. Our research group is carrying out parallel experiments concerning the direct utilization of zirconium silicate (zircon). On the basis of thermodynamic analysis by Taylor [20] who reports that on zircon plasma spraying, SiO_2 evaporates, leaving liquid zirconia which solidifies on the substrate. The plasma separation appears a viable method, for coating purposes, using direct plasma deposition of zircon as a ZrO_2 precursor.

3. EXPERIMENTAL

The sol preparation method has been investigated by Vesteghem et al[14]. A Zr-acetate solution was obtained by mixing Zr-n-propoxide and acetic acid in a molar ratio of 1:4. An excess of acid was added to insure full reaction. The mixture was stored at 60 °C for 24 h. A fine white zirconium oxiacetate precipitate was obtained, which was filtered, washed with 1-propanol and then dried. The Zr-oxiacetate powder was used diluted in water and submitted to a hydrothermal treatment in an autoclave system up to 260 °C (heating rate 2 °C min^{-1}) and a pressure of 45 bar, for 30 min.

The stainless steel substrates, AISI 316L (DIN 1.4404) used for the coating processes were discs of 16 mm diameter or sheets of 20x15x1mm. The substrates were prepared in four different ways to determine the effects of the passivation film on oxidation/corrosion resistance and on coating adhesion.

The samples as received, (SAR) were cleaned in an ultrasonic bath with acetone and the surface was left in the as received state. The samples polished (SP), were subjected to surface grinding and final polishing with Al_2O_3 powders. The samples polished, cleaned and passivated (SP C-P) were similar to substrate SP, but followed ASTM standards for cleaning and descaling stainless steel [21]. After polishing and cleaning with acetone, the samples were immersed for 30 min in a HNO_3 solution (40 weight-%) kept at 80 °C and then rinsed in water to remove the residual acid from the surface. The HNO_3 solution causes a corrosive surface cleaning, and promote the formation of a thin passivated film on the surface. This film, consisting mainly of $FeCr_2O_4$ (chromit), protects the steel from further oxidation by air and water in a similar way to the natural passivation layer. The substrate SS C-P (Sample Sandblasted and Cleaning Passivated) have been sandblasted with steel abrasive powder, to receive a surface roughness of approximately 3-5 μm and then cleaned-passivated as previously described for sample SP C-P.

For the wet-spraying process a commercial spray gun (same type as used for painting or glazing applications) was used. A fine beam was obtained with dry compressed air at a pressure of 5-6 bar and a spray-distance of about 10 cm. The prepared substrates were coated by spraying a fine layer on both sample sides. The samples were then dried at 100 °C for about 15 min. To obtain thick coatings, the spraying and drying processes was repeated up to four times. The densification of wet-sprayed samples was performed by heating at 10 °C min^{-1} up to 500°C in air and then sintering in argon atmosphere at different final temperatures between 600 and 900 °C during 3 h.

The thermal spraying was carried out by injecting the ZrO_2 sol into an oxyacetylene flame using a commercial powder spray gun (Terodyn 2000 from Eutectic & Castolin). The samples were coated at a spray distance of 15 cm from the spray gun nozzle, orienting the flame at 90° to the surface. After pre-heating the substrate up to about 200 °C, the solution was injected, using the on-off control. The flame temperature lied above 1200 °C, so overheating of the steel substrate during the coating process was prevented, by cooling with dry compressed air at 3 bar. The maximum surface temperature attained during the deposition process was 650°C.

For scanning electron microscopy (SEM) analysis of the sample surfaces, they were left in the "as delivered" state. For cross-section observation, samples were cut with a diamond saw, embedded in epoxy resin, surface grinded, polished to specular finish and covered with carbon. In some cases chemical etching was used to achieve a better contrast at the interface region.

The corrosion resistance of coated and uncoated samples was tried in a 20 weight-% H_2SO_4 solution. The samples were immersed in the 80 °C hot solution for 15, 30, 60 and 90 min and the weight-loss was registered, measuring the samples weight before and after immersion. Reported data represented the mean value of three samples. Coated and uncoated samples were also oxidized at 800°C in air during 15 h and charaterized by XRD, using CuK radiation, at a scanning speed of 2° min^{-1} between 25 and 70°.

4. RESULTS AND DISCUSSION

Fig.1 shows representative SEM micrographs of wet sprayed coatings heat treated at 550 °C for densification. Due to the high content of solvent and

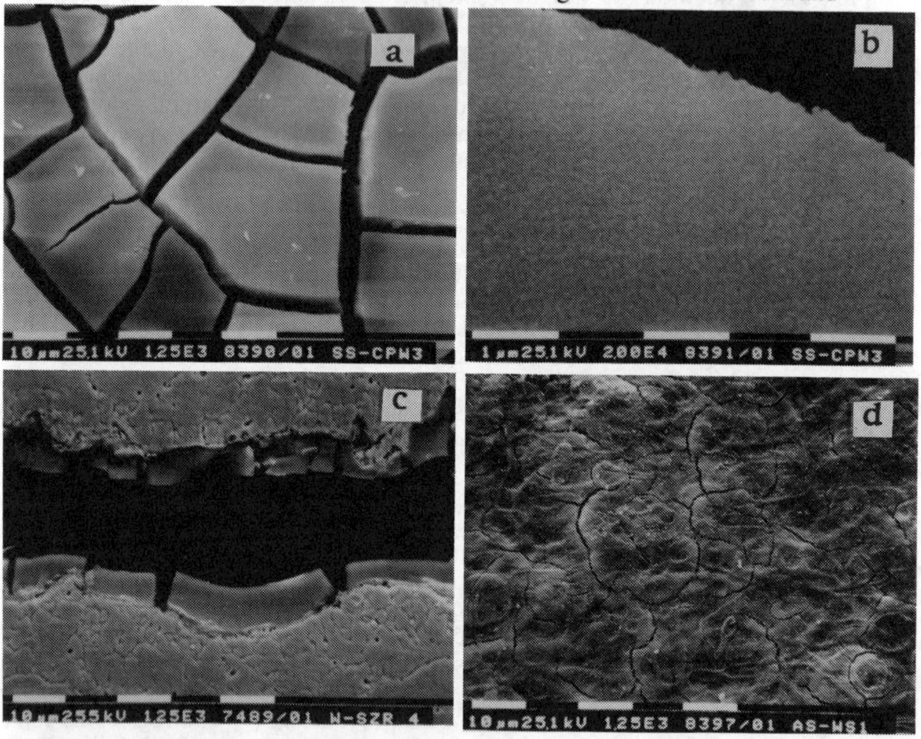

Figure 1 Wet-sprayed coatings. a) Surface of coating sintered at 550°C, showing cracked feature. b) Detail of 1.a micrograph showing dense and homogeneous nanostructure of the cracked platelets. c) Unetched cross-section d) Reduced cracking behavior achieved using sol of increased particle size.

organic groups the ZrO_2 coatings showed cracks produced during the drying and densification process. The coating cracks into platelets of approximately 30-40 μm (Fig.1.a), however such plates consisted of uniform nano-grains arranged in a dense and regular structure (Fig. 1.b). The same coating morphology was observed in samples sintered at higher temperatures. The unetched cross-section of wet sprayed samples (Fig.1.c) shows that the cracks are reaching to the substrates' surface, but very strong adhesion was achieved since no spalling of the cracked coatings occurred during manipulations for cross-section preparation. Fluctuation of the coating thickness between 5 and 10 μm was observed in this sample, due to the roughness of the substrate (SS C-P). The Fig.1.c also shows the oxide scale formed in the steel cross-section during storage of the cut and polished samples. This shows the passivation interlayer existing between ZrO_2 coat and substrate.

According to previous investigations [22,23] that evaluated the drying behavior of films obtained by wet deposition methods, using submicronic size alumina dispersions, a critical cracking thickness was obtained, characteristic of the substrate and coating properties as well as of the deposition conditions. The critical cracking thickness increases with particle size or when the suspension is flocculated. This behavior is confirmed in Fig 1.d that shows a much lower tendency to crack due to the formation of aggregates by flocculating the sol, however a less dense and uniform nanostructure was obtained. Current experiments to obtain dense, crack-free coatings by wet deposition, focus on manipulating the spray conditions (temperature, pressure, sol flow and spray distance) to promote the solvent evaporation before the ZrO_2 particles reach the substrate.

Figure 2. Thermal sprayed coatings. a) Typical surface microstructure showing interconnected aggregates. b) Detail of nanostructure of such aggregates.

The surface microstructure of thermal-spray coatings is illustrate on Fig.2. The coating surface shows high porosity and is made up of irregular and interconnected aggregates, some of them being hollow and nearly spherical. These aggregates consist of nanometric particles arranged in a relatively dense structure. To achieve high density coatings the flame must be capable of drying the solvent, burning the organics materials, and at least superficially melting the ZrO_2 particles, before they reach the substrate surface. Since the torch used was designed for powders with defined rheological and fusion properties, the deposition process, by spraying a solution was not evident. Although low quantities of solution were injected into the flame, full combustion of solvent and organics, respectively was not always achieved, so for further experiments a torch with solution flow regulator will be used to obtain a smaller and constant solution flow.

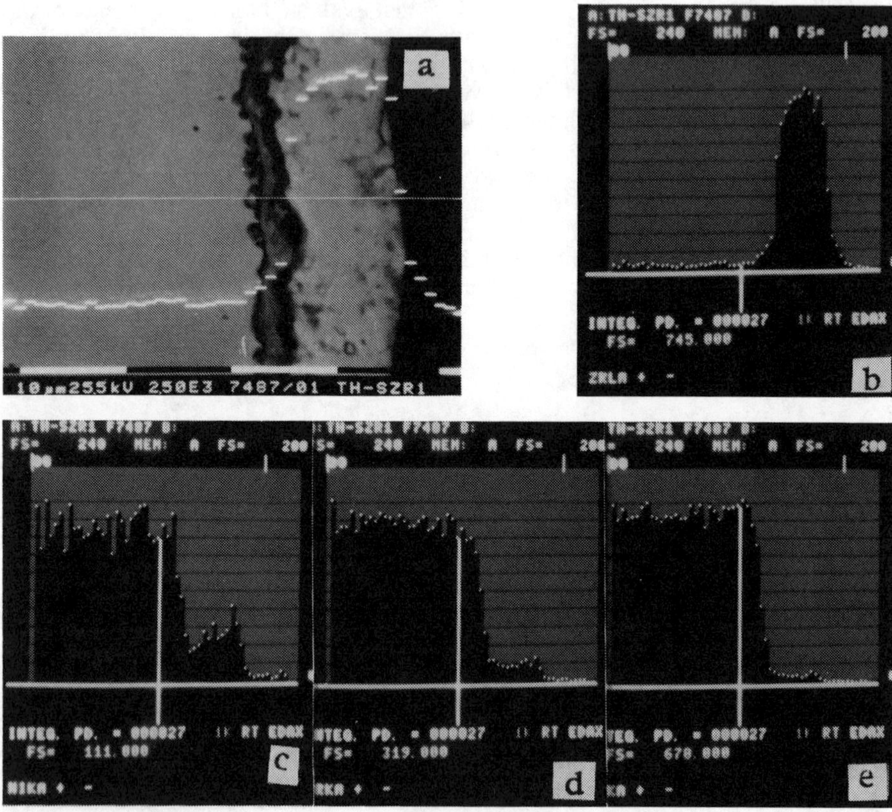

Figure 3. Cross-section analysis of thermal sprayed coatings. a) SEM image showing Zr profile by X-ray spectroscopy. b) Zr profile. c) Ni profile, d) Cr profile, e) Fe profile

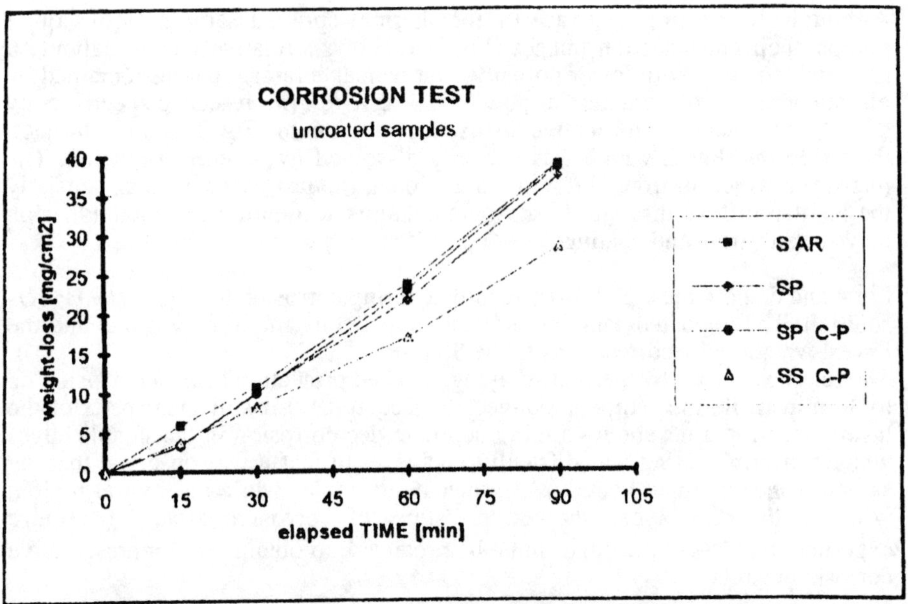

Fig 4 Corrosion behavior of uncoated substrates

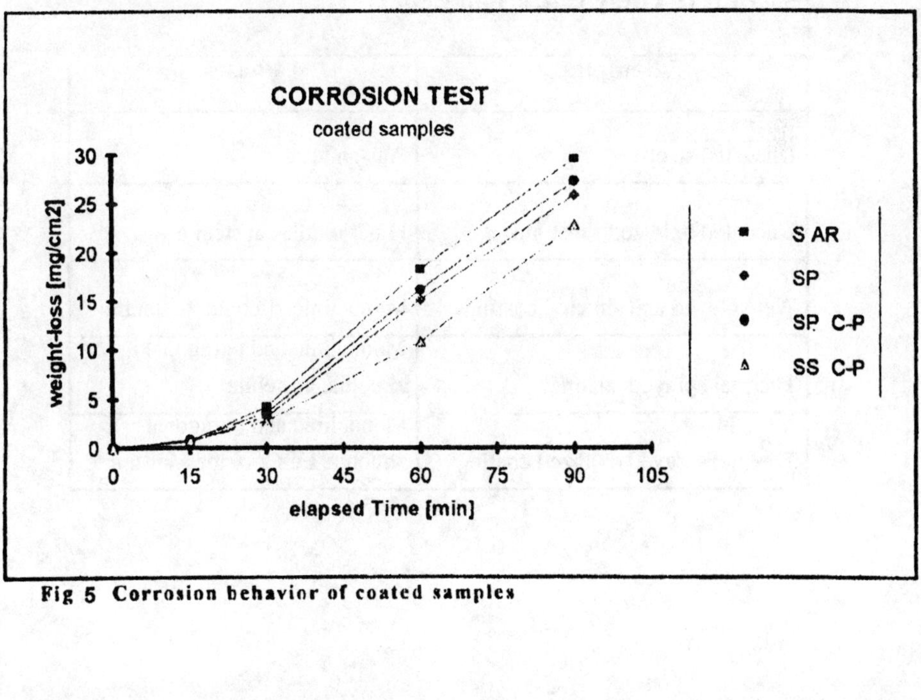

Fig 5 Corrosion behavior of coated samples

Although the coating's surface of the thermal sprayed samples looks quite porous, their cross-section images (Fig. 3.a) shows a relatively dense, about 10 μm thick coating, with lower porosity than typical ceramic coatings obtained by plasma spraying of commercial powders. The X-ray fluorescence spectroscopy (EDS) indicates no significative diffusion of zirconium (Fig. 3.a and b) through the oxide interlayer, which was partially dissolved by chemical etching. The interlayer is rich in iron, chromium and nickel, the last two elements, specially the Ni also diffuse through the zirconia coatings accounting for good adhesion between substrate and coating.

Fig 4 and 5 show the weight-loss recorded during corrosion tests in 20 % H_2SO_4 solution. The uncoated samples showed almost constant corrosion rate and the lowest weight gain corresponds to the SS C-P.
The samples coated by thermal spraying, showed protection against corrosion up to 30 min while the samples retained their coatings. Loss of great parts of the layers occurred after about 45 min due to crevice corrosion in the sample edges where uniform coating was difficult to achieve. In fact it was observed that the sample edges bubbled heavily, like uncoated samples while the central region, with good ZrO_2-layers showed no signs of corrosive attack. In future experiments, the sample edges must be protected, to obtain more representative corrosion results.

X-ray diffraction analysis was used to characterize the crystallization and oxidation behavior of the ceramic coatings and compare them with the uncoated steel substrate

TABLE I: X-Ray Diffraction Results

Samples	Phases
Uncoated steel	Austenite
Uncoated oxidized substrate	Fe-Cr oxides, austenite
Wet sprayed and sintered coatings	Monoclinic zirconia, austenite
Thermal sprayed coatings	Monoclinic and tetragonal zirconia, austenite
Thermal sprayed oxidized coatings	Monoclinic and tetragonal zirconia, Fe-Cr oxides, austenite

The characteristic peaks of the austenitic stainless steel were obtained for the AISI 316L substrate. They appear likewise for coated and uncoated samples, because the coatings are thin enough for the substrate to diffract through. The uncoated samples after heat treatment in air at 800 °C showed strong peaks belonging to the minority phase. Due to the thick oxide layer, which arises during the thermal treatment, the peaks for the steel substrate appear relatively weak.

The X-ray data from wet-sprayed samples after densification, showed that crystallization of the ceramic coatings started at 550 °C producing the monoclinic stable phase, and that the peaks got much stronger and sharper when increasing the densification temperatures.

The coatings deposited with the flame-spray technique showed monoclinic ZrO_2 and additional peaks of the tetragonal metastable form as a minor phase. No peaks from Fe-chromate were detected in the XRD indicating that no oxidation of steel substrate during the deposition process occurred. The flame-sprayed coatings submitted to oxidation test at 800 °C during 15 h showed peaks corresponding to the monoclinic and tetragonal phases of zirconia and small peaks of Fe-chromate indicating that a slight oxidation of the specimen's substrate, during the thermal treatment occurred. No significative variation in the relative intensity of monoclinic/tetragonal peaks was observed after the oxidation heat treatment, indicating that the tetragonal phase remains in the metastable form, probably due to the small grain size.

5. CONCLUSIONS

Thick zirconia coatings made up of non transformable phases, suitable for thermal barrier applications can be obtained by sol wet spraying at room temperature and sol flame-spraying, using colloidal ZrO_2 suspensions.
For applications that involve temperatures lower than 1170 °C, undoped monoclinic coatings could perform successfully.
Sol-gel derived nanometric particles, instead of commercially available micronic powder, can be deposited by flame spraying. In this way no dopants that decrease the corrosion resistance of ZrO_2 are required, and the formation of non equilibrium phases as those produced by plasma deposition, can be avoided.
Optimizing the deposition parameters, and redesigning the torch, the process proposed in this work could be very promising for the deposition of tick and dense coatings of zirconia and similar ceramics, providing good oxidation protection and thermal insulation for steel substrates.

Acknowledgments
The authors would like to thanks Mr. H. Ruiz for his helpful assistance in the experiments.

6. REFERENCES

1. G. Geiger, "Cover Story, Ceramic Coatings" Am. Cer. Soc. Bull. Vol 71 [10], (1992), 1470-1481
2. M. Shane, M.L. Macertney, "Sol-gel synthesis of Zirconia Thermal Barrier Coatings"J. Mater. Sci. 25 (1990) 1537-1544
3. M. L. Thorpe, "Thermal Spray Industry" Advanced Mater. & Proc., 5 (1993), 50-61
4. L. Lelait, S. Alpérine, C. Diot, M. Mévrel "Thermal Barrier Coatings: microstructural Investigation after Annealing", Mater. Sci. Engin. A, 121, (1989) 475-482
5. P. Vincenzini, "Zirconia Thermal Barrier coatings for Engine Applications" Industrial Ceramics, Vol. 10, N° 3 (1990), 226-239
6. A. R. Di Giampaolo Conde, M. Puerta, H. Ruiz, J. Lira-Olivares, " Thick Aluminosilicate Coatings on Carbon Steel Via Sol Gel" J. Non-Cryst. Sol., 147 &148 (1992) 467-473
7. T. Ogihara, K.. Wada, T. Yoshida, T. Yanagawa, N. Ogata, K. Yoshida "Preparation of Monodispersed Zirconia -Alumina and Yttria-Alumina Particles" Ceramic International, 19 (1993) 159-168
8. R. Guinebretière, A. Dauger, A. Lecomte, H.Vesteghem, "Tetragonal zirconia powders from the zirconium n-propoxide-acetylacetone-water-isopropanol system" J. Non-Cryst. Sol., 147 &148 (1992) 542-447
9. A. J. Lecloux"Texture evolution of zirconium oxide xerogel prepared by sol-gel process" J. Non-Cryst. Sol., 147 &148 (1992) 389-39
10. N.M. Ghoneim, S.B. Hanna"Sintering and Microstructure of Ultrafine yttria-Zirconia Compacts" J. Mater. Sci. 25 (1990) 5192-5198
11. D. L. Bourell, Perimal, W. Kaysser "Sol Gel Synthesis of nanophase Ytria-Stabilized Tetragonal Zirconia and Densification Behavior below 1600 K" J. Am. Ceram. Soc. 76 [3] (1993) 705-711
12. K. Izumi, M. Murakami, T. Deguchi, A. Morita, N. Tohge, T. Minami "Zirconia Coatings on Stainless Steel Sheets from Organozirconium Compounds" J. Am. Ceram. Soc. 72 [8] (1989) 1465-1468
13. M. Atik, M.A. Aegerter "Corrosion Resistant Sol-Gel ZrO_2 Coatings on Stainless Steel" J. Non-Cryst. Sol., 147 &148 (1992) 813-819
14 . H. Vesteghem, A. Lecomte, A. Dauger "Film Formation and Sintering of Colloidal Monoclinic Zirconia" J. Non-Cryst. Sol., 147 &148 (1992) 503-507
15. G. Wuest, S. Keller, A.R. Nicoll, A. Donnelly "Plasma Spray Deposition Efficiencies" J. Vac. Sci. Technol. A3[6] Nov/Dec (1985) 24642468
16. E. J. Kubel Jr. "Thermal-Spray-Coatings, Powder Dictate Properties" Advanced Materials & Processes, 12 (1990) 24-32
17. R. A. Miller, J.L. Smialek, R. G. Garlick "Phase Stability in Plasma Sprayed Partially Stabilized Zirconia-Yttria" Advances in Ceramics, Vol 3, pp 241-253, The American Ceramic Society, Columbus OH (1981)
18. B. Rhouta, B. Pieraggi "Interface in plasma deposited MCrAlY-Zr02 cermets" J. Mater. Sci. Lett. 11 (1992) 278-280.
19. R.J. Bratton, S.K.Lau, "Zirconia thermal barrier Coatings" Advances in Ceramics, Vol 3, pp 227-239, The American Ceramic Society, Columbus OH

(1981)
20. P.R. Taylor, M.Abdel-Lafit, S.A. Pirzada, X. Hou, "Non-transfered Arc, Thermal plasma Cyclone Reactor" Thermal Plasma Applications in Materials and Matallurgical Processing, pp 193-207, Edited by N.El-Kaddah, The Mineral, Metals & Materials Society, 1992
21. ASTM Standards "Standard Practice for Cleaning and Descaling Stainless Steel Parts, Equipment and Systems" Iron and Steel Productspp 141-152 ,A 380-88, Vol. 01.03 (1992)
22. R.C. Chiu, T.J. Garino, M. J. Cima "Drying of Granular Ceramic Films: I, Effect of Processing Variables on Craking Behavior" J. Am. Ceram. Soc. 76 [9] (1993) 2257-2264
23. R.C. Chiu, M. J. Cima "Drying of Granular Ceramic Films: II,Drying Stress and Saturation Uniformity" J. Am. Ceram. Soc. 76 [11] (1993) 2769-2777.

INFLUENCE OF CERAMIC PARTICLE CONTENT ON THE CYCLIC FATIGUE AND FRACTURE BEHAVIOR OF AN ALUMINUM ALLOY METAL-MATRIX COMPOSITE

T. S. Srivatsan

Department of Mechanical Engineering; The University of Akron
Akron, Ohio 44325-3903, USA.

ABSTRACT

A study has been made to understand the influence of ceramic particle reinforcement phase on the cyclic fatigue and fracture behavior of aluminum alloy 2124. The composite microstructures had similar particle size distribution. The 2124/SiC_p composites were deformed using fully-reversed, total strain-amplitude control cyclic loading. The cyclic strain resistance and resultant low-cycle fatigue life degraded with an increase in the volume fraction of the discontinuous ceramic particle reinforcement phase in the metal matrix. The cyclic fracture behavior of the metal-ceramic composites is discussed in light of concurrent and mutually competitive influences of volume fraction of the discontinuous reinforcement phase, intrinsic composite microstructural effects, cyclic strain-amplitude and resultant response stress.

1. INTRODUCTION

Reinforced metal matrices, also known as metal-matrix composites (MMCs), have tremendous potential for the future. They are an important class of materials with a long and sustained history of commercial applications. They provide better mechanical properties than the unreinforced metal matrix and tailorable physical properties. For example discontinuous ceramic-particle-reinforced aluminum alloy metal matrices are preferred because they offer a number of advantages such as a 15-40% increase in strength and a 30-100% increase in stiffness compared to the unreinforced alloy [1,2], while generally maintaining receptiveness to processing and characterization techniques used for the conventional unreinforced counterparts.

From a design perspective the attractiveness in choosing a discontinuous carbide particle (SiC_p)-reinforced aluminum alloy MMC stems from an improvement in specific modulus, i.e., density-compensated increase in elastic modulus. The moduli obtained are greater than those of typical titanium alloys and only marginally less than those of steels. Associated with an improvement in modulus are a concurrent increase in yield and tensile strength of up to 50% [3,4].

Furthermore, the discontinuously-reinforced aluminum alloy metal matrices based on particulate reinforcements are attractive because they can be made with properties that are near isotropic in three orthogonal directions, or in a plane. Also, conventional secondary fabrication and processing methods can be used to produce a wide range of product forms. However, the primary disadvantages with these materials is that they suffer from low tensile ductility (elongation-to-failure), inadequate fracture toughness, and inferior fracture resistance compared to the unreinforced matrix alloy [5-9]. Many practical applications of structural components made from fine ceramic particulate-reinforced aluminum alloy MMCs involve cyclic loading and, therefore, the fatigue properties are of need and interest.

The objective of this paper is to document the low-cycle fatigue and fracture behavior of a ceramic particle-reinforced aluminum alloy-based MMC. Emphasis was given to the procurement of the MMC with controlled synthesis (having same matrix alloy composition) and reinforcement particle characteristics (type (SiC), size distribution and shape), while varying only the volume fraction of the particulate reinforcement in order to rationalize its influence on composite microstructure and thus cyclic fatigue and fracture behavior.

2. MATERIAL

The MMCs selected for investigation in this study were based on a powder metallurgy processed Al-Cu-Mg matrix alloy designated as 2124 by the Aluminum Association of America. The nominal composition (in weight percent) of the matrix alloy is given in Table I. The reinforcement used was fine silicon carbide particulates [SiC_p]. Three different volume fractions [20 vol.%, 25 vol.% and 30 vol.%] of the discontinuous SiC_p reinforcement phase were chosen. The material was produced and made available by DWA Composite Specialties Inc., [Chatsworth, California, USA].

The DRA composite materials were fabricated by blending atomized aluminum alloy powders with SiC particles. Blending facilitates a homogeneous distribution of the SiC_p, both at the macroscopic and microscopic levels, in the 2124 aluminum alloy powder. The homogeneous blend was compacted by hot isostatic pressing, vacuum degassed, and pressed to produce a fully-dense billet. The degassing step assists in removing adsorbed moisture from the carbide particle (SiC_p) surfaces.

The billet was then extruded to bars having a rectangular cross-section. Extrusion ratios are unknown. Precise details of the primary processing treatment are considered proprietory by the manufacturer. Heat treatment of the 2124/SiC_p MMCs consisted of solutionizing the as-extruded plates at 498° C [920°F] for 2 hours, quenching in cold water. Subsequently, the MMCs were artificially aged at 120°C (250°F) for 18 hours to give the T4 temper.

3. EXPERIMENTAL TECHNIQUES

Samples for microstructural evaluation were prepared using standard metallographic techniques. The polished and etched samples were examined in an optical microscope and photographed using a standard bright-field technique.

The cyclic fatigue tests were performed on a fully-automated, closed-loop, servohydraulic materials test machine (INSTRON) equipped with a 10,000 kg load cell. The tests were conducted in axial total-strain amplitude control mode. Fully-reversed, push-pull, tension-compression low-cycle fatigue (LCF) tests were performed. The test command signal was triangular in shape and the mean strain was zero. For each individual test, the test machine was programmed to maintain a constant total strain rate of 1×10^{-3}/sec. All cyclic fatigue tests were begun in tension. The tests were conducted in controlled laboratory air environment

Figure 1. Triplanar optical micrograph illustrating the microstructure of the 2124 composites: (a) 20 vol.% SiC_p, and (b) 30 vol.% SiC_p

[relative humidity of 55 percent; ambient temperature (27°C)]. An axial 12.7 mm gage length clip-on extensometer was attached to the fatigue test specimen, with rubber bands, to monitor the total strain during fully-reversed total strain amplitude-control cyclic fatigue tests, such that the magnitude of negative strain equals the magnitude of positive strain [$R_\epsilon = \epsilon_{min.}/\epsilon_{max.} = -1$]. The controlled variable is total strain-amplitude. The stress-strain hysteresis loops were periodically recorded on an X-Y recorder equipped with a pen plotter. The plastic strain range ($\Delta\epsilon_p$) and cyclic stress range ($\Delta\sigma$) were provided as a print-out by the computer control console of the INSTRON servohydraulic test machine. The number of cycles to failure or separation is taken as fatigue life (N_f).

Fracture surfaces of the deformed fatigue test specimens were comprehensively examined in a scanning electron microscope (SEM) to characterize the fracture mode and the fine-scale features on the fatigue fracture surface.

4. RESULTS AND DISCUSSION

4.1. Microstructure

The triplanar optical micrographs illustrating the microstructure of the heat treated 2124/SiC/xxp-T4 [where xx denotes the carbide particle content in volume percent) composites, for 20 vol.% and 30 vol.% of the discontinuous ceramic particle (SiC_p) reinforcement, are shown in Figure 1. The discontinuous carbide particulate reinforcement phase [SiC_p] was found dispersed randomly through the 2124 aluminum alloy matrix and weakly aligned along the extrusion direction. At regular intervals clustering or agglomeration of the SiC_p was observed. An agglomerated site consisted of a few larger SiC particles intermingled with smaller and more regularly-shaped particles. The agglomeration of the SiC_p was evident along all three orthogonal directions (extrusion, long-transverse and short-transverse) of the MMC. The non-uniform distribution of the SiC_p reinforcements along the three orthogonal directions results in an anisotropic microstructure of the 2124/SiC/xxp composites. The ambient temperature tensile properties of the 2124/SiC/xxp composites, for the three different volume fractions of reinforcement, are summarized in Table II.

4.2. Low Cycle Fatigue Properties

A measurable physical quantity during strain-controlled cyclic deformation is plastic strain. Cyclic plastic strain produces a number of damaging processes

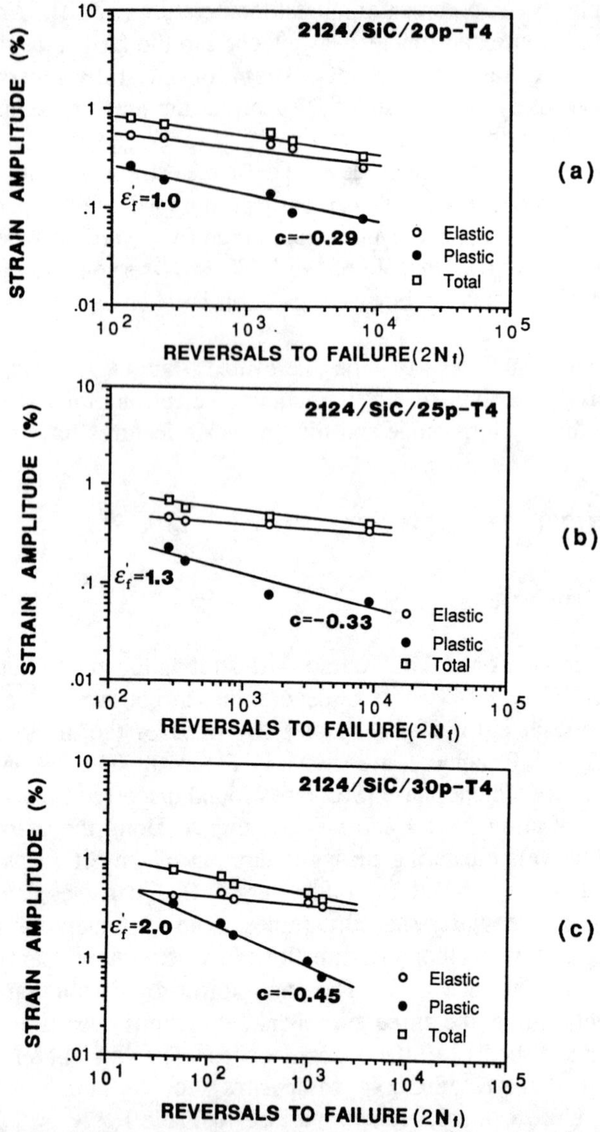

Figure 2. Variation of cyclic strain amplitude with reversals-to-fatigue life for the 2124/SiC/xxp composites:
(a) 2124/SiC/20p; (b) 2124/SiC/25p; (c) 2124/SiC/30p.

which affect the microstructure and resulting fatigue resistance. The progressive damage resulting from cyclic straining helps relate the low-cycle fatigue (LCF) life to cyclic strain-amplitude ($\Delta\epsilon/2$) in a completely reversed [$R_\epsilon = \epsilon_{min.}/\epsilon_{max.} = -1$] total strain-amplitude controlled, cyclic fatigue test.

The effect of strain cycling on LCF response of the 2124/SiC/xxp composite materials is shown in Figure 2. The plastic strain-amplitude ($\Delta\epsilon_p/2$)-reversals to LCF life ($2N_f$) curve can be viewed as an indication of the resistance of the composite material's microstructure to crack initiation and failure. Throughout LCF testing of the 2124/SiC/xxp MMCs, under fully-reversed total strain-amplitude controlled cycling, no evidence of specimen buckling was observed. Since the plastic strain varies throughout the life of the specimen, reaching a minimum value at maximum stress and a maximum value at minimum stress, the plastic strain range ($\Delta\epsilon_p$) corresponding to half-life is taken as the reference value. The plastic strain amplitude ($\Delta\epsilon_p/2$)-reversals to fatigue life ($2N_f$) curve is linear and the Coffin-Manson relationship:

$$\Delta\epsilon_p/2 = \epsilon_f' (2N_f)^c \tag{1}$$

can be used to determine the fatigue behavior in the LCF regime ($N_f < 10^4$ cycles) [10,11]. The value of c, the fatigue-ductility exponent, obtained from the plastic strain-amplitude-fatigue life curve is between -0.3 to -0.5. This value is close to those generally observed and reported {range of -0.7 to -0.1} for a large majority of monolithic aluminum alloys [12,13]. The fatigue-ductility coefficient (ϵ_f') increases with an increase in SiC_p reinforcement content in the 2124 aluminum alloy matrix. That is, the plastic strain amplitude ($\Delta\epsilon_p/2$)-reversals to fatigue life ($2N_f$) curve becomes steeper with an increase in SiC_p reinforcement content. The fatigue-ductility coefficients (ϵ_f') of the 2124/SiC/xxp MMCs do not correlate well with the monotonic ductility (ϵ_f), indicating that the cyclic deformation characteristics of the metal matrix is altered by the presence of the discontinuous particulate reinforcement.

Comparison of the plastic strain-amplitude-fatigue life curves of the 2124/SiC/xxp-T4 MMCs (Figure 3) reveals a degradation in low-cycle fatigue resistance with an increase in SiC_p content in the aluminum alloy matrix. The observed degradation in LCF life is more pronounced at low plastic strain-amplitudes and concomitant long fatigue life than at the higher plastic strain-amplitudes and resultant short fatigue life. The intrinsic mechanisms responsible for the observed degradation in cyclic strain resistance and resultant LCF life are discussed in the following sections on stress response characteristics and cyclic fracture.

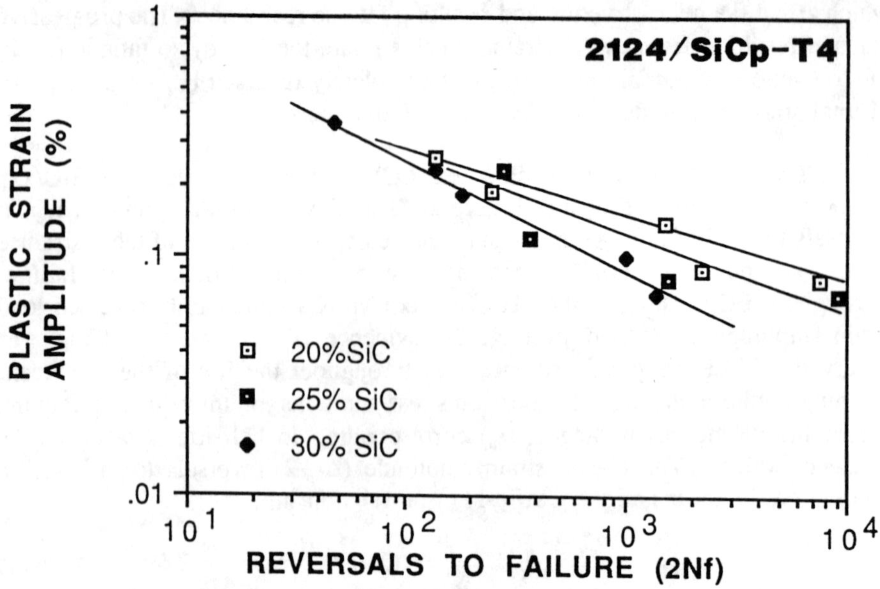

Figure 3. Influence of reinforcement content on plastic strain amplitude-reversals to fatigue life response of the 2124/SiC/xxp composites.

4.3 Stress Response Characteristics

The stress response curves or cumulative glide plots of variation of cyclic stress amplitude with number of cycles, at a fixed cyclic plastic strain-amplitude, illustrates the path by which the material arrives at its final flow stress level. The stress response curves of the 2124/SiC/xxp MMCs, for the different volume fractions of the discontinuous SiC_p reinforcement phase, are compared in Figure 4. The stress response curves reveal that at all cyclic plastic strain-amplitudes the 2124/SiC/xxp composite material progressively hardened and rapidly softened prior to failure. The hardening behavior dominated for most of fatigue life (80-90%) while the softening, or drop in load carrying capability, occurred following the initiation of multiple microscopic cracks, their growth through the composite matrix and eventual coalescence to form one or more macroscopic cracks and resultant growth of both the macroscopic and microscopic cracks. For a given volume fraction of the discontinuous SiC_p reinforcement phase the degree of hardening was greater at the higher cyclic strain-amplitude. That is, the tendency towards cyclic hardening decreased with a decrease in cyclic strain-amplitude.

The observed cyclic stress carrying capability and hardening of the 2124/SiC_p composites during fully-reversed, total strain-amplitude controlled deformation is responsible for the degradation in cyclic strain resistance and low-cycle fatigue life. Mechanisms contributing to the hardening or strengthening during fully-reversed strain cycling are discussed in detail elsewhere [14,15]. When the local stress concentration (τ^*) caused by dislocation build-up or pile-up at a ceramic particle (SiC_p) exceeds a critical value (τ_{cr}), microcrack initiation occurs through rupture of the hard and brittle particle. With continued cyclic straining, the microcracks, initiated at neighboring particles, rapidly link to form a macroscopic crack. Propagation of the macroscopic cracks along with the microscopic cracks results in softening or drop in load-carrying capability and culminates in failure.

4.5. Cyclic Fracture Behavior

To rationalize the influence of the ceramic reinforcement phase (SiC) on fracture behavior of the 2124/SiC/xxp MMCs, the LCF fracture surfaces were examined in a JEOL scanning electron microscope (SEM). Representative fractographs are shown in Figures 5-7.

Fatigue fracture features, on a macroscopic scale, were found to vary with particulate reinforcement content in the aluminum alloy metal-matrix. At high cyclic strain amplitudes, fracture on a macroscopic scale was essentially normal

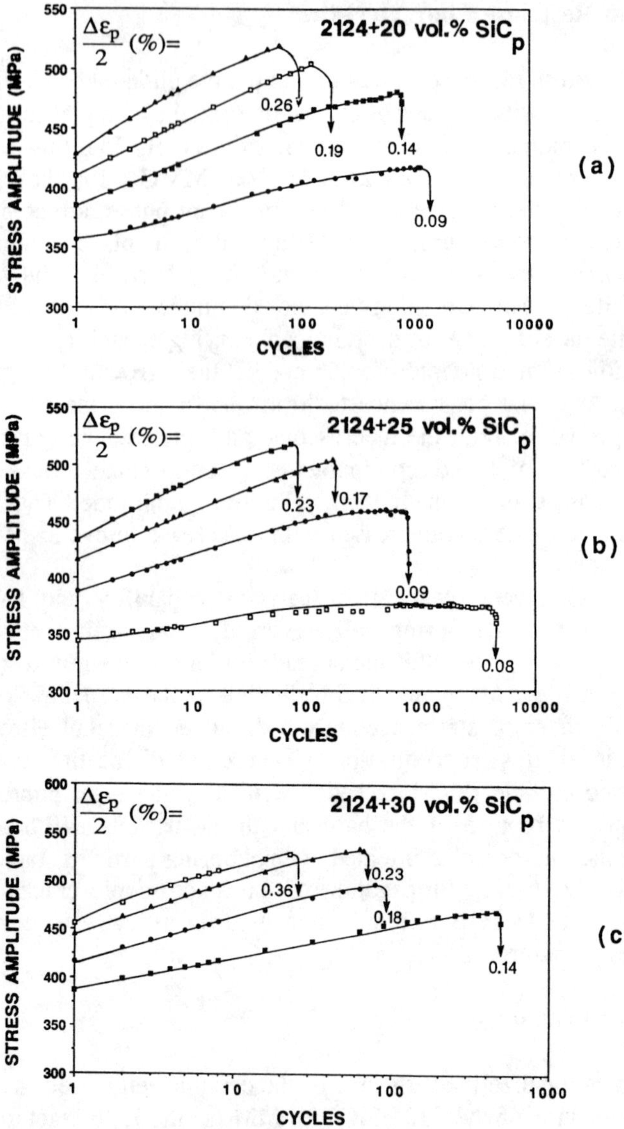

Figure 4. Stress response curves for the composites:
(a) 2124/SiC/20p; (b) 2124/SiC/25p; (c) 2124/SiC/30p.

to the stress axis. Macroscopic or low magnification observations revealed the fracture to be brittle (low cyclic ductility) with multiple microscopic cracks (Figure 5 a) and decohesion at the matrix-particle interfaces (Figure 5 b). The microscopic cracks were common in samples cyclically deformed at high total strain-amplitudes and resultant short fatigue life.

Microscopic observations of the fracture surface revealed microvoids of various sizes. The macroscopic and microscopic voids were distributed homogeneously throughout the fracture surface with evidence of microvoid coalescence. The voids were intermingled with shallow dimples indicating reminiscence of ductile failure. The constraints induced by the brittle carbide particle (SiC_p) reinforcement phase in the aluminum alloy metal matrix and the concomitant development of matrix triaxiality influences the cyclic flow stress and ductile void growth.

For this 2124/SiC/xxp composite, the events involving nucleation of a void tend to dominate cyclic ductility since subsequent stages of void growth and coalescence are rapid once void initiation has occurred. Preferential void nucleation was observed in regions containing an agglomeration of the SiC_p (Figure 6) and in the composite matrix with a high volume fraction of the particulate reinforcement (30 volume %) [Figure 7], thereby, suggesting the important role played by local plastic strains. Once the voids are nucleated they progressively grow with continued cyclic straining and eventually coalesce to form microscopic and macroscopic cracks.

The growth of voids is essentially a cavity enlargement process which is enhanced by localized plastic deformation. The limited growth of the microvoids, during far-field cyclic loading, clearly indicates that the deformation properties of the 2124/SiC/xxp composite microstructure are significantly altered by the presence of discontinuous SiC_p reinforcements. Few of the microvoids coalesce and the halves of these voids are the shallow dimples visible on the fatigue fracture surface. Enhanced void nucleation coupled with failure of the reinforcing SiC particles suggests that local plastic strains dominate in regions containing a high volume fraction of SiC particulates and are responsible, in part, for the inferior cyclic strain resistance of the composite with 30 vol.% SiC_p than the 2124 metal matrix with 25 vol.% and 20 vol.%.

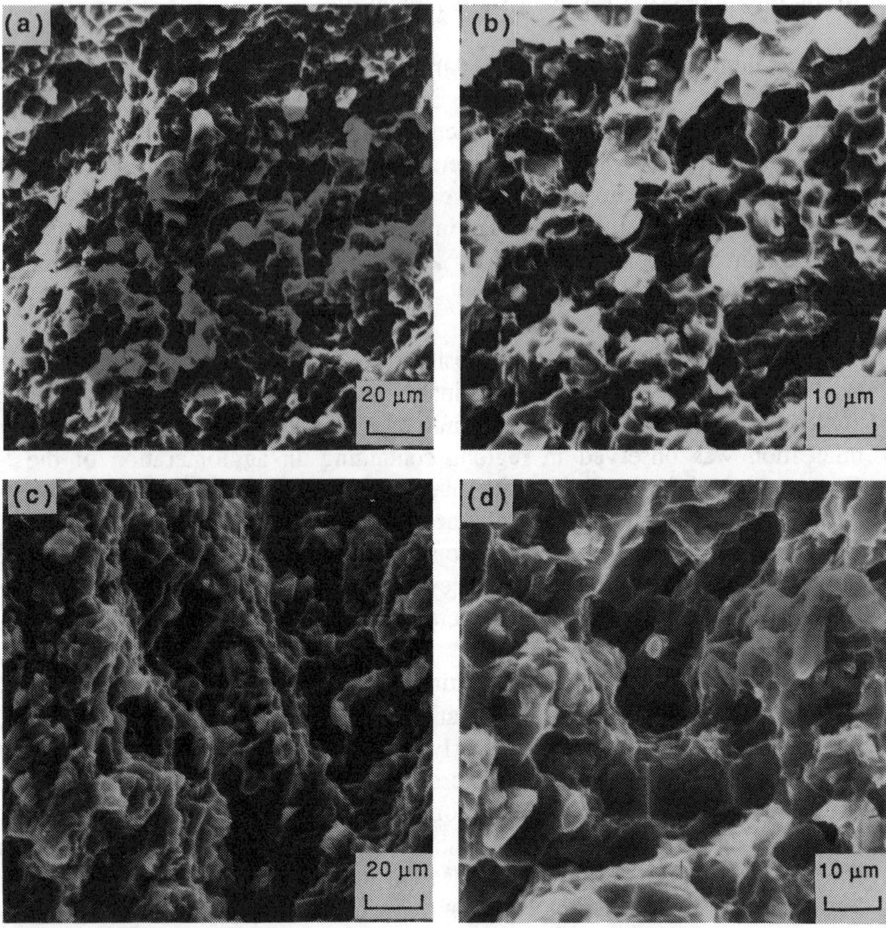

Figure 5. Scanning electron micrographs of the low-cycle fatigue fracture surface features of the 2124/SiC/20p MMC deformed at:
(a) and (b) plastic strain amplitude of 0.06%, N_f=3810 cycles
(c) and (d) plastic strain amplitude of 0.15%, N_f= 1120 cycles.

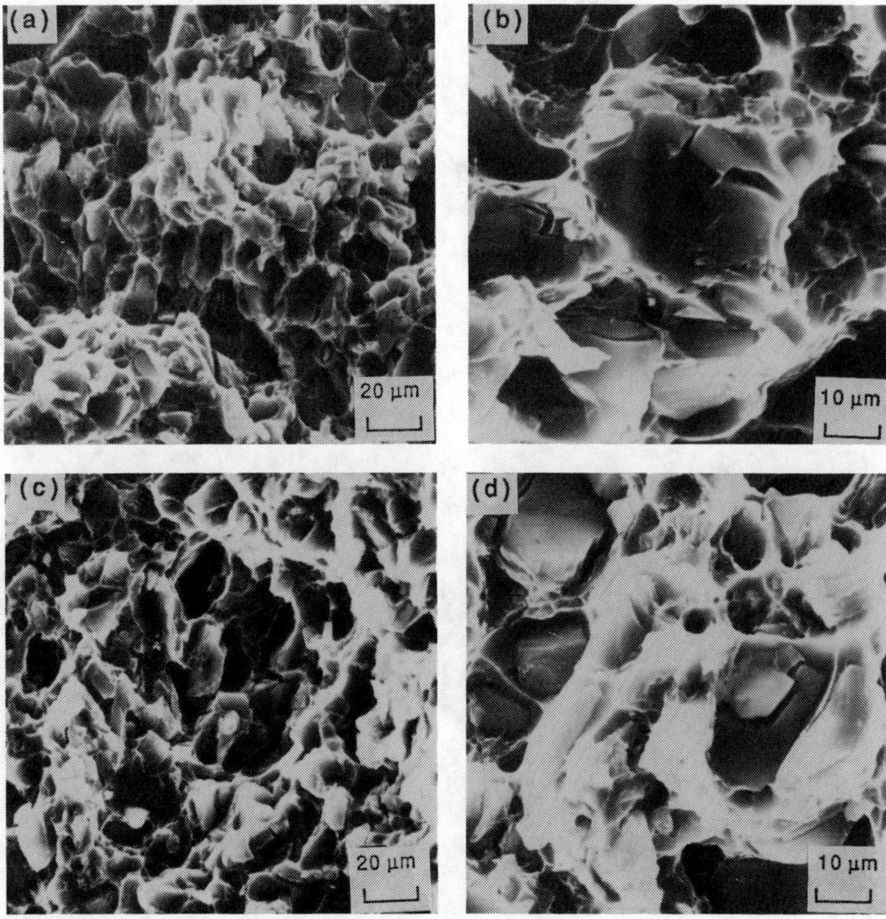

re 6. Scanning electron micrographs of the low-cycle fatigue fracture surface features for the 2124/SiC/25p MMC deformed at:
(a) and (b) plastic strain amplitude of 0.11%, N_f = 4510 cycles.
(c) and (d) plastic strain amplitude of 0.32%, N_f = 140 cycles.

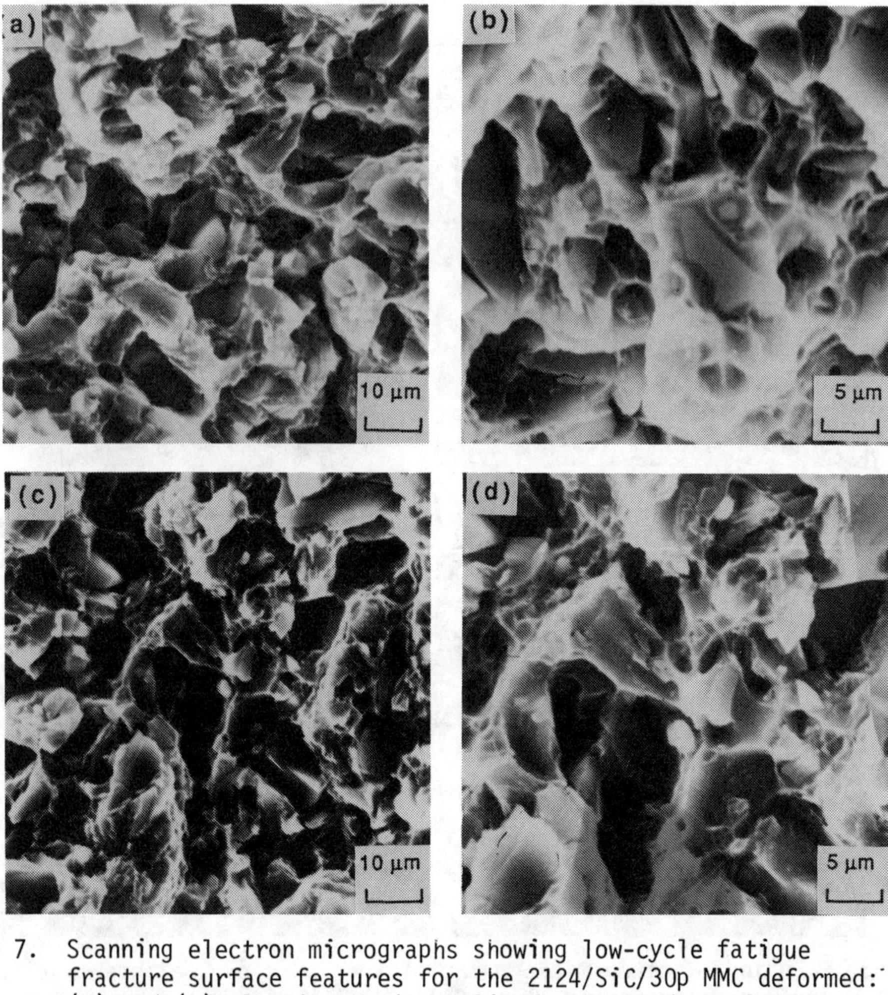

Figure 7. Scanning electron micrographs showing low-cycle fatigue fracture surface features for the 2124/SiC/30p MMC deformed: (a) and (b) plastic strain amplitude of 0.10%, N_f=3810 cycles (c) and (d) plastic strain amplitude of 0.17%, N_f=500 cycles

5. CONCLUSIONS

A study of the influence of ceramic particle reinforcement on the cyclic fatigue and fracture behavior of aluminum alloy 2124 discontinuously reinforced with silicon carbide particulates (SiC_p) provides the following observations :

1. The 2124/SiC_p composites revealed a non-uniform distribution of the SiC particulates in the three orthogonal directions of the as-received MMC plate. At frequent intervals an agglomeration of the particulates, of varying size, was observed.
2. Cyclic stress response of the ceramic particle (SiC_p) reinforced aluminum alloy revealed hardening for most of fatigue life followed by rapid softening to failure. The hardening response of the composite microstructure is dependent on mutually interactive influences of volume fraction of the discontinuous reinforcement phase and cyclic strain amplitude.
3. The 2124/SiC_p MMCs followed the Coffin-Manson relationship for variation of cyclic plastic strain-amplitude with fatigue life. A degradation in cyclic strain resistance and low-cycle fatigue life occurred with an increase in volume fraction of SiC_p reinforcement in the 2124 alloy matrix. The observed degradation in low-cycle fatigue resistance for the MMCs was noticeable at low cyclic strain-amplitudes and resultant long fatigue life than at the higher cyclic strain amplitudes and short fatigue lives.
4. Fracture morphology was found to be essentially similar at the different cyclic strain amplitudes. Fracture revealed an overall macroscopically brittle appearance but microscopically features reminiscent of ductile failure. With an increase in reinforcement content, cyclic fracture was dominated by crack initiation at the SiC particles. The constraints induced in the aluminum alloy matrix by the SiC particulates coupled with local stress concentration effects at the matrix-particle interfaces promotes failure through particle cracking and decohesion at the interfaces.

Acknowledgements

The author extends his thanks and appreciation to Aluminum Company of America (**ALCOA FOUNDATION**; Program Manager: Dr. A. Chakrabarti) for funding this research study under Grant 5-34392. Thanks are also extended to

DWA Composite Specialties Inc. [Chatsworth, CA; Program Manager: Dr. W. C. Harrigan, Jr.) for providing the material used in this study. Thanks and appreciation are extended to Ms. Pam Hoover for her timely assistance in the assembling of this manuscript.

TABLE 1:

NOMINAL CHEMICAL COMPOSITION OF ALUMINUM ALLOY 2124 (in Weight Percent)

Cu	Mg	Zn	Mn	Si	Fe	Al
4.65	1.65	0.01	0.9	0.04	0.30	Bal.

TABLE 2:

AMBIENT TEMPERATURE TENSILE PROPERTIES OF 2124/SiC$_p$-T4 METAL MATRIX COMPOSITE

Silicon Carbide [%]	E (GPA)	Yield Strength (MPa)	UTS (MPa)	Elong. (%)	R.A. (%)	Fracture Stress (MPa)
20	99	351	537	10.5	12.4	603
25	109	360	545	8.5	10.2	600
30	124	387	543	5.2	7.8	586

REFERENCES

1. D. F. Hasson and C. R. Crowe: in Strength of Metals and Alloys, ICSMA, Ocford, United Kingdom, Pergamon Press, 1985, pp. 1515-1520.

2. S. G. Fishman: Role of Interfaces in Material Damping, ASM Materials Park, Ohio, 1985, pp. 33-41.

3. J.F. Mandell, K.C. Hong and D.H. Grande: Ceramic Engineering Science Proceedings, Vol. 8 (7-8), 1987, pp. 937-940.

4. P. Niskanen, W.R. Mohn: **Advanced Materials and Processes**, Vol. 133(3), 1988, pp. 39-41.

5. S. V. Nair, J.K. Tien and R.C. Bates: **International Metals Reviews**, Vol. 30, No. 6, 1985, pp. 285-296.

6. D.L. McDanels: **Metallurgical Transactions**, Vol. 16A, 1985, pp. 1105-1115.

7. S. Dermarkar: **Metals and Materials**, Vol. 2, 1986, pp. 144-147.

8. T. S. Srivatsan and J. Mattingly: **Journal of Materials Science**, Vol. 28, 1993, pp. 611-620.

9. S. Norose, T. Sasdafa and M. Okabe: 28th Japan Congress on Materials Research, 1985, pp. 231-241.

10. L.F. Coffin, Jr.: **Transactions ASME**, Vol. 76, 1954, pp. 931-940.

11. S.S. Manson: NASA Technical Note, 1954, p. 2933.

12. B.Tomkins: **Philosophical Magazine**, Vol. 18, 1968, pp. 1041-1050.

13. R.E. Sanders, Jr., and E.A. Starke, Jr.: **Metallurgical Transactions**, Vol. 9A, 1978, pp. 1087-1095.

14. T.S. Srivatsan: **International Journal of Fatigue**, Vol. 14, No. 3, 1992, pp. 173-182.

15. T.S. Srivatsan: **International Journal of Fatigue**, Vol. 15, No. 5, 1993, pp. 355-367.

AlN-BN MACHINABLE CERAMICS

Jig-Sang Gam, Jung-Soo Ha*, Deock-Soo Cheong, and Kyong-Sop Han
Ceramics Division, Korea Institute of Science and Technology
P.O. Box 131, Cheongryang, Seoul, Korea

ABSTRACT

AlN-BN ceramics with BN contents in the range of 10 to 40 wt% and with or without additives (3 wt% Y_2O_3 or CaO) were prepared by hot pressing. Their machinability, strength, and microstructures were investigated. Both the main and radial cutting forces decreased with an increase in BN content in all samples. For the BN contents of 30 wt% or above, the cutting forces were lower than that of a mild steel tested at the same condition. Especially in the case of main forces, the values were less than a quarter of that of a mild steel, indicating excellent machinability. Bend strength (when the tensile surfaces of specimens were perpendicular to the hot pressing direction) also decreased with BN content mainly due to the much lower Young's modulus of BN compared to AlN. With the composition of 30 wt% BN at which the AlN-BN ceramics started to show better machinability than a mild steel, the bend strength was 155 MPa. With tensile surfaces parallel to the hot pressing direction, however, the bend strength obtained for the samples processed with the sintering aids showed low values (about 40 MPa), since most of the BN particles had such orientation that their cleavage planes (i.e., basal planes) were perpendicular to the pressing direction.

INTRODUCTION

In general, a ceramic material requires machining to achieve the precise tolerances of structural applications. However, most structural ceramics are too brittle and hard to allow cutting on a lathe like metals. Only lapping or polishing is applicable, which is not only inefficient but also is limited to simple shapes.

AlN-BN ceramics can be lathed by the aid of cleavage characteristic of the BN particles dispersed, inducing a crack produced at the tool-bit to follow the machining direction instead of propagating to the interior of the work piece [1]. This material, categorized as a machinable ceramic like the glass-ceramics based on a mica system [2,3], is already commerciallized in Japan. AlN-BN is also

*Present address: Department of Materials Engineering, Andong National University, 388 Songchun-Dong, Andong, Kyeongbuk 760-749, Korea (South).

known to be superior to the latter in refractoriness, strength, and thermal conductivity [1,4]. However, a detailed study on the machinability has not been reported in the literature. Only the fabrication method for producing a sintered body [5-8], and the effects of sintering aids and composition on the densification, microstructure, and strength [9] have been studied.

In this work, the machinability and strength of AlN-BN ceramics containing 10 to 40 wt% BN were investigated. Specimens were prepared using no additives, or with 3 wt% Y_2O_3 [10,11] or CaO [11-13]. The latter are common densification aids for AlN. The correlation of strength with microstructure was also studied.

EXPERIMENTAL PROCEDURE

The starting powders, AlN (Grade F, Tokuyama Soda, Japan) and BN (Grade A01, Herman C. Stark Berlin, Germany) were mixed in methanol for 4 h using an alumina ball mill. The resulting slurry was dried by stirring on a hot plate to produce mixtures of 10, 20, 30, and 40 wt% BN. The properties of the starting powders are given in Table I. In the mixtures involving 3 wt% Y_2O_3 or CaO (added as a form of $Ca(NO_3)_2 \cdot 4H_2O$) as a densification aid, the additive was dissolved in methanol prior to the addition of AlN and BN powders. This approach was used to disperse such a small amount of additives as homogeneously as possible. The prepared mixtures were hot pressed at 1900℃ for 1 h in nitrogen with a pressure of 30 MPa.

Table I. Characteristics of AlN and BN powders

	AlN*	BN†
Crystal structure	Hexagonal wurtzite	Hexagonal graphite
Agglomerate size (㎛)	<2.0	-
Specific surface area (m^2/g)	3-4	3-6
Purity	1.0 wt% O	<0.15 wt% B_2O_3

*Grade F, Tokuyama Soda Co., Japan.
†Grade A01, Hermann C. Starck Berlin, Germany.

The densities of the hot-pressed specimens were measured by the Archimedes method and converted into relative values to theoretical densities, obtained from a rule of mixture using the following component densities: 3.26 g/cm^3 for AlN, 2.25 g/cm^3 BN, 5.01 g/cm^3 Y_2O_3, and 3.32 g/cm^3 CaO [14]. Turning machining was performed to test the machinability. Cutting resistance in the tangential and radial directions, named main and radial forces as schematically shown in Fig. 1, was measured using a tool dynamometer of a strain gauge type (TD-500KA, Kyowa Electric Co., Japan). The tool used was K10 bite with the size of 2 by 10 by 4 mm, and the cutting conditions were as follows: the outer diameter of the work piece=25 mm, cutting speed=21 m/min, feed rate=0.025 mm/rev., and cutting depth=1 mm. The flexural strength was measured in four-point loading (inner span=10 mm, outer span=20 mm, and crosshead speed=0.5 mm/min) at room temperature on rectangular bars. The tensile surfaces were perpendicular to the hot-pressing direction in most cases, but for the samples of 30 wt% BN containing Y_2O_3 or CaO the strength for parallel cases also was investigated. The microstructure was characterized by scanning electron microscopy (SEM) of the fracture surfaces near the tensile surfaces.

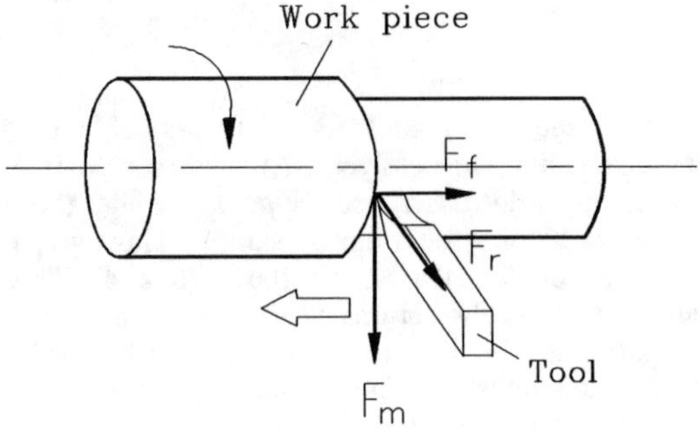

Figure 1. Schematic showing three components of cutting resistance in turning: main force (F_m), radial force (F_r), and feed force (F_f).

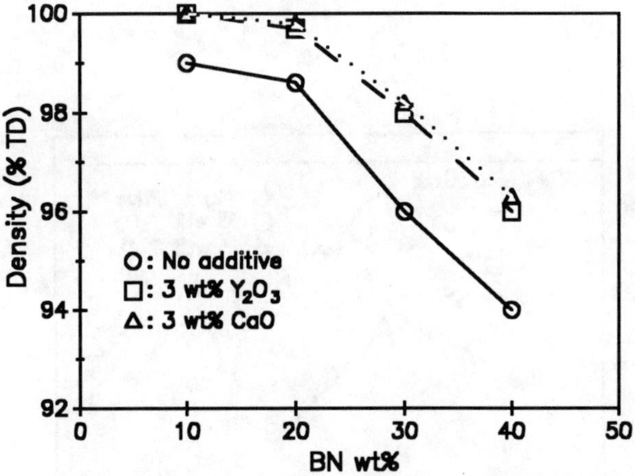

Figure 2. Densities of three kinds (no sintering additive, 3 wt% Y_2O_3 or CaO additives) of AlN-BN ceramics containing 10-40 wt% BN, hot pressed at 1900℃ for 1 h under 30 MPa and N_2 atmosphere.

RESULTS AND DISCUSSIONS

Densities

The densities of the hot-pressed specimens are shown in Fig. 2. They decreased with an increase in the BN content, and 1 or 2 % higher densification was obtained for the samples with sintering aids. There was little difference in density between the Y_2O_3- and CaO-added samples. Both showed almost full densification with BN contents to 20 wt%, and 98 % and 96 % densities at 30 wt% and 40 wt% of BN contents, respectively. Such a density decrease with an increase in the amount of dispersoid is common in ceramic composites. The decrease is attributed to the dispersed phase producing hydrostatic tensile stresses upon sintering, which inhibits the shrinkage of the matrix [15].

Machinability

Figures 3 and 4 show the main and radial forces of the AlN-BN ceramics measured on turning machining, respectively. At each composition, the main forces of the specimens showed little difference (Fig. 3), while the radial forces had slightly different values depending on type of sample. However, the difference was small for the compositions of 30 wt% BN or above (Fig. 4). These results indicate that the machinability of AlN-BN ceramics is almost entirely determined by the amount of BN. Compared to the results of a mild steel (SS41) tested under the same conditions (the dotted lines in Figs. 3 and 4), the two cutting forces of all the samples were smaller. The main forces were less than a quarter of that of the steel results, indicating good machinability. Thus, taking into account the strength decrease with increasing BN content as discussed in the next section, it can be seen that the optimum composition is 30 wt% BN.

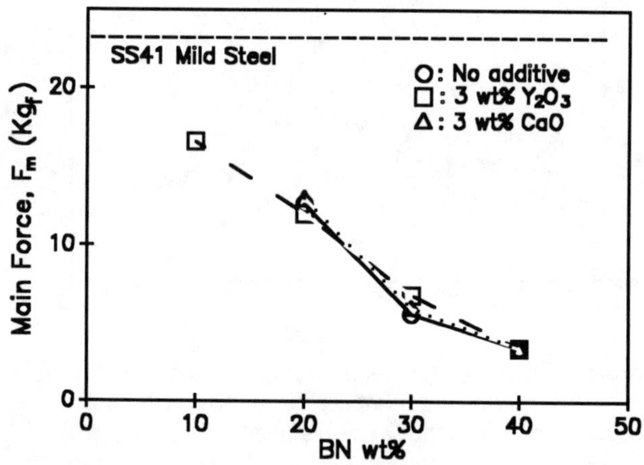

Figure 3. Results of machining tests by turning for three kinds (no sintering additive, 3 wt% Y_2O_3 or CaO additives) of AlN-BN ceramics containing 10-40 wt% BN: plots of main force vs. BN content. The horizontal dotted line denotes the main force value of a SS41 mild steel tested at the same conditions for comparison.

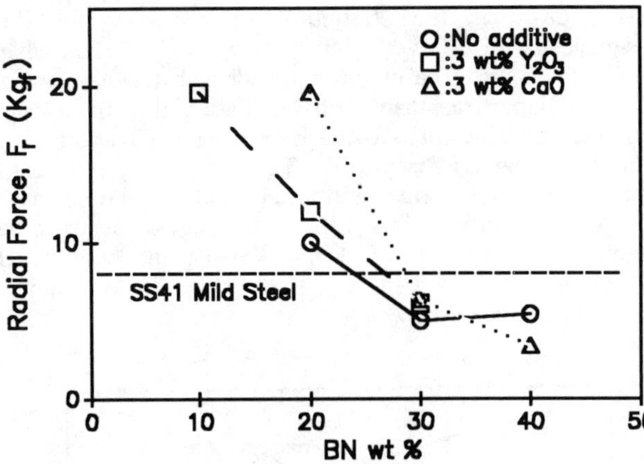

Figure 4. Results of machining tests by turning for three kinds (no sintering additive, 3 wt% Y_2O_3 or CaO additives) of AlN-BN ceramics containing 10-40 wt% BN: plots of radial force vs. BN content. The horizontal dotted line denotes the radial force value of a SS41 mild steel tested at the same conditions for comparison.

Strength

The results of flexural tests are given with the predictions of a rule of mixture in Fig. 5. A rule of mixture is widely used to predict the properties of a composite such as density, Young's modulus, and strength based on the property information of components. The following relation [16] is used for strength:

$$\sigma_c = \sigma_m V_m + \sigma_r V_r$$

where σ and V are strength and volume fraction, and the subscripts c, m, and r denote the composite, matrix, and reinforcement, respectively. This equation is derived assuming an isostrain condition (i.e., the strains of the composite, matrix, and reinforcement being equal), involving perfect bonding between the components, and the same values of Poisson's ratio for the components [16]. The strength values of AlN and BN used for the calculation were 300 MPa [17] and 110 MPa (for tensile surfaces perpendicular to the hot-pressing direction) or 50 MPa (for tensile surfaces parallel to the hot-pressing direction) [18], respectively.

In the case of perpendicular tensile surfaces, the strength of all of the three kinds of samples decreased with BN content, mainly due to a low Young's modulus of BN. The values at each composition were generally higher for the samples with sintering aids than for those without. It seemed that there was no notable difference between the samples with additives. The strength obtained with the composition of 30 wt% BN, begining to show better machinability than a mild steel, had values around 155 MPa. However, in the case of tensile surfaces parallel to the hot-pressing direction, the strength measured only for the Y_2O_3- or CaO-containing samples with 30 wt% BN was 40 or 50 MPa, which was less than one third of that for the cases of perpendicular tensile surfaces. This is because most of the BN particles had such orientation that their cleavage planes (i.e., basal planes) were perpendicular to the hot-pressing direction, as discussed in the following section on the microstructure. The experimental values were much lower than the predictions of the rule of mixture, and the differences were larger for the cases of tensile surfaces parallel to the hot-pressing direction. Such discrepancy probably results from the fact that the bonding between the AlN and BN might have been imperfect and the strength values of AlN and BN used for the calculation are not real values.

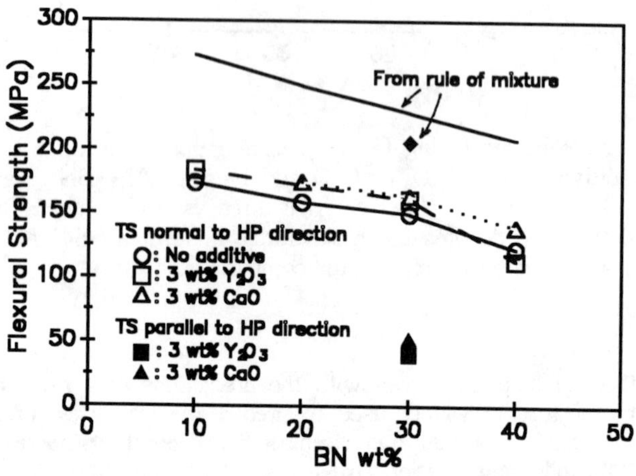

Figure 5. Results of four-point flexural tests at room temperature for three kinds (no sintering additive, 3 wt% Y_2O_3 or CaO additives) of AlN-BN ceramics containing 10-40 wt% BN, as compared with the predictions of rule of mixture: TS and HP denote tensile surface and hot pressing, respectively. Only the case of tensile surface normal to the hot-pressing direction was investigated for all samples, except the Y_2O_3 or CaO-containing sample with 30 wt% BN, for which both the cases, normal and parallel, were investigated.

Microstructure

Figure 6 shows the microstructures of the fracture surfaces of samples flexural-tested with tensile surfaces perpendicular to the hot-pressing direction. In the samples containing sintering additives, Fig. 6(a) and (b), BN particles were mostly oriented perpendicularly to the pressing direction with the cleavage planes, while in the samples without the additives, Fig. 6(c), such orientation was scarcely observed. It is presumed that in the former the sintering additives forming a liquid phase at the processing temperature provided fluidity for the rearrangement of the BN particles to relieve the pressure, resulting in the oriented microstructure. The

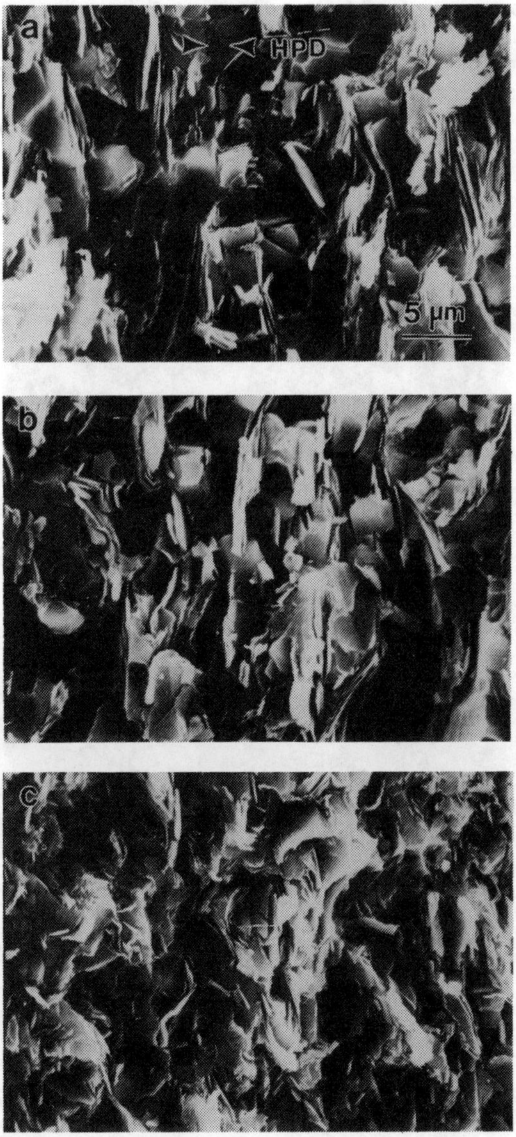

Figure 6. Scanning electron micrographs of the fracture surfaces of AlN-30 wt% BN ceramics after bend tests with tensile surfaces perpendicular to the hot pressing direction (denoted by HPD): (a) 3 wt% Y_2O_3 additive, (b) 3 wt% CaO additive, and (c) no additive. Note that the plate-shaped BN particles were perpendicular to the pressing direction in (a) and (b).

fracture of BN particles occurred mostly along the main crack plane, and some particles were pulled-out suggesting the bonding between AlN and BN was weak. The main fracture mode of AlN was intergranular for the samples containing additives, and grain sizes of 3 to 4 μm were observed. In contrast the samples without additives exhibited mostly transgranular fracture so that observation of the grain size was not possible. Figure 7 shows the microstructures of fracture surfaces of samples with tensile surfaces parallel to the hot-pressing direction. It can be seen that most areas of the fracture surfaces consisted of plates of BN, indicating that the fracture occurred along the cleavage planes.

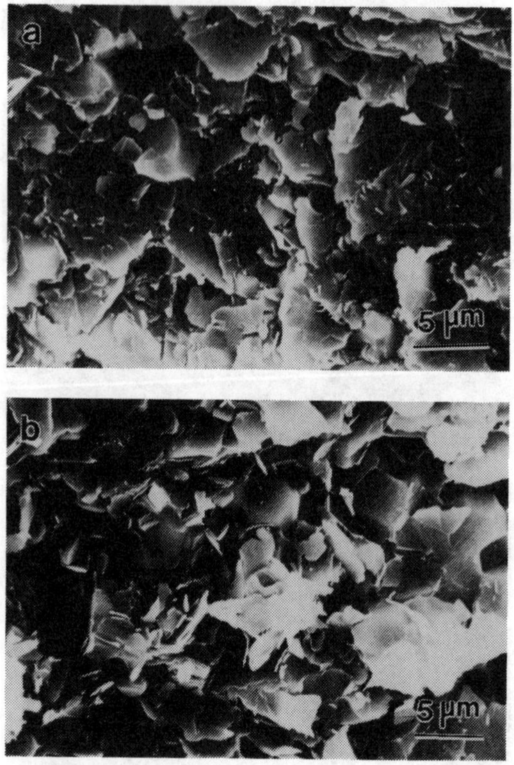

Figure 7. Scanning electron micrographs of the fracture surfaces of AlN-30 wt% BN ceramics after bend tests with tensile surfaces parallel to the hot pressing direction: (a) 3 wt% Y_2O_3 additive and (b) 3 wt% CaO additive.

CONCLUSIONS

The cutting forces and strength of AlN-BN ceramics in the case of tensile surfaces parallel to the hot-pressing direction decreased with an increase in the

BN content, and the dependence on the addition and type of a sintering aid was insignificant. With the composition of 30 wt% BN, the machinability was superior to that of a mild steel and an average strength of 155 MPa was obtained. However, in the case of tensile surfaces parallel to the pressing direction, the strength measured for the samples with sintering aids was just 40 MPa. This low strength resulted from a microstructure having cleavage planes of BN particles mostly oriented perpendicularly to the pressing direction.

REFERENCES

1. K. Takada, "Machining of Machinable Ceramics," *Machinery Technology*, 35(12),130-134 (1989).

2. D.G. Grossman, "Machinable Glass-Ceramics Based on Tetrasilic Mica," *J. Am. Ceram. Soc.*, 55(9), 446-449 (1972).

3. C.K. Chyung, G.H. Beall, and D.G. Grossman, "Microstructures & Mechanical Properties of Mica Glass-Ceramics," pp. 1167-1195 in *Electron Microscopy and Structure of Materials*, Ed. by G. Thomas, R.M. Fulrath, and R.M. Fisher, University of California Press, Berkeley, 1972.

4. Tokuyama Soda Co., Technical Bulletin No. 6103: Shapal-M, 1989.

5. N. Kuramoto, K. Takada, and Y. Numata, "Fabrication Method of a Composite Sintered Body," Japan Patent, No. 63572 (1986).

6. K. Takada and Y. Numata, "Fabrication Method of a Composite Sintered Body," Japan Patent, No. 56377 (1987).

7. J.R. Morris, Jr. and R.A. Tanzilli, "Aluminium Nitride-Boron Nitride Composite Article and Method of Making the Same," US Patent 4,666,873, May 19, 1987.

8. N. Kuramoto, K. Takada, and Y. Numata, "Composite Nitride Sintered Body," US Patent 4,642,298, Feb. 10, 1987.

9. K.S. Mazdiyasni, R. Ruh, and E.E. Hermes, "Phase Characterization and Properties of AlN-BN Composites," *Am. Ceram. Soc. Bull.*, **64**(8), 1149-1154 (1985).

10. K. Komeya, H. Inoue, and A. Tsuge, "Role of Y_2O_3 and SiO_2 Additions in Sintering of AlN," *J. Am. Ceram. Soc.*, **57**(9), 411-412 (1974).

11. K.A. Schwetz, H. Knoch, and A. Lipp, "Sintering of Aluminum Nitride with Low Oxide Addition," pp. 245-252 in *Progress in Nitrogen Ceramics*, Ed. by F.L. Riley, Martinus Nijhoff, The Hague, Netherlands, 1983.

12. K. Komeya, A. Tsuge, H. Inoue, and H. Ohta, "Effect of $CaCO_3$ Additions on the Sintering of AlN," *J. Mater. Sci. Lett.*, **1**, 325 (1982).

13. S.F. Horvath, S.R. Witek, and M.P. Harmer, "Effects of Carbon and Calcium Oxide on the Sintering Behavior of Aluminium Nitride," pp. 121–132 in *Advances in Ceramics, Vol. 26: Ceramic Substrates and Packages for Electronic Applications*, Ed. by M.F. Yan, K. Niwa, H.M. O'Bryan, Jr., and W.S. Young, The American Ceramic Society, Westerville, Ohio, 1989.

14. *CRC Handbook of Chemistry and Physics*, Ed. by R.C. Weast and M.J. Astle, CRC Press, 1979.

15. C.-H. Hsueh, "Sintering of Whisker-Reinforced Ceramics and Glasses," *J. Am. Ceram. Soc.*, **71**(10), C442–C444 (1988).

16. K.K. Chawla, *Composite Materials: Science and Engineering*, pp. 177–179, Springer-Verlag, New York, 1987.

17. E.S. Dettmer and H.K. Charles, Jr., "AlN and SiC Substrate Properties and Processing Characteristics," pp. 87–105 in *Advances in Ceramics, Vol. 26: Ceramic Substrates and Packages for Electronic Applications*, Ed. by M.F. Yan, K. Niwa, H.M. O'Bryan, Jr., and W.S. Young, The American Ceramic Society, Westerville, Ohio, 1989.

18. J.M. Blocher, Jr., "Nitrides," pp. 379 in *High-Temperature Materials and Technology*, Ed. by I.E. Campbell and E.M. Sherwood, John Wiley & Sons, New York, 1967.

Poster Session

GRAIN BOUNDARY CONTROLLED MECHANICAL AND DIELECTRIC PROPERTIES OF $BaTiO_3$-BASED NANOCOMPOSITE

H.J. Hwang, S.Ueda and K. Niihara
The Institute of Scientific and Industrial Research, Osaka University
8-1 Mihogaoka, Ibaraki, Osaka 567, Japan

ABSTRACT

Dense $BaTiO_3$ and $BaTiO_3$-based nanocomposites with nano-sized SiC particulate were successfully fabricated by a conventional ceramic processing method. Two types of ceramic capacitors, reoxidized and grain-boundary barrier layer(GBBL), were also prepared by oxidizing the sintered specimens in air. The crystal structure was changed from ferroeletric tetragonal to paraelectric cubic phase with an increase in SiC content. This phase change appears to be due to the grain size decrease by SiC addition and/or much oxygen vacancy which is caused by the reduced sintering atmosphere. Some mechanical properties were remarkably improved by the addition of SiC. The relative dielectric constant of monolithic $BaTiO_3$ after oxidation exhibits almost the same value of commercial $BaTiO_3$, whereas it drastically changed for $BaTiO_3$-SiC nanocomposites that had different microstructure and grain boundary structure. The relationship between grain boundary structure and mechanical/ dielectric properties will be discussed.

INTRODUCTION

The perovskite structure-type compound $BaTiO_3$ is widely used for PTC thermistors, capacitors of high dielectric constant, gas sensors and orther electronic devices. Its electrical properties and the grain boundary structure, including the PTCR (Positive Temperature Coefficient of Resistivity) phenomenon, have been studied extensively [1]. Much effort has been concentrated on improving the electrical characteristics [2-3].

Recently, electrical devices have been miniaturized and are used under quite severe atmospheres. Such environments often lead to the initiation and propagation of cracks due to thermal shock and fatigue. Hence, the improvement of mechanical properties and reliability is required in addition to electrical properties. Thus, the mechanical properties of $BaTiO_3$ have also been investigated together with the electrical properties [4-5]. These studies have been limited to internal stress and the resulting fracture strength change [6]. According to Pohanka et al [7], fracture strength of perovskite compounds is influenced by an internal stress developed in the transformation from the cubic to tetragonal phase after the sintering process. Bruce et al. [8] investigated the subcritical crack growth(SCG) phenomenon of PZT ceramics, and reported that the SCG in electroceramics strongly influences the mechanical strength and electrical properties. However, there are only a few reports about the relationship between microstructure and mechanical properties of $BaTiO_3$-based ceramics.

In the present study, dense and fine-grained $BaTiO_3$ and $BaTiO_3$-based nanocomposites with nano-sized SiC particle as a secondary phase were fabricated. The effects of particulate SiC addition on mechanical properties, such as fracture strength and toughness, as well as the microstructure evolution were investigated in detail. SiC is known as a good grain boundary controller and microstructure modifier in various fields of ceramic systems [9]. Therefore, it is thought to be possible to fabricate $BaTiO_3$-based composites having different microstructures if the nano-sized SiC particulate can be dispersed into the $BaTiO_3$ matrix grains. Subcritical crack growth(SCG) phenomena were also discussed based on data obtained from the controlled surface flaw(CSF) method. In addition, two types of ceramic capacitors, reoxidized and grain-boundary barrier layer(GBBL), were prepared by oxidizing the sintered specimens in air.

EXPERIMENTAL PROCEDURES

Material Preparation

Monolithic $BaTiO_3$ and $BaTiO_3$-based composites with nano-sized particulate SiC were prepared using a conventional powder metallurgical method for fabrication of ceramics. The starting materials were $BaTiO_3$(Sakai Chemicals Co. Ltd.) and β-SiC(Mitsui Toatsu Chemicals Co. Ltd.) powders. Ba/Ti atomic ratio of the starting $BaTiO_3$ powder was 0.999 and the crystal phase was cubic. $BaTiO_3$ with the required amount of SiC was wet milled in ethyl alcohol using a polyethylene pot and ZrO_2 ball for 24h. The wet milled slurry was dried, then dry milled and sieved through a 320 μm mesh screen. The mixed powder was hot pressed at 1100 to 1400°C under the pressure of 30MPa for 1h in Ar atmosphere. After sintering, in order to fabricate the reoxidized and grain-boundary barrier layer(GBBL) capacitors some specimens were oxidized in air at the prescribed temperature and time. Details of the composition and the fabrication process for the specimen preparation was described in Ref. [10]. Sintered bodies were cut with a diamond cutting machine. The section surface was polished with diamond paste for various evaluations and observations, as will be mentioned later.

Evaluation

The bulk density was determined by Archimedes method in toluene. X-ray diffraction pattern was taken by Ni filtered CuKα radiation. The grain size of specimens was determined using a linear intercept method. Fracture strength was measured on polished rectangular specimens with 3 by 4 by 40 mm in size using a three-point bending method at a crosshead speed of 0.05 mm/min. Fracture toughness was measured using microindentation technique using a micro Vicker's diamond indentor and controlled surface flaw(CSF) method for the optically polished surfaces at room temperature [11].

For the former method, the indentation load of 9.8N was applied for 15 second. For the latter method, the polished specimens were indented with a Knoop indentor at load levels of 9.8, 19.6, 49, 98 and 196N. Three-point bending tests were used for CSF experiments. Prior to the bending test, the indented surfaces were ground and polished to 50 μm to remove any possible surface-stressed layer caused by the indentation. The bending test was performed using an Instron Universal Testing machine at a constant crosshead speed of 0.05 mm/min. For all specimens, the fracture initiated at the site of the controlled surface flaw. Fracture surfaces were observed with a scanning electron microscope (SEM).

RESULTS AND DISCUSSION

Microstructure and Mechanical Properties

X-ray diffraction profiles show that $BaTiO_3$ nanocomposites with particulate SiC were mainly composed of $BaTiO_3$ and β-SiC. The addition of particulate SiC up to 3 vol% gives no unwanted phases such as $Ba_4Ti_{13}O_{30}$, $Ba_6Ti_{17}O_{30}$, and $BaTiO_3$-xBaO compounds. Polarization domain structures with a large quantity of 90° twins can be clearly seen for monolithic $BaTiO_3$. Dark and bright fringes were found in turn. This indicates that a lattice periodicity is changed and the orientation of polarization to each other is 90° as pointed out by an other investigation [12]. However, there are no domain structures for the $BaTiO_3$-5vol%SiC nanocomposite. In addition, most of the SiC particles were uniformly distributed within the matrix grain, with some bigger particles located at the $BaTiO_3$ grain boundaries, as shown in Figure 1.

Nearly 100% of theoretical density was achieved for monolithic $BaTiO_3$ after sintering at the temperature as low as 1200° C. On the other hand, the relative density of $BaTiO_3$-based nanocomposites with particulate SiC strongly depended upon the sintering temperature. The density gradually decreased, as the SiC content increased, to 98, 97 and 96% for $BaTiO_3$-1, 3 and 5 vol%SiC nanocomposite sintered at 1300°C for 1h, respectively.

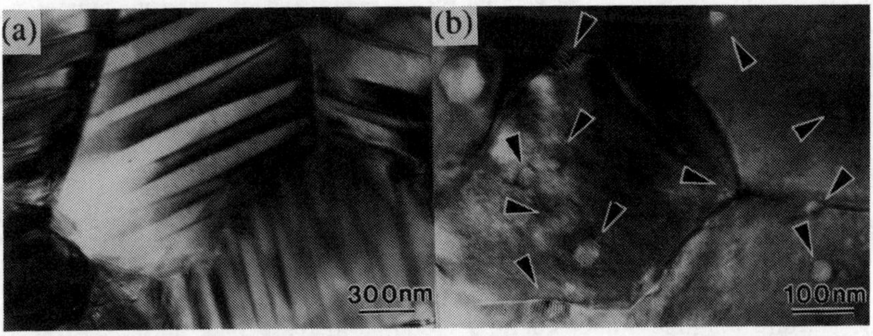

Figure 1. TEM photographs of monolithic $BaTiO_3$(a) and $BaTiO_3$-5vol% SiC nanocomposite(b) showing the typical domain structure of $BaTiO_3$ and intergranular SiC particle, respectively. Arrow indicates SiC particle.

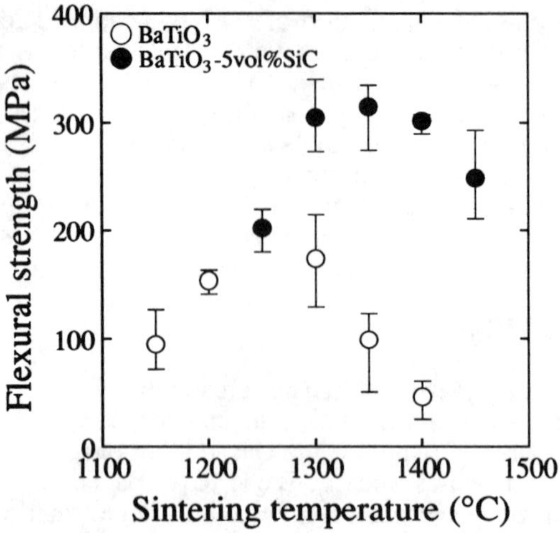

Figure 2. Flexural strength of monolithic $BaTiO_3$ and $BaTiO_3$-5vol%SiC nanocomposite as a function of sintering temperature.

Figure 2 shows the fracture strength variation as a function of sintering temperature for monolithic $BaTiO_3$ and $BaTiO_3$-5vol%SiC nanocomposites. The fracture strength increased with the sintering temperature up to about 1300°C for monolithic $BaTiO_3$. It began to decrease when the sintering temperature raises up to 1300° C. These fracture strength change observed for monolithic $BaTiO_3$ is mainly due to the densification and the gradual grain growth with respect to the sintering temperature. However, there is no change in fracture strength for $BaTiO_3$- 5vol%SiC composite even when the sintering temperature was increased up to 1400°C.

Some physical and mechanical properties of $BaTiO_3$ and $BaTiO_3$ nanocomposites with SiC sintered at 1300°C for 1h are summurized in Table I. Monolithic $BaTiO_3$ remained in the ferroelectric tetragonal phase regardless of sintering temperature. The crystal structure of $BaTiO_3$ gradually changed from tetragonal to cubic phase with an increase in SiC content. It was found that the crystal structure of $BaTiO_3$-3 and 5 vol%SiC composite, in particular, was not tetragonal any more, so we called this phase pseudo-cubic. Mechanical properties significantly changed with the addition of SiC. Fracture strength, in particular, increased with an increase in particulate SiC content. This was attributed to inhibition of grain growth by the SiC addition. The average grain size gradually decreased with increasing SiC content, which shows that SiC is very effective for controlling the grain structure of $BaTiO_3$ ceramics. It is also suggested that the increase in fracture toughness and Young's modulus contributed to the fracture strength improvement. As can be seen in Table I, the fracture toughness and Young's modulus increase with an increase in SiC content. This value are higher than those expected from the rule's of mixture. It must to be due to crystal structure change by the SiC addition.

Table I. Physical and Mechanical Properties of $BaTiO_3$ and $BaTiO_3$ Nanocomposites with Paticulate SiC.

SiC content (vol %)	Sintering temp. (°C)	Crystal# structure	Fracture strength (MPa)	Grain size (μm)	Fracture tougness (MPa $m^{1/2}$)	Young's modulus* (GPa)
0	1300	Tet	174	1.35	0.86†	94.0
	1400	Tet	46	---		
1	1300	Tet	301	0.84	0.99†	129.1
	1400	Hexa, Tet	45	Bimodal struc.		
3	1300	P-Cub	350	0.48	1.22§	141.2
	1400	P-Cub, Hexa	107	---		
5	1300	P-Cub	305	0.35	1.19†	141.4
	1400	P-Cub	248	---		

Tet=tetragonal; P-Cub=pseudo-cubic; Hex=Hexagonal and Hex, Tet=major phase is hexagonal
† CSF(Controlled Surface Flaw) method § Indentation Fracture(IF) method
* Resonance method

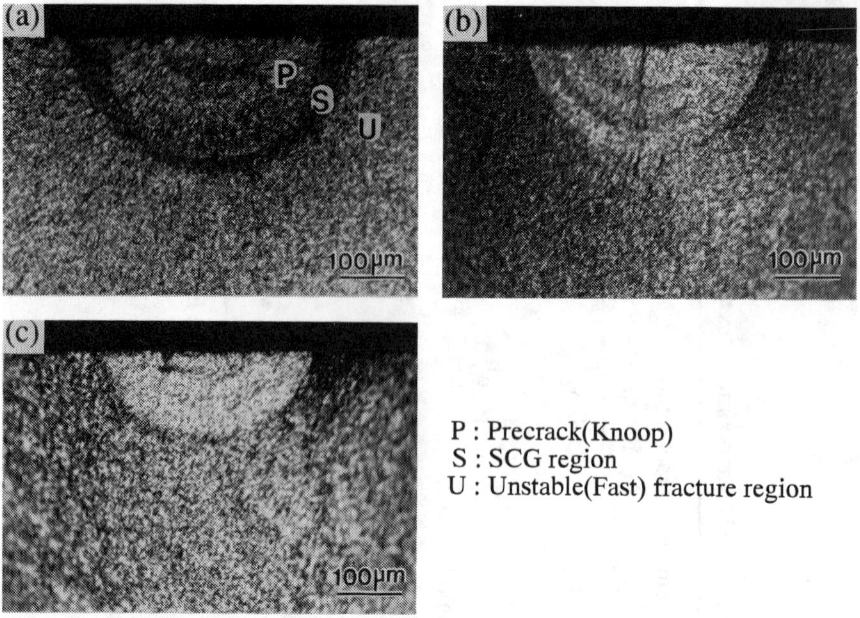

Figure 3. Fracture surfaces of the monolithic $BaTiO_3$(a), $BaTiO_3$-1vol%SiC (b) and $BaTiO_3$-5vol%SiC(c) nanocomposite precracked at 19.6N by knoop indentor.

Figure 3 shows the fracture surface of the specimens indented at 19.6N for monolithic $BaTiO_3$(a), 1 vol%SiC(b) and 5 vol%SiC(c) nanocomposites. The precracked region, as can be seen in Figure 2, becomes narrow as the SiC content increase. The dark region, which is thought to be subcritical crack growth(SCG) region, was observed between the precracked and unstable fracture regions for monolithic $BaTiO_3$ and $BaTiO_3$-1 vol%SiC. This SCG region also became smaller with an increase in the SiC content, and consequently was not observed for $BaTiO_3$-5 vol% SiC.

Generally, the fracture mode of monolithic $BaTiO_3$ has been known to be intragranular. In the present study, the fracture was intragranular regardless of the sintering temperature and the SiC dispersion. But intergranular fracture was observed at and near fracture origins for monolithic $BaTiO_3$ and $BaTiO_3$-1 vol%SiC nanocomposite. Figure 4 shows the high magnified fracture surface near fracture origin(a) and the boundary between the inter- and intragranular fracture(b) for monolithic $BaTiO_3$. In Figure 4, the arrows indicate the boundary between the intergranular and intragranular fracture. The intergranular fracture observed in Figure 3(a) is, however, limited near the fracture origin. It can not be seen in any other places except for the fracture origins. Thus, it is suggested that the intragranular fracture occurred before a brittle(fast) fracture in monolithic $BaTiO_3$.

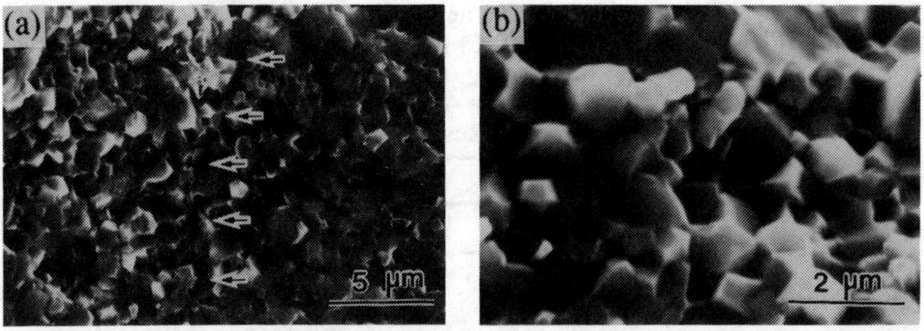

Figure 4. Typical fracture surfaces of monolithic $BaTiO_3$ showing the boundary of intergranular and intragranular fracture(a) and the high-magnified image of intergranular fracture observed at the fracture origins.

Dielectric Properties

As will be expected, there must be numerous oxygen vacancies in the sintered $BaTiO_3$ and $BaTiO_3$-based nanocomposites due to the reduced atmosphere sintering in the present study, resulting in semiconductive $BaTiO_3$ ceramics having a resistivity of 10^{-1} to $10^{-2}\Omega$ cm regardless of particulate SiC content, as shown in Table II. From the Hall constant measurement, the sintered body was a n-type semiconductor, and the conducting carrier was electron, which was caused by oxygen vacancies.

In order to fabricate dielectrics from the semiconductive monolithic $BaTiO_3$ and $BaTiO_3$-based nanocomposites, the oxidization of the sintered bodies was carried out. Their resistivity results are shown in Table II together with results for as-sintered specimens. Dielectrics having a higher resistivity than $10^{10}\Omega$ cm can be obtained. After oxidization, monolithic $BaTiO_3$ exhibits a bright yellow color which can usually be seen in typical and dense $BaTiO_3$ ceramics. In addition, the oxidization progressed throughout the sintered body. On the other hand, $BaTiO_3$-based nanocomposites have a thin oxidized layer on the surface. Consequently, the reoxidized type capacitor was fabricated.

The temperature dependence of the relative dielectric constant for $BaTiO_3$ and $BaTiO_3$ nanocomposites oxidized at 1050°C for 3h is shown in Figure 5. As can be seen in Figure 5, the monolithic $BaTiO_3$ exhibits the same behavior as found in small-grained $BaTiO_3$ ceramics [13]. The relative dielectric constant is 4000-6000 at room temperature. For $BaTiO_3$-based nanocomposites, the dielectric constant is lower than that of monolithic $BaTiO_3$. Moreover, the dielectric constant at room temperature gradually decreased with SiC content. Another aspect to take into account is the temperature independence of the relative dielectric constant. For pure $BaTiO_3$ ceramics, the tetragonal to cubic transformation around 120° C causes a sharp peak value in the dielectric constant at this Curie temperature. However, fairly good temperature independence of the dielectric constant is found for all of $BaTiO_3$-based nanocomposites.

Table II. Resistivity of $BaTiO_3$ and $BaTiO_3$ nanocomposites for the specimens sintered at 1300° C in Ar atmosphere and oxidized at 1050°C, respectively.

SiC content (vol%)	Resistivity (Ω cm)			
	As-sintered§	Sinter- and oxidized#		
		0	1	3†
0	0.113	4.74×10^{12}	4.30×10^{12}	7.82×10^{11}
1	0.021	3.16×10^{10}	7.00×10^{9}	6.16×10^{10}
3	0.015	3.32×10^{10}	2.02×10^{10}	1.15×10^{10}
5	0.013	5.48×10^{10}	3.69×10^{10}	2.54×10^{10}

§ Hall constant(Van der Pauw method) # Two probe method(applied voltage:25V) † oxidization time(hour)

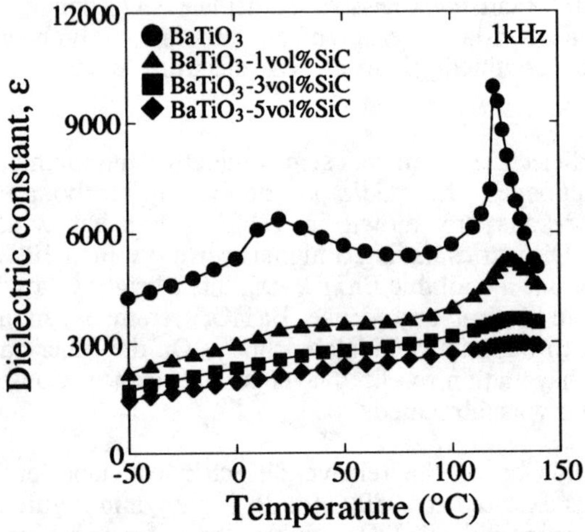

Figure 5. Temperature dependence of the relative dielectric constant for monolithic $BaTiO_3$ and $BaTiO_3$-SiC nanocomposites sintered at 1300° C in Ar atmosphere and oxidized at 1050°C for 3h in air.

CONCLUSION

Dense and small grained $BaTiO_3$ and $BaTiO_3$-based nanocomposites with particulate SiC were fabricated. In addition, two types of ceramic capacitors, reoxidized and grain boundary barrier layer(GBBL), were also prepared by oxidizing the sintered specimens in air. Mechanical properties, particularly fracture strength of $BaTiO_3$, were significantly improved by adding nano-sized SiC particulate. This may be mainly attributed to the depression of grain growth of $BaTiO_3$ and crystal structure variation. The SCG phenomenon was observed at room temperature in monolithic $BaTiO_3$ and $BaTiO_3$-1vol%SiC composite. By increasing SiC the SCG phenomenon disappeared, and the fracture type was completely intragranular. It is thought that the depression of SCG phenomenon by SiC addition is another reason for fracture strength improvement.

Acknowledgement
The authors gratefully acknowledge A.Nakahira and T.Sekino at Osaka University for their discussion and help. A part of this work was supported by the Japan Ministry of Education under a Grant-in-Aid for Science Research No. 03555157 and 04205099, and by the cooperative project "Advanced Materials Creation and Their Limit State Prediction for Environmental Preservation" between ISIR and WRI in Osaka University.

REFERENCES

1. Y.M.Chiang and T.Takaji, "Grain Boundary Chemistry of Barium Titanate and Strontium Titanate:II, Origin of Electrical Barriers in Positive-Temperature-Coefficient Thermistors,"J. Am. Ceram. Soc., 73(11) 3286-91 (1990).
2. H.Kanai, O.Furukawa, S.Nakamura and Y.Yamashita, Effect of Stoichiometry on the Dielectric Properties and Life Performance of $(Pb_{0.875}Ba_{0.125})[(Mg_{1/3}Nb_{2/3})_{0.5}(Zn_{1/3}Nb_{2/3})_{0.3}Ti_{0.2}]O_3$ Relaxer Dielectric Ceramic : Part I, Dielectric Properties," J. Am. Ceram. Soc., 76(2) 454-58 (1993).
3. H.Emoto and J.Hojo, "Sintering and Dielectric Properties of $BaTiO_3$-Ni Composite Ceramics," J. Ceram. Soc. Jap., 100(4) 555-59 (1992).
4. S.Tashiro, H.Igarashi and K.Okazaki, "Mechanical Strength of Several Kinds of Piezoelectric Ceramics," J. J. Appl. Phys., Supplement 20-4, 197-200 (1981)
5. S.K.Nag and D.C.Agrawal, "Piezoelectric and Mechanical Properties of Ceria-Doped Lead Zirconate Titanate Ceramics," J. Mat. Sci., 27(8) 4125-30 (1992).
6. R.C.Pohanka, S.W.Freiman and B.A.Bender, "Effect of the Phase Transformation on the Fracture Behavior of $BaTiO_3$," J. Am. Ceram. Soc., 6(1-2) 72-75 (1978).
7. R.C.Pohanka, R.W.Rice and B.E.Walker.JR, "Effect of Internal Stress on the Strength of $BaTiO_3$," J. Am. Ceram. Soc., 59(1-2) 71-74 (1976).
8. J.G.Bruce, W.W.Gerherich and B.G.Koepke, "Subcritical Crack Growth in PZT," in Fracture Mechanics of Ceramics, vol 4. edited by R.C.Bradt, A.G.Evans, D.P.H.Hasselman and F.F.Lange, Plenum Press, New York, 687-709 (1985).
9. K.Niihara, "New Design Concept of Structural Ceramics-Ceramic Nano-Composites," J. Ceram. Soc. Japan, 79(10) 974-82 (1991).
10. H.J.Hwang and K.Niihara," Dielectric Properties of $BaTiO_3$-based nano-composites, " in contribution.
11. J.J.Petrovic and M.G.Mendiratta, "Fracture from Controlled Surface Flaws," pp 83-102 in Fracture Mechanics Applied to Brittle Materials. ASTM STP 678. Edited by S.W.Frieman. American Society for Testing and Materials, Philadelphia,

PA (1979).
12. K.Koumoto, H.Tagawa, T.Nakano, S.Takeda, H.Yanagida, "TEM Observation of the Grain Boundaries in a Semiconducting Barium Titanate Thick Film," J. J. Appl. Phys., 23(5) L305-L307 (1984).
13. A.Yamaji, Y.Emoto, K.kinoshita and T.Murakami, "Preparation, Characterization and Properties of Dy-Doped Small-Grained $BaTiO_3$ Ceramics," J. Am. Ceram. Soc., 60(3-4) 97-101 (1977).

EFFECTS OF NANO-SIZED FILLER DISPERSION ON MECHANICAL PROPERTIES OF CARBON FIBER REINFORCED EPOXY COMPOSITES

M.Hussain, Y.Oku, A.Nakahira, S.Ueda and K.Niihara
The Institute of Scientific and Industrial Research, Osaka University
8-1 Mihogaoka, Ibaraki, Osaka-567.

ABSTRACT

A filament winding technique was applied for fabricating carbon-fiber unidirectionally-reinforced epoxy composites (CFRP). Nano-sized Al_2O_3 ceramic particles were dispersed into the epoxy matrix to fabricate carbon-fiber hybrid-reinforced epoxy composites (CFHRP). Their mechanical properties were evaluated to investigate the effect of nano-sized filler dispersion on the mechanical properties of CFRP. Young's modulus, flexural strength, fracture toughness, and interlaminar shear strength were significantly improved by incorporating 10 vol% of nano-sized Al_2O_3 particles. Fracture surfaces were examined by scanning electron microscopy to observe the role of filler particles into the matrix/carbon fiber interface.

INTRODUCTION

In order to improve the mechanical properties of epoxy composites the incorporation of carbon fiber into the matrix has attracted much attention in various fields[1-3], specially in the automobile and aerospace industries, because of their high fracture toughness, heat resistance, and environmental stability[4-5]. Addition of continuous carbon fiber into the matrix improves not only the brittleness of epoxy materials but also remarkably improves mechanical properties such as the fracture toughness, fracture strain, and Young's modulus. However, there have been few studies on the hybrid effect of incorporation of ceramic filler into the carbon fiber reinforced composite on the mechanical properties and fiber matrix interface. In the previous work [6] we succeeded to disperse nano-sized ceramic particles into the matrix to significantly improve the mechanical properties of the particulate filled epoxy composite.

In this experiment we succeeded to disperse ceramic filler particles into the epoxy matrix and fabricated carbon-fiber hybrid reinforced epoxy composites (CFHRF) which exhibited higher mechanical properties compared to carbon fiber reinforced composites without dispersed filler particles (CFRP).

EXPERIMENTAL PROCEDURE

Materials and Methods

The epoxy matrix used in the fabrication of CFRP was N.N.N'N'-tetra-glycidyl-methoxy-di-amine (TETRED-X) produced by Mitsubishi Gas Chemical Co.Ltd. 1,2-cyclo-hexane-dicarboxylic anhydride (HHPA) from Wako Jyunyaku Company, Japan was used as curing agent. Carbon fiber manufactured by Toho Rayon Co. Ltd. was selected as a fiber reinforced component in this experiment. γ-Al_2O_3 particles with an average particle size of 25 nm from Asahi Chemical Company were used as filler particles.

CFRP composites were fabricated by the filament-winding technique. A simple picture of the filament winding method is shown in Figure 1. Unidirectional fibers were pulled through the resin bath (mixture of epoxy and hardner) and wound onto a mandrel (100 mm in diameter) at a speed depending upon the volume fraction of fiber content into the composite. After winding was completed, the fiber/epoxy flat composite was cut into 100x100mm segments and stacked one upon another. The fiber composite was then pressed at 1.3 MPa and pre-cured at 90°C for 2 hr. After pre-curing, the press mold was post-cured at 180°C for 2 hr. Composites were then cut into 2x8x80mm pieces and ground for mechanical tests. During the CFHRP composite preparation the resin bath was filled with a mixture of epoxy resin/hardner and 10 volume percentage of Al_2O_3 filler in this experiment. Following procedure was same as CFRP composite fabrication. The mixing method used to obtain a homogeneous dispersion of 25 nm Al_2O_3 fillers is described in our previous work [6].

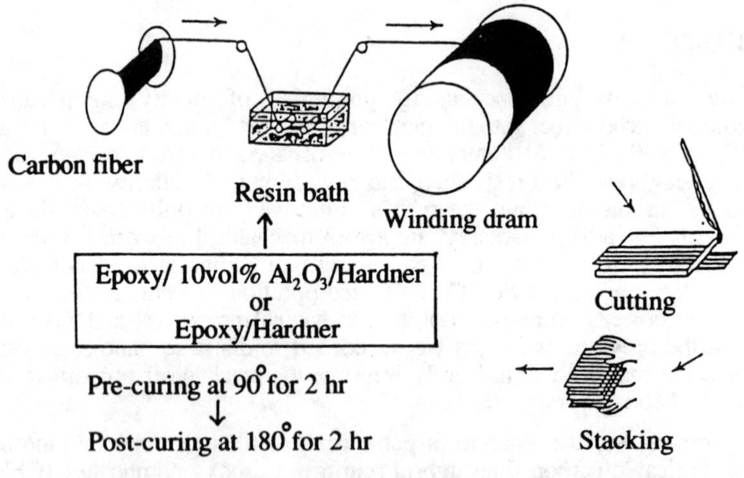

Figre 1 Fabrication procedure of carbon fiber reinforced composites with and without cermic fillers (CFHRP and CFRP) by a filament winding method.

Characterization of composites

The mechanical properties of CFRP and ceramic particle dispersed CFHRP composites were evaluated by measuring Young's modulus, flexural strength, interlaminar shear strength, and fracture toughness. Young's modulus was estimated using the flexural vibration resonance method at room temperature. The flexural strength tests were done using a 3-point bending method at a cross head speed of 2mm/min at room temperature. Interlaminar shear strength was performed according to ASTM D-2344. Fracture toughness was measured according to ASTM E 399 using the SENB method. Five specimens were used for determination of each mean value. SEM observation was performed to study the fracture behavior, fiber matrix adhesion/bonding, fiber matrix interface, fracture nature of carbon fiber, and fiber pullout.

RESULTS AND DISCUSSIONS

Figure 2 shows the variation of flexural strength of the composites with the fiber content. It was found that in each case the flexural strength of CFHRP composite was higher than those of CFRP composites without dispersed Al_2O_3 particles. The maximum flexural strength 1619 MPa was obtained for 49 vol% carbon fiber content CFHRP composites. On the other hand 1500 MPa was obtained for 50 vol% carbon fiber content CFRP composites.The load deflection patterns are shown in Figure 3. The crack arrest area was larger in ceramic filler dispersed CFHRP composites. Addition of Al_2O_3 particles into the epoxy matrix creates a roughness on the fiber surfaces due to mechanical binding with epoxy which later acts as a mechanical interlock between the fiber and the epoxy resin. When a load

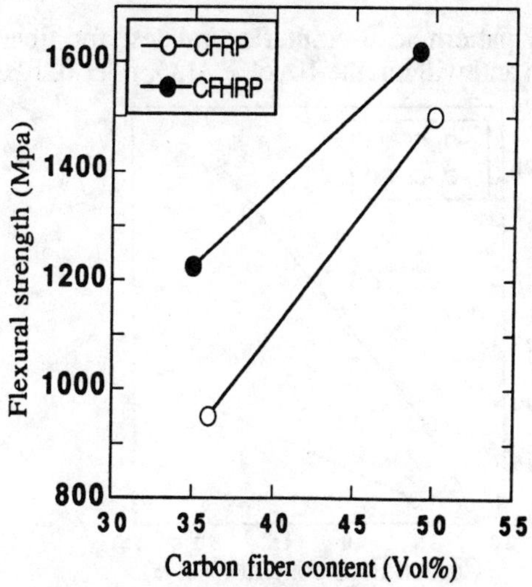

Figure 2 Flexural strength as a function of carbon fiber content for fiber reinforecd composites with and without the 10vol% Al_2O_3 filler (CFHRP and CFRP).

is applied, Al_2O_3 particles prevent carbon fiber sliding in the loaded direction and thus the applied load must be increase which contributes to a higher mechanical strength. This high strength is probably due to the strong interfacial adhesion bonding with the carbon fiber and the presence of Al_2O_3 particles. The particles prevent crack propagation and thus increases crack arrest area.

Interlaminar shear strength (ILSS) as a function of fibre content is shown in Figure 4. The ILSS increases when nano-sized Al_2O_3 particles are added into the CFRP composite. The interlaminar shear strength is applied on CFRP composite to determine the strength of fibre/resin adhesion bonding. The method is designed to cause the composite to fail in shear at the fibre/resin interface [7]. For the composite with 50 vol% fibre in addition of Al_2O_3 particles, the ILSS increases about 20%. Thus it is thought that the incorporation of ceramic filler into the CFRP composite increases the fibre/epoxy interfacial adhesion bonding

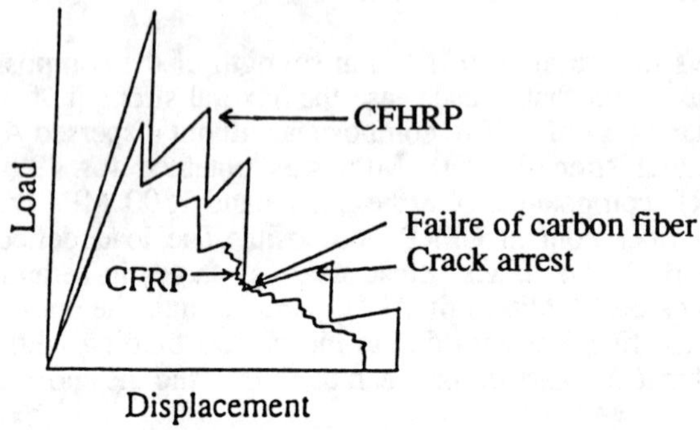

Figre 3 Load difflection pattern in 3-point flexural test for fiber reinforced composites with and without the 10vol% Al_2O_3 filler (CFHRP and CFRP).

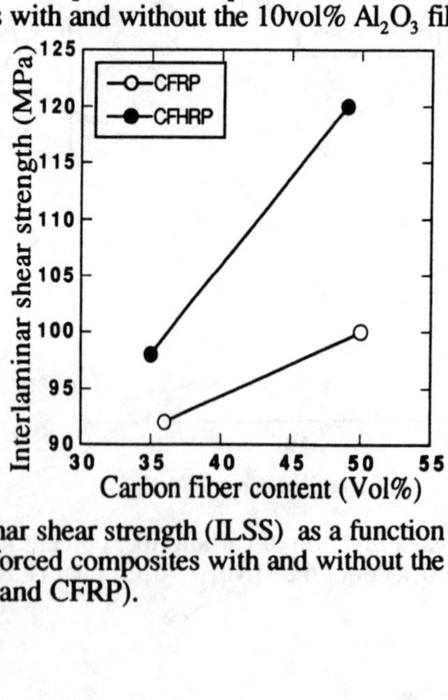

Figure 4 Interlaminar shear strength (ILSS) as a function of carbon fiber for the fiber reinforced composites with and without the 10vol% Al_2O_3 filler (CFHRP and CFRP).

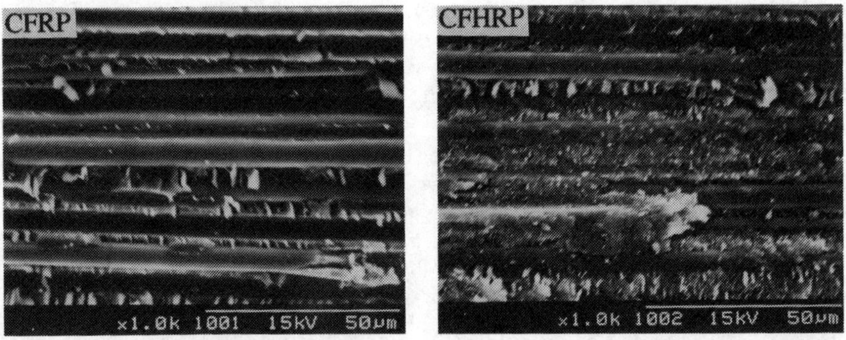

Figure 5 SEM micrographs of fracture surfaces after ILSS tests for the fiber reinforced composites with and without the 10vol% Al_2O_3 filler (CFHRP and CFRP).

which leads to an increasing ILSS in the composite. Figure 5 shows the SEM micrographs for both composites with and without dispersed Al_2O_3 particles after ILSS tests. It is observed clearly observed that the surface of carbon fibers in CFRP is almost smooth, which suggests that the fiber/epoxy adhesion bonding is not strong. On the other hand, there is a high degree of epoxy/particle adhesion bonding on carbon fibers in CFHRP as the rough and epoxy/Al_2O_3 rich area is observed on the surfaces of carbon fibers. The rough surface area of the fiber enhances the mechanical interlocking and thus the flexural and ILSS values increase for Al_2O_3 dispersed CFRP composites. Brejnick et al [8] reported that the surface topography of fibers played a very important role in ILSS of fiber reinforced composites and the 20% ILSS improvement was observed for roughened fibers.

Figure 6 shows the fracture toughness values of composites as a function of carbon fiber content. It is observed that toughness value increases more than 150% when Al_2O_3 particles are added to CFRP composites for 49 vol% fiber content. Fracture toughness also increases with the addition of Al_2O_3 particles for 36 vol% fiber content. It is well known that the fracture toughness (K_{1c}) increases with increasing filler volume fraction. For the poor filler-matrix adhesion/bonding, a linear variation of K_{1C} versus volume fraction is observed, whereas a maximum K_{1C} occurs for good adhesion [9-11] . Matrix toughness also plays an important role in determining the mechanical properties, especially the toughness, of polymer composites. Recently Gupta et al [12] studied the energy absorbing mechanism during fracture of fiber reinforced composites and suggested that the fracture toughness was a result of the debonding of the interface, the sliding or pullout of fibers, and plastic deformation of the matrix. Toughening methods and the various factors which affect the enhancement of toughness is recently given by Kim et al[13]. Figure 7 shows the surface morphology of the carbon fiber . The fiber surface contains an Al_2O_3/epoxy adhesion component in Al_2O_3 dispersed CFRP composites. On the other hand, the pullout fiber in CFRP composites is located individually containing almost no epoxy matrix on the surface. Therefore higher K_{1C} value for CFHRP composites is due the combination of higher K_{1C} value of epoxy by the addition of nano-sized Al_2O_3 particles and the followings strong bonding and mechanical interlocking between filler/epoxy and carbon fiber.

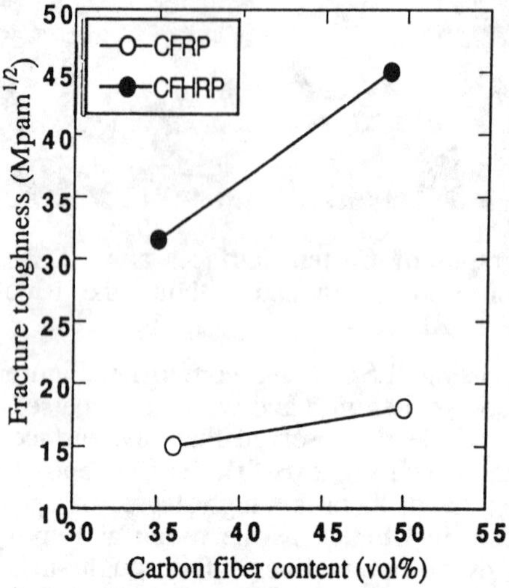

Figure 6 Fracture toughness as a function of fiber content for the fiber reinforced composites with and without the 10vol% Al_2O_3 filler (CFHRP and CFRP).

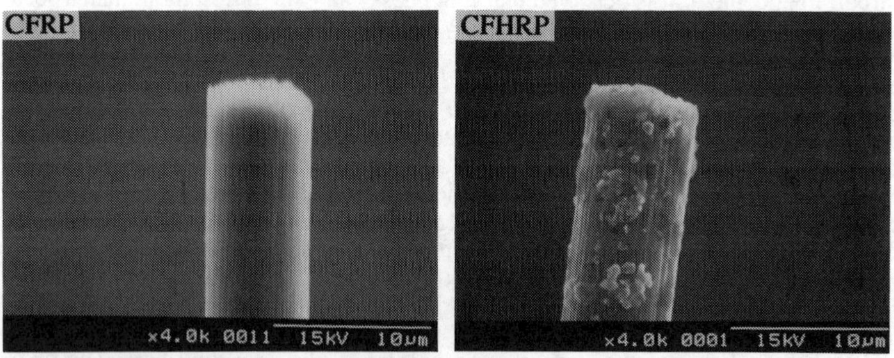

Figure 7 SEM micrographs of surface morphology on carbon fibers for the fiber reinforcd composites with and without the 10vol% Al_2O_3 filler (CFHRP and CFRP).

CONCLUSION

Effects of addition of 10 vol% nano-sized Al_2O_3 particles into CFRP composites on the mechannical properties have been studied. Mechanical properties increased significantly by the addition of filler particles into the CFRP composites. These improvements were mainly rationalized by the observed strong mechanical interlocking between filler/epoxy and carbon fiber, which was induced by the Al_2O_3 filler particles.

Acknowledgement
The authers gratefully acknowledge T.Sekieno at Osaka university for their discussion and help. A part of this work was supported by the Japan Ministry of Education under a Grant-in-Aid for Science Research No.03555157 and 04205099, and by the cooperative project " Advanced Materials Creation and Their Limit State Prediction for Environmental Preservation" between ISIR and WSI in Osaka University.

References

1] J.A.Rodosta and N.C.Trivedi, in "Handbook of Fillers and Reinforcements of Plastic", edited by H.S.Katz and J.V.Milewski (Van Hastrand Reinhold, New York,1987.
2] Z.D.Jastrzebski, " The Nature and Properties of Engineering Materials" 2nd Edn, (Wiley, New York,1977) ch.14.
3] R.B.Seymour and G.S.Kirshenbaun, " High Performance Polymers, Their Origin and Development " (Elsevier,New York,1986)
4] G.Lubin, " Handbook of Composites " (Von Nostrand, London,1982).
5] K.M.Gill, " Carbon Fibers in Composite Materials" (The Plastics Institute,London,1972)
6] M.Hussain, Y.Oku, A.Nakahira, K.Niihara " Study On the Dispersion of Ceramic Fillers and the Effect of Particle Size on Mechanical Properties of Particulate Filled Epoxy Composites " in preperation.
7] J.B.Sturgeon " Specimen and Test Method for Carbon Fiber Composites" Royal Air Craft Establishment, UK TRT1026(1971).
8] M.Breznick et al " Surface Treatment Technique for Aramid Fibers" Polym.Commun 28(1987)55.
9] A.C.Moloneg,H.H.Kausch, T.Kaiser and H.R.Bear, " Parmeter Determining the Strength and Toughness of Particulate Filled Epoxide Resins"J.Mater.Sci 22(1987)381.
10] J.Spanoudakis and R.J.Young, " Crack Propagation in Glass Particle-Filled Epoxy Resin, Part-1" J.Mater.Sci 19(1984)473.
11] Idem ibid, " Crack Propagation in Glass Partcle-Filled Epoxy Resin, Part-2,19(1984) 487.
12] V.B.Gupta, R.K.Mittal and Malligoel," Energy Absorbing Mechanism in Short-Glass-Fiber Reinforced Polypropylene" Compos.Sci .Technol.37(1990)353.
13] J.K.Kim and Y.W.Mai " High Strength, High Fracture Toughness Fibre Composite with Interface Control", Compos.Sci.Technol 41(1991).

EFFECTS OF SECOND PHASE PARTICLES ON INTERFACE AND GRAIN BOUNDARY STRUCTURE IN MgO-BASED CERAMICS

Y.H. Choa, S. Ueda and K. Niihara
The Institute of Scientific and Industrial Research, Osaka University
8-1 Mihogahoka, Ibaraki-shi, Osaka 567, Japan

ABSTRACT

Novel MgO/SiC nanocomposites, in which nano-sized SiC particles were incorporated mainly into the matrix grains, were fabricated by the normal powder metallurgical method. The micro/nano structure, including grain boundary and interface structure, was investigated by transmission electron microscopy and scanning electron microscopy. Observations showed that the matrix grain size, grain boundary and dislocation motion were controlled by the addition of nanometer-sized SiC particles. Significantly improved strength and toughness were observed at room to high temperatures for this MgO/SiC nanocomposites. These significant improvements of mechanical properties were attributed to the dispersion of nanometer-size second phase particles.

INTRODUCTION

There are various important variables in the microstructure that influence the mechanical properties of polycrystalline ceramics [1, 2, 3], but this paper mainly covers the effects of interface and grain boundary structures. Grain boundaries and/or interfaces are well known to be critically important in determining the mechanical properties of ceramic materials. Dependent on their structure and composition, for example, they are often preferred to act as fracture paths [4, 5], may control the creep behavior through interfacial sliding [6], and often provide paths for rapid diffusion [7]. In ceramic matrix composites, the grain boundary and interface structures are also well known to be crucial in fiber-matrix debonding, crack deflection, and microcracking [8, 9], as well as controlling fiber and whisker pull-out during eventual failure [10].

A number of room and high temperature fracture studies have been conducted on polycrystalline MgO. A.G.Evans et al. [11, 12] reported the effects of grain size, grain boundary, surface flaws and dislocation motion for MgO ceramics. They explained that the importance of grain boundaries by measuring the magnitude of the surface flaws produced by abrasion. When the existing flaws are small, the fracture is initiated by dislocation pile-ups according to the Stroh model. K.Yasuda et al. [13] investigated the effect of grain size and temperature on fracture toughness for MgO ceramics. They found that the increase in toughness with increasing temperature for the fine grained structure corresponded to the macroscopic ability of plastic deformation and the grain boundaries controlled the growth of microcracks at high temperatures. However, there seems to be no studies concerning the effect grain boundary/interface structure on mechanical properties for the MgO based composites.

In this paper, firstly the influence of second phase particles on the grain boundary and interface structure of MgO with SiC dispersoids was discussed. Secondly, the formation of microcrack, fracture morphology and mechanical properties at room and high temperatures were investigated for MgO monolith. Finally, the microstructure of MgO/SiC nanocomposites was observed and their mechanical properties were investigated.

EXPERIMENTAL PROCEDURE

Specimen Preparation

The following starting materials have been used: MgO powder (Ube Industries Ltd.) with an average particle size of 0.1 µm and SiC powders (Mitsui Touatu Chemical Ltd) with an average particle size of 0.15 µm. MgO and MgO/SiC specimens containing 3～50 vol% SiC were fabricated using hot press sintering route. MgO and MgO/SiC powders were mixed in n-butanol by the conventional ball milling method using SiC ball in the plastic pot for 24 hr. Then the dried powders were also mixed by the ball-milling for 40 hr. The mixed powers were hot-pressed in a graphite die with a 44 mm in diameter. Hot pressing was performed at temperatures ranging from 1400°C to 1900°C under 30 MPa applied pressure for 60 min in Ar atmosphere. Sintered materials were cut and ground into rectangular bar specimens (3 x 4 x 36 mm). For bending tests and microscopic observation, the samples were polished using diamond paste up to a 1/2 µm finished.

Evaluation

Densities of hot-pressed specimens were measured by the Archimedes method using toluene. The phase analysis was made by X-ray diffraction. The microstructure was observed using scanning electron microscopy (SEM) and transmission electron microscopy (TEM). Fracture strength was estimated by the three point bending test (span size; 30 mm) under the conditions of crosshead speed 0.3 mm/min at room temperature up to 1400° C in air atmosphere. The fracture toughness was estimated by the indentation fracture (IM) [14] and controlled surface flaw (CSF) methods[15, 16] at the same conditions as the fracture test. IM and CSF cracks were induced at a load of 49 N and loading duration of 15 second using a Vickers hardness tester. In the case of the CSF method, the careful surface polishing was carried out to remove the tensile residual stresses associated with indentation zone .

RESULTS AND DISCUSSION

General

The results of X-ray analysis revealed that the components of MgO/SiC composites hot-pressed at 1500°C to 1900°C were only MgO and β-SiC without any reaction phases. For monolithic MgO, highly dense (>99 %) and fine grained (average grain size; ~4 μm) MgO ceramics were obtained at 1400°C. However, MgO/5 vol% SiC composite hot pressed at 1400° C showed much lower density than monolithic MgO. Fully dense (>99 %) MgO/5vol%SiC composites were obtained at 1750° C. This depression of MgO sintering may be caused by : the limitation of diffusional transport paths, grain boundary sliding and plastic deformation during hot press sintering by addition of SiC particle.

Table I shows the variation of density and grain size with volume fraction of SiC content. As expected, the lowest density was obtained for samples containing the highest amount of particulates at 1850° C. The SiC presence in the MgO matrix appears to prevent grain boundary migration and to limit the grain growth. Table I also indicates that the grain size of the MgO matrix drops drastically from ~60 μm for pure MgO to less than ~ 2 μm for the 5 vol% SiC addition. This drastic reduction in grain size with the addition of SiC particles revealed that the SiC particles play a strong inhibiting role for the MgO grain growth.

Table I. Relative density and matrix grain size for MgO and MgO/SiC composites sintered at 1850°C.

SiC content (vol%)	Relative density (%)	Matrix grain size (μm)
0	99.95	64.5
0 *	99.9	4.2
3	99.8	2.4
5	99.7	2.0
10	99.3	1.6
30	98.3	0.51
50	95.3	---

* sintered at 1400°C

Nanostructure and MgO/SiC Interface

TEM observations revealed that most of the SiC particles were uniformly distributed within the matrix grain, with some larger particles located at the MgO grain boundaries, as shown in Figure 1. The closed pores were not observed at grain boundaries or within MgO matrix grains for dense MgO/SiC nanocomposites, whereas the intragranular pores were trapped in the MgO grains for dense monolithic MgO. These results imply that the grain boundary rapidly moved for monolithic MgO during the final stage of sintering. Figure 2 shows the formation process of the nanostructure for these composites. This observation means that the

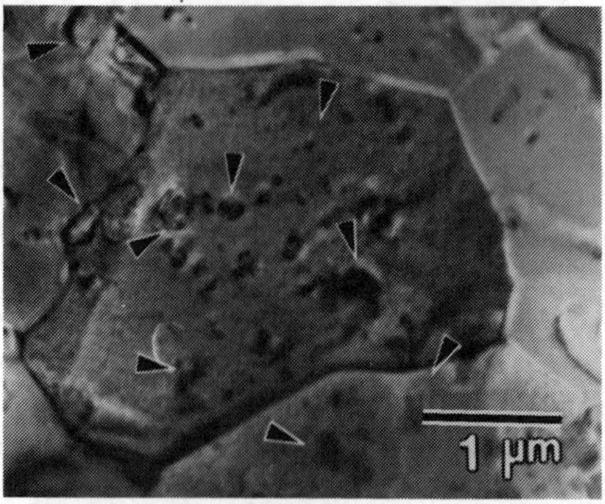

Figure 1. TEM photograph of the MgO/5vol%SiC nanocomposite prepared at 1850°C (Allows indicate the SiC particles).

Figure 2. Formation process of nanostructure of the MgO/SiC nanocomposite. (Allows indicate the SiC particles and moving direction of grain boundary).

intragranular SiC particles within the MgO matrix grains are caused by the break-away of the grain boundary from the pinning force of SiC particles like the pores trapped within grains. Kibbel and Heuer [17] discussed the break-away condition for particles located at the grain boundary, based on the Zener equation. The critical size of the intragranular SiC particle will depend on the grain boundary mobility, the grain boundary diffusion coefficient, the curvature of the matrix grains, the grain boundary thickness, and the temperature as given by Hsueh and A.G. Evans [18]. It was, in fact, observed that the fine particles were more easily trapped within MgO matrix grains at lower hot-pressing temperatures than the larger particles. Figure 3 is a TEM lattice image of the interface for the MgO/SiC nanocomposite, showing good compatibility between MgO and SiC and little sign of any reaction, except for the regular existence of dislocations at the interphase.

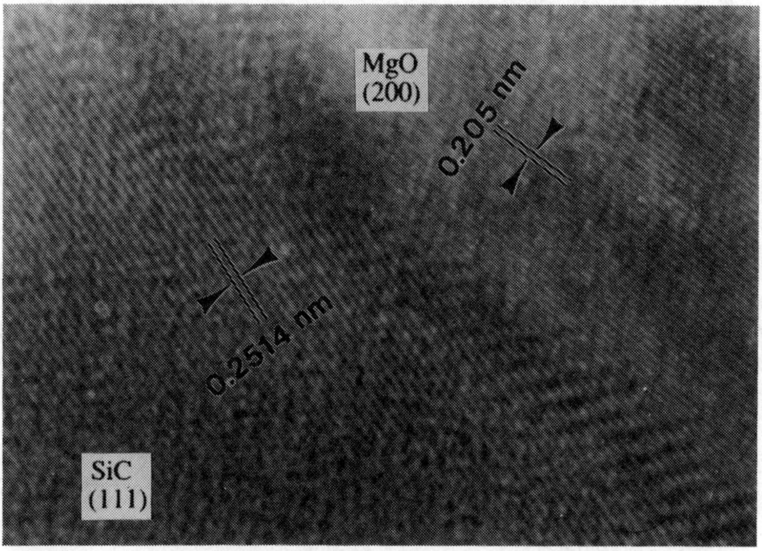

Figure 3. High resolution transmission electron micrograph of the MgO/SiC interface.

Mechanical Properties of Monolithic MgO

The bending strength as a function of grain size on $G^{-1/2}$ axis is plotted in Figure 4 for monolithic MgO ceramics. The fracture stress decreases with an increase in grain size and a linear relationship extrapolating to 150 MPa at infinite grain size is obtained. This suggests that the fracture is initiated from flaws produced by plastic flow (dislocation motion) as reported by Evans et al.[11].

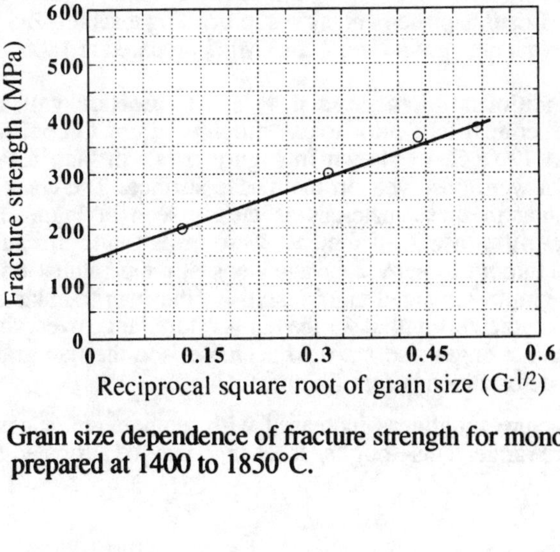

Figure 4. Grain size dependence of fracture strength for monolithic MgO prepared at 1400 to 1850°C.

Figure 5 shows typical load/displacement curves at various test temperatures for fine-grained MgO (average grain size; 4 μm) . Yielding and nonlinear behavior were observed at temperatures above 800° C. The larger-grain-sized (~ 60 μm) MgO, on the other hand, exhibited the apparent brittle behavior up to 1100° C. At higher temperatures, slight yielding was observed before fracture.

Fracture strength slightly decreased up to about 800°C with increasing temperature, as shown from the load/displacement plots of Figure 5. Above 800° C, however, a significant and rather sharp decrease in strength was observed, almost coincident with the observed plastic deformation before fracture. These observations suggest that the high-temperature strength behavior of polycrystalline MgO is strongly related with the dislocation movement, as reported by Evans et al. [12]. Above about 800°C, therefore, the plastic deformation can occur throughout the specimen in favorably oriented grains, before the necessary stress to extend inherent flaws is reached.

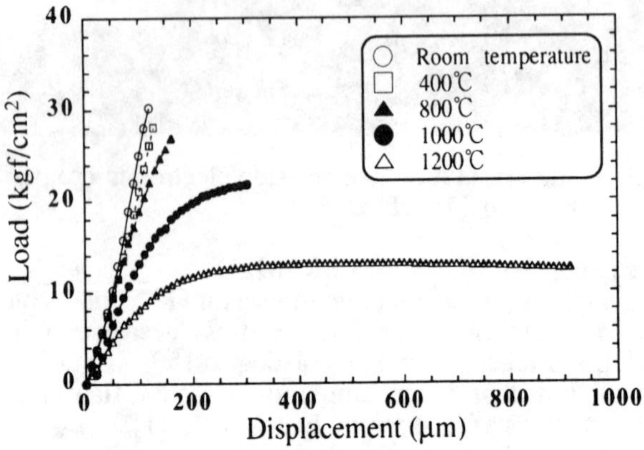

Figure 5. Load/displacement curves in bending tests for fine grained (average grain size; 4 μm) MgO prepared at 1400°C.

Under these conditions, microcracks that lead to fracture will be formed and crosslinked to the critical flaw prior to catastrophic failure. On polished specimens before fracture at 1200° C, as shown in Figure 6 (a), in fact, a large number of grain-sized cracks were observed on the tensile surface. The crack propagation is mainly intergranular in these materials as clearly seen in Figure 6 (b). Fracture surfaces for the 4μm MgO specimens were rough and irregular to suggest transgranular-fracture with many cleavage steps at temperatures up to 800° C, but the fracture topography changed significantly to the intergranular mode at higher temperatures. It is noteworthy that, in the temperature range over which the fracture mode changed, the strength also starts to decrease, and the fine-grained specimens exhibited plastic deformation before fracture.

The apparent fracture toughness increased with temperature at approximately the same temperature range. This toughness change may be attributed to the observed

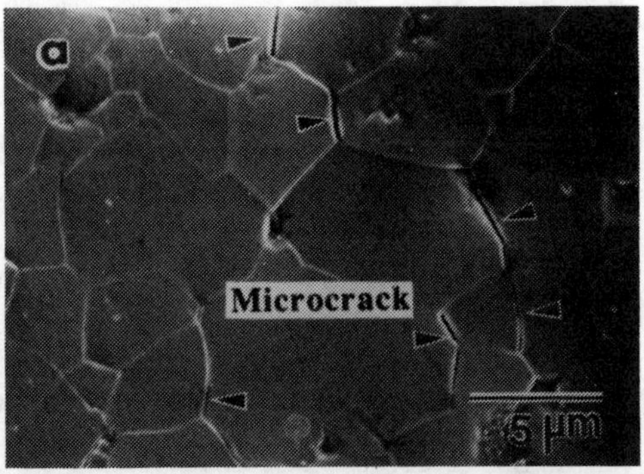

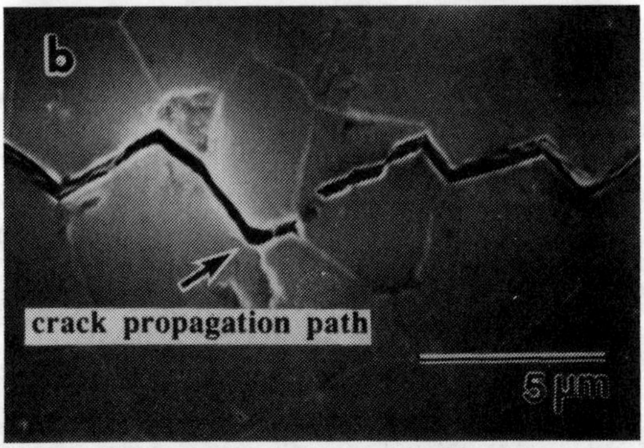

Figure 6. Formation of intergranular microcracks (a) and propagation of indentation -induced cracks (b) at 1200°C for monolithic MgO ceramics.

macroscopic plastic deformation. These observations suggest that plastic flow influences the fracture mode and high temperature strength of MgO [13]. Based on these results, both room and high temperature mechanical properties of MgO must be significantly improved when hard secondary phases are located at grain boundaries or are located within the MgO matrix grains.

Mechanical Properties Improvement by Nano-Sized SiC Dispersions
The fracture strength and toughness as a function of temperature are shown in Figure 7 for 5 vol% SiC dispersed MgO/SiC composite. At room temperature, the fracture strength of MgO was remarkably improved to 700 MPa by addition of only 5 vol% SiC particles. The high strength for this material may be attributed to two reasons. The fracture stress of MgO based ceramics will be given by the following modified Stroh equation ;

$$S_f = S_o + K_S\ G^{-1/2}\ , \tag{1}$$

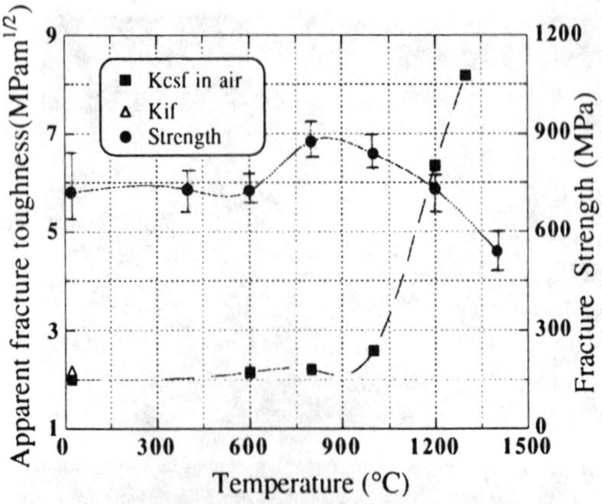

Figure 7. Effect of temperature on fracture strength and toughness (obtained by CSF method) for MgO/5vol%SiC nanocomposite.

where K_S is a constant that depends on the energy for crack-initiation, G is grain-size, and S_O is the flow stress for primary slip and independent of grain-size. In fact, as clearly seen in Table 1, the grain size of MgO drastically decreases by the SiC dispersions. Therefore, the first reason for the fracture strength increase by the SiC dispersions should be attributed to this observed decrease in grain size of MgO matrix.

As SiC particles dispersed within the MgO grains will prevent dislocation movement, the flow stress S_O must increase by the addition of SiC particles into the MgO matrix grains. A.G. Evans et al. [12] investigated the flow stress of MgO containing a higher concentration of impurities within the grains using the indentation wing technique. They found that the MgO material containing impurities exhibited higher flow stress than that for fully dense and pure MgO material. A similar increase in flow stress must be expected for the MgO/SiC nanocomposites. Based on the equation (1), therefore, it is concluded that the drastic increase in fracture strength for the MgO/SiC nanocomposites is due to both the decrease in grain size(G) and the increase in the flow stress (S_O) by the nano-sized SiC dispersions.

The bending strength of MgO/5vol%SiC composite, shown in Figure 7, indicated the temperature independence at low temperatures below 600°C. In this temperature range, as SiC particles dispersed within the MgO grains prevent primary slip, the fracture occurs by the extension of the inherent flaws in the composite. Above approximately 600°C, the strength of this composite increased with temperature, and then at 800°C and 1000°C the strength reached to about 850 MPa. In this temperature range, the dislocation motion within favorably oriented grains is confined to a small region around the crack tip. It reduces the stress concentration at the crack tip. Thus, the localized plastic deformation induced the higher strength. At elevated temperatures (>1000°C), the strength of this specimen gradually decreased with increasing temperature. In this temperature range (>1000° C), the strength is lower than the necessary stress to propagate grain-sized cracks. Thus, the linking of cracks to form the critical flaw size is required for catastrophic fracture. In spite of the decreasing strength with temperature, however, the strength of this material is higher than that at room temperature up to 1200°C.

Fracture surfaces for the MgO/SiC composite specimens were very rough, irregular and transgranular up to 1000°C, but at higher temperatures the fracture surfaces become smoother. It is noteworthy that, in the temperature range over which the fracture mode changed, the strength began to show the negative temperature dependence, and also the composite specimen started to exhibit plastic deformation before fracture.

The toughness of this composite was not changed with increasing temperature up to approximately 1000℃. But above this temperature, the toughness increased dramatically with increasing temperature as shown in Figure 7. This increase in toughness with temperature will be caused by macroscopic plastic deformation. These observations also suggest that the plastic flow may influence the fracture mode, high temperature strength and toughness of composites.

CONCLUSIONS

TEM observation revealed that the intragranular SiC within MgO matrix grain was caused by the break-away of the grain boundary from the pinning force by SiC particles. High resolution TEM lattice images of the interface of these composites showed good compatibility between MgO and SiC and little sign of any reaction. The room and high temperature fracture strength of MgO was remarkably improved by addition of only 5 vol% SiC particles. The drastic increase in fracture strength was attributed to the change of grain size and the increase of the flow stress by the intragranular SiC dispersions. The high-temperature mechanical properties were a result of the interaction between plastic flow and second phase particles. i.e, the observed significant improvements of mechanical properties at high temperatures were mainly rationalized by the nano sized SiC dispersions.

ACKNOWLEDGEMENT

The authors gratefully acknowledge A. Nakahira and T. Sekino at Osaka University for their discussion and help. Part of this work was supported by the Japan Ministry of Education under a Grant-in-Aid for Science Research No. 03555157 and 04205099, and by the cooperative project" Advanced Materials Creation and Their Limit State Prediction for Environmental Preservation" between ISIR and WRI in Osaka University.

REFERENCES

(1) A. G. Evans, " Structural Reliability: A Processing-Dependent Phenomenon," J. Am. Ceram. Soc., 65 [3] 127 - 137 (1982).
(2) M. P. Harmer, H. M. Chan and G. A. Miller, "Unique Opportunities for Micro structural Engineering with Duplex and Laminar Ceramic Composites,"

J.Am.Ceram. Soc., 75 [7] 1715 - 1728 (1992).
(3) J. Crampon and B. Escaig, "Mechanical Properties of Fine-Grained Magnesium Oxide at Large Compressive Strains," J. Am. Ceram. Soc., 63 [11 - 12] 681 - 686 (1990).
(4) T.F. Page, G. R. Sawyer, O.O. Adewoye and J. J. Wert, "The Hardness and Wear Behavior of SiC and Si_3N_4 Ceramics," Proc. Brit.Ceram. Soc, 26, 192 - 208, 1978.
(5) J. N. Ness and T. F. Page, "Microstructural Evaluation in Reaction-Bonded Silicon Carbide," J. Mater. Sci., 21, 1377 - 1397 (1986).
(6) M. H. Lewis, G. R. Heath, S.M.Winder and R. J. Lumby, "High Temperature Creep and Fracture of $\beta' Si_3N_4$," Material Science Research, 18, 605 - 616, 1984.
(7) H. Suzuki and T. Hase, "Boron Transport and Change of Lattice Parameter During Sintering of β-SiC," J. Am. Ceram. Soc., 63, 349 - 350 (1980).
(8) R.W. Rice, "Mechanisms of Toughness in Ceramic Matrix Composites," Ceram. Eng. Sci. Proc., 2, [7 - 8] 661 - 701, 1981.
(9) K. T. Faber and A. G. Evans, "Crack Deflection Process-I. Theory," Acta. Metall., 31, 565 - 576 (1983).
(10) T.N.Tiegs, P. F. Becher and L. A. Harris, "Interface in Alumina-SiC whisker Composites," Material Science Research, 21, 911 - 917 (1987).
(11) A. G. Evans and R. W. Davidge, "The Strength and Fracture of Fully Dence Polycrystalline Magnesium Oxide," Phil. Mag., 20, 373 - 388 (1969).
(12) A. G. Evans, D. Gilling and R. W. Davidge, "The Temperature-Dependence of the Strength of Polycrystalline MgO," J. Mater. Sci., 5, 187 - 197 (1970).
(13) K. Yasuda, S. D. Kim, Y. Kanemichi, Y. Matsuo and S. Kimura, "Influence of Grain Size and Temperature on Fracture Toughness of MgO Sintered Body," J. Ceram. Soc. Japan., 98, [10] 1103 -1108 (1990).
(14) K. Niihara, R. Morena, and D. P. H. Hasselman, "Evaluation of K_{IC} of Brittle Solids by the Indentation method with Low Crack-to-Indent Ratios," J.Mater. Sci. Letters, 1,13 - 16 (1982).
(15) J. J. Petrovic and M. Mendiratta, "Fracture from Controlled Surface Flaws," Fracture Mechanics Applied to Brittle Materials, S. W. Freiman, ed., ASTM STP 678, ASTM, Philadelphia, p 83 (1979).
(16) R. K. Govila, "Indentation-Precracking and Double-Torsion Methods for Measering Tracture Mechanics Parameters in Hot-Pressed Si_3N_4," J. Am.Ceram. Soc., 63, 319 (1980).
(17) B. W. Kibbel and A. H. Heuer, "Exaggerated Grain Growth in ZrO_2 -Toughened Al_2O_3," J. Am. Ceram. Soc., 69, 231 - 36 (1986).
(18) C. H. Hsueh and A. G. Evans, "Microstructural Development During Final Stage Sintering: I, Pore, Grain boundary Separation," Acta Metall., 30, 1269 - 1279 (1982).

MICROSTRUCTURE AND MECHANICAL PROPERTIES OF $MoSi_2$/SiC COMPOSITES

Yoshikazu SUZUKI, Atsushi NAKAHIRA, Tohru SEKINO and Koichi NIIHARA
The Institute of Scientific and Industrial Research, Osaka University, 8-1 Mihogaoka, Ibaraki, Osaka 567, Japan

ABSTRACT

Microstructure effects on mechanical properties were investigated for SiC particulate reinforced $MoSi_2$ composites. These composites were prepared by hot-pressing mixtures of sub-micron $MoSi_2$ and nano-sized SiC powders at 1300 to 1800° C in Ar atmosphere. The microstructure of $MoSi_2$-based composites consisted of $MoSi_2$, SiC and small amounts of Mo_5Si_3 and Mo_2C. Transmission electron microscopic observation revealed that the nano-sized β-SiC particles were dispersed at the grain boundaries and also within the $MoSi_2$ matrix grains. The fracture strength and toughness were considerably improved by the SiC dispersions. The $MoSi_2$ /15vol%SiC composite hot-pressed at 1500° C gave the maximum strength of 1215 MPa. The enhancement of fracture strength in this system was mainly caused by the grain growth inhibition by the SiC dispersion. Moreover, the intergranular SiC particles played important roles to control the grain boundary structure by connecting directly with the matrix $MoSi_2$ grains without impurity phases.

INTRODUCTION

Molybdenum disilicide ($MoSi_2$) is an excellent matrix material for application as high temperature structural components[1], mainly because of its high melting point of 2030° C and excellent oxidation resistance. However, its fracture strength and toughness are not high enough for use as structural materials. Petrovic et al.[2-5] reported that the mechanical properties of $MoSi_2$ could be improved by the addition of SiC whisker, CaO-PSZ, and carbon.

Recently, significant improvement of mechanical properties was achieved by using nanocomposite technology in ceramic/ceramic systems[6-12]. Nanocomposites are a new type of materials in which nano-sized particles are dispersed within the matrix grains and/or at the grain boundaries.

The purposes of the present work are to fabricate $MoSi_2$-based composites reinforced with nano-sized SiC dispersions for improving mechanical properties, and to clarify the correlation between microstructure and mechanical properties for $MoSi_2$ /SiC composites.

EXPERIMENTAL PROCEDURE

Fabrication of $MoSi_2$/SiC composites

Starting materials were $MoSi_2$ powder (Japan New Metals CO., Ltd) with mean particle size smaller than 0.98μm and β-SiC powder (Ibiden CO., Ltd) with primary particle size smaller than 0.07μm and average particle size smaller than 0.27 μm. The SiC content was changed in the range of 5 to 15 vol % in this experiment. $MoSi_2$ /SiC powder was mixed by wet ball milling in n-BuOH with ZrO_2 balls, and then the dried mixture was dry ball milled. The mixture was hot pressed at 1300 to 1800° C under applied pressure of 30MPa for 1h in argon atmosphere to obtain disks of 44 mm in diameter and 5.0 mm in thickness. The sintered bodies were cut, ground and polished into rectangular bars, 4×3×36 mm in size, with mirror-like surfaces.

Evaluation

The phase identification was carried out by X-ray diffraction analysis. Bulk density was measured by the Archimedes method in toluene. Young's modulus was determined by the resonance vibration method. Transmission electron microscopy with energy dispersive X-ray spectrometer (TEM-EDS) was used for micro- and nanostructure analyses. More than five rectangular specimens were subjected to three-point bending tests (span: 30 mm) to determine the fracture strength. Testing load was applied parallel to hot-press axis with crosshead speed of 0.5mm/min. The fracture toughness was evaluated by indentation fracture (IF) method with Vickers indentation of 98N-load and loading duration of 15 s.

RESULTS AND DISCUSSION

Figure 1 shows the X-ray diffraction patterns of $MoSi_2$ powder, SiC powder, monolithic $MoSi_2$, and $MoSi_2$ /SiC composite hot-pressed at 1500°C. As seen from these patterns, the main components of composites were $MoSi_2$ and SiC. But small amounts of Mo_5Si_3 and Mo_2C were also identified as reaction phases. It is considered that Mo_5Si_3 was formed because of Si vaporization from the $MoSi_2$ matrix. The Mo_5Si_3 depositing reaction is as follows[13];

$$5MoSi_2(s) = Mo_5Si_3(s) + 7Si(g)$$

Mo_2C may be formed from the reaction between impurity carbon and $MoSi_2$ matrix.

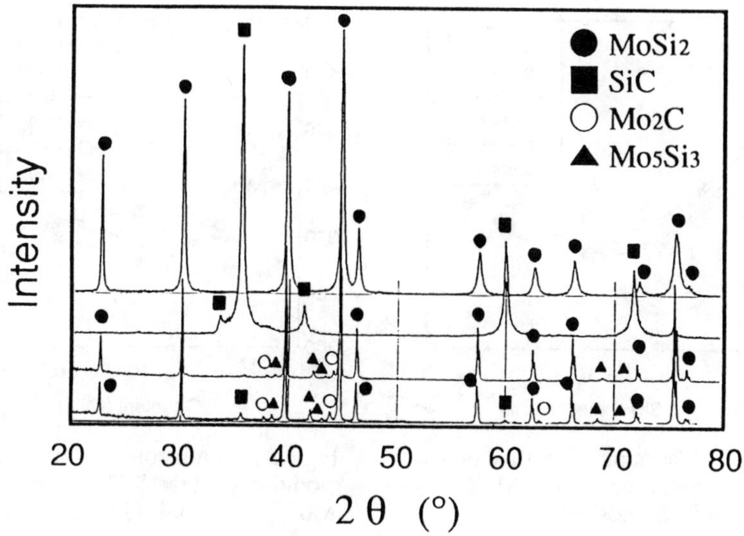

Figure 1 X-ray diffraction patterns for $MoSi_2$ powder (a), SiC powder (b), monolithic $MoSi_2$ (c), and $MoSi_2$/SiC composite(d) hot-pressed at 1500°C.

Figure 2 shows the variation of relative density with the SiC content. Although the relative densities of monolithic $MoSi_2$ and $MoSi_2$/SiC composites increased with the sintering temperature below approximately 1700° C, their increasing rates were not significant, far from the temperature dependence of usual ceramics. On the other hand, the relative densities decreased owing to overfiring above around 1800° C [14]. The density of monolithic $MoSi_2$ was approximately 96%. In the composite system, the densities decreased with the SiC addition owing to lowering of sinterability. As shown in Figure 2, the densities of $MoSi_2$/SiC composites reached a constant value of approximately 94% of theoretical value, which was calculated from densities of $MoSi_2$ and SiC without reaction phases.

The variation of Young's modulus with the SiC content is shown in Figure 3. The Young's modulus of monolithic $MoSi_2$ was approximately 380 GPa. The Young's modulus of SiC is reported to be 450 GPa[15]. Thus, the low Young's modulus observed for these composites must be attributed to the high porosity caused by the SiC addition. In fact, the variations of density and Young's modulus with the SiC content indicates the quite similar trend, as clearly seen in Figure 2 and Figure 3.

TEM photographs for the monolithic $MoSi_2$ sintered at 1300° C and the $MoSi_2$ / 15vol%SiC composite sintered at 1500° C are shown in Figure 4 and Figure 5, respectively. Homogeneously dispersed submicron-sized pores are observed in Figure 4. As expected from their low densities (Figure 2), similar sub-micron sized

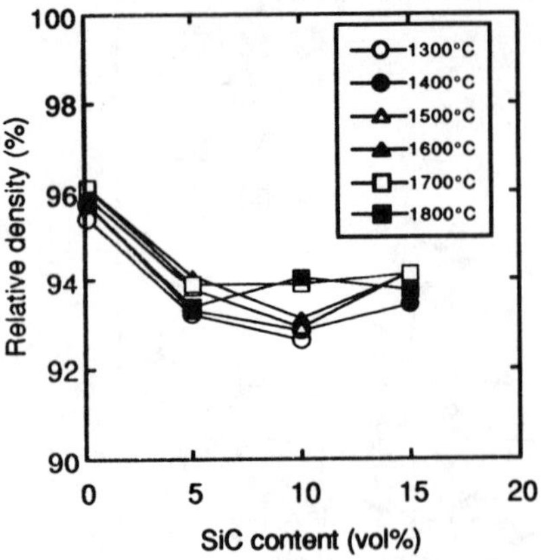

Figure 2 Variation of relative density with the SiC content for $MoSi_2$ and $MoSi_2$/SiC composites.

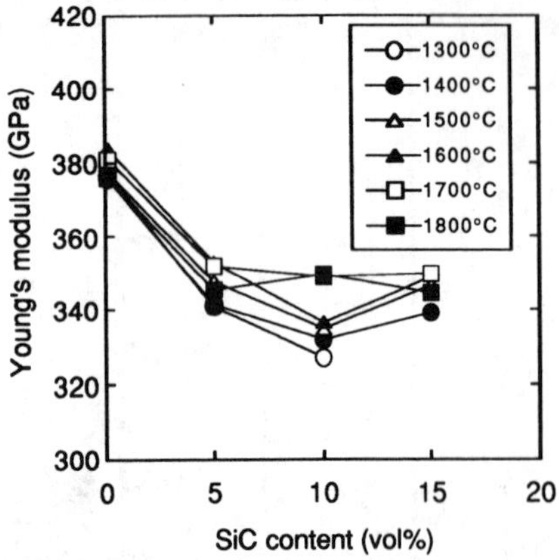

Figure 3 Variation of Young's modulus with the SiC content for $MoSi_2$ and $MoSi_2$/SiC composites.

pores are more frequent for the $MoSi_2$/SiC composites. As shown in Figure 5(a), in composites, the SiC particles less than approximately 100 nm were dispersed mainly within the $MoSi_2$ matrix grains, and the SiC particles larger than about 200 nm at the grain boundaries. EDS analysis indicated that Mo_5Si_3 particles were located within the $MoSi_2$ matrix grains. As clearly seen in Figure 5(b), however, SiC polycrystals were often observed in this composite. Detailed TEM observations showed that these SiC polycrystals were highly dense and included the nano-sized $MoSi_2$ dispersions at the grain boundaries.

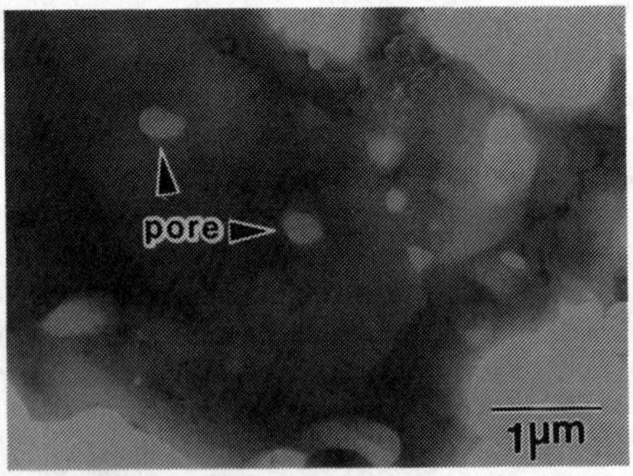

Figure 4 TEM photograph for monolithic $MoSi_2$ sintered at 1300°C.

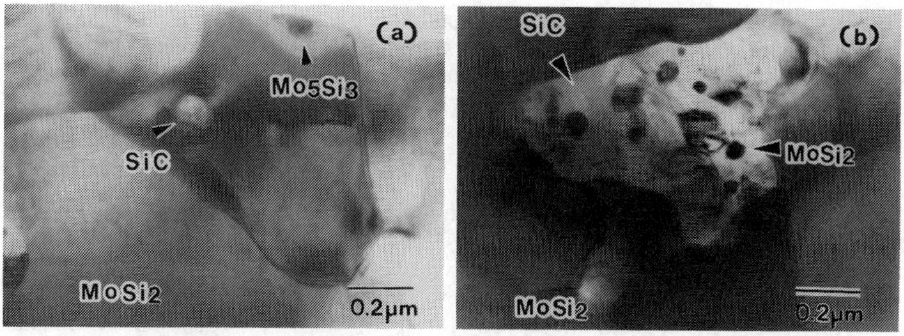

Figure 5 TEM photographs for the $MoSi_2$ /15vol%SiC composite sintered at 1500°C.

Figure 6 shows the high-resolution TEM image of the $MoSi_2$ /SiC interface. This SiC particle of about 50 nm in diameter was within the $MoSi_2$ matrix. There was no reaction phase at the interface between $MoSi_2$ and SiC. From this observation, it is reasonable to think that $MoSi_2$ does not react with SiC under sintering conditions used in this experiments.

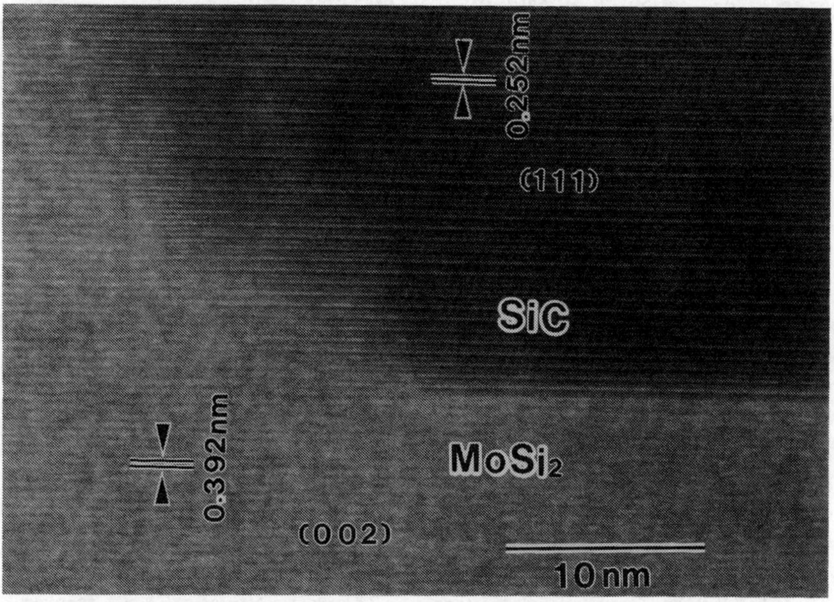

Figure 6 High-resolution TEM image of the $MoSi_2$ /SiC interface.

The variation of hardness with the SiC content is given in Figure 7. The hardness decreased with an increase in sintering temperature. On the other hand, it increased with an increase in the SiC addition, and a maximum value of 13.0 GPa was obtained for the $MoSi_2$ /15vol%SiC sintered at 1400° C. These results suggest that SiC dispersion is increasing the hardness of $MoSi_2$ ceramics regardless of the decreasing density.

As shown in Figure 8, the fracture strength increased with an increase in the content of SiC at sintering temperatures over 1500°C. SEM observations made clear that the monolithic $MoSi_2$ showed rapid grain growth. However, the grain growth of $MoSi_2$ in composites was strongly depressed by the SiC addition, and the maximum strength, 1215 MPa, was observed for the $MoSi_2$ /15vol%SiC composite hot-pressed at 1500°C. This maximum strength for $MoSi_2$/SiC composite is two to three times higher than those of the monolithic $MoSi_2$ in this experiment and more than five times higher than that in the literatures[3]. This improvement in strength will be mainly due to the grain growth depression by the SiC dispersions. To make clear strengthening mechanisms in this composite system, however, further studies of the effects of nano-sized SiC and Mo_5Si_3 dispersions are required.

The dependence of fracture toughness on the SiC content is shown in Figure 9. The influence of sintering temperature upon the fracture toughness was negligible. However, the toughness increased with an increase in the SiC content, and $MoSi_2$/15vol%SiC composite hot-pressed at 1800° C gave the maximum toughness of about 3.3 $MPa \cdot m^{1/2}$. This maximum fracture toughness is approximately one and half times higher than that of monolithic $MoSi_2$. The increase in fracture toughness may be attributed to the residual stress owing to the difference of thermal expansion coefficient between $MoSi_2$ and SiC.

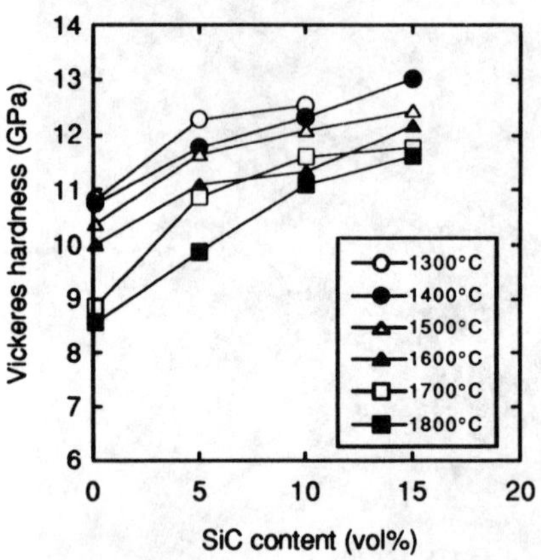

Figure 7 Variations of hardness with SiC content for $MoSi_2$ and $MoSi_2$ /SiC composites.

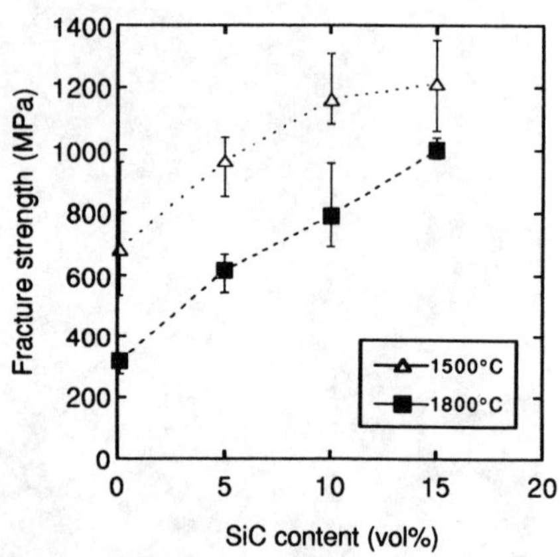

Figure 8 Variations of fracture strength with SiC content for $MoSi_2$ and $MoSi_2$ /SiC composites.

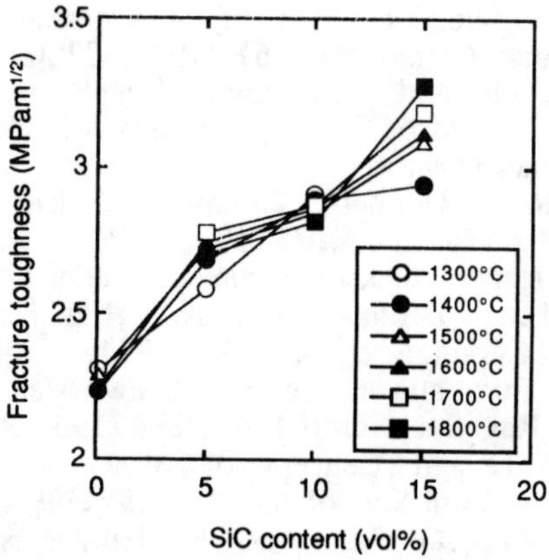

Figure 9 Variations of fracture toughness with SiC content for $MoSi_2$ and $MoSi_2$ /SiC composites.

CONCLUSIONS

(1) Fine SiC particle dispersed $MoSi_2$-based composites were successfully fabricated through powder metallurgy. The microstructure of $MoSi_2$ /SiC composites consisted of $MoSi_2$, SiC particles, and small amounts of Mo_5Si_3 and Mo_2C.
(2) Mechanical properties were strongly enhanced, especially the fracture strength. The strength reached 1215MPa for the $MoSi_2$ /15vol%SiC composite sintered at 1500°C. Hardness and toughness were also improved in this composite system.
(3) The enhancement of fracture strength in this system is mainly due to the grain growth inhibition caused by the SiC dispersion. The improvement of fracture toughness is due to the residual stresses from the difference of thermal expansion coefficient between $MoSi_2$ and SiC.

Acknowledgement

The authors gratefully acknowledge S. Ueda at Osaka university for their discussion and help. A part of this work was supported by the Japan Ministry of Education under a Grant-in-Aid for Science Research No.03555157 and 04205099, and by the cooperative project "Advanced Materials Creation and Their Limit State Prediction for Environmental Preservation" between ISIR and WRI in OSAKA University.

References

[1]P. T. B. Shaffer, Plenum Press Handbooks of High-Temperature Materials, (Plenum, New York, 1964).

[2]F. D. Gac and J. J. Petrovic, "Feasibility of a Composite of SiC Whiskers in an $MoSi_2$ Matrix", J. Amer. Ceram. Soc., **68** (1985) C200.
[3]D. H. Carter, W. S. Gibbs, and J. J. Petrovic, "Mechanical Characterization of SiC Whisker-Reinforced $MoSi_2$", Proceedings of Ceramics Materials & Components for Engines (1988).
[4]J. J. Petrovic and R. E. Honnell, "Partially Stabilized ZrO_2 Particle-$MoSi_2$ matrix composites", J. Mater. Sci., **25** (1990) 4453.
[5]S. A. Maloy, A. H. Heuer, J. J. Lewandowski, and J. J. Petrovic, "Carbon Addtions to Molybdenum Disilicide: Improved High-Temperature Mechanical Properties" J. Amer. Ceram. Soc., **74** (1991) 2704.
[6]K. Niihara, A. Nakahira and T. Sekino, "New Nanocomposite Structural Ceramics", Mater. Res. Soc. Symp. Proc., **286** (1993) 405.
[7]K. Niihara, "New Design Concept of Structural Ceramics -Ceramic Nanocomposites-" J. Ceram. Soc. of Jpn., **99** (1991) 974.
[8]K. Niihara, A. Nakahira, G. Sasaki and M. Hirabayashi, "Development of Strong Al_2O_3/SiC Composites", Proc. MRS Int'l Mtg. on Adv. Composites, (Plenum, New York, 1988) 129.
[9]K. Niihara, A. Nakahira, "Structural Ceramic Nanocomposites by Sintering Method; Roles of Nano-size Particles", Ceramics: Toward the 21st Century, Japan, (1991) 404.
[10]F. Wakai, Y. Komada, S. Sakaguchi, K. Izaki, K. Niihara, "A Superplastic Covalent Crystal Composite", Nature, **344** (1990) 421.
[11]K. Niihara, K. Izaki, T. Kawakami, "Hot-Pressed Si_3N_4-32% SiC Nanocomposite from amorphous Si-C-N Powder with Improved Strength above 1200°C", J. Mater. Sci. Lett., **10** (1990) 112.
[12]A. Nakahira, T. Sekino, Y. Suzuki, and K. Niihara, "High-Temperature Creep and Deformation Behavior of Al_2O_3 /SiC Nanocomposite", Ann. Chim. Fr., **18**(1993) 403.
[13]A. W. Searcy and A. G. Tharp, "The Dissociation Pressures and the Heats of Formation of the Molybdenum Silicides", J. Phys. Chem., **64** (1960) 1539.
[14]W. D. Kingery, H. K. Bowen and D. R. Uhlmann, Introduction to Ceramics, Second Edition, (New York, Chichester, Brisbane, Toronto, 1976).
[15]M. F. Ashby and D. R. H. Jones, Engineering Materials (Pergamon Press, 1980).

EFFECTS OF ADSORBED SPECIES ON MICROSTRUCTURE AND MECHANICAL PROPERTIES OF Al_2O_3

Hiroaki KITAYAMA, Masahiro INOUE* and Koichi NIIHARA
The Institute of Scientific Industrial Research, Osaka University
8-1 Mihogaoka, Ibaraki, Osaka 567, JAPAN
*Department of Materials Science and Engineering, National Defense Academy,
1-10-20 Hashirimizu, Yokosuka, Kanagawa 239, JAPAN

ABSTRACT

The structural change of the powder surface caused by adsorption of organic species was investigated using several analytical techniques. The powders with organic adsorbed species were obtained by ball milling the α-Al_2O_3 powder in acetone dispersant and subsequent microwave drying. Surface analyses suggested the surface structures of Al_2O_3 powders were strongly modified during drying. Microstructural observation revealed that the sinterability of Al_2O_3 powders and the microstructure development of sintered bodies were significantly changed by the powder surface modification. Based on these observations, the effect of organic adsorbed species on the fracture strength was also discussed.

INTRODUCTION

Milling oxide powders in such a polar organic medium as alcohol can change mechanochemically the characteristics of powder surfaces because the surface hydroxyl groups are substituted with organic ones. Usually this surface treatment is made to control the water adsorption on powder surfaces, which has serious consequences on the drying and sintering processes [1-2]. Thermal analysis indicated that Al_2O_3 and isopropanol, used as a dispersant, reacted to produce the surface species during milling [3]. It is also reported that the ethanol adsorption on Al_2O_3 produces the surface ethoxide, and at high temperatures the adsorbed species react to form surface acetates [4]. These surface reactions containing adsorption or desorption of organic species on the powder will have a strong effect on the sintering process.

In this study, the surface analyses were conducted using several techniques for observing the structure change of powder surfaces, and the influences of organic species adsorption on Al_2O_3 powder surfaces during powder processings were discussed in respect to both microstructure and fracture strength.

EXPERIMENTAL PROCEDURE

Powder Preparation and Characterization of Powder Surface

High-purity alumina powders, α-Al_2O_3 (AKP-30, 99.99%, Sumitomo Chem. Co.) and γ-Al_2O_3 (99.99%, Asahi Chem.Co.), with mean particle sizes of 0.4μm were used in this study. These powders were dispersed in acetone (reagent-grade,Wako Chem. Co.), and wet ball milling was carried out for 24h. After 1h microwave drying (2450MHz, 100V) of the slurry, the resultant soft agglomerates were crushed by dry ball milling for 12h to obtain starting powders. These powders surfaces were analyzed by several techniques, which were based on different principles. Table I shows the instrumentation used in the present experiment.

Table I. Powder Characterization Techniques

Technique	Comments
Thermogravimetric analysis (TGA) and differential thermal analysis (DTA)	Analyses of thermal (decomposition) reaction process of samples
Fourier transform infrared spectroscopy (FTIR)	Identification of adsorbed molecular structure by measuring characteristic vibrations and rotations of atoms in molecules
X-ray photoelectron spectroscopy (XPS)	Elemental analysis of layers 5nm thick, lateral resolution 1nm, depth profiles by ion milling, chemical bonding information

The thermal analysis of these powders was performed using a thermobalance (Rigaku, Thermoflex TG 8110). Adsorbed species on powder surfaces were analyzed using a Fourier Transform Spectrophotometer (Analect RFX-65). The analysis of surface and near-surface material was performed using a X-ray

Photoelectron Spectroscopy (Shimadzu / Dupont 650B). Irradiation of samples was performed under high vacuum conditions (1.38×10^{-10}Pa) with MgKα X-ray source ($h\nu$=1253.6eV). The Au4$f_{7/2}$ reference, whose binding energy was fixed at 93.8eV, has allowed the determination of the binding energies of the O1s, Al2p, C1s levels.

Microstructural Observation

Microstructural evolution of Al_2O_3 was observed not only by scanning electron microscopy (SEM) but also by liquid immersion microscopy (LIM) [5]. The LIM technique involves an adequate immersion liquid, which makes alumina green bodies transparent for optical microscopic examination. In this study, diiodomethane (CH_2I_2) (n=1.749) was used as an immersion liquid because its index was nearly the same as that of Al_2O_3 (n=1.77). The internal structures of green bodies were examined with an optical microscope of transmission mode on thinned specimens. The Al_2O_3 powders with different pre-treatments were hot-pressed at 30MPa in flowing N_2 at 1300~1500℃ for 1h. Microstructural observation of sintered Al_2O_3 was performed using SEM and transmission electron microscopy (TEM).

RESULTS AND DISCUSSION

Change of Powder Surface Structure by Adsorbed Species

Thermal analysis results showed a trace of the presence of organic adsorbed species on these powders' surfaces. In TGA/DTA, the γ-Al_2O_3 powder pretreated in acetone shows more weight loss than the as-received powder. The two specific peaks (650℃ and 1240℃) indicate the desorption of adsorbed species from the powder surface or the combustion of adsorbed species (Figure 1).

The adsorption of ethanol onto Al_2O_3 was investigated using infrared spectrometry. The alcohol is reported to adsorb in such fragments as -OH, -CH_2CH_2- and -OCH_3, bound directly to an Al atom on the surface to form Al-O-CH_2CH_3 [6]. In this study, the adsorption of acetone onto Al_2O_3 was investigated. The following IR spectra (Figure 2) indicate the acetone adsorption, CH asymmetric and symmetric stretching vibrational bands in the region 2970~2850cm^{-1}, unidentified peaks in the region 2100~1700cm^{-1} and at 1400cm^{-1}, etc.

After ball milling of as-received α-Al_2O_3 powder in acetone, the 45min microwave irradiation made the slurry completely dry. This powder was designated as powder AW. After 1h microwave drying, the powder was uniform in color. This powder was designated as powder AC. Thermal analyses and infrared spectra of acetone adsorbed on α-Al_2O_3, however, did not show any difference between powder AW and AC. The α-Al_2O_3 powder had even a smaller specific surface area than γ-Al_2O_3 (α-Al_2O_3 / γ-Al_2O_3 = 7 / 125), so had even fewer adsorbed species in proportion to the specific surface area. Namely, the acetone adsorbed on α-Al_2O_3 was beyond the measuring abilities of these instruments.

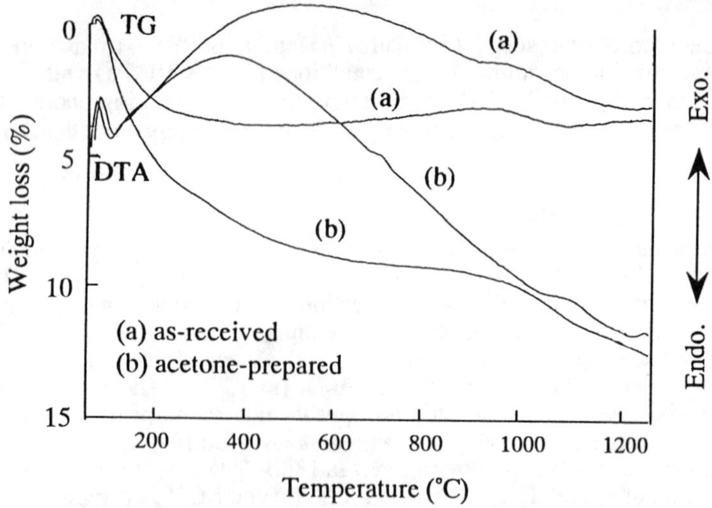

Figure 1. TGA and DTA curves for as-received γ-Al_2O_3 powder (a) and acetone-prepared γ-Al_2O_3 powder (b).

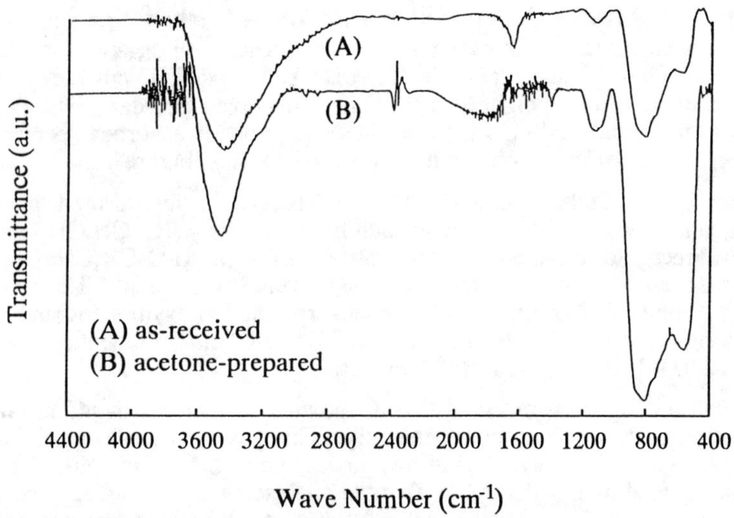

Figure 2. Transmission infrared spectra of as-received γ-Al_2O_3 powder (A) and acetone-prepared γ-Al_2O_3 powder (B).

It was then attempted to analyze these adsorbed species by XPS, which should be efficient for characterization of surface and near surface material. A high-resolution scan over the narrow energy range around Al2p, O1s, and C1s was performed to obtain information on the chemical environment of the surface elemental composition. Figure 3 shows XPS spectra of two kinds of powders (powders AW

and AC). Powders AW and AC showed a strong C1s peak in the XPS spectra. In the high resolution scan, each XPS signal of the C1s photoelectron consisted of two peaks. Literature data concerning the C1s binding energy of impurity hydrocarbon is 284.7eV, and then the lower binding energy peak was identified as impurity hydrocarbon in the analyzer chamber and the higher one indicated adsorbed species, carbonyl groups.

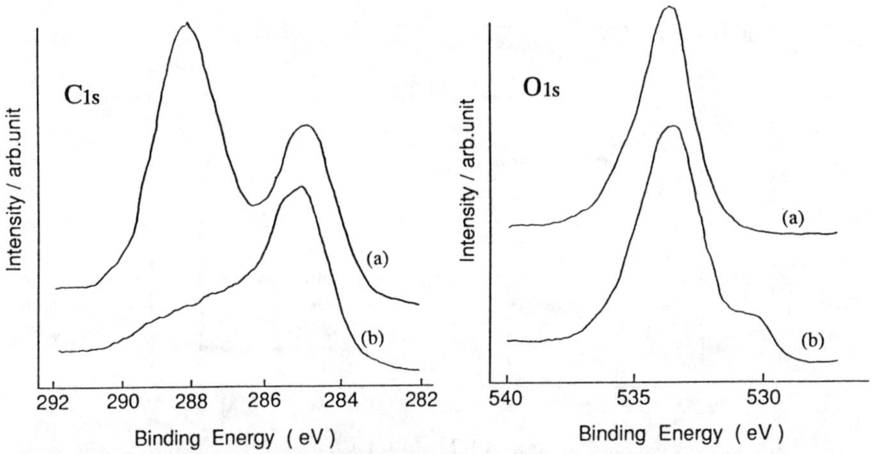

Figure 3. XPS spectra of powder AW (a) and AC (b)

The XPS signal of the O1s photoelectrons for the as-received α-Al_2O_3 powder clearly indicated a peak at 533.3eV, which was believed to be not due to the Al-O bonding in the Al_2O_3 (literature data; O1s: 531.6eV) but adsorbed water or hydroxy groups on the powder surface. Only O1s peak for powder AC showed another peak at 530.5eV. There are two possibilities on identification of this peak. One is that this peak will be ascribed to the reaction product O^{2-} of acetone with the Al_2O_3 powder surface, because it was reported that adsorbed organic species such as NO_x on metal surface decomposed to produce O^{2-} [7]. Another possibility is that this peak can be ascribed to the production of oxygen vacancies in Al_2O_3 powder by the decomposition of acetone. At present, it is difficult to determine which possibility is reasonable, but the observed change of the surface condition might have some influences on the sinterability of the powder.

Effects of Powder Surface Condition on Sinterability of Al_2O_3

Liquid immersion microscopic observation for Al_2O_3 green bodies from as-received, powders AW and AC, heated up to 500℃ and 900℃ indicated the retardation in the growth of highly-densified regions due to the change of surface conditions. The densification behavior for Al_2O_3 fabricated in this study is shown in Figure 4. Density of sintered Al_2O_3 was measured by Archimedes method in toluene. For preparation of the Al_2O_3 samples, the Al_2O_3 powder was used without any special treatment such as dispersion, milling, crushing, etc.

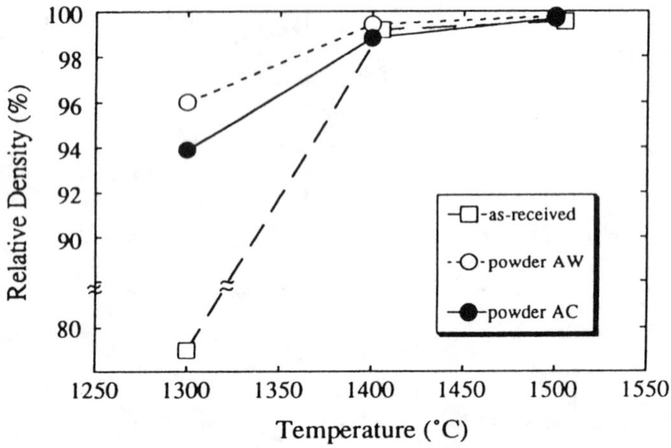

Figure 4. Densification behavior of Al_2O_3 fabricated from the powders with different pre-treatment.

(a) (b)

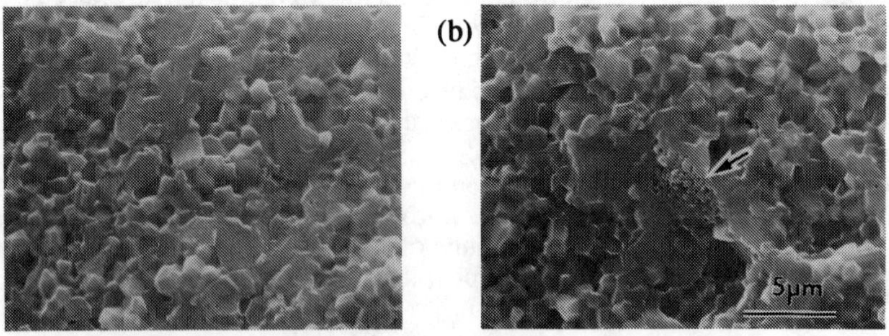

Figure 5. SEM micrographs for fracture fracture surfaces of Al_2O_3 sintered at 1400℃ from powder AW (a) and powder AC (b).

As clearly seen in Figure 4, Al_2O_3 samples sintered to nearly full density at 1400℃. However, the samples fabricated from powder AC reached only up to 98.5% theoretical density at this temperature. In addition, the microstructural difference could be observed between the samples fabricated from powder AW and AC. The samples from powder AW showed relatively uniform microstructure, while in the samples from powder AC, the presence of unsintered regions (as indicated by an arrow in Figure 5 (b)) could be observed in part. These unsintered regions were observed among highly contracted regions, and from the results of liquid immersion microscopy, they were ascribed to sudden formation of highly-densified regions and contraction of unsintered Al_2O_3.

Sintering at 1500℃ is enough to obtain the fully-densified samples even from powder AC. Figure 6 shows the SEM micrographs of sintered Al_2O_3 from powder AW and powder AC. Microstructural observation showed severe inhibition of grain growth of sintered Al_2O_3 made from powder AC.

(a)

(b)

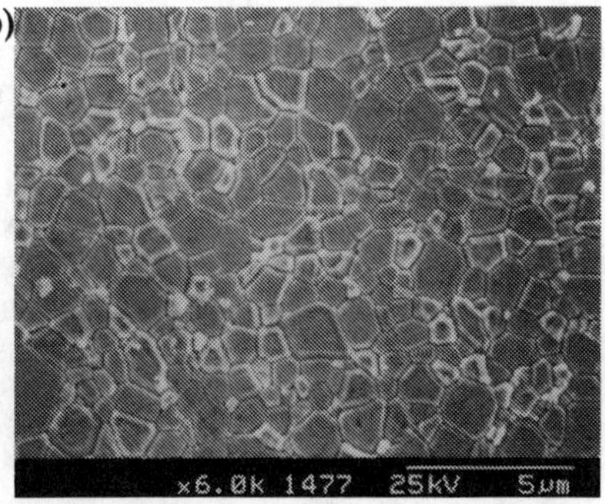

Figure 6. SEM micrographs for sintered Al_2O_3 at 1500℃ from powder AW (a) and powder AC (b)

Effects of Microstructure Change on Fracture Strength

As mentioned in the previous section, it was revealed that the organic species adsorbed during the powder treatment had a strong effect on the sinterability and hence microstructure of sintered Al_2O_3, as summerized in Table II. The fracture strength was also changed drastically as seen in Table II.

Table II. Fracture strength change of Al_2O_3 sintered from powders AW and AC.

Sintering Temperature (℃)	Relative Density (%)		Average Grain Size (μm)		Fracture Strength (MPa)	
	AW	AC	AW	AC	AW	AC
1400	99.4	98.5	0.80	0.60	960	564
1500	99.6	99.8	2.82	0.86	728	815

CONCLUSION

Various surface analysis techniques was used to analyze the organic adsorbed species on Al_2O_3 powder surfaces. The adsorbed species were introduced by wet ball milling Al_2O_3 powder in acetone and subsequent drying. XPS analysis suggested that acetone adsorbed on Al_2O_3 powder surface could react during drying due to high reactivity of the powder surface. These surface reactions changed the powder surface, resulting in a change in the sinterability of Al_2O_3 powder, and hence a change in the microstructure and mechanical properties of sintered Al_2O_3.

ACKNOWLEDGEMENTS

The authors gratefully acknowledge A.Nakahira and T.Sekino at Osaka University for their discussion and help. A part of this work was supported by the Japan Ministry of Education under a Grant-in-Aid for Science Research No.03555157 and 04205099, and by the cooperative project "Advanced Materials Creation and Their Limit State Prediction for Environmental Preservation" between ISIR and WRI in Osaka University.

References

1. M. S. Kaliszewski and A. H. Heuer, "Alcohol Interaction with Zirconia Powders", J. Am. Ceram. Soc., **73**, [6] 1504-509 (1990)

2. M. J. Readey, R. R. Lee, J. W. Halloran, and A. H. Heuer, "Processing and Sintering of Ultrafine $MgO-ZrO_2$ and $(MgO,Y_2O_3)-ZrO_2$ Powders", J. Am. Ceram. Soc., **73**, [6] 1499-503 (1990)

3. F. W. Dynys and J. W. Halloran, "Reactions During Milling of Aluminum Oxide", J. Am. Ceram. Soc., **64**, [4] C-62-63 (1981)
4. R. G. Greenler, "Infrared Study of the Adsorption of Methanol and Ethanol on Aluminum Oxide", *J. Chem. Phys.*, **37**, 2904-2100 (1962)
5. K. Uematsu, J. Y. Kim, M. Miyashita, N. Uchida, and K. Saito, "Direct Observation of Internal Structure in Spray-Dried Alumina Granules", *J. Am. Ceram. Soc.*, **73**, [8] 2555-57 (1990)
6. A. A. Babushkin, and A. V. Uvarov, *Dokl. Akad. Nauk SSSR*, **110**, 581 (1956)
7. C. R. Brundle, "XPS and UPS studies of the interaction of nitrogen-containing molecules with nickel: The use of binding energy patterns and relative intensities to diagnose surface species", *J. Vacuum Sci. Technol.*, **13**, 301-309 (1976)

INTERFACE STRUCTURE OF DIRECTIONALLY SOLIDIFIED COO-ZRO$_2$(Y$_2$O$_3$) EUTECTIC

Jun-ichi Echigoya* and Shigeyuki Hayashi**

*Department of Materials Processing
Faculty of Engineering, Tohoku University
Aoba, Aobaku, Sendai 980, Japan
**Institute for Materials Research
Tohoku University
2-1-1 Katahira, Aobaku, Sendai 980, Japan

ABSTRACT

The interface structure of directionally solidified CoO-ZrO$_2$(Y$_2$O$_3$) eutectic was investigated by transmission electron microscopy. The interface of the lamella was found to have the orientation relationship of (001)cubic ZrO$_2$ // (111)CoO and steps with periodic spacing. Precipitation of the Co$_3$O$_4$ or Co phase was observed at the interface between cubic ZrO$_2$ and CoO under experimental atmospheric conditions during directional solidification and successive cooling.

INTRODUCTION

The directional solidification of eutectics is a method used to produce an anisotropic composite directly from the melt. Attention has been focussed on mechanical properties of directionally solidified eutectics at high temperatures due to high structural stabilities up to the melting temperature [1-3]. The morphology and orientation relationships between the two phases have been investigated in directionally solidified oxide eutectic systems and determined by factors of volume fraction of the second phase, impurities, growth direction, and the ionic charge balance at the interface [3, 4]. The recent development of high resolution electron microscopy has made it possible to analyze the interface structure at the atomic level [5]. In oxide eutectic systems, such analysis has revealed the atomic fit at the interface to be extremely good and the sequence of atomic plane parallel to the interface to be metal/oxygen/metal/oxygen/metal/oxygen stacking across the interface [6].

Investigation of the $CoO-ZrO_2(Y_2O_3)$ eutectic system, in which the eutectic composition was around CoO-20 mol%ZrO_2, revealed its lamellar morphology by directional solidification [7]. However, since the composition of the compounds in the Co-O system changes in form from Co to Co_3O_4 through CoO due to oxygen partial pressure because of the valence fluctuation of Co, it may be possible to affect the interface structure by the oxygen partial pressure during solidification and successive cooling. The aim of this study was to investigate the interface structure of a directionally solidified $CoO-ZrO_2(Y_2O_3)$ eutectic.

EXPERIMENTAL PROCEDURE

Starting materials were prepared for solidification by blending weighed powders of CoO (99.99% purity), ZrO_2 (99.999% purity) and Y_2O_3 (99.99% purity), pressing the resulting mixture into a rod 5 mm in diameter and 80 mm in length and sintering it at 1470K for 3.6ks. The amount of Y_2O_3 added to the $CoO-ZrO_2$ eutectic was 8 mol%Y_2O_3 in the ZrO_2 phase when Y_2O_3 dissolved only in the ZrO_2 phase. Directional solidification was conducted by a floating zone melting method using a single ellipsoid image furnace in an atmosphere of stagnant air or flowing dry air.

Directionally solidified eutectics were cut and polished perpendicular to the direction of growth. Thin foils were prepared by sectioning the directionally solidified eutectics. After grinding down to 0.1 mm using a diamond polisher, the sections were ion-beam thinned in argon at 5 kV at an angle of 30 degrees. The foils were examined using JEOM-3010 and JEOL-1250 transmission electron microscopes.

EXPERIMENTAL RESULTS AND DISCUSSION

Figure 1 shows a typical example of a sample perpendicular to the direction of growth of a directionally solidified $CoO-ZrO_2(Y_2O_3)$ eutectic grown at a growth rate of 10 mm/h. The growth morphology of this system is a lamellar structure. Figure 2 is the electron diffraction pattern of a lamella in a directionally solidified $CoO-ZrO_2(Y_2O_3)$ eutectic grown in stagnant air. This reveals the orientation relationship between cubic ZrO_2(c-ZrO_2) and CoO to be

[100]c-ZrO_2//[110]CoO,

(001)c-ZrO_2//(111)CoO (interface).

This is the same relationship observed in a directionally solidified $MgO-ZrO_2$ eutectic [5]. Reflections from a Co metal were found in this electron diffraction pattern shown in Fig. 2. Since these reflections are just outside of the CoO reflections, Co seems to nucleate from the CoO matrix near the interface with the same orientation of {hkl}CoO // {hkl}Co. When the directional solidification takes place in stagnant air, this may be a reduced atmosphere for CoO as we later discuss.

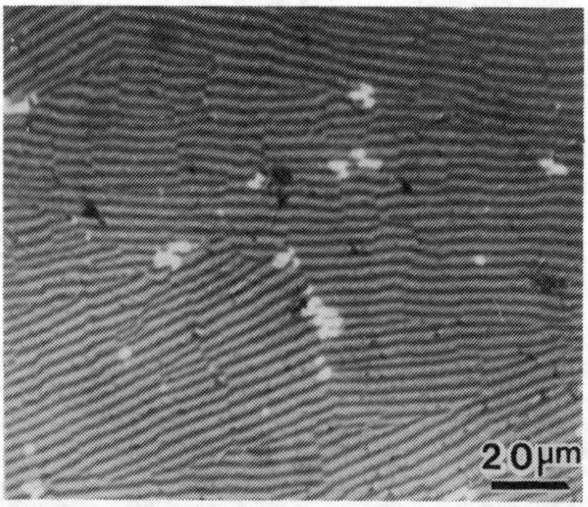

Fig. 1 Optical micrograph of a sample perpendicular to the direction of growth of a directionally solidified CoO-$ZrO_2(Y_2O_3)$ eutectic grown at a growth rate of 10 mm/h.

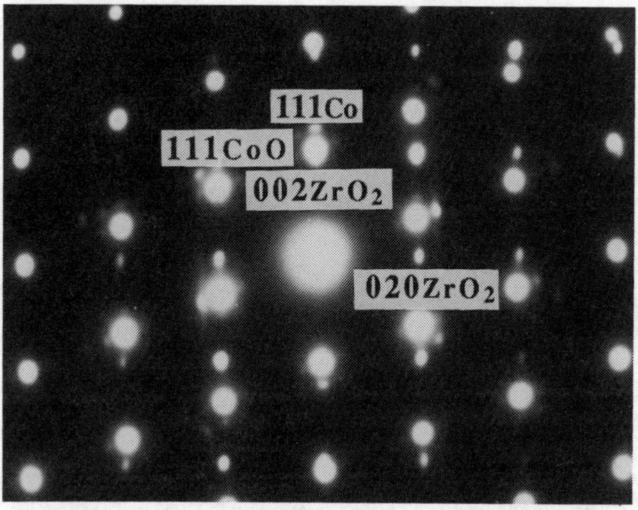

Fig. 2 Electron diffraction pattern of lamella in a directionally solidified CoO-$ZrO_2(Y_2O_3)$ eutectic grown in stagnant air. (Growth rate: 20mm/h)

The high resolution lattice image of the interface between CoO and c-ZrO_2 in the sample grown at a growth rate of 20 mm/h is shown in Fig. 3. The interface is formed by (111)CoO // (001)c-ZrO_2, microscopically, and has steps with periodic spacing. The misfit of atomic distance between [112]CoO and [010]c-ZrO_2 directions, which is along the interface, is about 1.9%. A good fit of oxygen atoms is realized by an atomic height which is one step up. This is one methods of accommodating the misfit strain at the interface [5]. The disorder of lattce fringes in the CoO phase close to the interface seems to be due to the precipitation of Co.

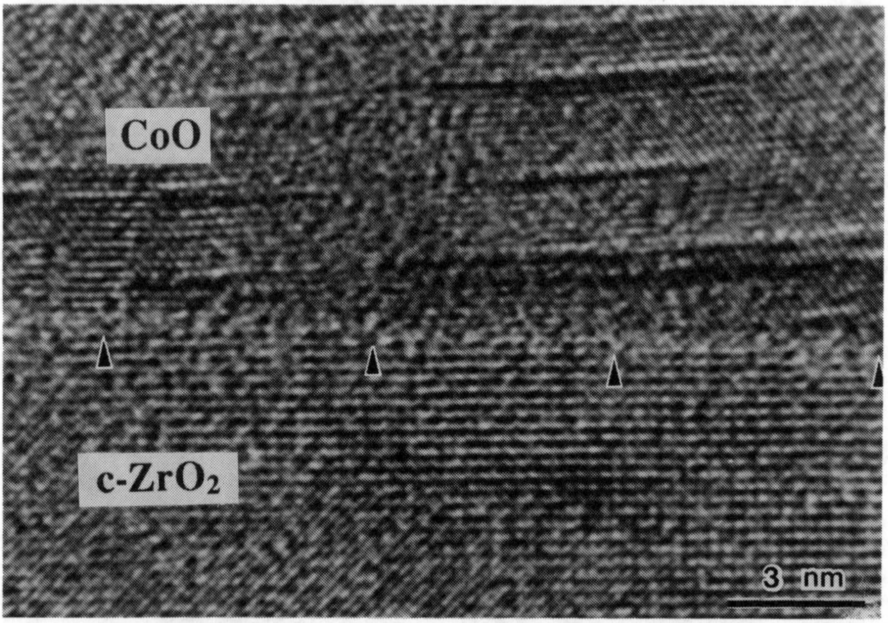

Fig. 3 High resolution lattice image of the interface between CoO and cubic ZrO_2 in a sample grown in stagnant air. (Growth rate: 20 mm/h)

When the directional solidification was performed in flowing dry air at a flow rate of 1 liter/min, the structure around the interface was different from that grown in stagnant air. Figure 4 shows the electron diffraction pattern of the CoO / c-ZrO_2 interface in a sample grown in flowing dry air. Fundamentally, this consists of CoO and c-ZrO_2 phases with an orientation relationship of (001)c-ZrO_2 // (111)CoO, though several faint reflections can be seen in this pattern. These faint reflections can be indexed by the reflections from the Co_3O_4 phase. The orientation relationship between CoO and Co_3O_4 is (111)CoO // (111)Co_3O_4, i.e., the relationship of {hkl}CoO // {hkl}Co_3O_4. Figure 5 shows a high resolution electron micrograph of the CoO / c-ZrO_2 interface in a sample directionally solidified at a growth rate of 10 mm/h in flowing dry air. Careful

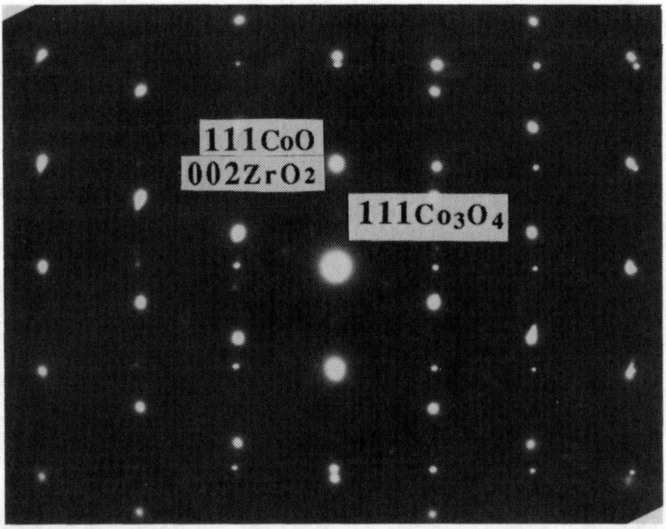

Fig. 4 Electron diffraction pattern of the CoO / c-ZrO_2 interface in a sample grown in an atmosphere of flowing dry air. (Growth rate: 10mm/h)

inspection of this micrograph indicates that the lattice image near the interface in the CoO phase is different from that of CoO. The thickness of the different contrast layer is about 3 nm, but not homogeneous along the interface. This different phase is estimated to be a Co_3O_4 phase based on the electron diffraction pattern of Fig. 4. Figure 6 shows a magnified image of Fig. 5. A lattice fringe spacing parallel to the interface close to the c-ZrO_2 phase on the CoO side is in good agreement with that of (111) Co_3O_4. This is further evidence that a Co_3O_4 compound at the interface in the CoO phase, when the directional solidification was carried out in flowing dry air. Since the interface has the steps already observed in Fig. 3, the Co_3O_4 phase is nucleated and grown at the interface during cooling after solidification. The (111) lattice fringe of the Co_3O_4 phase was found to rotate about 3 degrees to that of (111)CoO, though the lattice fringe of (111)CoO is parallel to that of (001)c-ZrO_2. The rotation of the (111)Co_3O_4 lattice plane to the (111)CoO and (001)c-ZrO_2 lattice planes introduces extra half planes parallel to the interface at both interfaces of c-ZrO_2 / Co_3O_4 and Co_3O_4 / CoO. The Co_3O_4 phase has a spinel-type cubic structure with a lattice parameter of a=0.808 nm, which causes a 5.3% misfit in comparison with that of the CoO phase in the case of {hkl}CoO //{hkl}Co_3O_4. Since the step structure of a c-ZrO_2 surface seems to be stable at the formation temperature of Co_3O_4, nucleation and growth of Co_3O_4 affect the existence of surface steps at the interface. This may be the reason for the rotation of the (111)Co_3O_4 lattice plane to the (111)CoO lattice plane.

We were able to estimate formation temperature of the Co_3O_4 phase in a CoO matrix from the phase diagram made by Aukrust and Muan [8]. When the

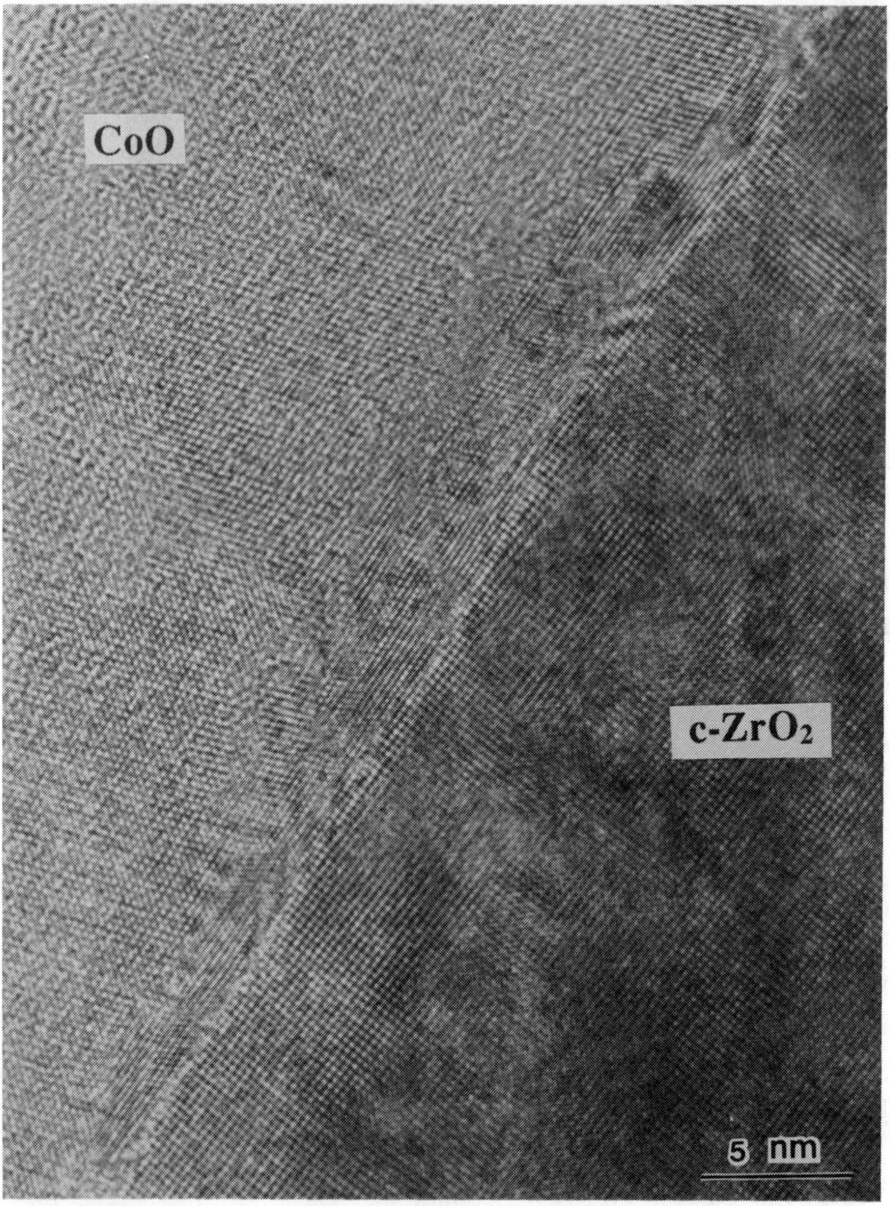

Fig. 5 High resolution electron micrograph of the CoO / c-ZrO_2 interface in a sample directionally solidified at a growth rate of 10 mm/h in flowing dry air.

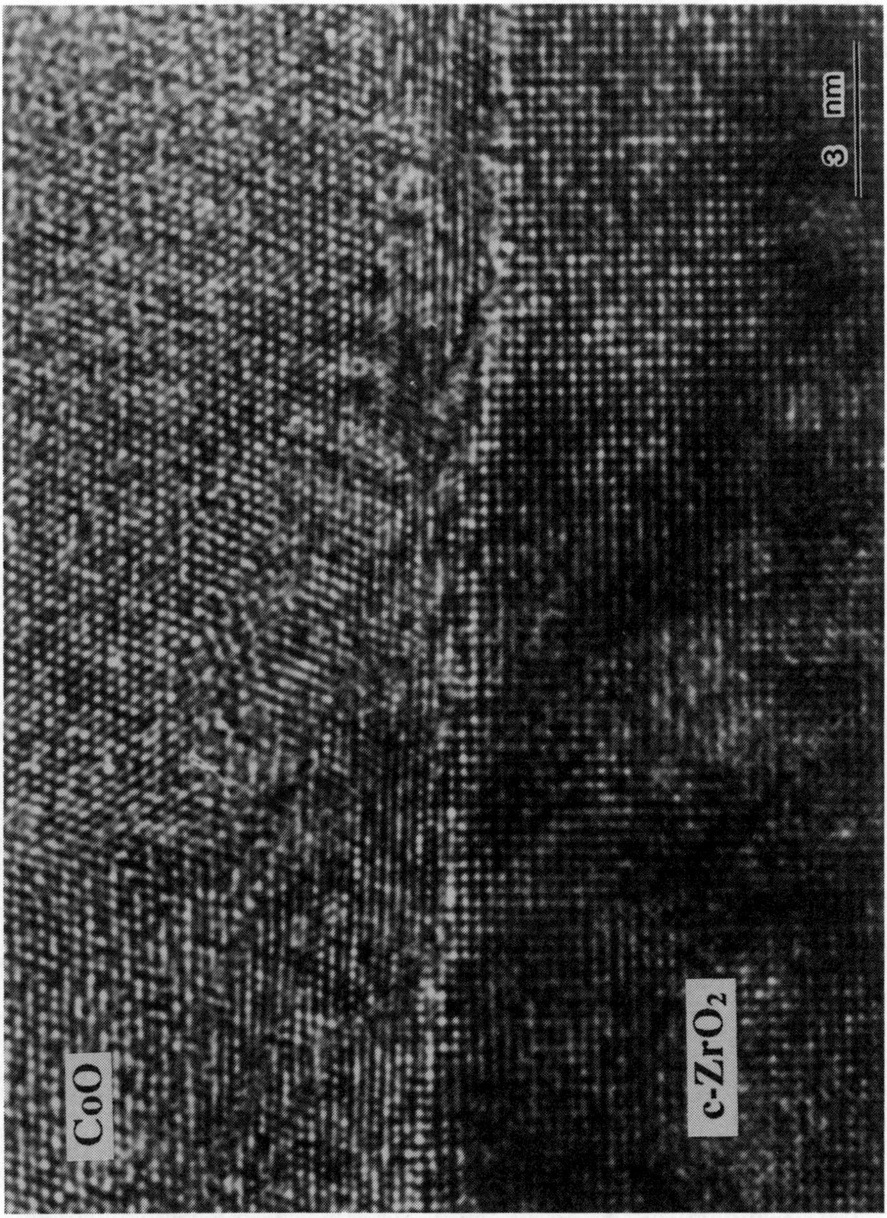

Fig. 6 Magnified image of a part of Fig. 5.

directional solidification was perfomed in flowing dry air, namely, the condition in which the partial pressure of oxygen, Po_2, is equal to one, the temperature coexisting in both the CoO and Co_3O_4 phases is estimated about 1240 K, which is high enough to allow diffusion of oxygen atoms in the c-ZrO_2 phase and the interface.

The existence of Co metal in the CoO phase of a sample grown in stagnant air is also explained by the thermodynamical equation of $\Delta G = RT \ln Po_2$, where ΔG, R, and T are the standard free energy of formation, gas constant and temperature, respectively, which are applied to the thermodynamic data listed in reference 9. At a temperature of 1800 K, which is below the eutectic temperature, a partial pressure of oxygen, Po_2, less than 10^{-6} results in the decomposition of CoO to Co and O_2, which may be a possible value in the present experiment. The formation of Co metal or the Co_3O_4 phase in the CoO matrix at the interface is due to the following factors: 1) cubic ZrO_2 is a fast ionic conductor of oxygen, and 2) the valence of Co is easily changed by the partial pressure of oxygen.

CONCLUSIONS

The interface structure of directionally solidified CoO-ZrO_2(Y_2O_3) eutectic was investigated by transmission electron microscopy and the following results were obtained.

(1) The interface of the lamella has an orientation relationship of (001)cubic ZrO_2 // (111)CoO and steps with periodic spacing due to the accommodation of mismatch from the near-coincidence relation.

(2) When the directional solidification was performed in an atmosphere of stagnant air, Co metal was observed at the interface in the CoO matrix. On the other hand, in an atmosphere of flowing dry air, Co_3O_4 precipitate was detected at the interface. All these materials have the orientation relationship of {hkl}CoO // {hkl}precipitates, though a small angle rotation of lattice was seen in Co_3O_4 to CoO.

(3) The precipitation behavior at the interface due to the atmosphere during solidification and successive cooling can be understood from thermodynamic considerations of the partial pressure of oxygen.

REFERENCES

[1] R. L. Ashbrook, "Directionally Solidified Ceramic Eutectics," J. Amer. Ceram. Soc. , 60 (1977) 428.

[2] W. J. Minford, R. C. Bradt and V. S. Stubican, "Crystallography and Microstructure of Directionally Solidified Oxide Eutectics," J. Amer. Ceram. Soc., 62 (1979) 154.

[3] V. S. Stubican and R. C. Bradt, "Eutectic Solidification in Ceramic Systems," Ann. Rev. Mater. Sci. 11 (1981) 267.
[4] J. Echigoya, H. Suto and S. Hayashi, "Growth Morphology and Orientation Relationships in Directionally Solidified $MgO-ZrO_2$ Eutectic," Trans. JIM, 26 (1985) 895.
[5] J. Echigoya and H. Suto, "Structure of Interphase Interface and Solute Segregation in Directionally Solidified $MgO-ZrO_2$ Eutectic," Proceedings of JIMIS-4 (1986) 213.
[6] J. Echigoya, G. Dhalenne and A. Revcolevschi, "TEM Study of Interfaces of directionally solidified $NiO-Gd_2O_3$ Eutectic Structures," Science of Ceramics, 14 (1988) 437.
[7] J. Echigoya and S. Hayashi, "Directionally Solidification of $CoO-ZrO_2$ Eutectics," J. Crystal Growth, 129 (1993) 699.
[8] E. Aukrust and A. Muan, "Thermodynamic Properties of Solid Solution with Spinel-Type Structure. I, The System $Co_3O_4-Mn_3O_4$," Trans. AIME, 230 (1964) 379.
[9] Thermo-Chemical Properties of Inorganic Substances, Ed. by O. Knacke, O. Kubaschewski and K. Hesselmann, Springer (1991) Berlin.

Acknoledgement

The authors express their thanks to Mr. E. Aoyagi for his assistance in the electron microscope experiments.

ESTIMATION OF DIFFUSION COEFFICIENTS FOR ZIRCONIA

Michiko Kusunoki, Kyoko Yonemitsu, Yukichi Sasaki, Yuichi Ikuhara and Yukio Kubo
Japan Fine Ceramics Center
Nagoya, Japan, 456

ABSTRACT

Grain growth and sintering phenomena of zirconia (ZrO_2) ultra-fine particles were observed by in situ heating up to 1200°C in a 400 kV transmission electron microscope[1]. The rates of grain boundary migration and sliding of lattice planes on the surface of ZrO_2 ultra-fine particles were measured by the in situ recorded video-system. With these values, the grain boundary diffusion coefficient D_{gb} was calculated to be 1.4x $10^{-17}m^2/s$ at 1200°C in accordance with Turnbull's theory[2] and surface self-diffusion coefficient D_s was estimated to be $2.2x10^{-17}m^2/s$ at 1200°C according to Herring-Mullins' theory[3].

INTRODUCTION

For ceramics, in situ observation at high temperature (>1000°C) by high resolution electron microscopy is considerably important to understand high temperature behavior of these materials. In the case of metals, not only volume diffusivity constants but also short circuit diffusivity constants obtained by tracer methods with radioactive isotope and other ways have been reported. Regarding surface diffusion in particular, a blunting phenomenon for the field emission cathode of a tungsten tip was directly observed[4,5,6] to determine the constants. The diffusion constant of the migration of tungsten atoms in blunting process was obtained by observing the change rate of the radius on each crystal plane of the tungsten tip during heat treatment.

In the case of ceramics, however, there are few experiments for surface diffusion measurement. We believe that direct observation of diffusion behavior on an atomic level should be a useful method to estimate the diffusivity constants in locarized area.

We have already shown the results of grain growth and sintering phenomena on zirconia ultra-fine particles obtained by in situ high resolution microscopy at 1200°C in the previous paper[1]. Rates of grain boundary motion and sliding of lattice layers on the surface of particles were measured by this experiment. The purpose of this paper is to estimate the grain boundary diffusion coefficient and the surface self-diffusion coefficient on the basis of these experimental results.

EXPERIMENTAL PROCEDURE

The zirconia powder used in the present experiment was a commercial material (Nippon Shokubai Co.,Ltd.) with 99.7 % purity. The average particle size was about 30 nm. Each particle is a single crystal with no remarkable lattice defects and is already coalesced a little by calcination. Electron diffraction patterns showed that the particles are of the monoclinic phase at all temperatures below 1200°C. A method for mounting the specimen was shown in the previous paper[1] in detail. The electron microscope used was a JEM 4000-FX operated at 400 kV, and the heating sample holder was GATAN model 628TA. In situ heating observations were carried out up to 1200°C in a vacuum of 10^{-5}Pa.

RESULTS AND DISCUSSIONS

Grain Boundary Migration

Figure 1 is a series of images for the coalescence between two ZrO_2 particles at 1200°C. In Fig. 1, the pictures from (a) to (d) were taken at a time interval of 3.7 s. These pictures show that the larger particle B became a seed for growth, and the smaller particle A was absorbed into the particle B, that is , the grain boundary migrated as indicated by the arrows. The driving force for the migration is considered to be due to a grain boundary interfacial free energy. The migration rate of grain boundary v can be presented[2] by:

$$v = - M \cdot d\mu / dx \qquad (1)$$

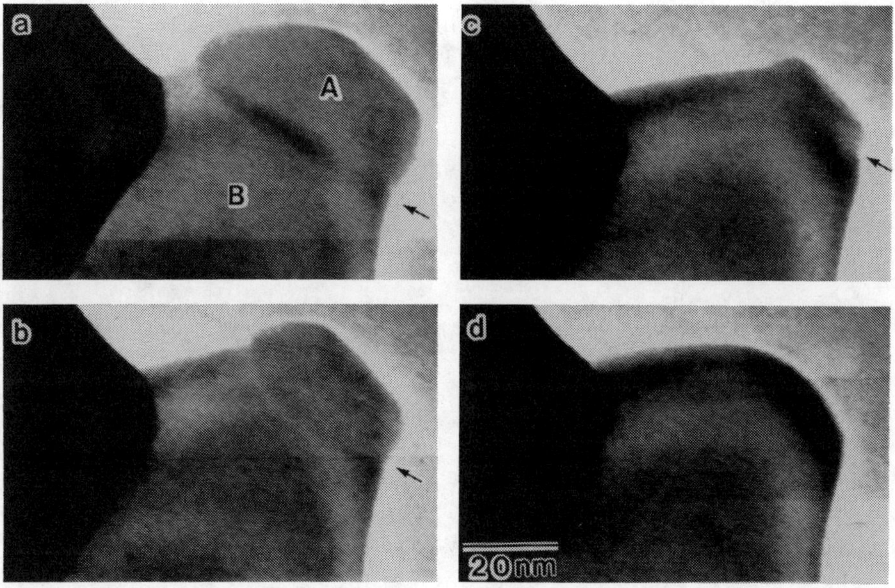

Fig. 1. Series of images of a coalescence of two small ZrO_2 particles at 1200°C. The images were taken in going from (a) to (d) at a time interval of 3.7 s.

where M is the atomic mobility, μ is the chemical potential per atom, and x is the coordinate measured in the direction of the grain boundary movement. Here, it is assumed that the mobility M is constant irrespective of the radius of the particle r. According to the Nernst-Einstein equation and Turnbull's expression, the basic relation between v and grain boundary diffusionc oefficient D_{gb} is presented by

$$v = D_{gb} \Delta F / \lambda RT \quad (2),$$

where ΔF is the total free energy difference (J/mol) on the two sides of the boundary, R is a gas constant, T is the absolute temperature, and λ is the thickness of the boundary. ΔF is the driving force and is proportional to the curvature of the particle, 2/r. Accordingly ΔF is presented by the next equation:

$$\Delta F = 2 \gamma V/r \quad (3),$$

where γ is boundary energy (J/m^2), V is molecular volume and r is the radius of the particle. Substituting eq 3 into eq 2 gives the basic relation between D_{gb} and v:

$$v = \frac{2 \gamma V}{\lambda RTr} D_{gb} \quad (4)$$

From the result of the in situ observation shown in Fig. 1, the value v was obtained as 6.7×10^{-9}m/s. Consequently, if γ can be known, the grain boundary diffusion coefficient is given by eq 3. However, the surface energy for ZrO_2 has not been reported yet. So, γ was estimated at roughly 1 J/m^2 comparing with the value obtained for MgO[7], V is the molecular volume of ZrO_2, $2.1 \times 10^{-5} m^3/mol$, the radius of the particle was used as the average radius of the two particles A and B in Fig. 1, 2.5×10^{-8}m, heating temperature T = 1473K. Substituting these values, D_{gb} for ZrO_2 was determined as $1.4x10^{-17}m^2/s$. There are no other data for pure ZrO_2 for comparison with this value. However, in the case of $ZrO_2(CaO)$[9], $6.6x10^{-20}m^2/s$ for Zr atom and $4.1x10^{-11}m^2/s$ for O atom are roughly estimated at 1200°C.

Surface migration

Figure 2 represents another series of images for grain growth selected from a videotape recorder. In this area, two particles A and B, were joined together, though the neck area was not observed. These pictures show the sliding of lattice planes during coalescing between particle A and B. The pictures in Fig. 2 were

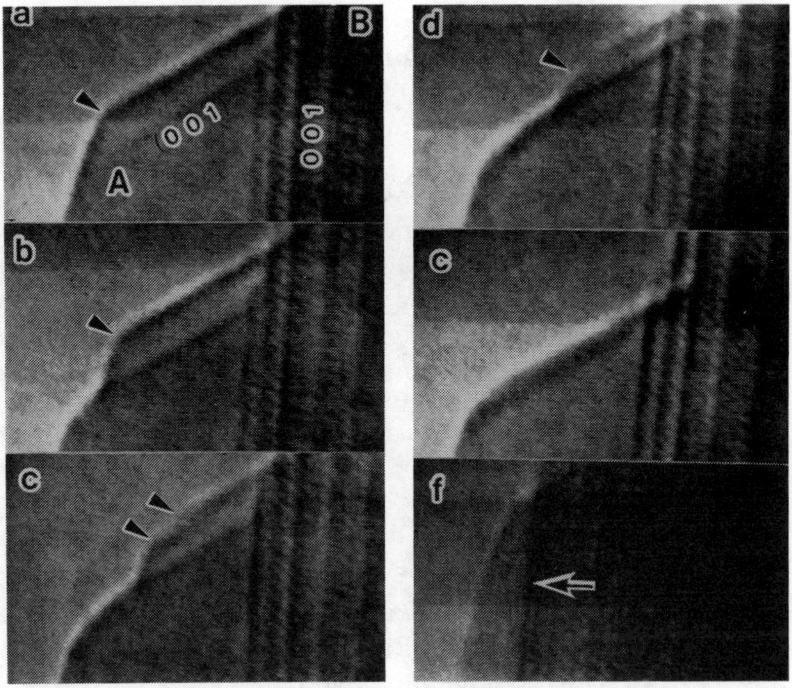

Fig. 2. Series of images of movements of (001) lattice planes during sintering at 1200°C. This phenomenon directly corresponds to surface self-diffusion.

taken in going from (a) to (e) at a time interval of 7 s, two or three (001) lattice planes were slided on the surface of particle A. The small triangles in Fig. 2 indicate the corresponding points of the sliding plane. In succession, during 10 s, (001) planes on the surface of the particle B increase gradually in the [001] direction on the left in Fig. 2(f). This phenomenon directly corresponds to surface self-diffusion.

For determination of surface migration constants of various materials, the fact that other transport phenomena such as volume self-diffusion are mixed with it, creates important problems. In this experiment, the effect of only surface migration was perfectly extracted. Moreover, using the methods of FIM (field ion microscopy) and FEM (field electron microscopy), surface migration constants for tungsten and some other metals were skillfully determined[4,5,6]. However, these methods are not generally applicable to ceramics because some ceramics such as ZrO_2, are insulators. Therefore, no experiments have been performed to determine surface self-migration of ceramics.

Herring[3] proposed a criterion for the geometrical equilibrium of the solid-vacuum interface at the surface of an isothermal single crystal of arbitrary shape. The chemical potential μ is presented as the following expression:

$$\mu = \mu_0 + \gamma \Omega K \qquad (5)$$

μ_0 is an arbitrary constant which represents the value below a stress-free plane surface. γ is the surface tension, K is the principal curvatures of the solid surface, Ω is the volume per atom.

If the flux of atoms is only due to surface migration, and surface curvature K varies along the surface as a function of curve co-ordinate, s, the migration rate of surface atoms, v_s, is yielded by the following expression:

$$v_s = -\frac{D_s}{kT}\frac{d\mu}{ds} = -\frac{D_s}{kT}\frac{dK}{ds} \qquad (6)$$

where D_s is the surface self-diffusion coefficient, T is the absolute temperature, and k is the Boltzman consrant.

In order to estimate dK/ds, if the shape of the cross section of the particle A is approximated as a part of the following equation:

$$y = a \sin \omega x, \qquad (7)$$

the migration rate v_s is yielded roughly approximately as follows:

$$v_s = -\frac{D_s \gamma \Omega}{kT} a\omega^3 \cos \omega x. \qquad (8)$$

Here, a is the amplitude of the surface's shape for the cross section and ω is the wave number as shown in Figure 3.

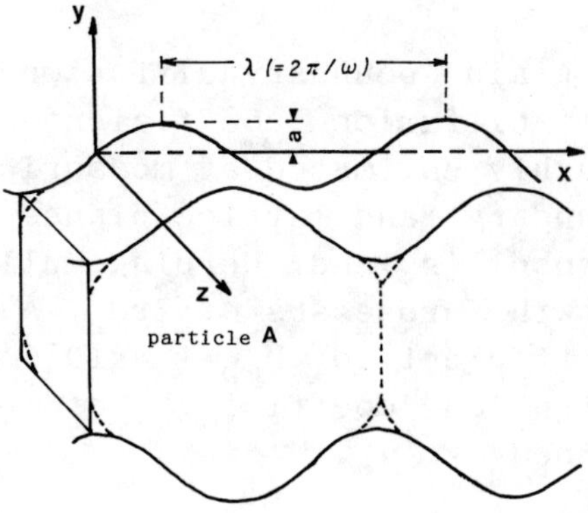

Fig. 3. Surface's shape of the cross section for two joined particles A and B was approximated by a part of sine curve.

From the videotape recording of the in situ observation as shown in Fig. 2, the migration rate of surface atoms v_s was measured as 38.3×10^{-10}m/s. Moreover, the series of pictures in Fig. 2 indicate that atoms move with the mass size of a few lattice planes, not individually. So, Ω was estimated as the volume of an unit cell, 1.41×10^{-28}m^3. γ was estimated at 1 J/m^2. Substituting these values and the other experimental data, T=1473K, a= 1.6×10^{-9}m, $\omega = 2.5 \times 10^8$m^{-1}, the surface self-diffusion coefficient of ZrO_2 at 1200°C was estimated as 2.2×10^{-17}m^2/s. There are no data of surface self-diffusion coefficient for ZrO_2 and other ceramics to compare with this value.

The results show that the two values D_{gb} and D_s obtained in this experiment were almost the same. However, in general, the surface diffusivity constant is rather larger than the volume diffusivity constant. The surface diffusivity constant of a dimer of tungsten[9] or molybdenum[10] obtained by FEM, were reported as 10^{-16}m^2/s. This value is considerably smaller than the value of an atom of W or Mo, 10^{-7}m^2/s. The present experiment showed that a zirconium atom or an oxygen atom did not jump individually, but moved with the mass of lattice planes. This fact suggests that the diffusion coefficient of ZrO_2 is presumed to be a rather small value.

CONCLUSIONS

A grain boundary diffusion coefficient and surface self-diffusion coefficient of ZrO_2 at 1200°C were roughly estimated by measuring migration rates of grain boundary and lattice planes on the surface, applying Turnbull's and Herring-Mullins' theories to grain growth processes of ZrO_2. A grain boundary diffusion coefficient of $D_{gb} = 1.4 \times 10^{-17}$m^2/s, and a surface self-diffusion coefficient of $D_s = 2.2 \times 10^{-17}$m^2/s were obtained.

By in situ observation using high resolution electron microscopy, it becomes possible to determine the diffusion coefficients, in particular the surface diffusion coefficients of ceramics. These measurements will be helpful to elucidate sintering behavior of ceramics.

ACKNOWLEDGEMENTS

This work was conducted in the program; "Advanced Chemical Processing Technology," consigned to the Advanced Chemical Processing Technology Research Association from the New Energy and Industrial Technology Development Organization, which is carried out under the Large-Scale Project enforced by the Agency of Industrial Science and Technology, the Ministry of International Trade and Industry.

Reference

[1]M. Kusunoki, K. Yonemitsu, Y. Sasaki and Y. Kubo, "In Situ Observation of Zirconia Particles at 1200°C by High-Resolution Electron Microscopy," J. Am. Ceram. Soc., 76[3], 763-65(1993).

[2]D. Turnbull, "Theory of Grain Boundary Migration Rates," Trans. AIME, 191, 661-65(1951).

[3]W. W. Mullins, "Theory of Thermal Grooving",J. Appl. Phys.,28[3],333-339(1957).

[4]E. Sugata and S. Nakamura, "Study of Field Emission Cathode by Field Ion Microscopy," Jpn. J. Appl. Phys., 1[1], 50-59(1962).

[5]J. P. Barbour, F. M. Charbonnier, W. W. Dolan, W. P. Dyke, E. E. Martin and J. K. Trolan, "Determination of the Surface Tension and Surface Migration Constants for Tungsten," Phys. Rev. 117[6], 1452-59(1960).

[6]C. P. C. Bettler and F. M. Charbonnier, "Activation Energy for the Surface Migration of Tungsten in the Presence of a High-Electric Field," Phys. Rev. 119[1], 85-93(1960).

[7]A. R. C. Westwood and D. L. Goldheim, "Cleavage Surface Energy of {100} Magnesium Oxide", J. Appl. Phys., 34[11], 3335-39(1963).

[8]Jpn. Chem. Soc., Kinzoku Binran, 2nd,p65, Ed.,Maruzen(1984) in Japanese.

[9]W. R. Graham and G. Ehrlich, "Surface Self-Diffusion of Atoms and Atom Pairs," Phys. Rev. Lett., 31[23], 1407-08(1973).

[10]T. Sakata and S. Nakamura, "Surface Diffusion of Molybdenum Atoms on Tungsten Surface," Surface Sci., 51, 313-17(1975).

FORMATION OF YTTRIUM ALUMINATES IN THE PRESENCE OF TiN

A. Kuzjukevics[a] and K. Ishizaki
Nagaoka University of Technology
1603-1 Kamitomioka, Nagaoka, Niigata 940-21, Japan

ABSTRACT

The formation and stability of $YAlO_3$ (YAP) and $Y_3Al_5O_{12}$ (YAG) ceramics with and without addition of TiN at high temperatures were investigated.

YAP formation is enhanced by the addition of TiN powder. During the formation, dissolution of Ti^{+3} in the YAP lattice also occurs. This makes the formed compound stable at high temperatures. The amount of dissolved titanium increases with decreasing TiN particle size.

The interaction between the matrix and TiN also occurs in the YAG ceramic. It leads to substitution of Al^{+3} by Ti^{+3} in the YAG lattice and appearance of free alumina.

INTRODUCTION

$Y_3Al_5O_{12}$ (YAG) ceramics have been intensively investigated during the last decade [1-7] because of these materials' application as translucent windows in optical devices. Before that only a few attempts have been done to produce yttrium aluminate ceramics [8, 9].

The theoretical value of elastic modulus, E, of YAG calculated from the single crystal stiffness constants is 283 GPa [3, 4]. This means that YAG is stiffer than Y_2O_3 (E=177 GPa) but more compliant than Al_2O_3 (E=400 GPa). Consequently, the hardness of YAG ceramics is less than that of Al_2O_3 and higher than that of Y_2O_3 [3]. The fracture toughness and strength of YAG ceramics depend on the processing route and presence of dopants [2-4]. A typical K_{Ic} value of vacuum-sintered YAG doped with SiO_2 (1000-1500 wt ppm) or MgO (500 wt ppm) is 1.7 $MN{\cdot}m^{-3/2}$. An increase in K_{Ic} to 3.1 $MN{\cdot}m^{-3/2}$ was observed for the ceramics sintered in hydrogen. This is due to the absence of Al-rich inclusions observed in vacuum-sintered YAG. The increase in K_{Ic} also gives higher fracture strength of

---

[a] Present address: Institute of Inorganic Chemistry, LAS, 34 Miera str., Salaspils, Latvia, LV-2169.

the hydrogen-sintered ceramic - 640 MPa compared to 420 MPa of the vacuum-sintered one. Values of K_{Ic} and fracture strength as low as 1.5 $MN \cdot m^{-3/2}$ and 234 MPa respectively were reported for the non-doped YAG ceramic produced by hot pressing [7].

Another yttrium aluminate, $YAlO_3$ (YAP), has a crystal structure consisting of AlO_6 octahedrons. These are considered more stable than the AlO_4 tetrahedrons of the YAG structure. The density of YAP (5.351 g/cm^3 [10]) is higher than that of YAG (4.552 g/cm^3 [10]), $Y_4Al_2O_9$ (YAM) (4.518 g/cm^3 [10]), Y_2O_3 (5.031 g/cm^3 [10]), or Al_2O_3 (3.987 g/cm^3 [10]). YAP is the only compound of the Al_2O_3-Y_2O_3 system with higher density than that of its constituents. This is due to shorter Me-O bonding distances. Consequently, the E-modulus of YAP might be larger compared, for example, with YAG. This should be beneficial for the mechanical properties of YAP ceramics. In spite of this, almost no research has been done to obtain YAP ceramics or to evaluate their properties. The main reason is it is difficult to produce single phase YAP materials. In the present study, YAP ceramics are produced by the solid state reaction between alumina and yttria in the presence of TiN. To the best of the authors' knowledge, this is the first attempt to produce YAP composite ceramics.

Table I. Composition of the starting mixtures

Sample	$[Y_2O_3]/[Al_2O_3]$ ratio	TiN ultrafine	TiN coarse
P	1:1 (YAP)	-	-
PTu	1:1 (YAP)	15 wt%	-
PTc	1:1 (YAP)	-	15 wt%
G	3:5 (YAG)	-	-
GTu	3:5 (YAG)	15 wt%	-

EXPERIMENTAL

Al_2O_3 [b] and Y_2O_3 [c] powders in 1:1 (YAP) and 3:5 (YAG) molar ratios with and without addition of 15 wt% ultrafine TiN[d] or coarse TiN[e] powders were mixed for 24 h in ethanol by milling with Al_2O_3 balls (99.5 wt%) in a plastic pot. The powder slurries were dried in a vacuum furnace for 6 h at 60°C and then crushed. The resulting powders were hot pressed in nitrogen for 1 hour at 1500°, 1600°, and 1700°C at a pressure of 35 MPa in a graphite mold and a plunger coated

[b] average particle size, 0.3 μm; Sumitomo Chemical Co., Ltd., Tokyo, Japan.

[c] average particle size, 1.35 μm; Shin-Etsu Chemical Co., Ltd., Tokyo, Japan.

[d] average particle size, 0.04 μm; oxygen content, 2.2 wt%; made by the plasma chemical synthesis [11]; Institute of Inorganic Chemistry, Salaspils, Latvia.

[e] average particle size, 1.38 μm; oxygen content, 1.31 wt%; Nippon Shin-Kinzoku Co.,Ltd., Japan.

with BN. A heating rate of 1000°C/h was used for all experiments. After the hot pressing, some billets were annealed in nitrogen for 5 h at 1600°C. The lattice parameters were measured by X-ray diffraction using Si as an internal standard. Table I shows the composition of the starting mixtures.

RESULTS AND DISCUSSION

YAG and YAG-TiN ceramics

A commonly used technique to fabricate yttrium aluminates by the solid-state reaction of yttria and alumina requires temperatures about 2000°C [12-14]. The reaction between Al_2O_3 and Y_2O_3 powders in 2:4, 1:1, and 5:3 ratios results in formation of the monoclinic YAM at 900°-950°C, independent of mixture composition [13, 14]. YAP with a perovskite-type structure forms at ~1100°C; and at 1300°C the garnet YAG crystallizes. During the interaction, Al diffuses into Y_2O_3. This diffusion is slow, and consequently, the reactions for the formation of yttrium aluminates are also slow. For example, the yield of YAP is less than 70% after 40 h at 1400°C [13]. Moreover, the traces of YAG and YAM remain up to the melting temperature of YAP. As for YAG, single phase garnet requires a sintering temperature higher than 1700°C. Below this temperature, traces of YAP are present [13, 14].

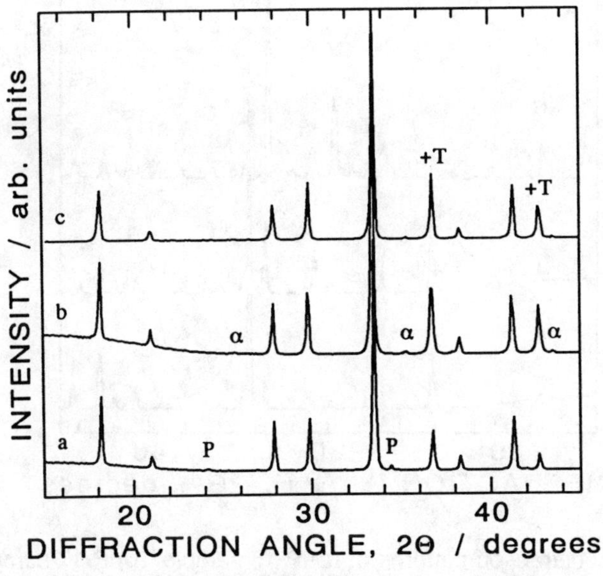

Figure 1. X-ray diffraction patterns of (a) G-15 sample, (b) GTu-15 sample, and (c) GTu-15A sample ("15" means that the sample was hot pressed at 1500°C, "A" - annealed sample). Identification marks: (P) YAP, (T) TiN, and (α) α-Al_2O_3. Non-marked diffraction maxima belong to YAG.

Results of the present work show that the G samples were single phase YAG when hot pressed at or above 1600°C. At lower temperatures traces of YAP were observed (Fig. 1 a). The presence of TiN in the GTu samples changed the phase composition (Fig. 1 b). Traces of the α-Al_2O_3 remained in the GTu samples hot pressed at 1500°-1700°C, as well as in the GTu samples annealed for 5 h at 1600°C. According to the Y_2O_3 -Al_2O_3 phase diagram [15], the YAG-Al_2O_3 mixture is stable at sintering temperatures used in this work.

The measured lattice parameter of YAG (a_{YAG}) of the G-16 sample (sintered at 1600°C) was 1.20061(14) nm and 1.20074(10) of the G-16A (annealed) sample. This is in good agreement with the reported value of 1.20089 nm [10]. However, a_{YAG} of the GTu-16 and GTu-16A samples are 1.20177(12) and 1.20171(8) nm, respectively, which is much larger than the reported value [10]. Therefore, the presence of free α-Al_2O_3 and the larger lattice parameter of YAG of the GTu samples clearly indicate that Ti^{+3} partially substitutes Al^{+3} in the YAG lattice. The ionic radius of Ti^{+3} is larger than that of Al^{+3} (0.076 and 0.051 nm, respectively, [16]), and consequently a_{YAG} increased as a result of the substitution of Al^{+3} by Ti^{+3}.

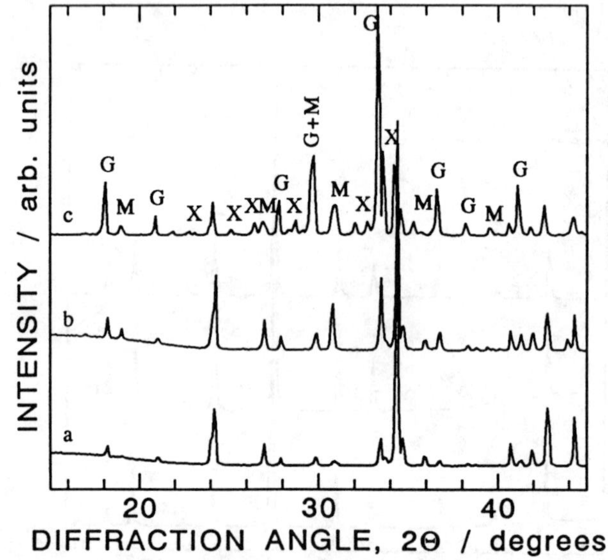

Figure 2. X-ray diffraction patterns of (a) P-15 sample, (b) P-17 sample, and (c) P-15A sample. Identification marks: (G) YAG, (M) YAM, and (X) unidentified X-phase. Non-marked diffraction maxima belong to YAP.

Lattice parameters of TiN are 0.42428(6) and 0.42429(6) nm for GTu-16 and GTu-16A samples, respectively. This coincides with the previously reported results of the changes of a_{TiN} for the Al_2O_3-TiN composites [17] caused by the substitution of nitrogen by oxygen in the TiN lattice. Therefore, the substitution of oxygen by nitrogen in the YAG lattice might also be expected since released aluminum was present as oxide and oxygen was dissolved in the TiN lattice. The substitution of oxygen by nitrogen in the titanium enriched aluminum titanate was reported by Perera [18].

YAP ceramics

According to Glushkova et al. [13, 14], YAP ceramics made by the solid state reaction of Y_2O_3 and Al_2O_3 contain traces of YAG and YAM up to the melting temperature of YAP. Similar results were observed for the P samples of the present study. As shown by X-ray diffraction, besides YAP, the P-15 sample contained also YAG and YAM (Fig. 2 a). The amount of these phases increased with increasing sintering temperature (Fig. 2 b). In contrast to the G and GTu samples, which remained unchanged after annealing, P-samples decomposed (Fig. 2 c). After annealing for 5 h at 1600°C, the amount of YAP decreased significantly. Beside increased amounts of YAG and YAM, an unknown phase appeared. This might be the same unidentified phase formed during the annealing of YAP single crystals [19]. The as sintered and annealed P-samples had an unusual marble-like microstructure consisting of large black grains in a light-grey matrix. However, no attempt was made to identify the black grains.

YAP-TiN ceramics

The phase composition of the PTu and PTc samples were different from those of the P samples. PTu with the ultrafine TiN and PTc with the coarse TiN powder showed no presence of YAM when sintered at and above 1500°C (Fig. 3 a and 4 a). The amount of YAG decreased with increasing sintering temperature from 1500°C to 1600°C. However, it remained almost constant at temperatures above 1600°C. Compared with the P samples (Fig. 2), the amount of additional phases was much smaller for the PTu and PTc samples. This implies that the presence of TiN particles in the mixture of Y_2O_3 and Al_2O_3 promotes YAP formation. Furthermore, the PTc samples with addition of coarse TiN contained less YAG (Fig. 4) than the PTu samples (Fig. 3). This is probably due to different amounts of Ti^{+3} which dissolved in the YAP lattice. The contact area between YAP and TiN is much larger for the PTu samples with ultrafine TiN. This must lead to a stronger interaction between the matrix and TiN and larger amount of YAG.

The interaction of YAP and TiN might be described by the following equations:

$$YAlO_3 + xTiN \rightarrow YAl_{1-x}Ti_xO_{3-1.5x}N_x[\ ]_{x/2} + (x/2)Al_2O_3 \qquad (1)$$

if the TiN substitutes Al_2O_3, or

$$YAlO_3 + (y/2)Ti_2O_3 \rightarrow YAl_{1-y}Ti_yO_3 + (y/2)Al_2O_3 \qquad (2)$$

if the Ti_2O_3 formed during hot pressing, substitutes Al_2O_3. In both cases a mixture of the YAP and Al_2O_3 forms and the composition shifts from a [Y]/([Al]+[Ti]) ratio of 1 to lower values or toward YAG. (For the [Y]/([Al]+[Ti]) ratio

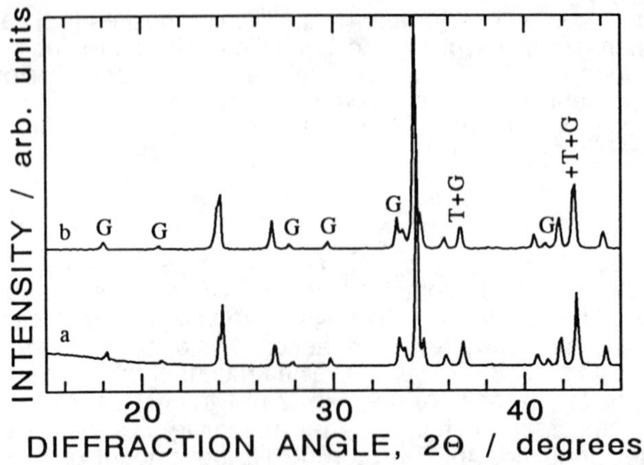

Figure 3. X-ray diffraction patterns of (a) PTu-16 and (b) PTu-16A. Identification marks: (G) YAG and (T) TiN. Non-marked diffraction maxima belong to YAP.

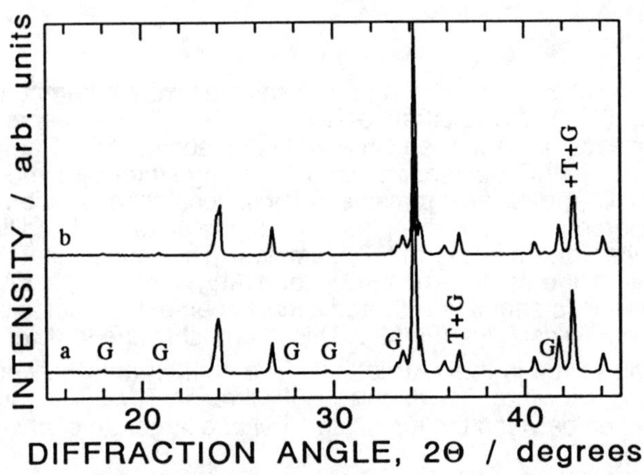

Figure 4. X-ray diffraction patterns of (a) PTc-16 and (b) PTc-16A. Identification marks: (G) YAG and (T) TiN. Non-marked diffraction maxima belong to YAP.

only the reacting part of Ti is considered.) Such composition is unstable at the sintering temperatures used and a mixture of YAP and YAG should form [15]. This is consistent with the experimental observations in this work.

As-sintered PTu and PTc samples differ from P samples by a larger amount of YAP and a lesser amount of additional phases. Another characteristic feature of them is stability at high temperatures. According to the X-ray patterns, no changes in phase composition occurred in the annealed sample (Fig. 3 b and 4 b). This indicates that the dissolution of Ti^{+3} and, probably, nitrogen in the YAP lattice made the formed compound more stable at higher temperature than pure YAP.

CONCLUSIONS

The formation and stability of $YAlO_3$ (YAP) and $Y_3Al_5O_{12}$ (YAG) ceramics with and without addition of TiN at high temperatures were investigated.

The addition of TiN promoted YAP formation. During the formation, dissolution of Ti^{+3} in the YAP lattice also occurred and made the formed compound stable at high temperatures. The amount of dissolved titanium increased with decreasing TiN particle size. The pure YAP ceramic without addition of TiN decomposes when annealed at 1600°C for a longer time. The interaction between TiN and YAG led to substitution of Al^{+3} of the YAG lattice by Ti^{+3} and appearance of free alumina.

REFERENCES

1. G. De With and H.J.A. Van Dijk, "Translucent yttrium aluminum oxide ($Al_5Y_3O_{12}$) ceramics", *Mater. Res. Bull.* Vol.19, No.12, pp.1669-74, (1984).
2. G. De With, "Translucent $Al_5Y_3O_{12}$ ceramics: something old, something new", *Mater. Sci. Monogr.*, Vol.38C (*High Tech Ceram., Pt.C*), pp.2063-2075, (1987).
3. G. De With and J.E.D. Parren, "Translucent yttrium aluminum garnet ($Y_3Al_5O_{12}$) ceramics: mechanical properties", *Solid State Ionics*, Vol.16, No.1, pp.81-86, (1985).
4. F.J.C.M. Toolenaar and G. De With, "Translucent $Al_5Y_3O_{12}$ The influence of process parameters", *Br. Ceram. Proc.*, Vol.37, No.8 (*Spec. Ceram.*), pp.241-246, (1986).
5. A. Ikesue and T. Kinoshita, "Synthesis of polycrystalline transparent YAG ceramics produced by solid-state reaction and its optical properties", *Kidorui*, No.18, pp.42-43, (1991).
6. J.W.G.A. Vrolijk, G.J. Bezemer and R. Metselaar, "Sintering of Yttrium Aluminium Garnet Ceramics", in: *European Ceramic Society Third Conference*, Madrid, Sept. 12-17, (1993).
7. K. Keller, T. Mah and T.A. Parthasarathy, "Processing and mechanical properties of polycrystalline $Al_5Y_3O_{12}$ (yttrium aluminum garnet)", *Ceram. Eng. Sci. Proc.*, Vol.11, No.7-8, pp.1122-1133, (1990).
8. L.P. Morozova, E.S. Lukin, T.V. Efimovskaya, and A.V. Smolya, "Preparation of ceramics from aluminum yttrium garnet", *Steklo Keram.*, No.8, pp.28-31, (1978).
9. E.S. Lukin, L.P. Kiryukhina, T.V. Efimovskaya, A.V. Smolya, and I.F. Panteleeva, "Study of the effect of zirconium dioxide, hafnium dioxide and magnesium oxide additives on the sintering, phase composition and microstructure of ceramics made from garnet yttrium aluminum", *Tr. Mosk. Khim.-Tekhnol. Inst. im. D.I. Mendeleeva*, No.100, pp.20-25, (1978).
10. JCPDS cards: 33-0041(YAP); 33-0040 (YAG); 34-0368 (YAM); 25-1200 (Y_2O_3); and 10-0173 (α-Al_2O_3).

11. J. Grabis, T. Millers, G. Heidemane, and Dz. Jankovica, "Interaction of titanium and nitrogen in high temperature gas flow", *Izv. AN LatvSSR. Ser. khim.*, No.1, pp.10-15, (1975).
12. D.R. Messier and G.E. Gazza, "Synthesis of $MgAl_2O_4$ and $Y_3Al_5O_{12}$ by thermal decomposition of hydrated nitrate mixture", *Amer. Cer. Soc. Bull.*, Vol.51, No.9, pp.692-694, (1972).
13. V.B. Glushkova, V.A. Krzhizhanovskaya, O.N. Egorova, Yu.P. Udalov, and L.P. Kachalova, "Reaction of yttrium and aluminum oxides", *Izv. Akad. Nauk SSSR, Neorg. Mater.*, Vol.19, No.1, pp.95-99, (1983).
14. V.B. Glushkova, V.A. Krzhizhanovskaya and O.N. Egorova, "Physicochemical study of compounds in the yttrium oxide - aluminum oxide system", *Dokl. Akad. Nauk SSSR*, Vol.260, No.5, pp.1157-60, (1981).
15. B. Cockayne, "The use and enigma of the Y_2O_3-Al_2O_3 phase system", *J. Less-Common Met.*, Vol.114, No. , pp.199-206, (1985).
16. D. Goldberg, "Des systemes formes par l'alumine avec quelques oxydes de metaux trivalents et tetravalents, en particulier, l'oxyde de titane", *Rev. Int. Hautes Temper. et Refract.*, Vol.5, No.1, pp.181-194, (1968).
17. A. Kuzjukevics, K. Ishizaki, J. Grabis, and T. Millers, "Hot pressing of TiN-alumina composites", *J. Ceram. Soc. Japan*, Vol.102, No.3 , pp.225-230, (1994).
18. D.S. Perera, "Phase relationships in the Ti-Al-O-N system", *British Ceramic Transactions Journal*, Vol.89, No.1, pp.57, (1990).
19. J.S. Abell, I.R. Harris, B. Cockaayne, and B. Lent, "An investigation of phase stability in yttrium oxide-aluminum oxide", *J. Mater. Sci.*, Vol.9, No.4, pp.527-537, (1974).

INTERFACES IN TUNGSTEN CARBIDE

Tetsuya Suzuki and Kunio Shibuki
Technical Research Laboratory, Toshiba Tungaloy Co., Ltd.
1-7 Tsukagoshi, Saiwai-ku, Kawasaki 210, Japan

Toshiyuki Suzuki and Yuichi Ikuhara
Materials Characterization Division, Research and Development Laboratory, Japan Fine Ceramics Center, 2-4-1 Mutsuno, Atsuta-ku, Nagoya 456, Japan

Abstract

Interfaces in tungsten carbide(WC) were observed by high resolution electron microscopy to determine energy-stable crystallographic relationships. It was found that the following crystallographic relationships were often observed between WC grains and considered to be stable.

$$(10\bar{1}0) \; // \; (10\bar{1}0)$$
$$[0001] \; // \; [1\bar{2}10] \text{ ------(A)}$$

The above simple relationship shows that the lattice spacings of the basal plane, (0002), and the prism plane, $(1\bar{2}10)$, are quite close with 2% misfit, that is a/c~1. The coincidence site lattice (CSL) method was applied for the basal and prism planes to simulate the results and predict another favorable relationship.

1. Introduction

Since the advent of high resolution electron microscopy, a number of ceramics have been investigated to determine the role of grain boundaries which play in controlling mechanical and electrical properties. These ceramics include Si_3N_4[1], SiC[2-4], AlN[5], Al_2O_3[6-8], CVD-diamond[9] and spinel[10].

Tungsten carbide(WC) is the major constituent of cemented carbide(WC-Co) cutting tools, although WC itself cannot be used for lack in fracture toughness. Hagege et al. first paid attention to the unique atomic configuration of the WC unit cell that the c/a ratio was identical[11]. They calculated and estimated that the most stable grain boundary in WC had the following crystallographic relationships in Σ 2 grain boundaries.

$$(10\bar{1}0) // (10\bar{1}0)$$
$$[0001] // [1\bar{2}10]$$

However, they have not fully shown experimental results that explain the atomic configuration and boundary directions on WC grain boundaries. In this study, we show the evidence of several Σ 2 grain boundaries originating from the identical c/a ratio through high resolution electron microscopy. Further, the coincidence site lattice (CSL) method was used to evaluate the interfaces of WC in detail, based on the experimental results.

2. Experimental Procedure

The commercially available cemented carbide(WC-Co) was used for HREM observations. The mixed powder was sintered at ~1400^oC and then cut into the required shape for cutting tools(12.7 x 12.7 x 4.5 mm^3).

The specimens for HREM observations were made as follows. The WC-Co insert, 12.7 x 12.7 x 4.5 mm^3, was cut into small pieces and subsequently mechanically thinned to a thickness of ~100 μm. Next, small pieces(2 x 2 x 0.1 mm^3) were dimpled to ~30 μm and then ion-beam thinned to electron transparency (5kV). Transmission electron microscopy was carried out with a Topcorn 002B microscope operating at 200 kV with a point-to point resolution of 1.8Å.

3. Results and Discussion

3-1 Most stable interface in WC

It is noted that the unit cell dimensions of WC are virtually identical, that is c/a ~ 1. The a-direction corresponds to the 1/3<11$\bar{2}$0> and the c-direction to the <0001>.The identical c/a ratio leads to the formation of the Σ 2 type grain boundary as shown in Fig. 1, as suggested by Hagege et al.[11].

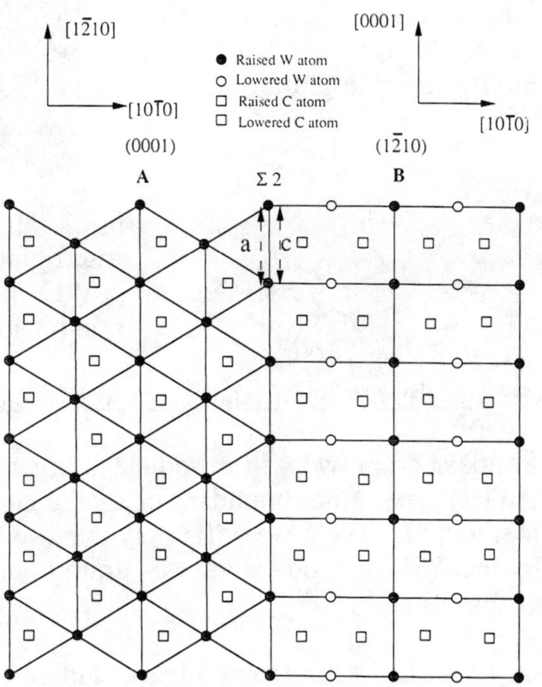

Fig. 1. Schematic illustration of Σ 2 grain boundary in WC.

Figure 2 shows a high resolution electron micrograph of WC-WC boundary, showing the evidence of a stable Σ 2 type grain boundary. The grains 1 and 2 are oriented to the [0001] and [$1\bar{2}10$] directions, respectively. The grain boundary was straight and two grains clearly had some crystallographic relationship. According to the micrograph, the following crystallographic relationship was obtained.

$(10\bar{1}0)$ // $(10\bar{1}0)$
[0001] // [$1\bar{2}10$] -----(A)

It was found that the (A) type boundary was the same as that of Fig. 1 and considered to be the most stable grain boundary.

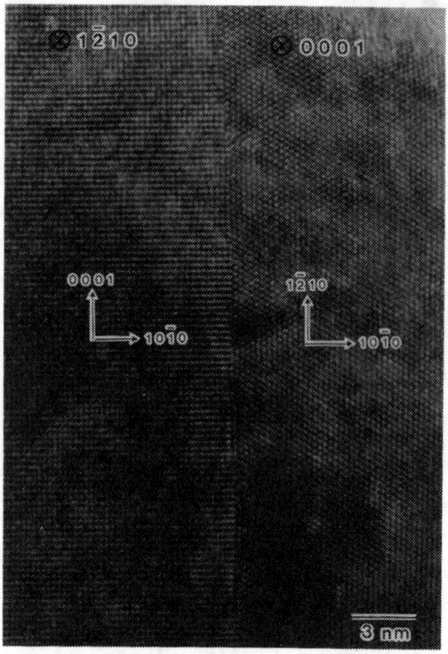

Fig. 2. High resolution electron micrograph of Σ 2 grain boundary in WC, indicating crystallographic relationship represented with arrows

3-2 Combinations of stable grain boundaries

In this section, we will introduce another stable boundary other than the (A) type grain boundary using the coincidence site lattice(CSL) method[12]. Then, we will apply the models which were calculated by the CSL method to various grain boundaries obtained from observation.

In the CSL model, the number of lattice sites of either grain contained by the primitive coincident lattice is defined as Σ. Thus, the basal twin boundary is referred to as the Σ 3 grain boundary since one-third of lattice sites on either side of the interface are common to both grains. Similarly, the (A) type grain boundary is a special case of $\Sigma 2$ grain boundaries.

It is generally accepted that two grains are superimposed to find the densest line in the number of overlapped atoms, along which grain boundaries should form[13]. Figure 3 shows CSL plots of the (0001) and $(1\bar{2}10)$ planes of WC, based on the result of Fig. 2. The densest line in the number of overlapped atoms is noted as (A) type grain boundary, which is the same type as seen in Fig. 2. The second densest line is noted as (B) which has the following crystallographic relationship.

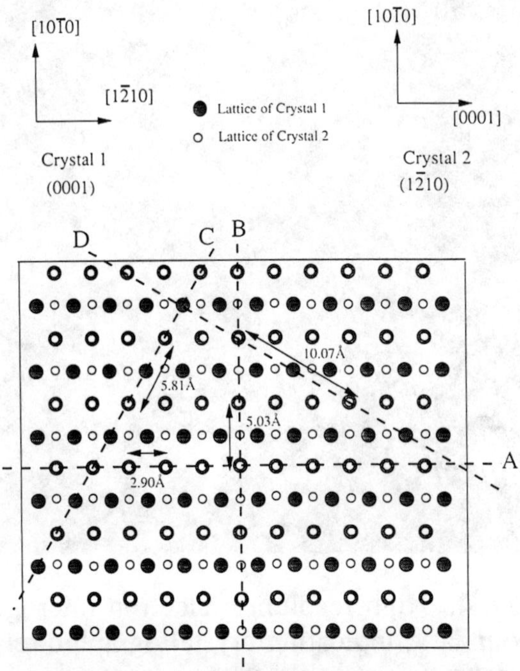

Fig.3. Coincidence site lattice (CSL) plots of the two WC grains oriented to the [0001] and $[1\bar{2}10]$ directions, showing four possible grain boundaries.

$(1\bar{2}10)$ // (0001)
[0001] // $[1\bar{2}10]$ ----(B)

Similarly, (C) and (D) type boundaries were defined in the same manner. Figure 4 shows a high resolution electron microgragh of grain boundary with the (D) type. This boundary runs in a zigzag path different from that in Fig. 2. However, it regularly turns as indicated by the broken line, reflecting the crystallographic

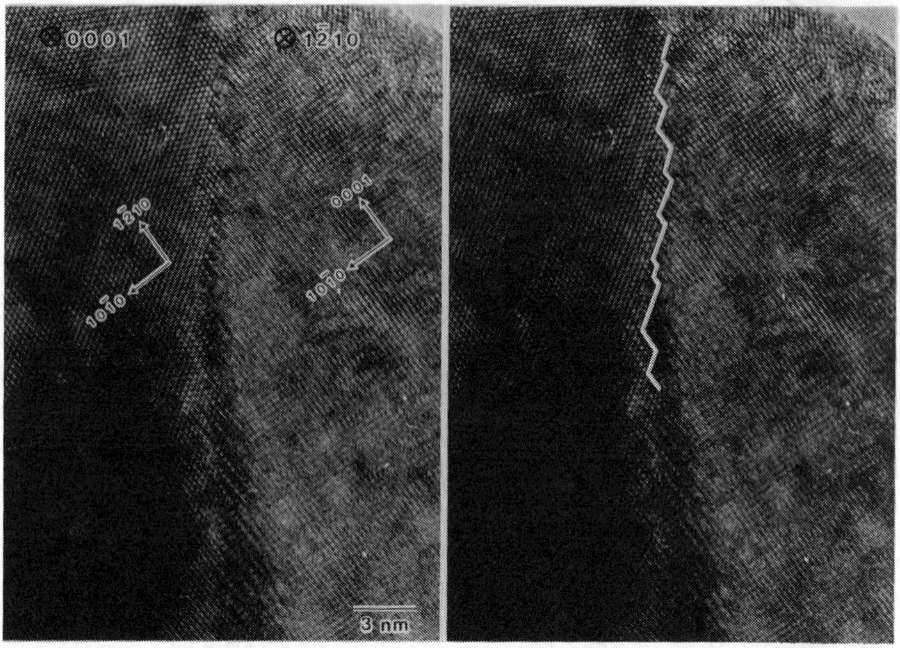

Fig. 4. High resolution electron micrograph of zigzag WC-WC boundary, indicating crystallographic relationship represented with arrows. (a) and (b) are the same region including the boundary.

relationships of (A) and (B) as schematically shown in Fig. 5. Thus, it is thought that the grain boundaries,which have low dense CSL directions,were divided into two stable boundaries following the crystallographic relationships of (A) and (B).

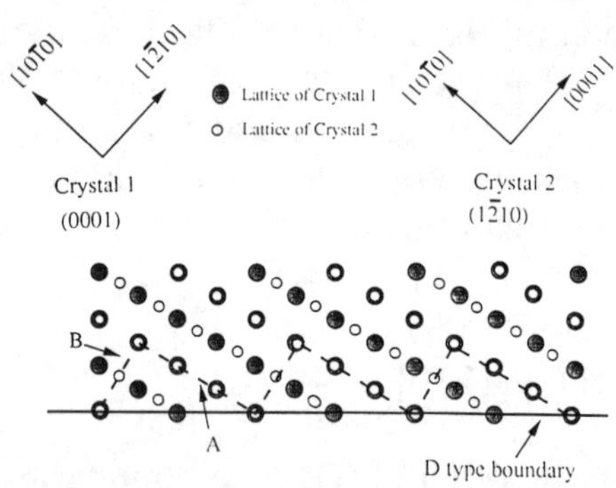

Fig. 5. Schematic illustration of zigzag boundary seen in Fig. 4. Note that the grain boundary runs following energetically stable (A) and (B) boundaries.

4. Summary

The interfaces in tungsten carbide were observed by high resolution electron microscopy. It was found that the unique interfaces were formed in WC, since the lattice spacing of (0002) and $(1\bar{2}10)$ are quite close with the identical a/c ratio.

4-1) The Σ 2 grain boundary with the following crystallographic relationship was often observed and considered to be the most stable grain boundary.

$$(10\bar{1}0) \; // \; (10\bar{1}0)$$
$$[0001] \; // \; [1\bar{2}10] \quad \text{-----(A)}$$

4-2) The coincidence site lattice (CSL) method predicted three more grain boundaries as well as the (A) type boundary. The second plausible grain boundary had the following crystallographic relationship and in fact was observed through electron microscopy.

$$(0001) \; // \; (1\bar{2}10) \quad \text{----(B)}$$
$$[1\bar{2}10] \; // \; [0001]$$

4-3) Even if the boundary directions favor a lower density of superimposed atoms such as the (C) and (D) types, the grain boundary is divided into the (A) and (B) type and runs in a zigzag path following the paths with lower interfacial energy.

Acknowledgement

The authors would like to thank Mr. Ueki, Mr.Kobayashi, Mr.Sasaki, Dr.Taniguchi and Mr.Yagi of Toshiba Tungaloy Co., Ltd. for their comments on the paper.

<References>

1. D.R.Clarke, "High Resolution Techniques and Application to Non-Oxide Ceramics," J. Am. Ceram. Soc., **62** [5-8] 236-46 (1979).

2. K.Hiraga, M.Hirabayashi, K.Niihara and T.Hirai, "Chemical Vapor Deposited Silicon Nitride and Silicon Carbide," Proc. IXth International Conf. on CVD, Ohio, 573-582 (1984).

3. Y.Ikuhara, H.Kurishita and H.Yoshinaga, "Grain Boundary and High Temperature Deformation of Sintered SiC," Yogyo-Kyokai-Shi, **95** [6] 638-645(1987).

4. Y.Ikuhara,H.Miyazaki, H.Kurishita and H.Yoshinaga, 'Grain Boundaries in SiC with No Additives," Proceedings of International Symposium, Sintering 87, Tokyo, (1987).

5. S.Hagege, S.Tanaka and Y.Ishida, "Structural Analysis of Planar Defects in Wurtzite Type Aluminum Nitride", Jpn. Institute of Metal, **52** 1192-98 (1988).

6. D.J.Barber and N.J.Tighe, "Electron Microscopy and Diffraction of Synthetic Corumdum Crystal II : Dislocation and Grain Boundaries in Impurity-Doped Aluminum oxide," Phil. Mag. **14** [129] 531-44 (1966).

7. C.C.Ahn and G.Thomas, "Microstructure and Grain Boundary Composition of Hot-Pressed Silicon Nitride with Yittria and Alumina," J. Am. Ceram. Soc., **66** [1] 14-17 (1983).

8. D.W.Susnitzky and C.B.Carter, "Structure of Alumina Grain Boundaries Prepared with and without Thin Amorphous Intergranular Films," J. Am. Ceram. Soc., **73** [8] 2485-93 (1990).

9. J.Narayan, "Dislocations, Twins and Grain Boundaries in CVD Diamond Thin Films : Atomic Structure Analysis," J. Mater. Res., **50** [11] 2414-23 (1990).

10. F.F.Lange and D.R.Clarke, "Morphological Changes of an Intergranular Thin Film in a Polycryslline Spinel." J. Am. Ceram. Soc., **65** [10] 502-506 (1982).

11. S.Hagege, G.Nouet and P.Delavignette, "Grain Boundary Analysis in TEM," Phys. Stat. Sol. (a), **61,** 97-107 (1980).

12. W.Bollmann, Crystal Defects and Crystalline Interfaces, Springer-Verlag, New York (1970).

13. C.B.Carter and S.L.Sass, "Electron Diffraction and Microscopy Technique for Studying Grain-Boundary Structure," J. Am. Ceram. Soc., **64** [6] 335-345 (1981).

INDEX